Chromosomes
—————— A SYNTHESIS ——————

Chromosomes
—————— A SYNTHESIS ——————

Robert P. Wagner
Life Sciences Division
Center for Human Genome Studies
Los Alamos National Laboratory
Los Alamos, New Mexico

Marjorie P. Maguire
Genetics Institute
Department of Zoology
The University of Texas at Austin

Raymond L. Stallings
Life Sciences Division
Center for Human Genome Studies
Los Alamos National Laboratory
Los Alamos, New Mexico

(W) WILEY-LISS

A JOHN WILEY & SONS, INC., PUBLICATION
New York • Chichester • Brisbane • Toronto • Singapore

Address all Inquiries to the Publisher
Wiley-Liss, Inc., 605 Third Avenue, New York, NY 10158-0012

Copyright © 1993 Wiley-Liss, Inc.

Printed in the United States of America.

Library of Congress Cataloging-in-Publication Data

Wagner, Robert P.
    Chromosomes : a synthesis / Robert P. Wagner, Marjorie P. Maguire,
Raymond L. Stallings.
      p. cm.
    Includes bibliographical references (p.   ) and index.
    ISBN 0-471-56124-X
    1. Chromosomes. I. Maguire, Marjorie P. II. Stallings, Raymond
L. III. Title.
QH600.W34 1992
574.87'322—dc20   92-25604
                                       CIP

The text of this book is printed on acid-free paper.

Contents

Preface

The existence of plant and animal chromosomes was first revealed in the 1870s. By the late 1880s some cell biologists realized that inheritance was associated with these physical bodies seen with the microscope. This observation was to establish the physical basis of inheritance as one of the greatest achievements of biology. In a sense the eukaryotic chromosomes and analogous structures in the prokaryotes and viruses are the most important objects in the living world, for they determine the existence and forms of organisms.

Since the time when chromosomes were first recognized, they have been studied extensively, first by cell biologists and then by geneticists, whose collaboration led to the emergence of the field now called "classical genetics." The demonstration that DNA is the genetic substance pushed chromosomes into the purview of biochemists, biophysicists, and molecular biologists, whose novel analytical techniques, added to the accepted cytological and genetic approaches, have greatly enhanced our understanding of gene structure and function.

Genes are the parts of chromosomes that we can now define structurally and functionally at the molecular level with a fair degree of precision. But what are those entities we call chromosomes? The aim of this book is to try to synthesize the relevant and significant classical and current findings by cytologists, cell biologists, cytogeneticists, geneticists, biochemists, biophysicists, molecular biologists, and evolutionists in their quest for a better understanding of what chromosomes are structurally and functionally. Chromosomes are obviously bearers of genes, but are they passive bearers, or do they represent a functional level above the gene level? A genome is constituted of genes that function in an integrated and harmonious fashion to create and maintain an organism. Do chromosomes represent a hierarchy of structure and function that governs more than the proper disjunction of genes? We are neither so presumptuous nor so naive as to believe that we have the definitive answer to these questions. Our goal has been to map a course through the many facts currently known about chromosomes as may eventually lead to consummate answers. We hope the reader can enjoy the excitement of the chase.

The study of chromosomes is framed within a continuum spanning over a hundred years. Any present understanding of chromosomes will be but a frame followed by a future frame, as in a strip of motion picture film. Notwithstanding this, we have attempted to describe a body of basic knowledge about chromosomes that has accumulated, and will probably not undergo fundamental change, with the hope that the reader can use it as a baseline to relate to future findings.

Nongeneticists and geneticists alike have difficulties with the jargon of genetics. Not all geneticists agree on the definitions of all terms in the extensive genetic vocabulary. We have attempted to follow the definitions given in the *Glossary of Genetics and Cytogenetics* by Rieger, Michaels, and Green (5th edition, 1991). Their definitions are based for the most part on those provided by the originators of the terms.

In order to make the contents of this book accessible to as many readers as possible, we have tried to follow the policy of introducing each topic with a rather fundamental introduction and building up from there. In addition, we have provided experimental details in boxes within the text to help readers unfamiliar with some of the important techniques used in chromosomal analysis.

There are two ways to approach a subject so broad as chromosomes. The first is to present the related material in the form of a general discourse that concentrates on generalities. The second is to use specific examples, drawing general conclusions from them. In this book we use a combination of both approaches. We present specific cases in some detail to illustrate important points. Having done this, we have made inferences and drawn conclusions that may or may not endure the test of time.

Robert P. Wagner
Santa Fe, New Mexico

Marjorie P. Maguire
Austin, Texas

Raymond L. Stallings
Los Alamos, New Mexico

Acknowledgments

We are indebted to the following for their interest in our endeavors to create this book, for their reading of chapters and their constructive criticisms, and for supplying significant information and references from relevent fields outside our immediate ken; from the Los Alamos National Laboratory: George Bell, E. Morton Bradbury, Joseph D'Anna, Deborah Grady, Amanda Ford, Carl E. Hildebrand, Jonathon Longmire, Juliann Meyne, Robert Moyzis, F. A. Ray, Richard Reynolds, James Spuhler, and David Torney; from the National Institutes of Environmental Health Sciences: Burke Judd; from the University of Texas at Austin: Mark Kirkpatrick, H. Eldon Sutton, and B. L. Turner; from The California Institute of Technology: Norman Horowitz; from Harvard University: Richard Lewontin; from Heidelberg University: Thomas Cremer.

Significant and excellent help in the preparation of the manuscript was provided by Barbara Judd and Monica Fink.

We thank Philip C. Wagner and Janet Young for their outstanding work in the preparation of the line drawings; Philip did the drawings for Chapters 1, 2, 3, 5, 6, 7, 8, 9, and 10, and Janet did the drawings for Chapter 4.

Introduction

Before one can understand the ground on which new ideas grow, one must understand how the old ones were modified or why they were eliminated.

E. Mayr, 1982

Chromosomes of the eukaryotes were not observed and described until about 100 years ago, but the groundwork for recognizing them started well over 300 years ago. The interval period was one in which the light microscope was developed and biology at the cellular level began to take form as an important aspect of the study of life. The beginnings of the understanding of the eukaryotic cell began in the last century with the development of this instrument, and with it chromosomes were first described. The story of the evolution of our perceptions about chromosomes can conveniently be partitioned into three time periods. The period 1850–1900 may be called "cells and chromosomes"; the period 1900–1940, "chromosomes and genes"; and the period 1940 to the present, "genes, nucleic acids, proteins, and chromosomes." A period before 1850 was important too, and we consider it first.

I. BEFORE 1850

A. Telescopes and Microscopes

The unaided human eye is a wonderful instrument. It can perceive objects as small as scales on a butterfly's wing and as far away as galaxies beyond the solar system.

This spectrum of seeing was the universe of humankind until a few centuries ago. Then, what were faint points of light in the sky or dust particles on earth became, in the seventeenth century, objects with details theretofore invisible, but henceforth visible with the development of two optical instruments: the telescope and its relative, the microscope. Who first conceived and constructed these instruments by the use of glass lenses is in dispute, but it is probable that the first practical telescope or "spyglass" was made in about 1608 by Dutch spectacle makers. It was by way of reports from Holland that the Florentine Galileo Galilei (1564–1642) first heard of the telescope, and with this knowledge he constructed his own instruments. With a telescope capable of magnifying 33 diameters he demonstrated the rotation of the satellites around Jupiter, observed the rotation of the sun, and established that Copernicus was probably right in affirming that the sun, not the earth, was the center of the solar system. This was one of the principal discoveries leading to the beginnings of the age of modern science.

Concomitantly with his work with telescopes, Galileo, along with contemporaries in Italy and Holland, began the construction of microscopes using knowledge gained from making telescopes. He did not himself make extensive observations with these instruments, but a fellow Italian from Bologna did. Marcello Malpighi (1628–1694) was probably one of the first to use the microscope to study animal and plant structure and became a founder of microscopic anatomy.

B. Beginnings of Microscopical Observations

Malpighi was not alone in his use of the microscope. Contemporary observers, such as Robert Hooke (1653–1703) and Nehemiah Grew (1641–1712) in England, and Antonie van Leeuwenhoek (1632–1723) in Holland, using microscopes with lenses they ground themselves, made extraordinary observations of living things theretofore unknown to humans. Hooke used the term **cell** to describe the compartments in thin slices of cork, and this was the beginning of the use of this term in biology. Leeuwenhoek looked at everything he could manage to put under his lens and disclosed a new world of organisms and parts of organisms to the attention of his contemporaries. It was in this period of the seventeenth century that the cell nucleus was first described by Leeuwenhoek and Marcello Malpighi, among others, and it may be said with good reason that what we now call **cell biology** started as an important field of biological science at this time.

Further developments in cell biology awaited the fashioning of more sophisticated microscopes. By the seventeenth century, the simple microscope invented in antiquity in crude form as a single lens was a valuable instrument, but its maximum useful magnification in the hands of most was only about 40 diameters. The early English compound microscopes consisting of several lenses used by Hooke and Grew magnified about 30 diameters. Leeuwenhoek was an exception; he was able to grind lenses that when mounted between two brass plates enabled him to obtain magnifications of several hundred diameters. But no one else was able to emulate him and repeat his performance for well over 100 years. It was not until the first part of the nineteenth century that compound microscopes began to be developed that were capable of the magnifications obtained by Leeuwenhoek with his single, simple lenses. And as the capabilities of these instruments increased, so did the abilities of the observers to see more about the cells of organisms, ranging from simple plants and animals to the more complex ones, including humans. By the 1850s the compound microscope had been developed to the point that microscopists were able to see some of the details of the cytoplasm and to determine that the nucleus was more than an empty space.

C. Cells as Living Morphological and Functional Units: The Cell Theory

The first part of the nineteenth century was important in the development of the concepts about cells. It was during this time that the so-called **cell theory** was enunciated. This theory stated that all plants and animals, although differing widely in external appearance, are fundamentally similar in their anatomical structure, since they are constituted of the same basic elementary units—cells. The botanists were the first to recognize this all-important generalization. It was perhaps most clearly stated by Meyen in 1830:

> Plant-cells appear either singly, so that each forms a single individual, as in the case of some algae and fungi, or they are united together in greater or smaller masses, to constitute a more highly organized plant. . . . Each cell forms an independent isolated whole; it nourishes itself, it builds itself up, and elaborates the raw nutrient materials, which it takes up into very different substances and structures. (Quoted from Hertwig, 1895)

This statement, made more than 150 years ago, could easily be used in a modern elementary biology text without change.

But even though it was becoming clear that cells were living morphological units, it was not at all clear how they arose. M. Schleiden and his friend T. Schwann, who are sometimes mistakenly identified as *the* founders of the cell theory, but were in reality only two contributors among many others to its formulation, did point out its universality, since it applied to animals as well as plants. They regarded nuclei as important constituents of cells, and Schwann (1839) suggested that new cells arose from formless organic matter by a process of crystallization. He was not invoking spontaneous generation as the explanation for the origin of new life by this, but trying to invoke a physicochemical explanation to supplant the mystical teleological explanations going back to Aristotle.

Many observers during this early period of cell biology regarded the cell membrane as the most important part of the cell. The contents inside the membrane were considered to be *slime*, with the implication that it might not be "alive" except perhaps for the nucleus. However, H. von Mohl in 1846 started using the term **protoplasm** for this slime and began to think of it as alive. (The term protoplasm was first used by J.E. Purkinje to describe the cell contents in 1839. It is derived from a theological term, *protoplast*, meaning the first man formed, Adam.)

At the end of this period M. Schultz and A. de Bary came to the conclusion that the important parts of cells, both plant and animal, was indeed protoplasm—*a living substance*. This was a significant step forward in the understanding of cell structure and function, and it introduced the developments of the second half of the nineteenth century.

II. CELLS AND CHROMOSOMES: 1850–1900

This period begins with the realization that "*the cell is a small mass of cytoplasm endowed with the attributes of life which contains in its interior a specially formed portion, the nucleus*" (Hertwig, 1895). However, it was still not clear where cells came from, although the more prescient of the observers were not at all of the opinion that they arose spontaneously. The plant cell biologists in particular were generally of the opinion that cells in some way came from preexisting cells, as Meyen, previously quoted, came very close to stating outright in 1830. It was during this period that the light microscope attained a stage of development that made it become possible to observe the changes in cells during the cell cycle. Even more significant, biology started to become a unified science. Previously there had been natural history and medicine, with its associated discipline of physiology. Now cell biology started to embrace medicine, and the seeds were sown for unification with natural history—an event that occurred in the next century with the gradual fusion of Darwinism and genetic and cellular biology.

A. Omnis Cellula e Cellula

With a celebrated passage from his book *Cellular Pathology*, Rudolph Virchow in 1858 ushered in a new period in biology and medicine. He wrote:

Where a cell exists there must be a preexisting cell, just as the animal arises from an animal and the plant only from a plant. The principle is thus established, even though the strict proof has not been produced for every detail, that throughout the whole series of living forms, whether entire animal or plant organisms, or their component parts, there rules an eternal law of continuous development.

With this statement Virchow enunciated the principle of life-continuity by cell division, and coined the celebrated aphorism, *omnis cellula e cellula*, "all cells come from cells."

The historic generalization was by no means original with Virchow. Meyen and von Mohl made closely similar statements in the 1830s with particular reference to plants. But the powerful influences of Schleiden and Schwann seem to have pushed these ideas to the background and replaced them with cell formation by a different kind of generation. It was not until the 1840s and 1850s that R. Remak, K. Naegeli, and K. Hofmeister among others began to restate it and, with Virchow's support, made it apparent to the general biological community that new cells indeed arose only from preexisting ones. This was a momentous step forward in the history of biology for it established the cell as the **unit of reproduction** as well as structure and function. It was the beginning of the concept of the cell as the basis of inheritance, leading E.B. Wilson to state what he called the **Law of Genetic Continuity**: *Life is a continuous stream.*

The recognition of the cell as a reproductive unit led to the demise of spontaneous generation as an explanation for the origin of current new life, and the recognition that the development of organisms from single cells was epigenetic; an animal or plant arises from an undifferentiated egg cell by successive stages of differentiation.

B. Omnis Nucleus e Nucleo

The cell nucleus first observed by the seventeenth century microscopists was considered to be an important cellular constituent. When it became evident that cells come from cells by division, a controversy arose as to the role and fate of the cell nucleus. One school held that the nucleus broke up during cell division and its contents diffused through the cytoplasm (Reichart, 1847); the other contended that it retained its identity in altered form, that it played an active role in cell division (Remak, 1852), and that two identical nuclei were formed from one: *omnis nucleus e nucleo*. The latter hy-

thesis was proved to be correct by a number of observers in the 1870s and 1880s. With their work began the study and understanding of **mitosis** or **karyokinesis**, and the recognition of the existence of chromosomes.

C. Fertilization

Actually the first studies leading to the recognition of continuity of nuclei came with studies on fertilization. Sexuality and fertilization in animals and plants were recognized thousands of years before the eighteenth century, but the details were far from understood, particularly with regard to the cellular basis of fertilization. Most of what people thought they knew was fancy rather than fact, as is well brought out by Aristotle's *De Generatione Animalium*. K.E. von Baer announced the discovery of mammalian (and human) eggs in 1828, but did not think of them as cells. This step was taken by Remak, who in 1852 described the frog's egg as a single cell. Koelliker had earlier, in 1841, proposed that sperm were also cells. The stage was thus set for the beginnings of the understanding of fertilization, but it was not until the 1870s that Schneider (1873) and Auerbach (1874), using material from platyhelminth worms, were able to conclude that the nucleus of the zygote was the result of the fusion of the egg and sperm nuclei. O. Hertwig (1878) advanced the idea that this fusion nucleus was one that became the progenitor of all subsequent nuclei in the developing sea urchin, *Paracentrotus*. In 1879 Fol confirmed this with a series of brilliant observations, and the fusion of sperm and egg and the fate of the subsequent fusion nucleus described by him and Hertwig became an accepted explanation for the beginnings of a new life. However, it should be recognized that at this time, in the 1870s, no one understood what was in those fusion nuclei. The identity of chromosomes was yet to be established and the process of meiosis was not even being thought about. The observation that there was a fusion nucleus, whatever its content, was nonetheless important because the conclusion that subsequent nuclei were derived from it focused attention on nuclei as being perhaps the most important visible entities in cells.

It should also be recognized that these significant advances were made possible by further developments in equipment and techniques. In 1870 oil immersion lenses were introduced. A few years previously the microtome was invented, making it possible to cut the extremely thin sections needed for high magnification of cell and tissue components. Finally new methods for fixing tissues were introduced and aniline dyes were used to stain cells and tissues, making it possible to see things previously invisible even with the best of microscopes.

D. The Nucleus as the Vehicle of Inheritance

The recognition of the nucleus as a self-perpetuating body led to the realization that it might be the vehicle of inheritance. However, only later was it hypothesized, almost simultaneously by O. Hertwig (1884) and Strasburger (1884), that the nucleus was the seat of heredity. At the same time that Strasburger and the others were drawing this important conclusion, the botanist Naegeli (1884) advanced his **idioplasm theory**. This was a theory based on pure speculation, but it was of great importance because it was an attempt to conceive heredity as being determined by the transmission of a specific physical substance, the **idioplasm**. He defined idioplasm as a substance that provided the nutritive elements necessary for the cell's survival and well-being. Notwithstanding this prescience, he failed to link the idioplasm with the nucleus. A second problem with Naegeli's theory was that he apparently conceived the idioplasm as a homogenous substance, although he did believe that each species had a somewhat different idioplasm.

(Naegeli achieved great eminence during his lifetime and was treated with reverence by many of his contemporaries, but he missed a third opportunity to advance the understanding of heredity in his time by his treatment of Mendel's work. He and Mendel corresponded over a 7-year period from 1866 to 1873. He failed to see the significance of Mendel's results with the pea plant, and advised him to try crosses with the hawkweed, *Hieracium*. This Mendel did and became quite discouraged as a result. This, coupled with his elevation to Abbot of his monastery, closed the door on future work with the pea plants. Unknown to him and other botanists of the time, *Hieracium* is apomictic, which means essentially that it reproduces asexually by a parthenogenetic process.)

E. Cell Division by Mitosis

Cell division was observed by a number of microscopists starting early in the nineteenth century, but its significance did not begin to be realized until it became known that cells came from preexisting cells. By the 1870s the resolving power of the light microscope was approaching the theoretical limit of 0.2 μ, and fixatives and stains were being used on tissues sliced with microtomes. It began to become evident that cells in many tissues underwent some sort of cycle of change. At one point in this cycle, bodies appeared that were intensely stained by aniline dyes. They were given the name **chromosomes** (chromo = color; soma = body) by Waldeyer in 1888. They were described as consisting of a substance called **chromatin** (literally colored material). As these bodies became visible, the nucleus as such disappeared because the nuclear membrane disappeared, and the chromosomes shortened and thickened and "split" longitudinally. The two members of the split pair then separated and moved to the opposite sides (poles) of the cell so as to make an equal division of the chromatic material. Meanwhile the cytoplasmic contents for the cell also underwent division and the chromatin in each became enclosed in new nuclear membranes. Thus two new daughter cells, each with its own nucleus, were formed. This process was called **karyokinesis** or mitosis. Both terms are still in use, karyokinesis being preferred to mitosis in continental Europe.

Fleming was one of the principals in mitotic studies. His drawings are even today noted for their clarity and accuracy. He and Ràbl (1885) made the important discovery that in *Salamandra* there are 24 chromosomes before and after mitosis, thus leading to the possible conclusion that chromosomes persist during the interphases between mitosis, albeit in masked form.

F. Omnis Chromosoma e Chromosoma

The question of the individuality of the chromosomes arose shortly after they were recognized beginning with the work on the nematode, *Ascaris*, by Van Beneden (1883). The question resolved itself to this: Does every chromosome that appears during the mitotic phase have some kind of direct connection with a corresponding chromosome that comes from a previous cell generation? Or to put the question in more modern phraseology: Is there a genetic continuity among chromosomes such that chromosomes come from chromosomes? A considerable number of the observers from the period of in the 1880s right up to the first decade of the twentieth century were of the opinion that there was no continuity. It was the eminent cell biologist Boveri who, starting in 1888, carried on a sustained program of research for 20 years and came to the following conclusion in 1909:

For every chromosome that enters into a nucleus there persists in the resting-stage some kind of unit which determines that from this nucleus come forth again exactly the same chromosomes that entered it, showing the same size relations as before and also the same grouping.

Also he concluded that in all cells of the offspring produced from the zygote or fertilized egg half the chromosomes are of maternal ancestry and half paternal. Thus **omnis chromosoma e chromosoma** became a fact, at least for some.

G. Chromosomes as Vehicles of Inheritance

The question of the identity of the hereditary substance in the nucleus was a burning one during the last part of the nineteenth century. One might assume that by the late 1880s there would be general agreement that it was the chromosomes. However, this was not to be until about 1920. A number of the observers in the 1880s did have the opinion that the chromosomes were indeed the hereditary vehicles, and some of them built elaborate hypotheses around this conclusion. Weismann in 1883 proposed in a lecture primarily concerned with arguments opposing the inheritance of acquired characteristics that all heredity depends upon the germ plasm of the germ cells. This germ plasm he conceived as a molecular substance similar to, but not identical to, Naegeli's idioplasm. In 1885 he expanded on his original ideas and stated that the germ plasm was the chromatin of the nuclei of the germ cells. Basically he built upon an observation of Roux in 1883 that mitosis makes sense only if one assumes that chromatin is not a uniform and homogenous substance, but differs qualitatively along the length of the chromatin, that is, chromosome threads. With this assumption in mind, Weismann (1892) proceeded to adopt some of the hypotheses of Hugo De Vries, who had postulated in 1889 the existence of hereditary particles he called **pangenes**. Weismann took the idea and changed the name of the particles to **biophores**. He hypothesized that biophores were invisible, self-replicating particles associated with the chromosomes that have the power to determine specific characteristics of the organism. They constituted a heterogeneous population, and the implication was that each biophore determined a different characteristic; a sort of one-biophore–one-function situation. He then proposed a hierarchy of determinants: Biophores were assembled to form ids and the ids in turn assembled to form **idants**, or chromosomes. The determinants were assumed to be arranged linearly along the chromosomes, and the stuff of the ids and their biophores was assumed to be chromatin. All this was, of course, guesswork, but it attracted a considerable amount of attention and came fairly close to describing the actual situation as we now know it.

However, another group of cell biologists led by Fol (1891), who had an excellent reputation because of his earlier work on mitosis, were of the opinion that the **centrosomes**, or **centrioles**, might be the hereditary material rather than the chromosomes. Centrosomes are bodies in animal cells around which the aster forms and the spindle arises. Fol and others thought they observed centrosomes arising from conjugation of sperm and egg. Hence they could conceivably carry the genetic information. But it became evident by the end of the last decade of the nineteenth century that, in some animals at least, the centrosomes of the zygote are descended from the sperm midpiece. In some cells, however, the centrosomes disappear and arise again *de novo*, so that it became improbable that they had genetic continuity. Also it was found that centrosomes are absent in higher plants. This left only the chromosomes as visible elements with a probable role of carrying the hereditary determinants.

H. Reduction of Chromosome Number

The development of the understanding of meiosis has been neglected up to this point because it was thought desirable to get the main ideas concerning the identity of chromosomes established first. Also some of the crucial details of the process of the formation of the gametes were not understood until after 1900. (It should also be mentioned that the term *meiosis* was not in use before 1904, when it was coined by Farmer and Moore as *maiosis*. Later writers changed the spelling to meiosis. (The term is from the Greek verb *meion*—literally, to make smaller.) Prior to 1905 the process was referred to as "reduction of the chromosomes" and the associated process of "maturation" of the germ cells in oogenesis and spermatogenesis.

In 1883 van Beneden described his observations on the conjugating nuclei of the gametes of *Ascaris*. He noted that these nuclei contained one-half the number of chromosomes found in the somatic cells of this parasitic nematode. He also recognized that the formation of the polar bodies during the maturation of the egg was part of the process of producing a single fertilizable egg with half the chromosome number of the unreduced egg. In a general review of the subjects of reduction and fertilization in 1902, Boveri showed that the process of maturation of the sperm and egg were essentially the same. It was now clear that the first polar body during maturation was comparable to a secondary spermatocyte in spermatogenesis, and the division of this secondary oocyte nucleus or first polar body gave rise to two more polar bodies with reduced chromosome numbers.

It soon became evident that the meiotic process was probably universal among sexually reproducing animals, and although the process varied in details, it always produced the same end result, that is, half the chromosome number. Meiosis was described in plants by Overton (1893), who showed that it occurred in sporophytic tissue and led to the formation of the haploid gametophytic generation.

I. Darwin, Mendel, and Miescher

The contributions of these three nineteenth century men have been either unmentioned or mentioned only in passing. Actually the impact of their findings on the understanding of the structure and functioning of chromosomes did not become fully appreciated until after 1900. However, some of these observers did recognize the possible significance of Miescher's work several years before 1900. For example, we find E.B. Wilson writing the following in 1895:

> Now chromatin is known to be closely similar to, if not identical with, a substance known as nuclein—which analysis shows to be a tolerably definite compound composed of nucleic acid (a complex organic acid rich in phosphorus) and albumin (protein). And thus we reach the remarkable conclusion that inheritance may, perhaps, be effected by the physical transmission of a particular chemical compound from parent to offspring.

These conclusions of Wilson were in large part derived, as he himself admitted, from the speculations of Claude Bernard (generally considered to be one of the principal founders of scientific medicine) and Albrecht Kossel. They both advanced the idea that the nucleus of the cell controlled the formation of the organic matter in the cytoplasm. Indeed in 1882 Kossel proposed that the nuclein (nucleoprotein) of the nucleus plays the central role in this control.

The first person to recognize the existence of this nuclein was a Swiss chemist, Friedrich Miescher (1844–1895). Work-

ing in the laboratory of the noted physiological chemist Hoppe-Seyler, Miescher succeeded in isolating nuclei from human pus cells. (These were readily available in those days from hospitals, since antisepsis was not in fashion.) From these nuclei he isolated a substance he called nuclein, which we now refer to as nucleoprotein. He published these results in 1871, and then turned to work with salmon sperm, which were readily available from the spawning salmon of the Rhine River. From these sperm he succeeded in separating what we now call nucleic acid from its associated protein, protamine. These historic findings supported the realization that chemical techniques applied to cell structure are a necessary adjunct to the study of cells with the microscope. Miescher was well aware of this and in 1871 in a letter to a friend he wrote:

> I believe that in the organic world each complicated case is built on simpler things, and these in turn on still simpler things . . . each case must be reduced to its simplest terms. (Greenstein, 1943)

Here we have one of the earliest statements of what has come to be called **reductionism** in biology. (See Mayr, 1982 for an in-depth discussion of reductionism.)

III. SUMMARY AND SIGNIFICANCE OF THE PERIOD 1850–1900

The period from 1850 to 1900 was one in which biology emerged from some of the last vestiges of Aristotelian medieval thought and began to become a unified science. A biological *Weltanschauung* little different from what we have today came into being.

One is struck, on reading in the biological literature of this period, by one overwhelming fact: The leaders in biology of that time were almost compulsive in seeking the physical basis of heredity. This is not something arbitrarily read into their writings, but is a clear pattern of intellectual activity expressed in many different ways directed toward this one goal. Even Darwin speculated about this matter, though he paid little attention to what the cell biologists across the channel were doing. He developed a theory called **pangenesis**, which postulated the existence of hereditary particles he named **gemmules**. It turned out to be wrong, but it did stimulate some thinking on the matter.

The fruits of the labors of the continental cell biologists culminated in identifying as the chromatin in the chromosomes material of heredity. This conclusion was admittedly based on indirect evidence, and definitive experimental underpinnings were to come later in the next century, but it was basically right on the mark. The following quotation from the first edition of E.B. Wilson's book *The Cell*, published in 1896, gives some idea of the kind of thinking current among some of the leading intellects of the time.

> In bringing the foregoing discussion into more direct relation with the general theory of cell-action we may recall that the cell-nucleus appears to us in two apparently different roles. On the one hand, it is a primary factor in morphological synthesis and hence in inheritance, on the other hand an organ of metabolism especially concerned with the constructive process. These two functions we may with Claude Bernard regard as but different phases of one process. The building of a definite cell-product, such as muscle fiber, a nerve process, a cilium, a pigment granule, a zymogen granule, is in the last analysis the result of a specific form of metabolic activity, as we may conclude from the fact that such products have not only a definite physical and morphological character, but also a definite chemical character. In its physiological aspect, therefore, inheritance is the recurrence, in successive generations, of like forms of metabolism; and this is effected through the transmission from generation to generation of a specific substance or idoplasm which we have seen reason to identify with chromatin. This remains true, however we may conceive the morphological nature of the idioplasm—whether as a microcosm of invisible germs or pangenes, as conceived by De Vries, Weismann, and Hertwig, as a storehouse of specific ferments (enzymes), as Driesch suggests, or as a complex molecular substance grouped in micellae as in Naegeli's hypothesis. It is true, as Verworn insists, that the cytoplasm is essential to inheritance, for without a specifically organized cytoplasm the nucleus is not able to set up specific forms of synthesis. This objection, which has already been considered from different points of view, both by De Vries and Driesch, disappears as soon as we regard the egg cytoplasm as itself a product of the nuclear activity; and it is just here that the general role of the nucleus in metabolism is of such vital importance to the theory of inheritance. If the nucleus be the formative center of the cell, if nutritive substances be elaborated by or under the influence of the nucleus while they are built into the living fabric, then the specific character of the cytoplasm is determined by that of the nucleus, and the contradiction vanishes. In accepting this view we admit that the cytoplasm of the egg is, in a measure, the substratum of inheritance, but is it so by virtue of its relation to the nucleus, which is, so to speak, the ultimate court of appeal. The nucleus cannot operate without a cytoplasmic field in which its peculiar powers may come into play; but this field is created and moulded by itself. Both are necessary to development; the nucleus alone suffices for the inheritance of specific possibilities of development.

It is a sad commentary that many textbooks of elementary biology used in our schools, and written as long as 80 years after the above was written, had not achieved either the perception of what was, or the clarity of thought to express what was, the **role** of cells and their component parts in the functioning and development of organisms. One can also go so far as to say that few more original thoughts have been uttered in the twentieth century up to the present writing beyond those expressed in the times that Wilson wrote. Since that time we have not so much been writing the text as the commentary.

The full significance of this statement by Wilson cannot be appreciated without some knowledge of the ideas, suppositions, and superstitions current among not only the lay people, but some biologists of the nineteenth century. It is in this context that the quotation from Mayr given at the head of this chapter assumes special meaning.

Here are some examples of ideas current in the nineteenth century, most of them carried over from preceding centuries and going back even to antiquity.

1. Living organisms possess vital factors that are divorced from physical factors. These are life essences and are not explicable by the known (and unknown) principles of physics and chemistry.

2. In some unknown way these vital essences are transmitted from one generation to the next. Inheritance therefore has no physical basis.

3. Whether spiritual or physical what is inherited

    (a) is contributed by one of the parents in toto, or by one of the parents more than the other.

    (b) is contributed by both parents equally, but their contributions are blended in the offspring (blending inheritance).

    (c) is the characteristics of the parents as such rather than the directions to develop these characteristics.

    (d) incorporates characteristics of the parents not necessarily inherited by them but acquired by them (Lamarckism).

4. Life can arise spontaneously out of inert matter. This is particularly true of lower forms such as slugs, worms, and insects.

5. Each organism has within it the miniature preformed members of its succeeding generations (preformationism).

Some of these concepts survive even today under the guise of vitalism and Lamarckism and other unproved assumptions for which no known scientific evidence exists. But the great majority of persons now accept that inheritance has a physical basis in chromosomes and that what a parent learns or experiences is not programmed in the chromosomes to be passed on to the next generation. This revolutionary idea was the fruit of the labors of hundreds of cell biologists of the nineteenth century only a few of whom have been mentioned in the foregoing. The formulation of this idea must be considered as one of the great intellectual achievements of the nineteenth century.

IV. CHROMOSOMES AND GENES: 1900–1940

This period began with the recognition in 1900 of the significance of Mendel's 1865 paper in contributions by Hugo DeVries, Karl Correns, and Erich von Tschermak. These three persons published papers describing experiments that verified Mendel's earlier neglected results and made it impossible to continue to ignore his conclusions. Mendel's "laws" as exemplified by the now familiar ratios obtained after making certain kinds of crosses were now accepted as being important in the analysis of the inheritance of diverse characteristics. The original ratios were discovered by use only of plants, but in 1902 William Bateson and Lucien Cuénot showed that the same kinds of ratios were obtained with chickens and mice. Hence the laws of Mendel appeared to be applicable for plants and animals in general.

A. Mendelism, Cell Biology, and the Chromosomes Theory of Inheritance

During this Mendelian renaissance the process of chromosome reduction or meiosis involved in the formation of gametes began to be understood better. The phenomenon of synapsis or pairing of chromosomes in the first prophase was something of a mystery until T.H. Montgomery showed in 1901 that the paired chromosomes forming the bivalents were each in fact derived from a different parent. That is, each bivalent contained a maternally derived and a paternally derived chromosome. Findings of this type and "hunches" based on observations with experimental material led T. Boveri in 1901 to write to his friend Hans Spemann:

> Taking everything into consideration, I believe that the essential point can finally be approached. I feel beyond doubt that the individual chromosomes must be endowed with different qualities and that only certain combinations permit normal development.

This set the stage for what only can be considered as one of the most momentous events in the history of genetics.

E.B. Wilson, his pupil W.S. Sutton, and Boveri united Mendelism and cell theory. In a series of two papers Sutton (1902, 1903) essentially founded eukaryotic genetics. He studied the meiotic process in the grasshopper *Brachystola magna*, which has 11 pairs of readily distinguishable chromosomes plus one unpaired element. In his second paper, he made the following generalizations:

1. The chromosome group of the presynaptic germ-cells is made up of two equivalent chromosome series, and strong grounds exist for the conclusion that one of these is paternal and the other maternal.

2. The process of synapsis . . . consists in the union in pairs of the homologous members . . . of the two series.

3. The first post-synaptic or maturation mitosis is equational and hence results in no chromosome differentiation.

4. The second post-synaptic division is a reducing division, resulting in the separation of the chromosomes that have conjugated in synapsis and their relegation to different germ-cells.

5. The chromosomes retain a morphological individuality throughout the various cell divisions.

Despite the errors in (3) and (4), Sutton goes on to state that

> many points were discovered which strongly indicate that the position of the equatorial plate of the reducing division is purely a matter of chance, that is, any chromosome pair may lie with paternal chromatids indifferently toward either pole irrespective of the positions of other pairs—and hence that a large number of different combinations are possible in the mature germ-product of an individual. (Sutton, 1902)

He then went on to estimate the total numbers of combination possible with various numbers of chromosomes pairs, and showed that when germ cells with the reduced number of chromosomes unite in fertilization, the result will be the ratios in offspring of the type that Mendel found.

With Mendelian ratios thus explained, the spotlight turned on chromosomes as the probable bearers of the Mendelian *merkmale* or **unit factors**, as they were beginning to be called. This did not mean that this association was universally accepted, however. A number of prominent, genetically oriented biologists did not accept the possibility of Mendel's particulate characteristics being part of chromosomes. W.L. Johanssen, who invented the terms **gene** defining it as a "little word . . . useful as an expression for the unit factors . . . in the gametes demonstrated by modern Mendelian researchers" as well as genotype and phenotype, referred in 1913 to the idea that genes were on chromosomes as a "piece of morphological dialectic." As late as 1916 Bateson, one of the most prominent geneticists of the first part of this century, who coined the term

genetics and wrote the first English textbook of genetics, made the following statement in a book review:

> It is inconceivable that particles of chromatin or any other substance, however complex, can possess those powers that must be assigned our factors. . . . The supposition that particles of chromatin . . . can by their material nature confer all the properties of life surpasses the range of even the most convinced materialism. (Bateson, 1916)

By "our factors" Bateson meant genes. Obviously he viewed genes in an almost mystical context. But despite the skepticism of these two prominent and influential geneticists, genes were put on chromosomes. Proof was provided by T.H. Morgan and his students A.H. Sturtevant, C.B. Bridges, and H.J. Muller, who worked together at Columbia University during the period 1910 to about 1920 using the fruit fly *Drosophila melanogaster* as the experimental organism.

Sutton's hypothesis, quoted in the preceding paragraphs, had made it clear that the segregation of genes and the process of independent assortment could be explained by assuming that the chromosomes are the bearers of the hereditary material. He and H. De Vries also recognized that if this is so, each chromosome must bear more than one kind of gene. In 1910 Morgan defined this association of two or more genes on the same chromosome as linkage, and he described linked genes as constituting a **linkage group** in 1911. His and his students' objectives were to show that linkage groups corresponded to the visible chromosomes. Over a period of several years they demonstrated, by making the appropriate kinds of crosses with *D. melanogaster*, a species with four pairs of chromosomes, that there were indeed four identifiable linkage groups. In the process they elucidated the process of crossing-over, demonstrated the linear arrangements of genes on the chromosomes,

and constructed **linkage maps**. Their work formed the experimental proof for the **chromosome theory of heredity**. They summarized this theory in 1915 in *The Mechanism of Mendelian Heredity*, one of the great classics in the literature of genetics. In 1916 Bridges provided the ultimate proof that chromosomes bear genes in his paper on nondisjunction in *Drosophila*. By about 1920, some 50 years after the first recognition of chromosomes, even the most skeptical of the nonbelievers among the geneticists had to admit that chromosomes are the bearers of the hereditary substance(s) made manifest by the results from genetic crosses.

B. Sex Determination

When cytologists first began to study the chromosomes of insects, they noted that some Hemiptera and Orthoptera had a chromosome that did not have a homologue with which to pair. At first it was thought to not be a proper chromosome and was designated with an X to indicate its questionable state. McClung in 1901 proposed that it be called an accessory chromosome, and that it determined that its possessor will be a male, or XO, and that the female should not have an X chromosome at all (1902). This was one of the first attempts to identify a specific chromosome with a specific trait, even though McClung proved to be wrong in assuming that the X was a male determining chromosome. N.M. Stevens straightened things out in 1905 when she showed that in the beetle *Tenebrio* the male has an X that pairs with a readily distinguishable smaller Y, while the female has two Xs. Later it was shown that species with XO males had XX females. And still later, in 1913, Bridges turned to the study of these sex chromosomes in *Drosophila* and showed in a summary paper published in 1925 that sex in *Drosophila* is determined by a balance between the ratio of the number of

Xs to the number of sets of autosomes. The Drosophila X has a female producing effect and the autosomes, a male determining effect. Out of this study came the important concept of genic balance. Bridges found that the Drosophila Y has no sex-determining effect. XO *Drosophila* are males in phenotype, but sterile. In mammals the XO individual is also sterile, but female.

C. Cytogenetics

Concomitantly with the *Drosophila* work by Morgan and his associates, experiments were being done with maize—in particular by R.A. Emerson and E.M. East (1913). Their early work showed that maize could be an extremely useful organism with which to study genetics. In a sense this was a rediscovery because Mendel had worked with maize crosses and found that he obtained the same type of results that he got with pea plants, but he never published his results (Rhoades, 1984). A.E. Longley (1924) was one of the first to do cytogenetic work with maize, and he published a series of papers in which he demonstrated that maize chromosomes were ideal for studies of certain aspects of chromosome structure. This was an important step forward because it led to the further development of maize cytogenetics, principally by L.F. Randolph (1928) and Barbara McClintock (1929). Although *Drosophila* was magnificent material for studying genetics by making crosses, it was not good for studying chromosomes visually during this early period in the 1920s. The only known workable chromosome preparations were from oogonial cells and larval brain cells, and these were small and showed little detail. But late prophase and early metaphase maize chromosomes from the first microspore mitoses stained by the acetocarmine smear technique showed enough detailed differences among the 10 pairs of chromosomes

to enable McClintock to identify the ten linkage groups and make the first idiogram. This important development enabled the maize cytogeneticists to visually identify translocations, inversions, and other aberrations.

In 1927 H.J. Muller announced what he called the "artificial transmutation of the gene." He had developed a technique to determine quantitatively the effects of X rays in producing sex-linked lethals in *Drosophila*. It soon became evident that, in addition to causing lethal mutations, "visible" chromosome breaks were increased in rate by X rays. The breaks led to rearrangements that, if they caused a change in the position of the centromere by pericentric inversion or translocation, could sometimes be identified cytologically in *Drosophila*. Otherwise tedious genetic methods had to be used. However, in 1933 the ability to do cytogenetics with *Drosophila* was increased enormously when T.S. Painter in Texas, and E. Heitz and H. Bauer in Germany, announced their discovery of the significance of the giant salivary gland chromosomes of *Drosophila* and *Bibio*. These dipteran polytene chromosomes, first described by Balbiani in 1881, showed distinct banding patterns when stained with appropriate stains, or even when unstained under the phase contrast microscope. The banding patterns were found to be distinctly different not only for each of the four chromosomes in the *D. melanogaster* genome, but even for different regions of the same chromosome. With metaphase chromosomes from brain cells only centromeres and chromosome ends would serve as markers. Now one had thousands of bands to work with as markers, and inversions (whether pericentric or paracentric), small deletions and duplications, and transpositions and translocations could be recognized and analyzed with relative ease. Thus a treasure trove was revealed that enabled *Drosophila* geneticists to make detailed cytological analyses to accompany their genetic analyses, and enabled the *Drosophila* evolutionists to make significant progress toward the analysis of speciation in this genus, which comprises close to 1,000 species distributed over most of the temperate and tropical regions of the world.

D. Cytological Analysis of Crossing-over

Crossing-over can be defined as the process leading to genetic recombination between linked genes. Perhaps no single process has received more attention and created more debate among cytologists and geneticists than this one. Eukaryotic crossing-over occurs in germ line cells presumably during the first prophase of meiosis, and also in mitotic cells. It is the meiotic process that has received the most attention, and despite about 75 years of study, starting with Janssens in 1909, it is still not understood in its basic details. Numerous hypotheses were advanced during the period up to 1940, but none were generally accepted. However, two landmark investigations in the 1930s demonstrated beyond doubt that recombinants arising from crossing-over are the result of physical exchange between cromatids. In 1931 H. Creighton and B. McClintock, using maize, and C. Stern, using *Drosophila*, showed that crosses between strains heterozygous for morphologically distinct chromosome markers, and marker genes between the cytological markers, produced offspring with recombined marker genes accompanied by recombination between the physical markers.

Early in the study of crossing-over there was considerable speculation about whether it occurred in the two-strand stage before sister strand formation, or in the four-strand stage between chromatids in the tetrads. It was E. Anderson, in 1925, and G. Beadle and S. Emerson, in 1935, who showed that crossing-over occurred in

Drosophila in the four-strand stage. This was also verified with tetrad analysis in *Neurospora* by C. Lindegren in 1933. The processes of genetic recombination are discussed in detail in Chapter 4.

E. Position Effects

The early studies of genes and chromosomes led most geneticists to the opinion that chromosomes were the passive bearers of genes, and that the position of a gene on its chromosome had no effect on its activity. However, in 1925 Sturtevant obtained results with the so-called Bar mutants of *D. melanogaster* that indicated that a gene's position might indeed be important in its functioning. He proposed that the effect of a gene may be dependent on its position with respect to neighboring genes, and that the phenomenon of **position effect**, as he termed it, might have considerable bearing on the organization of chromosomes. Later studies, summarized by Demerec (1940), showed that genes moved out of their euchromatic environment by inversion or translocation so that they were next to heterochromatin demonstrated an instability in their phenotypic expression. This reinforced the concept of position effect and made it evident that it is a general phenomenon.

F. Chromatin Chemistry

It has already been mentioned that cell biologists prior to 1900 were speculating that the chromatin of the chromosomes and the nuclein of Miescher were identical. R. Altmann (1889) showed that nuclein is composed of two substances: protein and nucleic acid. He was able to prepare relatively pure nucleic acid, free of protein. In the period 1900–1930 it was chemists who worked with nucleic acids; the substance was to a large extent ignored by geneticists and cell biologists. Indeed in the 1925 edition of his book *The Cell* E.B. Wilson had second thoughts about his earlier enthusiasm for nuclein and nucleic acid, and states that

> a large and increasing body of evidence shows that some of them (proteins) differ characteristically from species to species, and even indicate that they may constitute the fundamental chemical basis of heredity. (p. 644)

The possibility of protein being the hereditary substance was accepted by many geneticists up until the 1940s. There were cogent reasons for this assumption. First of all the chemists had shown that two general types of nucleic acid existed: ribonucleic, or plant nucleic acid, and desoxypentose, or animal nucleic acid. Plant nucleic acid, which we now know as RNA, was originally isolated from yeast, and animal nucleic acid, which we now know as DNA, was isolated from bovine thymus glands. For some years it was believed that RNA was confined to plants and DNA to animals. This was shown to be incorrect in 1924 by the biochemists R. Feulgen and H. Rossenbeck. In 1914 Feulgen had devised a test that could distinguish the so-called plant type from the animal type. By using this test, he and his collaborator then showed that wheat germ contained DNA.

Feulgen's test detects aldehydes by the use of fuchsin–sulfurous acid, or Schiff's reagent. By means of this reagent, cytogeneticists showed that the bands of the salivary gland chromosomes of the Diptera were Feulgen-positive. Some thought the genes were located in the bands, and this gave some credence to the idea that DNA was indeed the hereditary substance. There was one other difficulty, however, and this was the second hindrance to the acceptance of DNA as the genic material. It was generally believed that DNA always contained equimolar amounts of the nucleotides: adenylate, thymidylate, guanylate, and cytidylate, and that these occurred in the repeated sequence A-T-G-C over and over again. This was called the tetranucleotide hypothesis, and so long as it was embraced there was good reason to reject DNA as being of genetic significance. It was disproved in 1951 by Chargaff et al., which again made DNA a strong candidate for being the substance of the genetic material.

In the late 1930s a group of biophysicists working with T. Caspersson in Stockholm began to use ultraviolet light photomicrography as a means of studying the distribution of nucleic acids in cells. Since the nucleic acids by virtue of their purine and pyrimidine bases absorb strongly in the region of 260 nm of the spectrum, the concentration of both RNA and DNA can be measured with quartz optics and photoelectric cells in various parts of the cell. DNA and RNA are distinguishable from one another by the use of the Feulgen reagent and the use of hydrolytic enzymes specific for RNA and DNA, respectively. The work of Caspersson and his associates led to the renewal of interest in the possibility that the chromosomal DNA was possibly the genetic material, and in some way related to the synthesis of RNA and protein (reviewed in Caspersson, 1950).

V. SUMMARY AND SIGNIFICANCE OF THE PERIOD 1900–1940

This period started with a bang and ended in a whimper. Things did not look very exciting in 1940. Genetic biology was in a period of pause except for some applications in the realm of evolutionary theory. However the total period of 1900–1940 was extremely important in the history of chromosome studies, since it did unite the invisible Mendelian genetic elements, the genes, with the visible physical elements, the chromosomes. Genes became real things with a known location within the nucleus. By crossover recombination analysis they could even be mapped to *loci* on their chromosomes.

Also during this period a beginning was made toward trying to understand what genes did as components of chromosomes. As early as 1903 L. Cuénot concluded from his results with crosses with mice of different coat colors that genes, which he called **mnemons**, and enzymes, which he called **diastases**, were related. In modern terminology he stated that enzymes were the result of the activity of genes. About the same time, in 1902, A.E. Garrod, after considering the inheritance of albinism, alkaptonuria, and cystinuria in humans, concluded that these conditions were the result of inherited metabolic differences. In 1909 Bateson declared that he believed that mutant genes (which he termed **allelomorphs**) that resulted in the absence of a pigment were the result of the *absence* of a gene that would otherwise produce an enzyme involved in the synthesis of that pigment. Cuénot, Bateson, and Garrod can be considered to have been the first link to genes with enzymes (Wagner, 1989). In 1917 S. Wright, after analyzing the results of crosses with mammals of different coat color, concluded

first that melanin is produced by the oxidation of certain products of protein metabolism by the action of specific enzymes; second, that this reaction takes place in the cytoplasm of cells probably by enzymes secreted in the nucleus; third that various chromogens are used, the particular ones oxidized depending on the character of the enzymes present; and finally that hereditary differences in color are due to hereditary differences in the enzyme element of the reaction. (p. 230)

This was a clear statement of a possible gene–enzyme relationship. But little further attention was paid to this hypothetical role of genes until G. Beadle and B. Ephrussi did a series of experiments in the 1930s with eye color mutants of *D. melanogaster* and came to the conclusion that the wild-type allele of the vermilion gene, v^+, is responsible for the production of a

specific enzyme that converts the amino acid tryptophan to an eye pigment precursor shown by Butenandt et al. (1940) to be kynurenine (reviewed by Ephrussi, 1942). This finding set the stage for the work with *Neurospora* described in the next section.

VI. GENES, PROTEINS, AND NUCLEIC ACIDS: 1940–

Things began to come together in the 1940s. Chemists and physicists began to concentrate attention on the problems of biology, and one obvious place to start was to look at this business of genetics. Their attention was focused on genes. What were they and what did they do (Schroedinger, 1945)? It was accepted that they must be physical entities, but of what substance? There was some evidence that they had a relationship with enzymes, but how to demonstrate this experimentally? Attempts to answer these questions in the period 1941–1965 led to the transformation of genetics into a molecular science and hence to the further development of its status as the most reductionist of all biological sciences. An interlocking matrix of discoveries involving new techniques for studying proteins and nucleic acids, and use of the fungus *Neurospora*, and the bacterium *Escherichia coli* and various bacteriophages that infected it, led to the development of the experimental means to answer these questions, which were otherwise intractable to Mendelian techniques of analysis.

A. The Neurospora Revolution and What Genes Do

The experimental basis for answering the question of gene function was established by G. Beadle and E. Tatum in 1941, when they demonstrated that the fungus *Neurospora*, which in its native wild-type state would grow on a synthetic medium consisting of inorganic salts, carbohydrate, and one vitamin (biotin), could be mutated to produce strains that had to be supplied with specific supplemental nutrients in order to grow. A single mutation brought about the need for a specific amino acid or vitamin. This work by the biochemist Tatum and the geneticist Beadle ushered in the second revolution in genetic biology (Horowitz, 1991). It led to the development of the one-gene–one-enzyme hypothesis (Horowitz, 1948; Horowitz and Leupold, 1951) and to the conclusion that a relationship existed between genes and proteins, since enzymes are proteins. This was the beginnings of genetic molecular biology. Now genes could be linked to specific molecules, namely proteins.

B. Prokaryotic Genetics and New Horizons

About the time that these discoveries with *Neurospora* were being made, it began to be realized that much about the nature of inheritance could be learned by using bacteria and viruses as experimental material. Bacteria are classified as prokaryotes, since they do not have nuclei, in contrast to the nucleated eukaryotes such as *Neurospora*. It was shown that mutants of *Escherichia coli* could be obtained that required specific growth factors similar to those found in *Neurospora*. Also in this period during the forties fundamental observations were being made by members of the so-called Phage Group, led by S. Luria, M. Delbruck, and A. Hershey, among others, who proved that certain viruses called bacteriophage or phage replicated in *E. coli* and passed their genes on from generation to generation just as did cellular organisms while replicating in their bacterial host cells (reviewed in Portugal and Cohen, 1977). In the period from about 1945 to 1960 J. Lederberg, W. Hayes, J. Monod, F. Jacob, and E. Wollman worked out the fundamentals of inheritance in *E. coli* (reviewed in Hayes,

1968). These developments caused *E. coli* to become the workhorse of genetics, replacing *Drosophila* in that role for at least the next 20 years. Basic eukaryotic genetic mechanisms had been essentially worked out, and efforts were now focused on the inheritance patterns in the prokaryotes and their phages. The result was a much greater understanding of gene function and structure, which carried over to the eukaryotes. Also techniques were developed with the bacteria and their phages that allowed for extensive analysis of eukaryotic genomes, which would not otherwise have been possible. The progress in eukaryotic chromosomal structure and function analysis beyond the period of classical genetics was to a very large extent made possible by the exploitation of bacteria and their viruses.

C. DNA Is Finally Established as the Genetic Substance

Genes may be involved in protein synthesis, but what are they made of? This was the second big question of the 1940s. Although E.B. Wilson had speculated in 1886 that the genetic material was probably nuclein, he had changed his mind by 1925 and thought it might be protein. This possibility was generally accepted up to about 1945, principally because the nucleic acid deoxyribonucleic acid, which cytologists had identified as the principal nucleic acid of the chromosomes, was thought by chemists to have a structure of such a type as to render it incapable of carrying a message. A breakthrough came in 1944, when O. Avery, C. McLoed, and M. McCarty reported that the first experiments to demonstrate that the substance capable of transforming the genetic characteristics of the bacterium *Pneumococcus* was almost certainly DNA. Shortly thereafter chemists changed their original conclusions about DNA structure, and it became evident that it was indeed capable of carrying a message (Chargaff, 1951). In

1953 J. Watson and F. Crick, with the help of crystallographers and new chemical findings, were able to work out the details of DNA structure and to show that it was a molecule theoretically able to replicate itself. Not only was it able to carry a message, but it could replicate that message and pass it on to the very next generation by chromosomal divisions. Thus some 90 years after the discovery of chromosomes the dream of establishing the chemical nature of the hereditary substance in chromosomes was realized.

D. DNA, RNA, and Proteins

One of the first steps toward clarifying the gene–protein relationship came from the discovery of L. Pauling and associates in 1949 that showed that persons with the inherited disease sickle cell anemia possessed a hemoglobin that had to be different from normal hemoglobin. This was done by using the technique of electrophoresis, in which charged molecules are caused to move in an electric field. This technique was to revolutionize not only the study of protein differences but also nucleic acid differences. By using enzymes to digest hemoglobins from normal and sickle cell individuals to small polypeptides and then separating these by electrophoresis, V. Ingram in 1957 was able to show that of the 146 amino acids of the beta chains of the two different proteins only one is different. One of the glutamic acids at a specific place in the normal globin chain (position six counting from the amino end) is replaced by a valine in the sickle cell globin chain. This explained the results reported by Pauling et al., because the side chain of glutamate is negatively charged while that of valine is not. Thus Ingram established a definite connection between the effects of a base pair change in DNA and the substitution of a specific amino acid in a specific place in a polypeptide chain.

He states in his paper that "the sequence of base pairs along the chain of nucleic acid provides the information which determines the sequence of amino acids in the polypeptide chain for which the particular gene, or length of nucleic acid is responsible. A substitution in the nucleic acid leads to a substitution in the polypeptide" (Ingram, 1957, p. 328).

The burning question now became: How does the information in DNA become translated into specific proteins? It had long been known that besides DNA another nucleic acid, ribonucleic acid, existed in all cells. The embryologist J. Brachet suggested in the early 1940s that this RNA probably was involved in the synthesis of proteins in cells. Biochemists such as M. Hoagland and associates (1958) and Brenner et al. (1961) also became convinced that this was so and proceeded to work out the details. At least three types of RNAs existed in cells: transfer RNAs (tRNA), ribosomal RNAs (rRNA), and messenger RNAs (mRNA). It was shown that these forms of RNA form a direct link between DNA and proteins. The genetic information in DNA is **transcribed** into a complementary nucleotide sequence mRNA from which it is **translated** into a specific amino acid sequence to form a polypeptide in the presence of tRNAs and organelles, called ribosomes, that are constituted of rRNA and protein. The aphorism DNA→RNA→Protein became the dogma of molecular genetics.

It was obvious that DNA contained a code, but what was the nature of this code? The answer to this came very quickly, starting in the early 1960s. By using *E. coli* N. Nirenberg, H. Mattaei, H. Khorana, and C. Yanofsky and their associates showed that what was to become known as the genetic code was a trinucleotide code; each trinucleotide was described as a codon responsible for the determination of a specific amino acid except for three that acted as stop signals in the translation process. The connection between DNA, RNA, and proteins was established by showing that the code in DNA was transcribed into a complementary strand of mRNA, which then was translated into protein polypeptide in the presence of ribosomes containing rRNA and the tRNAs. With these findings it became apparent that the coding parts of genes are constituted of contiguous codons.

E. Advances in Chromosome Structure and Function

While molecular genetics was getting its start, things were beginning to stir in the field of chromosomes studies. A comprehensive history of some of these developments has been written by T. C. Hsu (1979), who also made significant experimental contributions in this field. We may delineate three phases starting with the period 1952–1959, during which time it became possible to demonstrate mammalian chromosomes in somatic cells in culture by the use of hypotonic solutions. The next period (1960–1969) saw the development of somatic cell cytogenetics and led, starting in about 1969, to the present period, in which the molecular and cytogenetic approaches are coming together to explore the structure and function of chromosomes.

Plant and *Drosophila* chromosomes had long been easily observed and counted in many species and, starting early in the twentieth century, their karyotypes had been determined. This was not true, however, for many animals, especially for humans and other mammals, until about 1952, when techniques were developed to demonstrate metaphase chromosomes in somatic cells in culture. In 1956 J. Tijo and A. Levan were able to demonstrate that the true chromosome number of humans was $N = 46$, correcting the previously accepted 48 that had been determined in spermatogonial cells. In 1959 J. Lejeune and colleagues made the important discovery that children with Down's syndrome were trisomic for one of the smallest of the human chromosomes. This was the beginnings of medical cytogenetics, which has now expanded to become an important part of human clinical genetics.

The next important development came with the technique of staining mammalian chromosomes with Giemsa stain, which is a mixture of methylene blue, eosine, and other components. The development of the C-banding procedure by F. Arrighi and T.C. Hsu in 1971, followed by other modifications of Geimsa staining, led to the ability to distinguish the different chromosome types in a genome. This has led to the development of techniques to separate chromosomes of the different types in sufficient numbers for detailed analysis of their chemical and genetic structure.

F. Chemical and Physical Structure of Chromosomes

The fact that eukaryotic chromosomes are constituted of chromatin consisting of about 50% nucleic acid and 50% protein was appreciated in the 1920s, but little was then known of the organization of these two basic structural elements until the 1960s, when the extensive application of chemical and physical techniques began to be used. The first major advance was the identification of the major protein components of chromatin: the five basic histones H1, H2a, H2b, H3, and H4, which are present in all eukaryotic chromosomes. Structurally all five of these histone types have related amino acid compositions whether plant or animal. The second major advance was the finding that four of these, namely H2a, H2b, H3, and H4, form octamers called **core particles**, about which DNA is wrapped to form structures associated with H1, which are called **nucleosomes**. These appear to be basic structural elements of chromosomes. By the use of X-ray analysis it was then established that the nucleosomes are connected to one another by non-histone-associated DNA, forming a fiber about 10 nm in thickness. This fiber is continuous from one end of a chromosome to the other but is highly convoluted and compacted even in interphase chromosomes.

G. Analysis of the DNA Structure of Chromosomes

Though the discovery of the code and the protein-encoding role of genes made it evident that genes are stretches of contiguous DNA codons, it also began to become apparent in the latter part of the 1960s that only a small part of the total DNA in the cells of the multicellular eukaryotes encoded polypeptides. Beginning in 1966, R. Britten and D. Kohne showed by kinetic analysis of the reassociation of short segments of single-stranded DNA that a large part of the DNA of plants and animals consists of repetitive sequences with no apparent genetic function. It became evident that the parts of chromosomes that can be identified with genetic changes that by mutation result in base substitutions in codons to cause amino acid substitutions in polypeptides constitute only a very minor part of their total DNA. In mammalian genomes this genic part is generally considered to be at most 4–5%. These sorts of findings have led to the view currently held by some geneticists that a large part of the DNA of eukaryotic genomes is nonfunctional "junk."

About the same time that these kinetic analyses of DNA content of chromosomes were being conducted, a group of enzymes called **restriction endonucleases** were being isolated from *E. coli* and other bacteria. These restriction enzymes recognize specific tetra-, penta-, or hexanucleotide segments in duplex DNA molecules and

cut at these specific sites. Historically, it was the discovery of these enzymes and the elucidation of their activities by M. Meselson and R. Yuan (1968), and W. Arber and S. Linn in (1969) that gave birth to DNA manipulation as a science, and changed the whole course of the development of molecular genetics in its application to the study of chromosomes. Beginning in this period and extending to the present time, phages, plasmids, and bacteria whose life cycles and genetics had been worked out in the forties and fifties became essential tools for the analysis of eukaryotic chromosomes.

By means of electrophoretic and chromatographic techniques of various types it has been found possible to isolate DNA fragments with specific base sequences and insert them into phage or plasmid vectors. These chimeras containing both carrier vector DNA and the "foreign" DNA from whatever nonvector source are **recombinant DNA molecules**. They are introduced into bacterial or eukaryotic host cells and allowed to replicate in these cells. The "cloned" foreign DNA is separated from the vector DNA by using restriction enzymes. Thus, sufficient specific DNA is provided to be sequenced or use in a variety of other ways.

H. Mapping of Chromosomes

Linkage data from a large variety of plants and animals obtained from the results of crosses have generated genetic maps with considerable detail for most eukaryotes with the exception of humans (O'Brien, 1990). Up to 1968 only one gene (the Duffy blood type) had been assigned to a human autosome. This lack of information about autosomal gene loci changed rapidly beginning in the 1970s, however, principally as a result of the discovery of Barski and associates in 1961 that somatic cells from different lines could fuse to form hybrids and of the subsequent discovery by Weiss and Green, in 1967, that man/mouse hybrid cells could be used to assign human marker genes to their proper chromosomes. Since about 1980 the introduction of molecular techniques using labeled specific DNA probes has tremendously increased the rate at which new loci are being identified, especially in the human genome. Consequently it is now possible to map chromosomes without making crosses. As a result of the use of these two techniques about 2,000 genes have been assigned to the chromosomes of the human genome. Similar increases in the development of plant and other animal linkage maps have also been made.

VII. SUMMARY AND SIGNIFICANCE OF THE PERIOD 1940 TO THE PRESENT

The present period started with a second revolution, namely the fusion of classical genetics with molecular biology. How it will end is not certain, but it has already been demonstrated that great advances have been made in the last 50 years toward an understanding of the function and structure of eukaryotic chromosomes; we are in fact approaching the level of understanding at which structure and function become one. No greater achievement in biology can be imagined.

1 Basic Chromosome Structure

The phenomena included here under the name of karyokinesis are an integral part of the behavior of easily stained threadlike structures occasionally visible during the division of the cell nucleus.

. . . I propose that those objects, so often identified by Boveri with one of the important stages of karyokinesis as "chromatic elements," be given the specific technical name "chromosomes."

Waldeyer, 1888

With these words, translated from the German, W. Waldeyer coined the term **chromosome** for those occasionally visible "threadlike structures" or "chromatic elements" so well known to his fellow cell biologists. Although the term did not gain immediate acceptance among cell biologists in the last century, it is now generally indiscriminately applied to any structure known to bear genes. Viruses, (whether the RNA- or DNA-type), bacteria and other prokaryotes, mitochondria, and chloroplasts are all considered to have chromosomes such as the nuclear elements described by Waldeyer. It is well established, however, that the eukaryotic structures described by him are significantly different from the analogous structures found in bacteria, mitochondria, plasmids, chloroplasts, and all known viruses with the exception of a few animal ones. Typical eukaryotic chromosomes have centromeres and telomeres, engage in mitosis and meiosis, and have a complex association between histones and other proteins and DNA, but these other "chromosomes" do not show this high degree of organization, although they may have some protein associated with them (Pettijohn, 1988). It has been suggested, therefore, that the name chromosome be assigned only to the nuclear structures first described by Waldeyer. This is the policy that will be followed in this book. The other gene-bearing structures will be called **genophores**. Waldeyer's article of 1888 was a review and an appraisal of the work of a number of European biologists who had begun shortly after the 1850s to use

the light microscope to observe plant and animal cells that had been stained with aniline dyes. The bodies seen in the cells during certain stages of the cell cycle took stain readily and so were first called **chromatic** (colored) bodies or elements. The term chromosome (colored body) given them by Waldeyer came into popular use in the first decade of the twentieth century when it was becoming evident that they were the bearers of the hereditary determinants—Mendel's **Merkmale** or Johannsen's **genes**. This revolutionary finding was to establish genetics as a science.

I. CHROMOSOMES ARE CHROMATIN

Chromatin is the name applied to the collective substances chromosomes are made of. It is an organized aggregate of nucleic acids and protein. When chromosomes are in the mitotic and meiotic stages of the cell cycle, chromatin stains readily and is easily visible in some cases even without the aid of a microscope. When in **interphase** between these active stages of cell division only that chromatin called **heterochromatin** is readily visible with stains (Fig. 1.1). The remainder of the chromatin is called **euchromatin**, except for some other chromosome parts to be discussed in the next chapter. Heterochromatic regions of chromosomes are generally referred to as positively **heteropycnotic** (pycnotic = dense) by the cytologist. They may vary in intensity from one part of the cell cycle to another. Two types of heterochromatin are classified: **constitutive** and **facultative**. Constitutive heterochromatin is present at identical positions on all chromosomes in all cell types. Facultative heterochromatin is variable in its expression. It varies with the cell type and may be manifested as condensed, or heavily stained, chromatin in only certain differentiated somatic cells in the same organism.

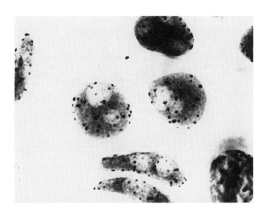

Fig. 1.1. *Cells in interphase showing hetero-pycnosis in heterochromatic regions of chromosomes. Onion cells stained with Giemsa. Reproduced from Stack and Clark (1974) with permission of the publisher.*

Basically most of the above was what was known about chromatin in the early part of this century. Now a great deal more is known. We know that the DNA, which constitutes about 50% by weight of the chromosomes is the genetic material, and we understand something about its structure and its relationship with the protein components. In this chapter we describe some of the characteristics of the chromosomal DNA and protein of the chromatin as an introduction to the basic structure of chromosomes.

II. DNA EXISTS IN A NUMBER OF FORMS IN CHROMOSOMES

A. Right-Hand–Left-Hand Conformations

The generally accepted model of a chromosome prior to its replication is that of a flexible rod composed of a single molecule of DNA millions of **base pairs** (bp) long associated with a variety of proteins as a nucleoprotein complex (Kavenoff et al., 1973; Comings and Okada, 1973). The DNA is usually described as the B-form originally proposed by Watson and Crick

(1953). However, the biophysicists and physical chemists have since described a number of different forms or conformations; it is now evident that chromosomal DNA may have considerable flexibility of form and that the different forms may be in equilibrium with one another. Chief among these are the **A-, B-, C-,** and **Z-forms**. The A-, B-, and C-forms are generally assumed to be right-handed helices, while the Z-form is left-handed. All four forms are distinguishable by their X-ray diffraction patterns. Table 1.1 gives some of the characteristics differentiating them. It must be recognized that these are properties of purified DNA fibers and crystals and that they will not necessarily obtain *in vivo*.

It is generally thought that the forms most likely to be found in intact cells are the B- and Z-forms, although it should be recognized that others, such as A- and C-forms and even forms now unknown, cannot be completely ruled out. The B-form may indeed be an important constituent of chromosomes because it is found in the presence of high humidity and low ionic strength—conditions that are prevalent in living cells. The Z-form, on the other hand, requires high salt concentrations *in vitro*, but it has been found that methylation of the cytosines in DNA stabilizes the Z-form and allows it to exist at a low salt concentration. It is maybe a natural form under certain conditions (reviewed by Rich et al., 1984). Left-handed DNA has also been shown in *Escherichia coli* cells (Jaworski et al., 1987).

The differences between the B- and Z-forms are quite pronounced. Figure 1.2 shows van der Waal's models of Z-DNA and B-DNA. Note the zigzag phosphate diester backbone of Z as compared with the smooth one of B-. Z-DNA has a single groove that extends to the helix axis as compared to the two shallower grooves of the B-form. The zigzag organization of Z is a consequence of the conformation of its

TABLE 1.1. Characteristics of Some of the Known Forms of DNA

DNA form	Base pairs per turn	Helical diameter (nm)	Relative humidity percent	Salt concentration required	Helical handedness
A	11.0	230	75	High	Right
B	10.4 ± 0.1[a]	200	92	Low	Right
C	7.9–9.6	190	66	High	Right
Z	12.0	180		High	Left

[a]As generally assumed but not proved for intact native DNA in chromosomes; 10.4 when in solution under physiological conditions; 10.0 when in fibrous form.

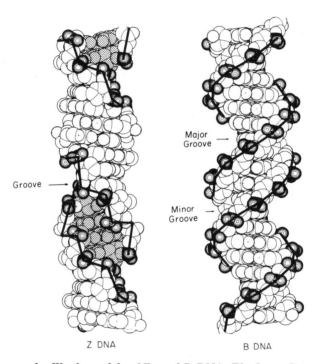

Z DNA   B DNA

Fig. 1.2. *van der Waals models of Z- and B-DNA. The heavy lines follow the phosphate backbones. Reproduced from Rich et al. (1984), with permission of the publisher.*

nucleosides. In B-DNA all nucleosides have the *anti* configuration, as shown in Figure 1.3 for deoxyguanosine, whereas in Z-DNA the purine nucleosides have a *syn* conformation alternating with the pyrimidine nucleosides in the *anti* form, as shown diagrammatically in Figure 1.4. The pyrimidines rotate with their sugars, whereas the purines rotate around the glycosidic bonds (Fig. 1.3). Hence the pyrimidine nucleosides remain in the *anti* form. Z-DNA is most likely to be found in sequences that have an alternation of purine and pyrimidine bases. The geometry of the AT

B-DNA

Z-DNA

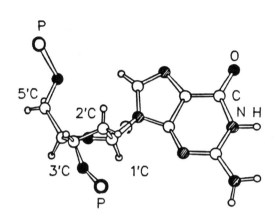
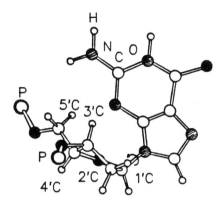

Fig. 1.3. *Conformation of deoxyguanosine in B-DNA* (**a**) *and Z-DNA* (**b**). *The sugars are oriented so that the plane defined by C1'−O1'−C4' is horizontal. Atoms lying above their plane are in the* endo *conformation. In B-DNA C2' is* endo *whereas in Z-DNA C3' is endo. B-DNA has guanine in the* anti *position while Z-DNA has guanine in the* syn *position. Reproduced from Rich et al. (1984), with permission of the publisher.*

B-DNA

Z-DNA

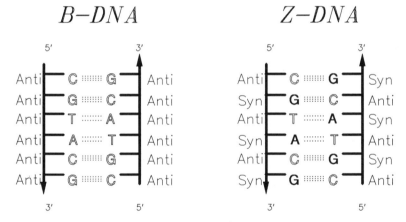

Fig. 1.4. *Comparison of the B- and Z-forms of DNA with an identical sequence of nucleotides. The Z-DNA has the* syn *forms of the purines while the pyrimidines remain in the* anti *form. Reproduced from Rich et al. (1984), with permission of the publisher.*

base pairs is similar to that of CG, with the adenines in the *syn* form and the thymines in the *anti* form.

A further consequence of the *syn* conformation of the purines is that the nucleotide pairs near the center of the B-helix will be near the periphery in the Z-helix, as shown by the end views of the helices in Figure 1.5. Topological analysis of these two forms of DNA indicates that

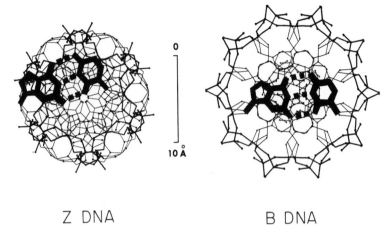

Z DNA   B DNA

Fig. 1.5. *Looking down the helices of Z- and B-DNA. Note the difference in the position of a guanine–cytosine pair; it is near the center of B-DNA, but near the periphery of Z-DNA. Also the phosphate backbone shown with the heavier lines is less exposed to the outside environment in the Z-DNA. Reproduced from Rich et al. (1984), with permission of the publisher.*

they may easily undergo interconversion by the flipping or rotating of the nucleoside pairs. The B $\rightleftarrows$ Z equilibrium point will be determined by environmental conditions and by methylation of the cytosines, which is conducive to pushing the B $\rightleftarrows$ Z equilibrium to the right.

Other differences between the B- and Z-forms and their biological consequences are discussed in the following chapters in their relevant contexts. For the present it should be emphasized that we can by no means be confident that we understand the conformational status of the DNA in chromosomes. It can be different from one part of a single chromosome to another, and from one nucleus to another. The flexibility of the various states may in fact be quite great and directly related to cell differentiation and the functioning of chromosomes in their native state.

B. Other Topological Variations

In addition to a variety of forms such as B and Z, DNA also assumes different states associated with the coiling of the duplex helix itself. This is called **supercoiling**, which is essential to the organization of the chromosome and its functioning in the replication, transcription, and the recombination of the DNA. The different topological states associated with this supercoiling are controlled by a group of enzymes within the nucleus and in intimate association with the chromosomes. These enzymes collectively are called **topoisomerases**.

Supercoiling of DNA occurs in both prokaryotes and eukaryotes. Vinograd (Vinograd et al., 1968) was the first to point out the importance of supercoiling in DNA function on the basis of his pioneering studies with the polyoma virus. This virus, like other viruses and prokaryotes in general, has a circular genophore. The supercoiling of a circular DNA, such as is found in a phage, is illustrated in Figure 1.6c.

It is convenient to think of the DNA double helix as a twisted ribbon the edges of which have opposite or antiparallel directions $3' \rightarrow 5'$ versus $5' \rightarrow 3'$. A circular molecule of DNA such as illustrated in Figure 1.6a can be formed only if edges with the same polarity are joined to-

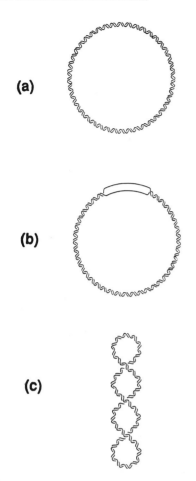

(a)

(b)

(c)

Fig. 1.6. *Configurations of circular DNA. a. Relaxed, zero supercoiling. b. An originally supercoiled molecule, as shown in (c), converted to a partially relaxed molecule by strand separation, reducing the number of times that one strand twists about the other. c. Negatively supercoiled DNA. Reproduced from Lewin, 1987, with permission of the publisher.*

ruption of some of the base pairs to give a region of unpaired DNA bases (Fig. 1.6b). The supercoil illustrated in Figure 1.6c is said to be **negative** because its formation involves a partial unwinding of the helix. On the other hand, if the duplex is twisted clockwise, it becomes overwound and the supercoil is **positive**.

The state of DNA supercoiling of a closed duplex DNA circle is given by the simple equation:

$$Lk = Tw + Wr, \qquad (1)$$

where Lk, the **linking number**, is the number of times that one DNA strand crosses the antiparallel strand, that is, the number of times the two strands are interlinked. Lk, therefore, must be an integer that is constant for a circle. It can be changed only by making single- or double-stranded DNA cuts, rotating the DNA ends, and religating to form another circle with a different integral value of Lk. A right-handed rotation is by definition positive, and the DNA end must be rotated through 360° before religating because of the polarity of the DNA strands.

Tw in Equation 1 is the **twisting number;** this is the number of helical repeats contained in the DNA circle. For a circle with N base pairs of B-form DNA with a helical repeat of 10.5 bp, $Tw = N/10.5$. Thus for most DNA circles, Tw is not an integer. A closed DNA molecule is often depicted as a planar circle as shown in Figure 1.6a. It is important to recognize that such a circle can exist only in the relaxed state, if it contains an integral number of helical repeats such that $Lk^0 = Tw^0$. This equality, however, holds only for a particular set of conditions. B-DNA is flexible and its precise helical repeat depends on the ionic strength, pH, and temperature of the solution medium it is in. A change in the solution conditions can cause a change in the precise helical repeat of the DNA with the result that Tw

gether. If, however, before a circle is formed, one end of the ribbon is twisted counterclockwise in the direction of unwinding the helix (assumed here to be right-handed), the DNA will be under strain, and no longer in the relaxed state. The strain can be relieved by the twisting of the duplex upon itself as shown in Figure 1.6c. Also the strain can be altered by dis-

differs slightly from Lk. The DNA circle then becomes slightly strained and deviates from the previously relaxed planar state to become superhelical. One response of a DNA circle to alleviate strain is called **writhing**, Wr in equation 1. The other response is a change in the twisting number Tw: i.e., a change in the DNA helical repeat in prokaryotic, but not eukaryotic cells. The introduction of significantly higher levels of superhelical strain into a prokaryotic DNA circle requires the action of an ATP dependent DNA gyrase that changes the linking number of the circle so that $Lk^0 > Tw^0$. The **writhing number**, Wr, is equal to the number of superhelical turns introduced into the circle. For a fully relaxed planar circle $Lk^0 = Tw^0$ and $Wr = 0$. If these values are subtracted from Equation 1 for a superhelically strained circle, then

$$(Lk - Lk^0) = (Tw - Tw^0)$$

$$+ (Wr - 0) \text{ or } \Delta Lk = \Delta Tw + Wr. \quad (2)$$

Thus any superhelical strain introduced into a DNA circle will be partitioned between a change in twist Tw and the introduction of writhe, Wr.

Both positive and negative supercoiling can be introduced into a DNA circle by a variety of methods. All DNA circles, however, are negatively supercoiled, in which case $\Delta Lk < 0$. This negative supercoil strain can be accommodated by the circle by the unwinding of the B-form DNA helix and the introduction of negative superhelical DNA turns. If in the DNA circle there is a sequence of alternating purines and pyrimidines with the potential of forming left-handed Z-form DNA, then high levels of a negative supercoil strain can be relieved by the transition in this region from the B- to the Z-form (or a left-handed B-form). This transition of one turn of B-form to Z-form DNA will relieve the strain of two negative superhelical turns of DNA. Another DNA structural transi-

tion that can relieve negative supercoiling involves inverted DNA repeats or palindromic sequences with the potential of forming DNA cruciforms. With the increasing levels of negative DNA supercoiling, there is a high probability of cruciform formation.

DNA supercoiling in eukaryotes is not as well understood as it is in prokaryotes and viruses, which have much smaller genophores. Chromosomes have extraordinarily long linear DNA molecules associated with proteins in very specific conformations. Another consideration is that the enormous mass of chromosomes in comparison with genophores could be expected to hinder their ends from rotating if a constrained internal DNA loop was subjected to supercoiling. So far a eukaryotic equivalent to a prokaryotic DNA gyrase has not been identified. The packaging of DNA in nucleosomes, discussed later in this chapter, introduces a singular aspect to supercoiling not found in prokaryotes.

III. CHROMOSOMAL DNA CONSISTS OF UNIQUE AND REPETITIVE SEQUENCES OF BASE PAIRS

A. Kinetics of Reassociation of Single-Stranded DNA Identifies Unique and Repetitive DNA

Much analysis of DNA structure has been done not only by crystallographic methods, but also by studying the kinetics of the reassociation of short segments of single-stranded DNA that have complementary sequences. The hydrogen bonds between the adenines and thymines, and between the cytosines and guanines, in double-stranded, native DNA are easily ruptured by elevating the temperature of a suspension of DNA to promote the formation of single strands. It is possible to do this without breaking the covalent phosphate bonds of the strand backbones or those in the other nucleotide components, because hydrogen bonds are weaker than covalent bonds by a factor of more than 10. DNA isolated from cells can first be sheared to short segments several hundred nucleotides long by passing it through a hypodermic needle, and then can be transformed to the single-strand state preparatory to allowing it to reform double strands as described in the following sections.

Double-stranded DNA is **disassociated**, **denatured**, or **melted** (its strands are separated) by being subjected to heat, high salt concentrations, or alkaline pH. Figure 1.7 shows the course of DNA denaturation by heat as measured by changes in optical density. As single strands are formed, the optical density at the 260-nm wave length of the DNA solution increases, a phenomenon known as hyperchromicity. The DNA sample used in this example denatured rapidly between 80°C and 90°C. The midpoint of this range is called the melting temperature and designated as **Tm**. The Tm value is a function of the ionic strength of the solution, which is determined by the salt concentration, and the base content of the DNA in solution.

The DNA of most mammals will have a Tm of about 86°C, or an average GC content of about 40%. But a DNA that is 60% GC will have a Tm of about 94° under the same conditions, and part of mammalian DNA does have this high GC content. Tm increases linearly about 0.4° for each 1% increase in GC content, other conditions being equal. Thus one can estimate the GC content by determining the Tm of a sample of DNA. The reason for the difference is that GC nucleotide pairs have three hydrogen bonds between the bases, whereas AT pairs have only two. Hence it takes a higher temperature to separate guanines from cytosines in a polynucleotide.

Nuclear DNA that is purified, sheared to various lengths, and heated to give a solution of single strands will renature, reanneal, or reassociate with lowering of the temperature, usually to 60–70°C. The complementary strands begin to reassociate to produce duplex (double-stranded) DNA again. The reassociation occurs over a period of time and can be measured by noting the reduction in absorbance at 260 nm, and the single-strand fraction can be separated from the double-stranded at any time during reassociation by passing the

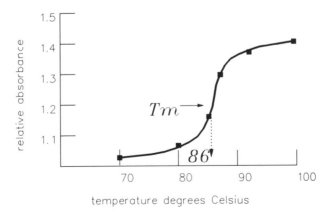

Fig. 1.7. *S-shaped curve obtained when a sample of native DNA is heated to temperatures above the normal physiological range sufficiently to separate the strands.*

solution through a column of hydroxyapatite. Hydroxyapatite is a complex form of calcium phosphate. Under the proper conditions it binds the duplex DNA but not the single strands. Measurements of single strands remaining after reassociation is started can be made by using DNA labeled with tritium or carbon-14 and determining the radioactivity that elutes from the hydroxyapatite. Since some of the reassociated DNA will be part single-stranded and part double-stranded, some of the single-stranded fraction will be retained on the column. This can lead to an underestimation of the amount of single-stranded DNA, but this can be corrected for in part by using the nucleases, such as S1 which hydrolyzes single-stranded, but

not double-stranded DNA. By using various combinations of these methods one can assay the amount of duplex DNA present after a period of reassociation with a considerable degree of accuracy.

When the temperature of a solution of single-stranded DNA is lowered to 60°, it does not renature instantaneously but over a period of time. The rate of reassociation depends upon the concentration of complementary sequences according to the law of mass action (Britten and Kohne, 1966; Davidson et al., 1973). The equation

$$C/C_0 = 1/(1 + kC_0t) \qquad (3)$$

describes the course of the reactions. Its derivation is given in Box 1.1, section A.

BOX 1.1 KINETICS OF DNA STRAND REASSOCIATION AND THE CALCULATION OF GENOME SIZES

A. Reassociation Kinetics

The reassociation of single complementary strands of DNA follows *second-order reaction kinetics* because the rate is proportional to the concentration of *two* reactants. Thus it is a *bimolecular reaction* involving two complementary strand molecules. The second-order rate equation is

$$\frac{dC}{dt} = -kC^2, \qquad (1)$$

in which C is the concentration of single-stranded DNA at time t and k is the rate constant. Integration of this equation between limits of C_0, the initial concentration of C at $t = 0$, and the concentration of DNA remaining single stranded, C, at time t gives a general expression (Equation 2) that describes the course of the reaction:

$$\frac{dC}{C^2} = -k \int_0^t$$

$$\left. \frac{1}{C} \right]_{C_0}^{C} = -kt$$

$$1/C - 1/C_0 = kt$$

$$C/C_0 = 1/(1 + kC_0t). \qquad (2)$$

C_0t is the product of the nucleotide concentration in moles/liter and time in seconds ($M \times s^{-1}$).

When the reaction is one-half complete,

$$C/C_0 = 1/2 = \frac{1}{1 + kC_0t_{1/2}}$$

$$1 + kC_0t_{1/2} = 2$$

$$C_0t_{1/2} = 1/k \qquad (3)$$

or

$$k = 1/C_0t_{1/2} \qquad (4)$$

B. Calculation of Genome Sizes

The size of the genome can be estimated from the kinetics of the renaturation. Given a genome of total size G bp with a slow unique nonrepetitive component U, then

$$U = \alpha G \text{ bp} \qquad (5)$$

in which α is the proportion of U in the total genome, G bp.

Assuming that this genome, G bp also contains an intermediate fast (repetitive) component R repeated F times per G bp, then

$$R = \frac{\beta G \text{ bp}}{F} \qquad (6)$$

in which β is the proportion of the component, and F is the number of times the sequence is repeated. R and U can be determined from the $C_0t_{1/2}$ values:

$U = \alpha C_0t_{1/2}$ (nonrepetitive)

$\qquad \times 4.2 \times 10^6/C_0t_{1/2}$ (*E. coli*);　(7)

$R = \beta C_0t_{1/2}$ (repetitive)

$\qquad \times 4.2 \times 10^6/C_0t_{1/2}$ (*E. coli*).　(8)

Substituting Equation 5 into Equation 7,

αG bp $= \alpha C_0t_{1/2}$ (nonrepetitive)

$\qquad \times 4.2 \times 10^6/C_0t_{1/2}$ (*E. coli*);

G bp $= C_0t_{1/2}$ (nonrepetitive)

$\qquad \times 4.2 \times 10^6/C_0t_{1/2}$ (*E. coli*).　(9)

Thus the size of any genome can be determined directly from the $C_0t_{1/2}$ value of its nonrepetitive component. This $C_0t_{1/2}$ value must, however, represent the total DNA, not the proportion α that it occupies in the total genome. Thus, if it occupies 50% of the genome and the $C_0t_{1/2}$ value is determined from the pure nonrepetitive component, this must be divided by 0.5 to get the $C_0t_{1/2}$ for the total DNA to be substituted in the Equation 9.

If now Equation 6 is substituted in Equation 8:

βG bp$/F = \beta C_0t_{1/2}$ (repetitive)

$\qquad \times 4.2 \times 10^6/C_0t_{1/2}$ (*E. coli*)

and if this is divided by Equation 9, then

$$F = \frac{C_0t_{1/2} \text{ (nonrepetitive)}}{C_0t_{1/2} \text{ (repetitive)}}. \qquad (10)$$

Thus, if the nonrepetitive component is unique ($F = 1$), then the F of the repetitive component can be determined.

C_0 is the concentration of the total DNA in the single-stranded sample in moles/liter at time $t = 0$, and C is the concentration of single strands at any time t, in seconds, after reassociation is started.

The course of reassociation of a sample of DNA is described by plotting the percentage of single-stranded DNA remaining after the reaction is started against the logarithm of $C_0 \times t$. Figure 1.8 depicts the results expected for an ideal second-order reaction when a sample of single-stranded pieces of DNA about 500 nucleotides (C_0) in length reassociates back to the 100% double-stranded form over time t. The concentration of single-stranded

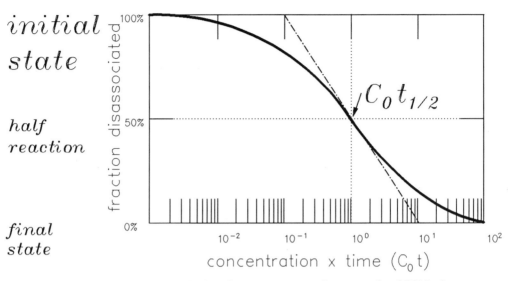

$$\text{association course:} \quad \frac{C}{C_0} = \frac{1}{1 + kC_0 t}$$

Fig. 1.8. *An ideal reassociation time course curve for a sample of DNA. See text and Box 1.1.*

DNA is plotted against time. Note that about 80% of the reassociation reaction approaches linearity over a 2-log interval of $C_0 t$. Curves of this general type are obtained whenever reassociation is carried out with samples of DNA from all organisms, whether prokaryotic or eukaryotic. The data obtained are most useful when the $C_0 t$ value at one-half completion ($C/C_0 = \frac{1}{2}$) is estimated from the linear portion of the curve. Ideally this is the inflexion point as shown in Figure 1.8. The substitution of $C_0 t_{1/2}$ in Equation 3 shows that $C_0 t_{1/2}$ is the reciprocal of the second-order rate constant k:

$$C_0 t_{1/2} = 1/k. \qquad (4)$$

If we consider for illustrative purposes an artificial copolymer consisting of U and A strands only: (UUUU)/(AAAA)n, the strands of which are reassociated following their dissociation, reassociation will

be rapid and give a $C_0 t_{1/2}$ of about 2×10^{-6} (Figure 1.9). If, on the other hand, a copolymer such as (ATATAT)/(TATATA)n is treated in a similar fashion, the $C_0 t_{1/2}$ value will be greater than this since two pairs of bases A:T and T:A must align rather than one, as for U:A. A repeating tetramer of the form (ATGC)$_n$ with its complement (TACG)$_n$ will take longer to reassociate and have an even higher $C_0 t_{1/2}$ value in the range of 4×10^{-6}.

In order to quantify the relationship between the length of repeating segments and $C_0 t_{1/2}$, the concept of complexity is introduced. The **complexity Xc** of a preparation of DNA segments is defined as the length in nucleotide pairs of the longest nonrepeating sequence that can be produced by reannealing together complementary fragments in the population. Hence, for the examples given above, U:A has a complexity of 1, AT:TA a complexity

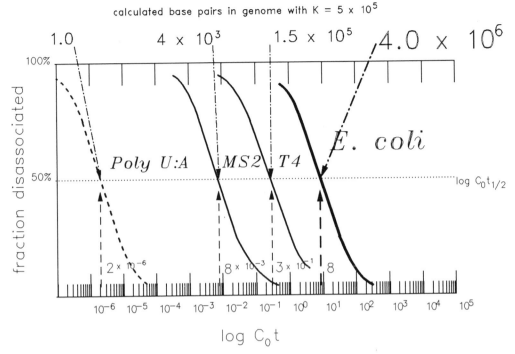

Fig. 1.9. C_0t *curves for a synthetic DNA (U:A), phages MS2 and T4, and the bacterium* E. coli. *For details see text.*

of 2, and $(ATGC)_n$ a complexity of 4. $C_0t_{1/2}$ is directly proportional to complexity. Therefore we can write

$$Xc = K\, C_0t_{1/2} \qquad (5)$$

Where K is a proportionality constant whose value depends on reaction conditions. Under the conditions generally employed with a 0.18N concentration and a 400 nucleotide fragment size, $K \simeq 5 \times 10^5$. Therefore we can write the more specific equation

$$Xc = (5 \times 10^5)\, (C_0t_{1/2}), \qquad (6)$$

from which we can calculate that U:A has a $C_0t_{1/2} = 2 \times 10^{-6}$. It follows that AT:TA will have a $C_0t_{1/2}$ of 4×10^{-6} and $(ATGC)_n:(TACG)_n$ a $C_0t_{1/2}$ of 8×10^{-6}.

B. Reassociation Kinetics Can Determine the Size of a Genome

The direct relationship between Xc and $C_0t_{1/2}$ allows one to determine the total number of base pairs (bp) in an unbroken length of DNA or RNA such as for a viral or bacterial genophore. Consider the data for reassociation of the RNA phage MS2, the DNA phage T4, and the DNA of *E. coli* (Figure 1.9). The calculated Xc values using the observed $C_0t_{1/2}$ values in Equation (6) are 4×10^3, 1.5×10^5, and 4.0×10^6, respectively. The number of base pairs represented in a genophore can be determined directly by spectrophotometric methods to give the **chemical complexity**. When this is done, MS2 is found to have 3.6×10^3 bp, T4 about 1.8×10^5 bp, and *E. coli* about 4.2×10^6 bp. Thus there is a striking correspondence between the

Xc values or **kinetic complexity** and the directly determined chemical complexity. The sizes of the genophores of viruses and bacteria are equal to their *Xc* values, or in other words their kinetic complexity is approximately equal to their chemical complexity. This is because their genomes (or genophores) contain few or no repeated sequences. Each genome has about only one representative of each type of segment. Sequences that do not repeat are generally called **unique**, that is, they have a frequency of 1, or very few, per genome. It is characteristic of viruses and bacteria that their genomes are made up of unique sequences of DNA. However, it should be noted that repeated identical sequences of less than about 10 cannot be distinguished from unique sequences; therefore the kinetic method has a resolving power of 10. Given the $C_0t_{1/2}$ value for the DNA of any prokaryote, its genome size can be determined by using the well-established value for the chemical complexity of *E. coli* (4.2×10^6 bp), by substituting in the following equation:

$$\frac{C_0t_{1/2} \text{ of any genome}}{C_0t_{1/2} \; E. \; coli}$$

$$= \frac{\text{Complexity of any genome}}{4.2 \times 10^6 \text{ bp}} \quad (7)$$

(The $C_0t_{1/2}$ of an *E. coli* sample should be determined under the same conditions as the $C_0t_{1/2}$ of the unknown sample, because these values vary with the temperature, ionic strength, and length of the reassociating strands.)

Equation 3 and its derivatives, Equations 4–6, are useful in the analysis of eukaryotic genomic DNA as well as the prokaryotic variety. However, unlike prokaryotes, eukaryotes have genomes consisting of repetitive DNA as well as unique DNA. The repeated DNA segments naturally reassociate more rapidly than the unique, since they are present at a higher concentration in a total genomic DNA sample. Many different kinds of repetitive segments may be present and constitute anywhere from a few percent in the fungi, to 50%–60% or more of the total genomic DNA in the higher eukaryotes. Figure 1.10 gives two examples, one for the DNA from the fungus *Aspergillus* and the other for DNA from Syrian hamster cells. The C_0t curve for *E. coli* is also given as a reference.

Note that the C_0t curves shown in Figure 1.10 for the fungus and hamster DNA are spread out over a much greater range than the approximately 2 logs for *E. coli*. The explanation given for this difference is that there are more different sequences in the population of DNA segments in the eukaryote samples. Some sequences are repeated at high frequency and some are unique. If in a mole of DNA 50% of the sequences are repeated and 50% are not, it is to be expected that the repeated sequences will reassociate more rapidly than the unique, since the repeated segments are present in higher concentration than the unique. Therefore, a curve such as shown by the solid line in Figure 1.11 should be obtained. When the early-reassociating (repetitive) strands are separated from the late-reassociating (unique) strands and these fractions are allowed to reassociate separately, curves like that shown by dashed lines are expected.

If the $C_0t_{1/2}$ for each of the separated pure components shown in Figure 1.11 are determined, a value of about 10^{-1} is obtained for the repetitive fraction and about 10^3 for the unique fraction. If the $C_0t_{1/2}$ values for the mixture shown by the solid line C_0t curve are determined, they differ from those observed for the pure forms. $C_0t_{1/2}$ for the repetitive fraction is about 2×10^{-1} and the unique fraction is about 2×10^3. The difference is the result of the fact that the mixture C_0 represents the total DNA concentration of both the unique and repetitive, whereas the isolated frac-

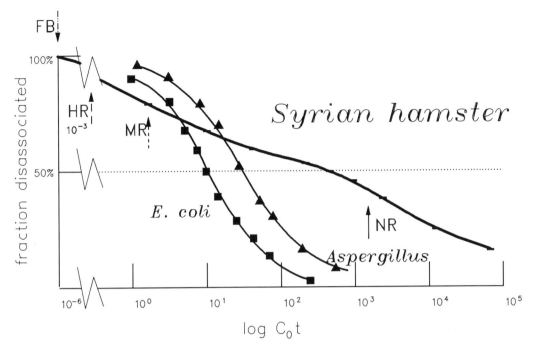

Fig. 1.10. C_0t *curves for the DNA from* Aspergillus nidulans, *the Syrian hamster, and* E. coli. *See text. FB = foldback; HR = highly repetitive; MR = moderately repetitive; NR = nonrepetitive or unique.*

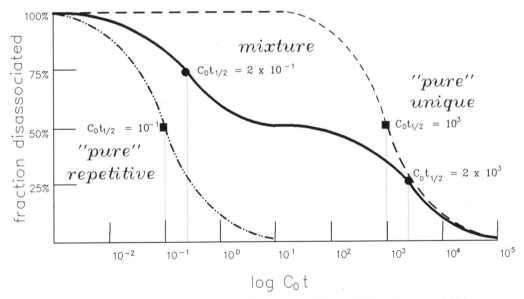

Fig. 1.11. C_0t *curves expected from "pure" repetitive and "pure" unique DNA fractions. Note that the mixture of the two has two flexion points that are different from the* $C_0t_{1/2}$ *values gotten with the pure fractions. The "pure" repetitive fraction corresponds to the MR and the "pure" unique to the NR fractions.*

tions have a C_0 for either the unique or the repetitive only. That is, a mole of mixture has 50% repetitive and 50% unique sequence while the "pure" fractions are either 100% repetitive or 100% unique. The pure fractions should therefore reassociate more rapidly, and they do.

The size of any genome, totally or partially unique, may be determined by kinetic analysis of the reassociation. Box 1.1, section B gives the details of the reasoning involved. It is sufficient to say here that the $C_0t_{1/2}$ of the **nonrepetitive component** can be used to determine the size of any genome. But this $C_0t_{1/2}$ value must represent the total DNA of the genome, not the fraction that the unique part occupies in the total genome.

The C_0t curve for a sample of total DNA from *Aspergillis nidulans* is similar to that of *E. Coli*, also included in Figure 1.10, but it has a higher $C_0t_{1/2}$. Single-strand fragments of 0.5 kb were reassociated and gave a $C_0t_{1/2}$ of about 24, which is about three times the $C_0t_{1/2}$ for *E. coli*. From these data the genome size for *Aspergillus* is calculated to be in the range of 10^7 bp, as compared with the 10^6 bp of *E. coli*. The shape of the curve for *Aspergillus* indicates very little if any repeated sequences. However, if a sample of single-strand total DNA is reassociated to a C_0t of 1 and this duplex DNA is separated from the remaining single-stranded DNA by passing

the sample through a hydroxyapatite column, it is found that single-stranded DNA derived from the fast-reassociated duplex fraction has a very low $C_0t_{1/2}$. This is a repetitive fraction calculated to be about 2%–3% of the total genome DNA (Timberlake, 1978). In general, fungi have about 98% unique DNA. The rest is in repetitive components. Since the $C_0t_{1/2}$ value determined for *Aspergillus* is for the total DNA, Equation 7 can be used to determine the kinetic complexity. The calculated value of 1.71×10^7 bp compares reasonably well with the chemical complexity (or total DNA/genome) of 2.1×10^7 bp. The ratio of chemical complexity to kinetic complexity should be about unity; in this case it is 1.2. Table 1.2 gives some examples of amounts of nonrepetitive DNA for a number of animals.

The C_0t curve for Syrian hamster DNA shown in Figure 1.10 is spread over 12 log units with the nonrepetitive component C_0t curve covering about 10^2 to 10^5. Moyzis et al. (1981) have fractionated the total DNA into four components: (1) foldback, (2) highly repetitive, (3) moderately repetitive, and (4) nonrepetitive. These fractions are indicated as FB, HR, MR, and NR, respectively, in Figure 1.10. Table 1.3 lists these with estimations of percentages of each type, their $C_0t_{1/2}$ values, and their frequencies. If the $C_0t_{1/2}$ of 2,300 for the nonrepetitive fraction is used together with

TABLE 1.2. Examples of Amounts of Nonrepetitive DNA Among Eukaryotes

	Total genome size, bp	Nonrepetitive comp. (bp)	Percentage nonrepetitive
Drosophila melanogaster	1.8×10^8	1.3×10^8	70
Musca domestica (house fly)	8.6×10^8	3×10^8	34
Bombyx mori (silkworm)	5.0×10^8	2.8×10^8	56
Antheraea pernyi (a moth)	9.7×10^8	3.4×10^8	35
Rattus norvegicus (rat)	3.0×10^9	2.0×10^9	67
Homo sapiens (human)	3.3×10^9	1.7×10^9	52
Xenopus laevis (toad)	3.1×10^9	1.7×10^9	55
Bufo bufo (toad)	6.7×10^9	1.3×10^9	19
Necturus necturus (salamander)	5.0×10^{10}	12×10^9	24

$C_0t_{1/2}$ of 8 and complexity of 4.2×10^6 bp for *E. coli*, the size of the hamster genome calculated from Equation 7 is 2.9×10^9 bp. This is quite close to the chemical complexity determination of 3.3×10^9 bp. This ratio of chemical to kinetic complexity is 1.1.

The estimated 3% of foldback DNA is considered to be single-strand segments of DNA with palindromic ends (Figure 1.12). These ends, being complementary, reassociate almost instantaneously. The $C_0t_{1/2}$ of 10^{-6} given in Table 1.3 is an estimate. The highly repetitive component has a great many copies per genome and a complexity, determined by Equation 6, of about 500. The moderately repetitive compound comprises about a third of the genome, with a complexity of 1.3×10^6. Equation 6 gives an Xc value of 1.2×10^9 for the nonrepetitive component.

In addition to determining the genome size, the $C_0t_{1/2}$ values can also be used to determine the frequency F of the occurrence of repetitive sequences of the genome. The equation

$$F = \frac{C_0t_{1/2} \text{ of nonrepetitive DNA}}{C_0t_{1/2} \text{ of repetitive DNA}} \quad (8)$$

can be used with the same strictures applying for its use as for Equation 5. (See Box 1.1B.) Thus, for the middle repetitive sequence in Table 1.3, $2,300/2.8 \simeq 820$, and for the highly repetitive one, $2,300/ <10^{-3} \simeq 2.3 \times 10^6$.

Evidence for the validity of the general application of Equation 7 is given in Figure 1.13. Kinetically determined genome sizes are plotted against chemically determined sizes for a variety of organisms ranging from *E. coli* to humans. The straight line drawn through the points has a slope close to 1, and the correlation coefficient $r = .98$.

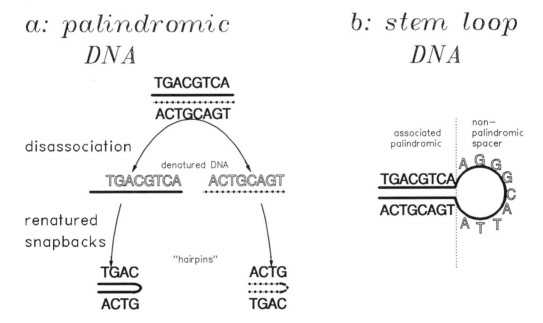

Fig. 1.12. *Hairpins and stem loops expected from palindromic DNA* **(a)** *and from stem loop DNA* **(b)**. *See text.*

TABLE 1.3. Reassociation Data for Syrian Hamster DNA: Total Nuclear DNA Sheared to 250 bp Lengths

Component	Percentage DNA in each component	Observed $C_0t_{1/2}$	No of copies per genome	Complexity (X_c)
Foldback	3	$<10^{-6}$	—	—
Highly repetitive	2	$<10^{-3}$	$>2.3 \times 10^6$	500
Middle repetitive	35	2.8	820	1.3×10^6
Nonrepetitive	60	2,300	1	1.2×10^9

Data from Moyzis et al. (1981a).

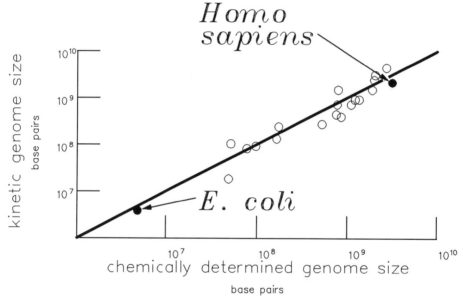

Fig. 1.13. *A plot of kinetically determined genome sizes against chemically determined sizes for DNA from a variety of organisms. See text.*

C. Eukaryotic Repetitive Sequences Are Heterogeneous in Structure

The C_0t curve for the DNA of the Syrian hamster shown in Figure 1.10 is characteristic of those found for mammals. C_0t extends from 10^{-6} to 10^5, a range of 10^{11}, in contrast to *E. coli* DNA, which has a range of about 10^2. The 37% of this DNA that is repetitive is far from homogeneous.

Values in Table 1.3 indicate that the 5% component reassociates extremely rapidly, with extremely low $C_0t_{1/2}$ values. In contrast, the middle repetitive sequences reassociate more slowly, with a $C_0t_{1/2}$ of about 2.8. It is this component that has been further analyzed by Moyzis et al. (1981a, 1981b) and shown to be heterogeneous in its composition. It is a mixture of many different kinds of segments, each with its own sequence.

If the middle repetitive region of the C_0t curve is isolated by fractionization on hydroxyapatite columns to obtain DNA from C_0t 10^{-6} to C_0t 50, this component can be further fractionated into subcomponents after it is sheared to 150 average fragment length (Mòyzis et al., 1981b). Fractionation is done by reassociating the single strands at a variety of temperatures ranging from 60°C to 80°C (about 25°C to 5°C below Tm, respectively). When only one temperature is used for reassociation, such as 25°C below Tm (the commonly used value), strands with sequences that have only 75%–85% complementarity will reassociate, thus obscuring any heterogeneity in sequence that may exist. By using a series of temperatures, curves such as those shown in Figure 1.14a are obtained. These curves have been arbitrarily separated for clarity by one log $C_0t_{1/2}$. Actually the $C_0t_{1/2}$ values for all six fractions center around 4.5. When duplex middle repetitive DNA is thermally eluted from hydroxyapatite at a

series of temperatures from 60° to 75°, the series of curves shown in Figure 1.14b are obtained. The Tm values are lower for each fraction than the Tm for the native (unprocessed) DNA, the 60°-eluted fraction having a Tm value a full 9° lower than the native DNA. A reasonable interpretation of these results is that the middle repetitive fraction of the hamster genome appears to contain "families" of different, but probably related, sequences of DNA that have different GC contents, since they have different Tm values. Furthermore, the sequences within a family appear to be closely or completely homogeneous. This conclusion is based on the observation that the increase in the temperature at which they reassociate does not change their $C_0t_{1/2}$ values significantly.

All mammals whose DNA has been analyzed by reassociation kinetics have numerous repetitive sequences such as are found in the Syrian hamster (Singer, 1982). It is also apparent that the repetitive DNA

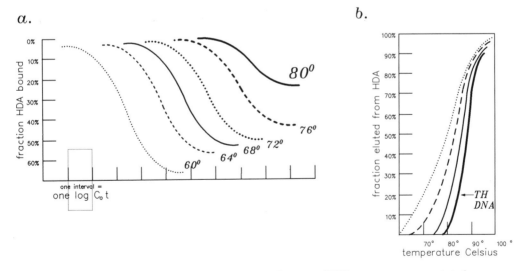

Fig. 1.14. a. *Isolated moderately repetitive hamster DNA segments reassociated at different temperatures as described in the text. The percentage bound to hydroxyapatite is the percentage of reassociated DNA.* **b.** *Repetitive duplex DNA eluted from hydroxyapatite at temperatures indicated. Data from Moyzis et al. (1981b).*

sequences are distributed at the centromeric and telomeric parts of chromosomes as well as dispersed throughout the unique regions of chromosomes, as described in the next chapter.

IV. DETERMINATION OF DNA NUCLEOTIDE SEQUENCES

The use of renaturation kinetics yields important information about the frequencies of occurrence of different types of repetitive and nonrepetitive DNA sequences. However, a knowledge of the actual base pair sequences is also important for a further understanding of basic chromosome structure. The following discusses two ways in which nucleotide sequencing is done. The primary structure of DNA is its sequence of base pairs represented by its nucleotides. The procedure for determining the nucleotide sequence for a stretch of DNA consists in cleaving the DNA into small segments, replicating

the segments by cloning in plasmids or phage, and then determining the nucleotide sequences of selected segments—usually about 300–400 kb long. Two procedures are currently being used to replicate and isolate specific segments for sequence determinations.

Box 1.2 describes the procedure for inserting and cloning DNA segments in the phage M13, preparatory to determining their nucleotide sequences. After the isolation of the phage with the desired inserted segment of DNA, one can proceed with the process of sequencing.

Box 8.4, Chapter 8, describes the procedure for replicating specific fragments of DNA by using the polymerase chain reaction (PCR).

Two basic methods have been developed to determine nucleotide sequences in DNA segments 300–500 bases in length: the dideoxy method (Sanger et al., 1977) and the chemical method (Maxam and Gilbert, 1977, 1980).

BOX 1.2 FORMATION AND CLONING OF RECOMBINANT DNA MOLECULES

The procedure (general references: Winnaker, 1987, and Sambrook et al., 1989) involves four basic steps that can be outlined as follows: (1) preparation of "foreign" DNA segments from a nuclear source and the selection of a **vector** to carry them; (2) ligation of the vector and foreign DNA to form a chimera DNA molecule (this is the **recombinant** DNA); (3) introduction of the vector and its recombinant molecule into a suitable host cell such as *E. coli.* (this process is referred to as transformation or infection); (4) replication of the host cells along with their contained recombinant DNA, which is also replicated

with the vector (the host cells that carry the recombinant molecules are selected for).

A. Vectors and Hosts

Either plasmids or viruses are generally used as *vectors* to carry the *foreign* or *insert* DNA. The phage M13 is used as an example in the following discussion, but the principals involved in forming recombinant DNA molecules are essentially the same for it as for any other *vector*. Vectors in general use are plasmids, cosmids, phage, animal viruses, and yeast artificial chromosomes. Most are highly modified from their "natural" state. They are "engineered" for specific roles as described in the following discussion for M13. The vectors with their foreign DNA included are cloned in host cells, which may be bacterial, yeast, or certain plant or animal cells.

The M13 Vectors: The genophore of M13 in its vegetative (nonreproducing) phase is a single stranded circle of DNA with 6,407 nucleotides. M13 is called a *filamentous phage* because the genophore is not confined in a protein capsule of limited size. Instead, the genophore takes on an extended linear shape, and when foreign DNA is inserted, the protein coat expands linearly to resemble a filament. Hence longer pieces of foreign DNA up to 2 kb can be inserted and still be maintained within the phage coat.

M13 infects the "male," *F* episome-bearing form of *E. coli* by attaching to a pilus and injecting its circular DNA. During injection the single-stranded (SS) DNA is converted to the double-stranded replicative (RF) form (Fig. 1). The RF form consists of a parental (+) SS and a complementary (−) strand. The (−) strand acts as a template for the generation of new (+) strands, and hence new nonreplicative genophores, which are maintained in the SS state by a single-stranded binding protein (SSB protein). The RF form also produces more duplex RF progeny within the host. These do not acquire protein coats. The SS(+) strands with SSB leave the host with a protein coat to continue the cycle. In addition to the advantage of being filamentous, M13 also has the advantage of not lysing its host cell. It just continues to reproduce phage particles that escape into the medium, leaving the host to maintain and replicate itself while producing more phage. In a liquid culture as many as 5×10^{12} phage particles per 1 ml may be obtained. Pure, uncontaminated SS(+) DNA is easily obtained from them.

The duplex RF M13 genophore has been modified by introducing into it

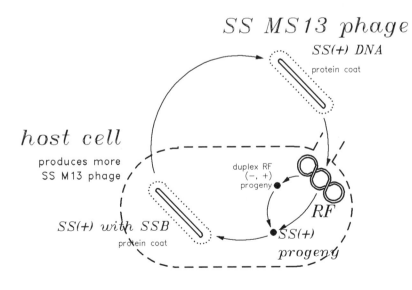

Fig. 1.

(Continued)

BOX 1.2 FORMATION AND CLONING OF
RECOMBINANT DNA MOLECULES *(Continued)*

parts of the *lac* operon of *E. coli*: specifically the *lac I* gene, the *lac Z* promoter, and the part of the *lac Z* gene that encodes the α-peptide of β-galactosidase (Fig. 2a). The α-peptide is not a functional enzyme; it consists of 146 amino acids of the complete functional β-galactosidase. Messing and his colleagues (Messing, 1981; Yanisch-Perron et al., 1985) constructed the modified M13 M13mp2, for example, by introducing the *E. coli* DNA segment (**a** in Fig. 2.) into M13 (**b** in Fig. 2.) at a specific intergenic region of the phage (**c** and **d** in Fig. 2).

The foreign DNA to be sequenced is introduced into the modified phage such as M13mp2 by the use of restriction endonucleases.

B. Restriction Endonucleases

Restriction endonucleases are enzymes that cleave DNA duplex strands. Many different kinds of them have been isolated from various species of bacteria. Three general types (I, II, III) are known. Of these the type II endonucleases are the most frequently used in recombinant DNA technology. They recognize specific nucleotide sequences in DNA. Generally, the sequence recognized is a tetra-, penta-, or hexanucleotide stretch with an axis of rotational symmetry. For example, a widely used endonuclease EcoRI recognizes the hexanucleotide sequence:

$$5'\text{-}G \downarrow pApApTpTpC\text{-}3'$$
$$3'\text{-}CpTpTpApAp \uparrow G\text{-}5'$$

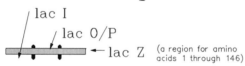

$a:$ *E. coli operon*

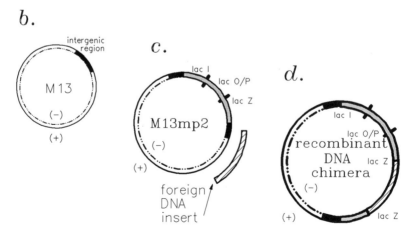

Fig. 2.

which reads the same way in the $5' \to 3'$ or $3' \to 5'$ directions on the complementary strands. The strands are therefore palindromic. The enzyme makes staggered cuts so that the result is that the cut ends have overlaps. Thus,

$5'$-GOH-$3'$ $\qquad$ $5'pApApTpTpC$-$3'$
$3'$-CpTpTpApAp $5'$ $\qquad$ $3'OHG$-$5'$

EcoRI produces protruding $5'$ ends. Other endonucleases such as PstI produce $3'$ protruding ends. Thus,

$5'$-CpTpGpCpA $\downarrow$ pG-$3'$
$3'$-Gp $\uparrow$ ApCpGpTpC-$5'$

$5'CpTpGpCpAOH$-$3'$ $\quad$ $5'pG$-$3'$
$3'$-Gp-$5'$ $\quad$ $3'OH$ ApCpGpTpC $5'$

Or the enzyme can cut to make "blunt ends" such as, for example, Hae III

$5'$-GpG $\downarrow$ pCpC-$3'$
$3'$-CpCp $\uparrow$ GpG-$5'$

$5'$-GpGOH-$3'$ $5'$-CpC-$3'$
$3'$-CpCp-$5'$ $3'OHGp$-$5'$

The great advantage of cuts made by endonucleases is that the ends generated by a given enzyme can form complementary base pairs with any other DNA molecules with similar ends. Thus, if a fragment of DNA cut with EcoRI is placed together with a genophase of a plasmid or phage that has also been cut with EcoRI, then both will have overlapping ends. The complementary base pairs reassociate and the phosphate bonds reformed with *ligase* enzyme as shown below:

The bases shown underscored are the contributions of the foreign DNA inserted into the genophore; those indicated as N can be any one of the four DNA bases.

Although the DNA fragments are generally produced by cutting with endonucleases, other methods for producing them—such as mechanical shearing, by RNA-directed synthesis, and chemical synthesis—are also used. These can be introduced into a cut vector by modifying the ends of both the vector and the foreign segment by various means (Winnaker, 1987).

C. Making, Cloning, and Rescuing Recombinant DNA Molecules

The RF MS13mp2 vector diagrammed in Figure 2c is cleaved with an appropriate restriction enzyme to form a linear molecule. It has a site recognized by EcoRI within the *lac Z* region. The foreign DNA with the appropriate sticky ends, also formed by EcoRI, is introduced, the ends of the vector are annealed with it, and a ligase is used to complete the phosphate bonds to form a circular duplex chimera (Fig. 2d). This recombinant molecule can be used to infect a strain of *E. coli* that carries the *F* episome. The *coli* host genophore is modified so that it lacks the entire *lac* operon, but its F factor carries the *lac* operon with a deletion in that part of the *lac Z* gene that encodes the α-peptide of galactosidase. This defect in the F plasmid can be complemented by the α-peptide, 146 aminoterminal fragment produced by the M13mp2, which *has not* been

Foreign Insert

Ligase $\qquad$ | Ligase
↓ $\qquad$ ↓ $\qquad$ ↓

.....NpGpApApTpTpCpN.....NpGpApApTpTpCpN.....
.....NpCpTpTpApApGpN.....NpCpTpTpApApGpN.....

↑ $\qquad\qquad$ ↑
Ligase $\qquad\qquad$ Ligase $\qquad$ *(Continued)*

BOX 1.2   FORMATION AND CLONING OF
RECOMBINANT DNA MOLECULES (Continued)

transformed by the foreign DNA. The phages that *have been* will not produce the complete α-peptide and hence will not encode active β-galactosidase even with the help of the F factor. This forms the basis for separating the phages that have incorporated the foreign DNA from those that have not. To do this separation the bacteria infected with the phages are plated on nutrient agar containing a *lac* inducer isopropylthiogalactoside (IPTG),

and a colorless compound, 5-bromo-4-chloro-indolyl-β-D-galac[34]toside (X gal), which is hydrolyzed by β-galactosidase to yield a dark blue dye, 5-bromo-4-chloro-indigo. Colored plaques are formed by untransformed phages, whereas the colorless plaques are produced by the transformed, recombinant phage. These can be isolated and used to provide the inserted single-stranded DNA for sequencing.

A. The Sanger Dideoxy Method

This is an enzymatic method involving the use of polymerase I of *E. coli* to synthesize a DNA segment complementary to the single-stranded DNA inserted in the phase vector shown in Figure 1.15a. To start the reaction a primer is added and annealed to the part of the M13 DNA adjacent to the insert (Fig. 1.15b). Polymerase is added along with the four deoxynucleotides (dNTP) needed for DNA synthesis, resulting in a strand complementary to the template strand (Fig. 1.15c). This synthesis is carried out in the presence of ^{32}P-labeled deoxynucleotides so that the newly synthesized strands can be identified. When a dideoxynucleotide (ddNTP) is present, synthesis proceeds in the presence of the polymerase until it incorporates a dideoxynucleotide. Chain elongation is terminated then because the chain now has no 3′hydroxyl group. Thus the reaction mixtures containing ddNTPs will have chain terminations at G, A, T, or C, depending on the ddNTP present. Each of the four reaction mixtures has one of the ddNTPs present at sufficiently low

concentration for only an occasional ddNTP to be inserted instead of the natural dNTP. The results of carrying out the different reactions with the template DNA sequence shown in Figure 1.15a are depicted in Figure 1.15d. The reaction products are added to a sequencing gel and separated electrophoretically. In this gel the DNA fragments are separated by size; the shorter the fragment, the greater its mobility in the electric field. The sensitivity is so great that fragments differing by one nucleotide in length can be separated. In Figure 1.15d, the direction of electrophoretic movement of fragments is from the bottom up so that the shorter fragments are toward the top. In the presence of ddGTP three fragments will be formed 5′TACG, 5′TACGG, and 5′TACGGTACATTCG, as shown in the first column of Figure 1.15d. The other three terminators produce fragments shown in the next three columns. The 3′ ends of the fragments are emphasized and give the sequence of the strand complementary to that shown in Figure 1.15d. With this method sequences up to 300 nucleotides can be determined with reasonable accuracy.

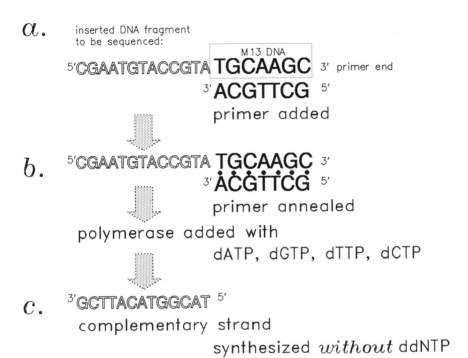

a. inserted DNA fragment to be sequenced:

5'CGAATGTACCGTA **TGCAAGC** 3' primer end

3' **ACGTTCG** 5'

primer added

b. 5'CGAATGTACCGTA **TGCAAGC** 3'

3' **ACGTTCG** 5'

primer annealed

polymerase added with

dATP, dGTP, dTTP, dCTP

c. 3'GCTTACATGGCAT 5'

complementary strand

synthesized *without* ddNTP

d. ddNTP products

resulting from polymerase actions
in the presence of ddNTP chain terminators

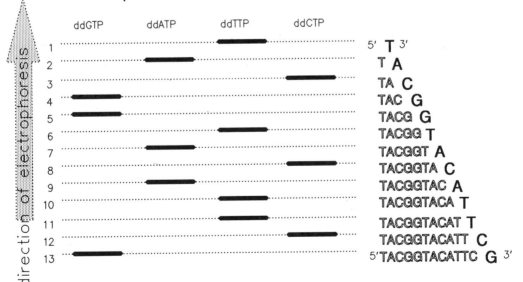

Fig. 1.15. *Diagram for an autoradiograph of a sequencing gel that determines the nucleotide sequence of DNA by the dideoxy enzymatic method of Sanger. The direction of electrophoretic movement of the fragments is indicated by the arrow at left. Therefore the smallest fragments are toward the top. The fragments with terminal ddNTP's are shown on the right.*

B. Chemical Sequencing Method

The chemical sequencing method of Maxam and Gilbert is based upon the modification of bases within the DNA segment to be sequenced. Hydrazine, formic acid and dimethylsulfate specifically modify the four bases; and when piperidine is added to the reaction mixture, it catalyzes strand breakage at the modified bases. Dimethylsulfate methylates the seven nitrogen of guanine (G) in Figure 1.16; formic acid protonates the purine ring nitrogens of adenine and guanine, weakening the glycosidic bonds so that piperidine replaces the purines (G + A in Fig. 1.16). Hydrazine splits the ring of thymine or cytosine, and the fragments can be re-

placed by piperidine (T + C). Cytosine reacts with hydrazine in the presence of NaCl, but not thymine. The specifically modified cytosine can then be displaced by piperidine (C in Fig. 1.16). Piperidine, in addition to displacing modified bases, also catalyzes phosphodiester bond cleaveage at the point where the modified bases are displaced. The result, then, is similar to chain termination with ddNTP in the enzymatic method, but the sequence of the original cloned segment is determined by the chemical method, not its complement. Figure 1.16 shows the results with the same DNA fragment that was sequenced by the enzymatic method. Note that the chemical method sequences the template strand itself rather than its complement.

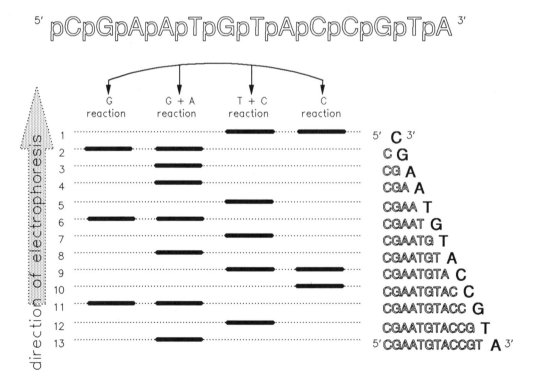

Fig. 1.16. *Determining the nucleotide sequence of DNA by the use of the chemical method of Maxam and Gilbert. The direction of electrophoretic movement of the fragments is indicated by the arrow at left. Therefore the smallest fragments are toward the top. The fragments resulting from the chemical treatment described in the text are on the right.*

V. CHROMATIN CONTAINS MANY DIFFERENT KINDS OF PROTEINS IMPORTANT TO ITS STRUCTURE AND FUNCTION

A. The Histones Form the Bulk of Chromosomal Protein

The histones form the bulk of the different kinds of proteins that are contained in chromatin. They consist of five basic types and constitute the fundamental structural proteins of chromatin along with a type II topoisomerase (TOPO II). These five are usually designated as histones H4, H3, H2A, H2B, and H1. Together, they constitute about 45% of the total mass of a chromosome. Histones organize the DNA double helices of the chromosomes, and probably play an important role in DNA replication and transcription.

1. Basic structures

The histones of a number of eukaryotes have had their amino acid sequences determined. Figure 1.17 summarizes some of the results of these determinations. Histones have a higher proportion of the basic

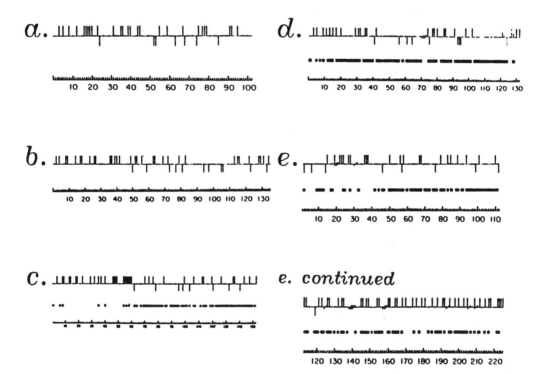

Fig. 1.17. *Diagram summarizing amino acid compositions of certain histones.* **a.** *Calf H4 distribution of basic, acidic, hydrophobic, and unchanged hydrophilic residues. An upward line indicates Arg, Lys, or His; a downward line shows Glu or Asp; and a dot represents Val, Met, Leu, Ile, Try, Phe, Pro, or Ala. H4 has 11% Lys; 14% Arg.* **b.** *Calf H3 distributions as in* **(a).** *H3 has 10% Lys; 13% Arg.* **c.** *Calf H2b distributions as in* **(a).** *Black bars below indicate identity of amino acid residues for calf, trout, sea urchin sperm, and Drosophila. H2b has ~ 16% Lys; 6% Arg.* **d.** *Calf H2a distributions, also trout and sea urchin. Symbols as in* **(a)** *and* **(c).** *Slashes indicate deletions. H2a has ~ 11% Lys; 9% Arg;* **e.** *H1 distributions for calf and trout. Symbols as in* **(a)** *and* **(d).** *H1 has ~ 29% Lys; 6% Arg. It has the highest content of Lys of the histones. Reproduced from Isenberg (1979), with permission of the publisher.*

amino acids, arginine, lysine and histidine, and a lower proportion of glutamate and aspartate than do other proteins. Also, they have a relatively high content of hydrophobic amino acids.

From amino acid sequences and conformational analyses it is now apparent that all five histones are multidomain proteins; histones H3 and H4 have well-defined, basic, flexible domains that extend from apolar globular domains; H2A and H2B have basic, flexible N-terminal domains and C-terminal tails flanking central apolar globular domains; and H1 has a similar conformation but with much longer N- and C-terminal domains.

Histone H4 is probably the most highly conserved protein among the eukaryotes. H4 from the pea plant differs from calf thymus H4 by only two amino acid residues. Although the indications are that there may be more divergence between other higher plant and animals and some of the *Protista*, the structure of H4 is fundamentally the same throughout the eukaryotic superkingdom in as much as all H4's have a high concentration of basic amino acids at the amino terminal end of the polypeptide, as illustrated for calf thymus in Figure 1.17a. Histone H3 is also a highly conserved protein; the difference between the pea plant and calf H3 is in only four residues. Histone H2B has a highly conserved sequence starting two-thirds of the way from the carboxyl end and going toward that end. The flexible N-terminal region is poorly conserved. Histone H2A, on the other hand, is relatively highly conserved, at least among the different animal groups.

Histone H1 is the least conserved in structure among the histones. Indeed it can have a number of subtypes in a given species that vary from tissue to tissue in the adult stage. Within each subtype, the globular domains are more conserved than the flanking flexible domains. The variations are in molecular weight as well as in amino acid sequence. The data in Figure 1.17e show how great a difference does exist between the two vertebrates rabbit and trout. Even so there are many conserved regions.

2. Histone subtypes

All of the histone types described in the preceding discussion have subtypes or variations on a theme. In animals, specific histone genes are turned on and off in the process of differentiation during development and gametogenesis. In the bony fish, Teleostei, histones are replaced during spermatogenesis by simpler proteins, the **protamines**. These are composed almost completely of arginine. In Echinodermata the sperm have different histones than the mature adults, and the embryos have still a third set. Similar degrees of variation are also found in Mammalia. Histone H5 largely replaces H1 in the nucleated erythrocytes of fish, amphibians, reptiles, and birds. Also the sperm of some marine invertebrates contain 01 histones that replace H1. H1, 01, and H5 have many similarities of sequence and may therefore be considered as variants with somewhat the same functions in chromatin. A mammalian variant of H1, called $H1^0$, is similar to H5 inasmuch as it has similar amino acid sequences in parts of its chain. All of these variants may play important roles in cell differentiation. $H1^0$, for example, is absent from the liver of embryonic mice, but it appears after birth. The histone genes exist in families constituted of various genic forms that are involved in the regulation of gene activity. Some of these encode various forms of histone and mRNAs at different times in development in different tissues of the same adult animal, and during the cell cycle. The chromosome replication process during the interphase of the cell cycle leads to a partial replacement of the normal histones observed in proliferating cells (D'Anna et al., 1984, 1985, 1986).

These differences could be either the cause or the effect of forms of differentiation.

To summarize the generalities about variability among the histones: H3 and H4 are strongly conserved in all parts; H2A and H2B are variable at the amino basic ends; and H1 is variable at the amino and carboxyl ends, but not in the apolar globular domains (Von Holt et al., 1979). The H1 variants and H1⁰ and H5 have conserved globular regions and variable N- and C-terminal domains. The histones are obviously highly important structural elements that participate not only in the maintenance of basic chromosome structure, but also in the regulation of chromosome function.

3. Posttranslational modifications of histones

In addition to the subtypes with different amino acid sequences that result from the action of different histone genes, even more variation is created through modification of the side chains of the constituent amino acids after translation. At least five types of modification are known: **methylation, acetylation, phosphorylation, poly(ADP) ribosylation**, and **ubiquination** (Bradbury, 1992). The methylation and ADP ribosylation of histones is apparently prevalent in all eukaryotes. The precise role of these modifications is unknown, but it is suspected that they may be involved in the interactions between the histones that occur during chromatin formation.

Nonreversible acetylation blocks the N-terminal serines of most of H1, H2A, and H4, and occurs shortly after their synthesis during the cell cycle. Reversible acetylation of all four histones: H2A, H2B, H3, and H4, occurs at the specific lysines shown in Figure 1.18. The acetylations clearly have the capability of modulating the interactions of these N-terminal domains with DNA in chromatin. Acetyla-

tion of all four of these histones, spread over all sites, occurs during the S phase of the cell cycle and is apparently associated with the requirements of DNA replication and transcription. Strong evidence supports the strict correlation of H3 and H4 tri- and tetraacetylations with transcriptionally active genes. Extensive acetylation also occurs during spermiogenesis when histones are replaced by protamines. Thus it appears that all aspects of DNA processing require histone acetylation.

H1 and H3 phosphorylations at sites indicated in Figure 1.18 have been associated with chromosome functions that are different from those associated with acetylation. Phosphorylation of serines and threonines in H1 is at a low level in the G1 phase of the cell cycle, but increases slowly in the S phase and rapidly in G2 to finally reach a hyperphosphorylated state at metaphase. H3 histones undergo a late phosphorylation of serines just prior to metaphase. These phosphorylations of H1 and H3 during the G2 and early M phases have been associated with the process of chromosome condensation (discussed later in this chapter).

The importance of **H1 kinase** activity in control of the cell cycle is indicated by the finding that the yeast cell cycle control genes, *cdc2*, encode a kinase subunit, $p34^{cdc2}$. This subunit forms a complex with **cyclins** A and B, which becomes activated as the H1 kinase by the dephosphorylation of a phosphorylated tyrosine during the G2 phase of the cell cycle. The cyclins are selectively degraded after metaphase but begin to be synthesized again with the beginnings of the next cycle.

Ubiquination involves the linkage of the free amino groups of histones 2A and 2B to the carboxyl terminus of the polypeptide **ubiquitin**. This is a 8,500-D polypeptide that links specifically with the free ε-amino group of lysine 119 of histone H2A (Hershka, 1983). The function of these

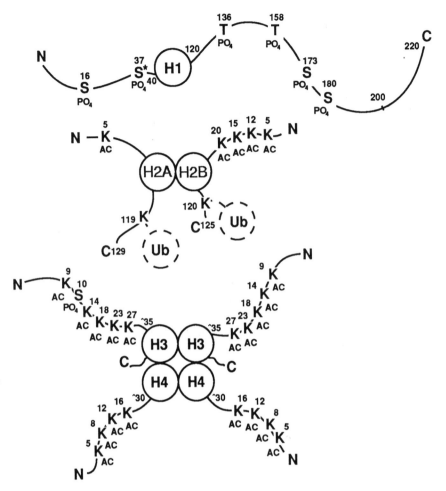

Fig. 1.18. *Diagram showing the sites of chemical modifications of the N- and C-terminals of the five basic histones. Reproduced from Bradbury (1992), with permission of the publisher.*

ubiquinated forms is not known. They constitute about 5% of H2A and 1% of H2B and are present during the cell cycle until just minutes before metaphase, when they are deubiquinated. Reubiquination starts immediately after metaphase.

B. Nonhistone Chromosomal (NHC) Proteins

1. NHC proteins in general

Nonhistone chromosomal proteins are sometimes referred to as acidic proteins of the chromosomes, since they do not con-

tain the high percentage of basic amino acids that are characteristic of histones (Elgin et al., 1973). They are generally referred to as the **DNA-binding proteins**. Some bind with DNA and apparently organize it so that it can be read by DNA and RNA polymerases. Their role in the regulation of transcription is extremely important and is discussed in Chapter 6.

It is not known how many kinds of proteins of the nonhistone variety there are. One of the problems in working with them is that they vary in amount from one species to another and from one cell type to another in the same species. When one considers all the different types of proteins involved in the repair, transcription, and replication of DNA, it is quite possible that the total number may range into the hundreds or even thousands, although not all will be present at once in the DNA from isolated chromosomes.

2. *The topoisomerases*

The **topoisomerases** are a class of DNA-binding proteins with structural and enzymatic functions that we consider separately from other nonhistone proteins because of their unique properties.

The topological form of a DNA molecule, or section of a molecule, is not assumed spontaneously and independently of outside influences. The states of relaxed or not relaxed and coiled are controlled in part by the topoisomerases. Two classes of topoisomerases are generally described: type I (TOPOI) and type II (TOPOII). Type I enzymes transiently break a single DNA strand, and type II enzymes break both strands transiently (reviewed by Wang, 1985).

The type II topoisomerases are distinguished from the type I enzymes in that they catalyze changes in DNA linking number through the transient introduction of double-stranded breaks. In prokaryotes such are *Micrococcus luteus* or *E. coli* these enzymes are called **DNA gyrases** because of their ability to catalyze the introduction of negative supercoils into closed circular DNA.

The eukaryotic type I enzyme(s), as found in mammals, are associated with chromatin and can relax both positively and negatively supercoiled DNA. At least some of them attach to the DNA covalently through 3'-phosphotyrosine side chains. The net effect of the enzymes is to cause single-strand breaks, allowing the DNA to assume the relaxed form.

The type II topoisomerases, some of which are also called DNA gyrases, can catalyze the conversion of relaxed duplex DNA to superhelical forms. They have been studied mostly in bacteria. The reaction requires energy, since the supercoiled form has a higher free energy than the relaxed has. The energy is supplied by ATP. The action of the enzyme is reflected in the reduction of the linking number. It is also characteristic of the gyrases of this group that they can relax negatively supercoiled DNA when ATP is not present.

Type II topoisomerases so far isolated from the eukaryotes appear to be primarily ATP-dependent enzymes that **relax** rather than promote supercoiling. But at least one of them appears to be part of a protein scaffold backbone to which DNA is attached, as discussed later in this chapter. This peculiarity of their action as compared with the bacterial enzymes is possibly related to nucleosome formation (discussed in the next section). In addition, topoisomerase activity has been found to be important in the processes of DNA replication, transcription, and recombination.

VI. CHROMATIN IS HIGHLY ORGANIZED

The structure of chromatin at the molecular level began to be intensively investigated in the 1970s by a number of biological and physical chemists using X-ray diffraction, neutron scattering, electron microscopy, and electric dichroism, as well as the standard biochemical techniques of enzymatic clipping of DNA into pieces followed by electrophoresis (described in the previous section on DNA sequencing). Although we are still far from understanding the functioning of the structures that have been elucidated up to the present time, it may be said that progress toward this understanding has been encouraging to the extent that reasonable questions can now be asked about chromatin function with some possibility that they can be answered. This was not possible until the present types of sophisticated physical and chemical analyses were employed.

A. Nucleosomes

A major advance in the understanding of chromatin structure was made by Hewish and Burgoyne (1973), who showed that when chromatin was treated with Ca^{++}-activated endonucleases and the DNA products were subjected to electrophoresis, a ladder pattern of DNA fragments was obtained of the type shown in Fig. 2. It was subsequently shown that these were products of a 200-bp DNA repeat and that this repeat was a constituent of an organized chromatin body now known as the **nucleosome**. This was a momentous discovery that profoundly advanced our understanding of chromosome structure.

Nucleosomes constitute 95% or more of the weight of chromatin and are considered to be the basic structural units of chromatin. They have the basic structure shown in Figure 1.19. In most cells of the higher eukaryotes, nucleosomes contain 195 bp of DNA, a histone octamer with $[(H2A, H2B)_2 (H3_2, H4_2)]$, and one H1 molecule. (The object illustrated in the Figure 1.19 is not a complete nucleosome as ordinarily defined, since it does not show the $\sim$27-bp DNA "tail" not directly associated with the histones. This tail connects with the next nucleosome in a string of nucleosomes, as discussed in the next section.) The DNA repeat wrapped around the octamer (Figure 1.19) may vary considerably in length, ranging from $\sim$170 bp in *Saccharomyces cerevisiae* to 241 bp in sea urchin sperm. Also certain tissue cells in birds and mammals may have lengths different from other cells in the same individual. The reasons for these differences in repeat lengths are not known, but it is conceivable that they may be a reflection of differences in the regulation of gene action.

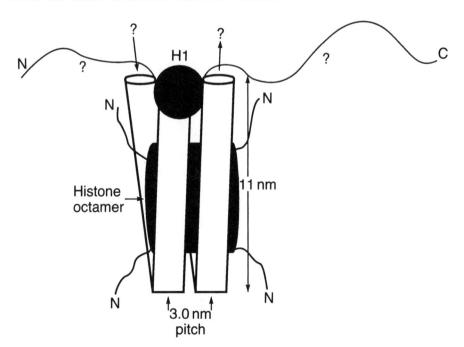

Fig. 1.19. *Diagram of a nucleosome particle around which is wrapped approximately two turns of duplex DNA. Reproduced from Bradbury (1992), with permission of the publisher.*

1. The nucleosomal core particle

The histone octamer has about 1.7 turns of DNA 146 bp long wrapped around it. This is known as the **nucleosomal** or **histone core particle** (Richmond et al., 1984). It is a nucleosomal subunit whose structure has been determined in solution by neutron scatter and in crystals by X-ray and neutron diffraction (reviewed by Bradbury and Baldwin, 1986; Klug et al., 1985; Burlingame et al., 1985). It has by these means been shown to be a flat disk 11.0 nm in diameter and 5.5–6.0 nm thick with 1.7 turns of DNA of pitch 3.0 nm coiled on its edge and wrapped around it.

The 146 bp of DNA of the core particle is not bent uniformly with constant radius around the histone octamer, but follows a path of straight segments and intervening tight bends. The tetramer ($H3_2$, $H4_2$) interacts with the DNA and coils the DNA around itself, as proposed in the model shown in Figure 1.20a. This tetramer also provides the interaction sites for the two (H2A, H2B) dimers in each face of the disk, as shown in Figure 1.20b, to complete the octamer of the core particle. Core particle shape structures have been isolated that contain only H3 and H4, indicating that they are the possible "kernels" of core particles. The fact that they are also the most highly conserved in structure of the histones is important evidence supporting their basic role in the structure of the core particle, and hence of chromatin itself. Strong surface interactions exist between the histones of the octamer (D'Anna and Isenberg, 1974). H3 and H4, H2B and H4, and H2A and H2B interact strongly, and these interactions presumably provide the adhesive forces that maintain the octamer's integrity.

The core particle is a subunit of the subject illustrated in Figure 1.19, which is called the **chromatosome**. It contains 168 bp of DNA and the fifth histone, H1, which is shown associated with the DNA entering and leaving the nucleosome. The cen-

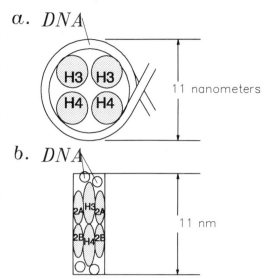

Fig. 1.20. *Model of the octamer structure of the core particle, showing the probable arrangement of the core histones: H3, H4, H2A, and H2B, which constitute the octamer. DNA interacts with the $(H3,H4)_2$ tetramer in* (**a**) *to form the kernel of the disk shown here (inside view).* **b.** *The disk shown in end view with the H2A and H2B histones associated on the outside of the kernel tetramer.*

tral globular region of H1 apparently interacts with the C-terminal half of histone H2A and thus binds it to the core particle (Boulikas et al., 1980). Since a significant portion of the histone electron density has not been identified in the crystal structure, it is thought that some of the basic N-terminal domains and C-terminal tails of the histones protrude from the chromatosomes (Fig. 1.19). It is these protruding arms that may be subjected to the reversible chemical modifications described in the preceding section on posttranslational modifications of histones. The flexible basic domains possibly interact with the DNA entering and leaving the nucleosome. These areas are indicated by the question marks in Figure 1.19.

In summary, neutron scatter and X-ray crystallography have demonstrated that chromatin is organized at three basic levels:

- the **core particle**, consisting of 146-bp DNA plus a histone octamer with two molecules each of H2A, H2B, H3 and H4;
- the **chromatosome**, which is a core particle plus 22 bp more of DNA to give a total of 168 bp and a molecule of H1; and
- the **nucleosome**, with a total of about 200 bp of DNA plus the histones found in the chromatosome.

HMG proteins may or may not be associated with the nucleosomes.

The interactions between DNA and the core histones are apparently electrostatic ones involving the highly basic amino side chains of arginine and lysine as well as the somewhat weakly basic histidine with the phosphate backbone of the DNA (Nieto and Palacian, 1988). The spacer DNA linking the chromatosomes is variable in length. It may be anywhere from close to 0 to 80 bp in the organisms so far studied. It is the chromatosome that is highly conserved, not the rest of the nucleosome. Portions of the linker DNA may, however, interact with histones (Morse and Cantor, 1985).

The properties of DNA are altered when it is wrapped on a protein surface such as a histone octamer (White et al., 1988). As indicated on p. 30, the lower the linking number, the more negative the supercoiling. The linking numbers of intracellular DNAs of the eukaryotes are generally at least 6% lower than what would be expected for relaxed DNA. This reduction has been attributed to nucleosome formation (Germond et al., 1975). In addition, it has also been shown that positive as well as negative supercoiling occurs in yeast (Graever and Wang, 1988). The simultaneous presence of both types has been ascribed to the occurrence of transcription (see Chapter 6).

2. *The 10-nm fiber*

When nuclei are suspended in an aqueous solution of low ionic strength (e.g., <2 mM NaCl), they swell and then burst, and their chromosomes fall apart, releasing fibers of chromatin that appear under the electron microscope as beaded strings (Fig. 1.21). These fibers, generally called **10-nm fibers** because this approximates their thickness, consist of a continuous strand of duplex DNA along which are distributed bead-like nucleosome chromatosomes serially connected by linker DNA not associated with histones. Treatment of these fibers with micrococcal nuclease initially cuts the DNA at one of the junctions of each bead to free the structures described above as nucleosomes.

B. The Chromatin Solenoid

A "higher form" of chromatin is generated when one increases the ionic strength of the medium in which chromosomes or 10 nm fibers are suspended (Thoma and Koller, 1977). At concentrations of 2 mm of dionic electrolytes such as $MgCl_2$ and in the presence of histone H1, a fiber of 25–30 nm thickness is produced. It is generally alluded to as the **30 nm fiber** in the literature (McGhee et al., 1983; Felsenfeld and McGhee, 1986). Analysis of this fiber by electric dichroism, electron microscopy, X-rays, and neutrons reveal a structure that is present among all Metazoans and Embryophytes whose chromatin has been analyzed. Since it is demonstrable in both interphase and metaphase chromosomes, this structure probably is a basic chromosomal element for all eukaryotes with the possible exception of some of the Protista and eukaryotic Thallophyta.

One model of a small portion of a 30-nm fiber is depicted in Figure 1.22. The chromatosomes are shown arranged in a left-handed helical ramp six to a turn with

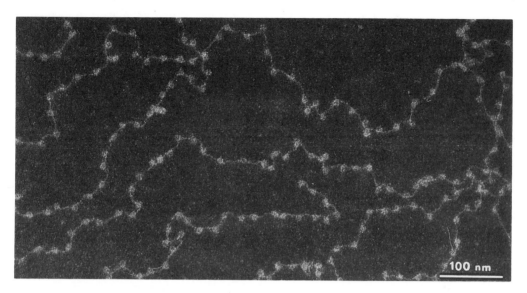

Fig. 1.21. *An electron microscope photo of a 10-nm fiber. The round bodies are nucleosomes. Duplex DNA of variabling length connects the nucleosomes. The dark-field electron micrograph of the edge of a chicken erythrocyte nucleus shows extended chromatin fibers composed of nucleosomes and linkers. The nucleosomes are ~10 nm in diameter and contain 146 bp of DNA and eight histones, two each of H2A, H2B, H3, and H4. The linkers are somewhat variable in length (~50 bp in chicken erythrocytes) and are thought to interact with H1 and H5 histones. Courtesy of A.L. Olins and D.E. Olins, Oak Ridge National Laboratory.*

a pitch of about 11 nm. A view down the axis of the helix, also called a **solenoid**, shows the chromatosomes arranged radially, the arrangement that best fits the X-ray data (Fig. 1.23). They are probably not arranged parallel to the solenoid axis, but at an angle of 20°–30° to that axis. The H1 histones are possibly located at or near the central core of the solenoid. There seems to be one molecule of H1 per chromatosome covering that part of each chromatosome where the DNA strands enter and leave the core particle (Fig. 1.19).

The DNA of the solenoid consists of at least two parts, a part attached to the histone core and a part referred to as the spacer or linker DNA, which connects the chromatosomes laterally in the solenoid. As already mentioned, these spacer segments vary in length. The question arises: How are they organized? One suggestion that is consistent with the data from dichroism measurements is shown in Figure 1.24 (McGhee et al., 1983). The spacer DNA in this model is a supercoil wound around the helical part passing through the chromatosome centers. The line AB passing through the centers of adjacent chromatosomes is shown in Figure 1.24 as being about 10° from the line CD, which is perpendicular to the solenoid axis. This determines the pitch angle r of the spacer DNA supercoil, and this in turn determines the distance between neighboring chromatosomes in the solenoid. If $r = 0°$, the spacer will be a ring, and if $r = 90°$, the spacer becomes a rod, as in the 10-nm fiber, and the chromatosomes lie far apart. This model has the advantage of accommodating chromatin with different spacer lengths. Figure 1.25 diagrams a DNA supercoil hypothesized for chicken erythro-

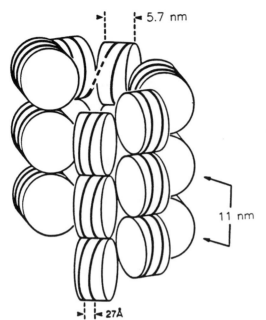

Fig. 1.22. *Diagram of a proposed solenoid model. Six nucleosomes per turn are assumed. The nucleosomes are shown as having a thickness of 5.7 nm and a diameter of 11 nm. Note the continuation of DNA from one nucleosome to the next, indicated by the broken line. Reproduced from Widom and Klug (1985), with permission of the publisher.*

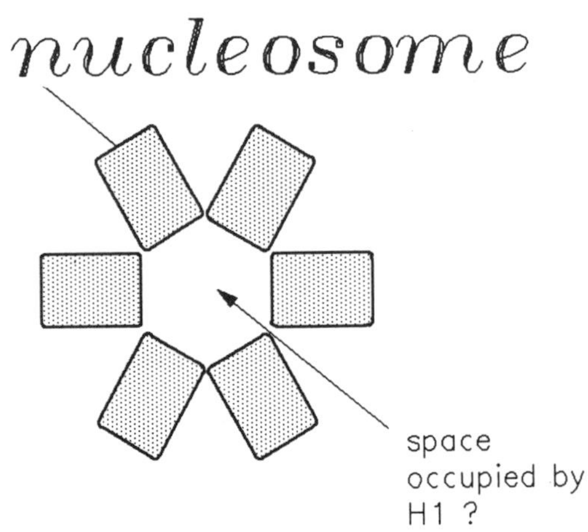

cyte DNA that has spacers known to be 44 bp in length.

Alternative, but not necessarily completely different, models of the structure of 30-nm fiber have been proposed (Hozier et al., 1977; Subirana et al., 1985). These models are generally based on observations made with the electron microscope rather than X-rays, and in some cases they lead to the conclusion that the fibers are the result of stacking of nucleosomes into disks or beads. Figure 1.26 is an electron microscope micrograph of a ~30-nm fiber. Parts of it clearly show stacking of nucleosomes that can be interpreted as the type of structure developed from X-ray data as illustrated in Figures 1.22 and 1.23.

Structures revealed at this level of chromatin organization are necessarily the result of the method of preparation by the observers. One therefore expects different results from different laboratories. All seem to agree that 30-nm fibers exist, but an open mind should be maintained with regard to the exact structure. At present, the solenoid structure revealed by X-rays, neutrons, and electric dichroism seems to have the most support. Artifacts of preparation are not ruled out, however. Frozen-hydrated fibers from an echinoderm, *Thyone*, and a salamander, *Necturus*, do not show a constant 30-nm diameter (Athey et al., 1990). A solenoid model with a variable diameter that increases with linker length has therefore been proposed as a possibility.

C. Beyond the Solenoid

1. Chromosome compaction

The higher-order structure of chromatin, as seen, for example, in metaphase chromosomes and the heteropycnotic (stainable and visible) regions of interphase chromosomes, is by no means well understood. But since metaphase chromosomes yield 30-nm fibers, it is probable

Fig. 1.23. *Radial arrangement of chromatosomes as viewed looking down the axis of the 30-nm solenoid. The central space may be occupied by histone H1.*

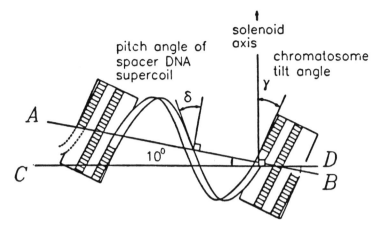

Fig. 1.24. *Proposed organizations of chromatosomes and spacer DNA. See text for details. Reproduced from McGhee et al. (1983), with permission of the publisher.*

Fig. 1.25. *DNA supercoil for chicken erythrocyte DNA that has spacers ~ 44 bp in length. The chromatosome faces are proposed to be tilted 25° from the solenoid axis. The spacer DNA between neighboring chromatosomes is supercoiled about the helix, as indicated by the broken line that passes through the chromatosome centers. Reproduced from McGhee et al. (1983), with permission of the publisher.*

that the condensation of chromatin in them consist of the looping and folding of these fibers. Take human chromosome 16 as a close to average example. It has about 1.1×10^8 bp, which translates to about 3.7×10^4 µm in length. At maximum compaction in metaphase it is ~ 3 µm long. The degree of compaction, or the linear **packing ratio**, is therefore about 12,000 (3.7×10^4 µm/3 µm). At early prometaphase, chromosome 16 is about 4 µm long, which gives a packing ratio of about 9,250. It has been suggested that the difference in the packing ratio between the condensed metaphase chromosome and its extended interphase state is about 7.5 (Manuelidis and Chen, 1990). The interface 16 should therefore be about 23 µm long, which represents a ratio of 1,600. How then do we go from 3.7×10^4 µm to 23 µm? Scanning electron microscope micrographs of mammalian metaphase chromosomes reveal the type of structure shown in Figure 1.27. Their surface appearance is in conformity with a compaction of chromatin fibers by some sort of folding to reduce the length of chromatin to metaphase length. The course of compaction

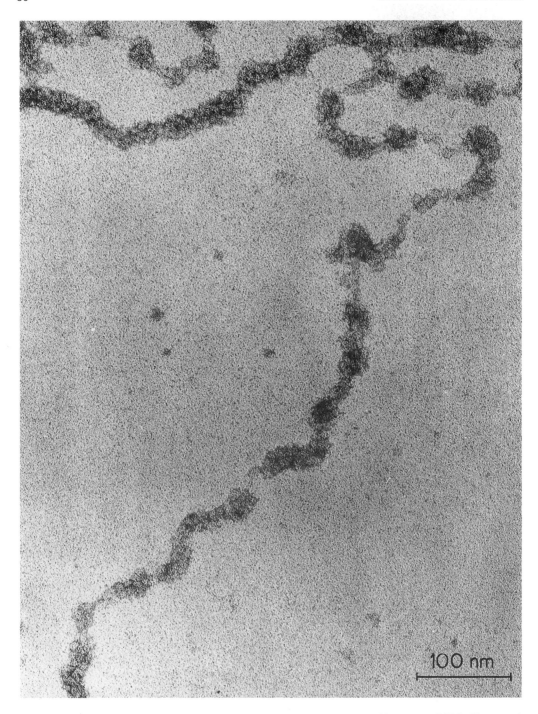

Fig. 1.26. *Bright-field electron micrograph clearly demonstrating close-packed nucleosomes in 20- to 30-nm chromatin fibers spilling out of a chicken erythrocyte nucleus. Isolated nuclei were washed in 0.2 M KCl and diluted 1:150 in 0.02 M KCl + 0.5 mM MgCl$_2$, centrifuged onto a glowed carbon-coated grid, and stained for 30 s with 0.1% aqueous uranyl acetate. Courtesy of A.L. Olins and D.E. Olins, Oak Ridge National Laboratory.*

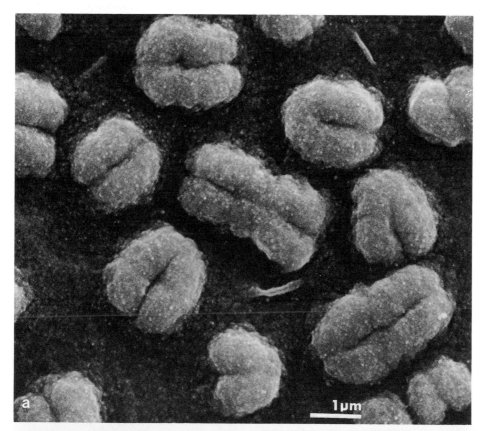

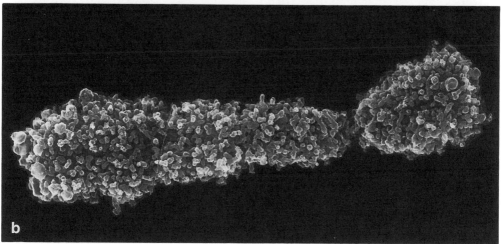

Fig. 1.27. a. *Dog metaphase chromosomes as seen by scanning electron microscopy (SEM). Note that some are acrocentric and some metacentric. Courtesy of T.M. Seed, Argonne National Laboratory.* **b**. *Scanning electron micrograph of a submetacentric Chinese hamster CHO metaphase chromosome isolated by means of nitrogen cavitation by the method of Wray and Stubblefield (1970). The chromosome was stabilized with 2% aqueous uranyl acetate, dehy-*drated in a graded series of acetone solutions, critical-point-dried, sputter-coated with gold-palladium prior to observation with a JEOL JEM-100CX scanning transmission electron microscope. (Courtesy Drs. Susanna M. Gollin, Department of Human Genetics, University of Pittsburgh and Wayne Wray, National Institutes of Health.) Copyright Susanna M. Gollin and Wayne Wray, 1983. All Rights Reserved.*

from solenoid to the metaphase level then is germane to our understanding of chromosome structure.

Part of this 1600-fold compaction is achieved by the wrapping of the DNA about the core particles and further spiralization in the solenoid. This gives a packing ratio of about 40. Obviously the rest of the compaction must be accounted for by other processes. Many models have been proposed, and in general they may be grouped loosely into two classes: helical coiling of the 30-nm fiber and repeated folding transversely and longitudinally. Actually there is no reason to eliminate one of these at the expense of the other. Depending on the observer, both types have been supported by direct observation of metaphase preparations.

2. Loops and folds

Laemmli and his associates have proposed that the DNA of the higher-order organization of chromosomes is folded in the 30-nm form into loops attached to a central protein core (Laemmli et al., 1977). One version of this is shown in Figure 1.28. Metaphase chromosomes freed of 99.1% of their histones still retain their DNA in approximately the same form as intact chromosomes, presumably because it is attached at many points to what is called a nonhistone protein **scaffold** (Paulson and Laemmli, 1977). This scaffold consists primarily of two proteins, designated Sc1 and Sc2, with molecular weights of 170 kD and 135 kD, respectively (Lewis and Laemmli, 1982). Sc1 has been identified as topoisomerase II (Earnshaw et al., 1985) and the radial loops of the DNA are attached to this protein at sites designated as **scaffolding attaching regions** or **SARs** (Earnshaw and Heck, 1985; Gasser et al., 1986; Gasser and Laemmli, 1986a). The scaffold structure itself is maintained by metalloprotein interactions and dissipates in the presence of chelating agents

Fig. 1.28. *DNA freed of histones apparently attached at points along its length to a protein scaffold to form loops. The dense material at the bottom of the EM photomicrograph is the scaffold. Courtesy of U.K. Laemmli.*

(Lewis and Laemmli, 1982). In *Drosophila melanogaster* the loops are attached at SARs in interphase chromosomes that are located in nontranscribed regions of the DNA (Mirkovitch et al., 1984). The SARs of three *melanogaster* genes (alcohol dehydrogenase (*Adh*), fushi tarazu (*ftz*) and *Sqs-4*, a gene encoding one of the glue proteins) co-map with enhancer-like regulatory sequences in DNA, which we discuss further in Chapter 6 (Gasser and Laemmli, 1986b). The implication is that the "free" loop domains contain actively transcribing regions regulated by the regions of the DNA associated with the SARs. The loops are in the range of 4–13 kb long and should contain far more information than is necessary to encode polypeptides.

The extended mammalian interphase chromatin fiber or chromosome has an average diameter of ~240 nm (Manuelidis and Chen, 1990). It is suggested that this transition from the 30-nm solenoid fiber to 240 nm is achieved primarily by the folding into loops of solenoid (Fig. 1.28) to give an interphase structure as shown in Figure 1.29a. It is then proposed that further compaction is achieved by coiling as shown in (Fig. 1.29b) to give a metaphase chromosome ~700 nm in thickness and 7.5 times shorter than its interphase counterpart.

3. Helices, spirals, and coils

The model of chromosome organization sketched in Figure 1.29 is accepted in varying degrees, or not at all, by other observers. In the first place, evidence does exist that histone H1 is necessary for the tight assembly of the 30-nm fibers (Labhart et al., 1982). Second, there is also evidence that 200-nm fibers arranged perpendicularly to the long axis of chromosomes exist (Taniguchi and Takayama, 1986). These spiralized fibers are in turn composed of 30-nm solenoid. The compacted chromosome is hypothesized to consist of an additional spiralization of the

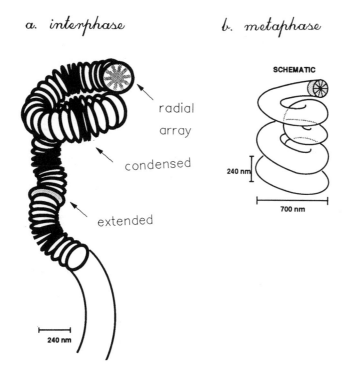

Fig. 1.29. *Models for 240-nm-thick interphase and 700-nm-thick metaphase chromosome structure.* **a.** *Model of an interphase chromosome in which the 30-nm solenoid is represented as a radial array to give a chromosome fiber structure about 240 nm thick, which is* hypothesized to be the thickness of the interphase chromosome. **b.** *Structure of a metaphase chromosome as attained by the coiling of the 240-nm structure in* (**a**). *Reproduced from Manuelidis and Chen (1990), with permission of the publisher.*

200-nm fibers as illustrated in Figure 1.29. Some support for this type of supra structure is found in the observations of various investigators who used electron microscopes with chromosomal material prepared in a variety of ways (Haapala, 1984; Sedat and Manuelidis, 1977; Zatsepina et al., 1983; Hozier et al., 1977; and Okada and Comings, 1979). Finally, it is quite possible that a combination of the various views of the compacted structure is nearest to the actual situation. Scanning electron microscope studies by Rattner and Lin (1985) have led them to posit the kind of organization depicted in Figure 1.29. This combines Laemmli's loops with the ideas of those who advocate spirals, coils, and helices.

VII. SUMMARY

Little knowledge about the basic structure of chromosomes was attained for some 80 years after they were first described. It was not until the demonstration in the 1940s and 1950s that their DNA was indeed the genetic material, and biochemists began to work on their protein components, that a picture began to emerge and make it possible to design acceptable models. Since then it has become apparent that chromosomal DNA is composed of both unique and repetitive sequence segments and that it can assume various forms such as the A-, B-, and Z-configurations. Furthermore, the histone proteins, which constitute the major protein fraction of the chromatin, are not constant in their qualitative properties at different times in the cell cycle and even during the different stages of development and differentiation. These and the nonhistone proteins are a major factor in maintaining the basic structure of chromosomes and play an important role in the functioning of their DNA. It is now apparent that the earlier view of chromosomes as very stable entities is quite wrong. This will be made increasingly clear in the subsequent chapters.

In summary, chromosomes of eukaryotes have several levels of organization:

1. Their DNA is predominantly in the B-form, the right-handed helical form.

2. Their DNA is bound to histone octamers consisting of histone types 2A, 3B, 3, and 4 to form nucleosome core particles, which require histone H1 to complete the nucleosomes.

3. At low ionic strength chromatin unfolds into a 10-nm fiber. This fiber forms a 30-nm fiber structure or solenoid with an increase in ionic strength beyond what is necessary to maintain the 10-nm fiber.

4. There is an interaction with nonhistone proteins that results in the formation of looped domains ranging in size from 10 kb to 100 kb. These domains are found in interphase and metaphase chromosomes, and when the histones are extracted from these chromosomes, a number of nonhistone proteins remain to which the DNA is attached. This protein residue, part of which has been identified as a topoisomerase II, is the scaffold. Other nonhistone proteins bind to the DNA and play a regulatory role.

5. A major factor in the functioning of DNA, as will be considered in subsequent chapters, is its topological state with regard to the winding of the helix. DNA exists in several different topological states: negatively supercoiled, positively supercoiled, relaxed, and a mixture of all three.

2 Chromosome Organization

Chromosomes are not merely aggregates of discrete genic units. To a certain extent they are units in themselves.

G.L. Stebbins, 1950

Chromosomes look simple under the microscope lenses. But they are highly organized entities—certainly among the most highly organized organelles in cells. They are the ultimate prime directors of what goes on in cells, their directives ranging from housekeeping metabolic activities to differentiation and beyond. They therefore have a higher order of structure superimposed upon the basic structure described in the preceding chapter. Two of the most important visible structures are the centromeres and the telomeres.

I. ALL FUNCTIONAL CHROMOSOMES HAVE CENTROMERES

Centromeres (kentron = center; meros = part) have been given a variety of other names. Synonyms commonly in use are kinetochore, spindle fiber attachment or locus, primary constriction, achromite, and just plain spindle attachment. There are more. Schrader (1953) lists 27 synonyms. All of these are descriptive of centromeres insofar as they describe the chromosome region where the spindle fibers attach, their appearance at metaphase and anaphase as constrictions in the chromatin, their poor uptake of stains, and their involvement in the movement of the chromosomes during cell division.

Many observers consider centromere and kinetochore as synonymous terms applying to precisely the same chromosomal

structures (see Rieder, 1982, for review). But it is of heuristic value to consider kinetochores as special entities within the more general regions, the centromeres. The centromere is chromatin like the rest of the chromosome, but has within its domain a proteinaceous entity, the kinetochore, that can be detected by antikinetochore antibodies found in the serum of patients with the autoimmune disease scleroderma CREST (Earnshaw and Rothfield, 1985; Brenner et al., 1981). In this book we shall use the general term centromere with the understanding that it refers to a region of chromatin that may also contain a kinetochore, even though there is a considerable degree of confusion in the literature as to the use of the two terms (Godward, 1985). A kinetochore can be simply defined as that region of the centromere to which spindle fibers attach (Ris and Witt, 1981). Hence it can be overwhelmingly important in the disjunction of chromosomes during cell division.

A chromosome without a centromere is described as acentric, and is like a ship without sails, motor, or rudder. It drifts aimlessly and, although capable of replicating, is lost after one or two cell divisions. It should be noted, however, that certain Protozoa, fungi, and algae do not undergo dispersal of the nuclear membrane during karyokinesis (Kubai, 1975). They are eukaryotes that may represent transitional states between the prokaryotic and the more typical eukaryotic forms. Failure of the nuclear membrane to dissipate during karyokinesis is accompanied by a peculiar relationship between what appear to be kinetochores and the spindle fibers that arises outside the nuclear membrane (see Dodge, 1985, for review).

A. Centromeres May Occupy Different Positions on Chromosomes

1. Monocentric chromosomes

Monocentric chromosomes have a single centromere and are characteristic of most Metazoan plants and animals and some of the unicellular forms such as yeast. The location of the single centromere can categorize a chromosome as **acrocentric** (acro = extremity) if the centromere is visible at or near one end, **metacentric** (meta = between) if near or at the middle, submetacentric if between the midpoint and chromosome end, and **telocentric** (telo = end) if it is terminal. True telocentrics probably do not exist, as discussed later in this chapter. The terms acrocentric (having one "arm") and metacentric or submetacentric (having two "arms") are descriptive of chromosomes that appear as rods, or V's, or J's, respectively, in metaphase and anaphase (Fig. 2.1) and are quite useful in describing the gross morphology of a chromosome. Frequently the shorter arm of a metacentric or submetacentric is referred to as the "p" arm and the longer one as the "q" arm.

Centromeres of monocentric chromosomes, whether they be acrocentric or metacentric, nearly always have stretches of

heterochromatin, sometimes described as **centric chromatin**, closely associated on either side. Indeed most of the heterochromatin of a chromosome is generally associated with its centromere, although in many cases it is also present at the telomeres, as discussed in the next section. During the replication of chromosomes in the S-phase of the cell cycle, the centromeric regions of the chromosomes generally replicate late, as described in Chapter 3.

2. Polycentric chromosomes and diffuse centromeres

In some animals such as the ascarid nematodes, insects in the orders Hemiptera, Homoptea, and Lepidoptera, and some centipedes, as well as plants in the genus *Luzula*, centromeric chromatin is present in multiple copies along the length of the chromosome (reviewed by Pimpinelli and Goday, 1989). These chromosomes are described as polycentric or **holocentric**, in contrast to those with single localized centromeres. In organisms with holocentric chromosomes spindle fibers attach along the whole length of the chromosomes, as shown in Figure 2.2. In the ascarid nematodes centromeric chromatin is separated by noncentromeric segments; in other, such as the coccid bugs, every point on the chromosomes shows spindle attachment ability (Hughes-Schrader, 1948). The latter are said to have **diffuse centromeres**. In some cases the meiotic chromosomes have localized centromeres, while the mitotic ones have nonlocalized centromeric activity, as in *Luzula* (Braselton, 1981). In other cases the reverse is true (Comings and Okada, 1972). Related to the holocentric type is a semilocalized centromere type in organisms such as *Pluerozonium* that has potentially multiple centromeres. In these the position of the active centromere may shift in meiosis although remaining localized in one place during mitosis (Vaarama, 1954). Mouse L

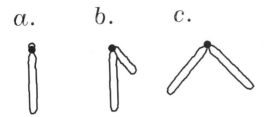

Fig. 2.1. *Forms of monocentric chromosomes.* **a.** *Acrocentric.* **b.** *Submetacentric.* **c.** *Metacentric.*

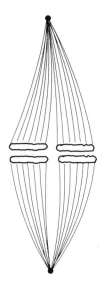

Fig. 2.2. *Spindle fibers and holocentric chromosomes with diffuse centromeres during mitosis in a coccid bug Statococcus, n-2. After Hughes-Schrader (1948).*

cells carried in culture may develop chromosomes with two or more centromeres (Vig, 1984). Ordinarily a chromosome with two centromeres (generally called a dicentric) will break at anaphase when the centromeres are pulled in opposite directions as described on page 191. However, these mouse multicentric chromosomes, which may have up to eight distinguishable centromeres, segregate normally. The reason appears to be that in the mouse cells only one of the centromeres retains kinetochore activity as determined by the use of the antikinetochore serum from patients with scleroderma CREST (Zinkowski et al., 1986). Rat cells of the B_1 line in culture also produce multicentric chromosomes but all centromeres retain kinetochore activity toward the antiserum. In some cases only one kinetochore per chromosome retains the ability to react with microtubules. In others, anaphase bridges are formed, as described on p. 191. Hence the chromosomes function as monocentrics. Loss of ability of one of the kinetochores of a dicentric to function actively at ana-

phase has also been noted with human chromosomes (Merry et al., 1985). This can have important evolutionary consequences.

B. Centromeres Have a Unique Molecular Structure

Centromeric chromatin contains both histone and other proteins and DNA. The DNA molecule of the chromosome is continuous from one end to the other, but the molecular organization of the centromeric region is different from the rest of the chromatin. It has already been noted that active centromeres contain kinetochores with unique antigenic properties, but in addition centromeres contain DNA sequences that are unique. We consider first the structure of the DNA in the centromeric regions of yeast chromosomes.

1. Centromeres of Saccharomyces

The molecular structure of the centromeres of the yeast *Saccharomyces cerevisiae* has been extensively investigated and the nucleotide sequences of the centromerically active regions of several of the chromosomes have been determined. DNA segments from five of the yeast's 17 chromosomes have been isolated and cloned in plasmids that replicate in yeast cells. These segments, all of which have centromeric activity, have been designated as CEN3, CEN11, CEN5, CEN6, and CEN4, being from chromosomes III, XI, V, VI, and IV, respectively. When present as recombinant DNA in autonomously replicating plasmids in yeast cells, they are stably maintained, and they segregate in mitosis and meiosis with a high degree of accuracy. All are functional as small segments of DNA about 1 kb or less in length. They, plus the plasmid vector DNA, constitute a "minichromosome" about 2 lm in length. When assayed for mitotic stability, a nondisjunction frequency of about 10^{-2} is found as compared with the normal yeast rate of about 10^{-5}. Normal centromeres in their natural positions on chromosomes are very efficient in maintaining proper segregation in yeast, but their functioning in the abnormal position in plasmids is still impressive even though 1,000 times less efficient. A plasmid without the recombinant centromeric DNA is quickly lost after a few cell divisions. The efficiency of the recombinant centromere DNA is also impressive in meiosis. Plasmids with a CEN segregate normally in 60%–90% of the tetrads. Again, this is a high rate of nondisjunction, but plasmids without the CEN do not disjoin properly at all (Koshland et al., 1988).

The nucleotide sequences of four of the isolated CEN segments have been determined, and all have a great deal in common, as shown in Table 2.1. Within each CEN there is a section that shows a high degree of sameness of sequence with the other three CENs. This section has been segmented into three elements (I, II, and III). Element I, consisting of 14 bp in each CEN, is identical in sequence in CEN3 and CEN11, and shows substantial homology with CEN4 and CEN6. Of some interest is the finding that 40 bp to the left of element I in CEN6 there is a segment with the sequence ATAAGTAAAA-TAAT that is more similar to the element I sequences of CEN3 and CEN11 than the element I sequence of CEN6 itself. Element II is 82–89 bp long and in all four cases is over 90% (A + T)-rich. The A + T sequences are somewhat variable, but

CEN3 and CEN11 show about 71% sequence homology between them for this region. Element III is 11 bp long and like element I is identical in CEN3 and CEN11; it is also obviously related although not identical in CEN4 and CEN6 (Bloom and Carbon, 1982).

These approximately 220-bp segments are believed to be the **centromere cores** (Fig. 2.3) for the CEN of chromosome 11. The core is 15–20 nm in diameter and is resistant to both micrococcal and DNA I nucleases. It contains the elements necessary for centromere function. A 219-bp deletion that removes the left 5 bp of element I, along with some 214 bp of the sequences to the left, has no effect on CEN11 activity. But a longer deletion that removes element II up to the boundary with III completely removes activity. Element III is also very important. Its removal, or even small deletions or single base changes within it, inactivate the centromeric activity of the CEN. Minor changes within element II also cause loss of activity (Bloom et al., 1983). The evidence indicates that only a single spindle fiber microtubule attaches to a yeast centromere (Peterson and Ris, 1976). Presumably its attachment is to the core, which then satisfies the definition of kinetochore. A microtubule has a diameter of about 20 nm, which is about the diameter of the core.

Of considerable interest is the finding that if a 627-bp segment containing centromere activity is deleted in one chromosome 3 in a diploid with one intact chromosome 3, extreme instability in mi-

TABLE 2.1. Sequences of Nucleotides of Four Yeast CENs

	Element I	Element II	Element III
CEN 3	ATAAGTCACATGAT ←	88 bp (93% A + T) →	TGATTTCCGAA
CEN 11	ATAAGTCACATGAT ←	89 bp (94% A + T) →	TGATTTCCGAA
CEN 4	AAAGGTCACATGCT ←	82 bp (93% A + T) →	TGATTACCGAA
CEN 6	TTTCATCACGTGCT ←	89 bp (94% A + T) →	TGTTTTCCGAA

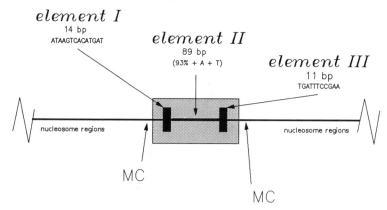

centromere core
220 basepairs of nuclease resisting core, about 15 to 20 nanometers diameter

element I
14 bp
ATAAGTCACATGAT

element II
89 bp
(93% + A + T)

element III
11 bp
TGATTTCCGAA

nucleosome regions

nucleosome regions

MC

MC

Fig. 2.3. *The centromeric core of chromosome XI of S. cerevisiae. MC, Sites at which micrococcal nuclease cuts. Nucleosomes are present outside the core DNA. Data from Bloom et al. (1983) and Carbon (1984).*

tosis results, and the acentric chromosome is quickly lost. But if, on the other hand, this segment is inverted instead of being deleted, or replaced by a 858-bp segment containing CEN11, neither mitotic nor meiotic activity is lost (Clarke and Carbon, 1983; Carbon, 1984; Clarke and Carbon, 1985). These results indicate that at least in yeast the orientation of the centromere is not of great importance and that the centromeres of chromosomes 3 and 11 are interchangeable. This lack of centromere specificity has also been found in Neurospora and maize. But lack of specificity does not necessarily extend to interspecific exchanges. Yeast centromeres introduced via plasmid vectors into other yeast species, Neurospora, or cultured animal cells do not always function. Even so, some sequences cloned from the yeast *Saccharomyces uvarum* function in *S. cerevisiae* (Huberman et al., 1986). The alga *Chlamydomonas*, and the nematode *Caenorhabditis elegans* also exhibit some CEN activity in yeast. Thus a degree of homology may exist in widely unrelated forms. Also element I shows some sequence ho-

mology with satellite DNAs from cattle (75%–93%), rat (77%), baboon (77%), and human (72%).

2. Centromeres of the higher eukaryotes

The Metazoan animals and the higher plants generally have larger chromosomes than the lower eukaryotes such as yeast. Fungi have chromosomes about 1 μ or less in length during the metaphase as opposed to plants and animals, whose chromosomes average at least 10 times longer. In addition the centromeric regions of higher plants and animals are much more complex. Their centromeres are usually associated with a constriction, the primary constriction, in distinction from a secondary constriction that may be located on the same chromosome but not associated with centromeric activity. A region of heterochromatin generally surrounds the primary constriction on either side. This constitutive centric heterochromatin can be made visible with the appropriate staining techniques as described in Box. 2.1.

Centromeres have been extensively studied in mammals, in which they develop special differentiations during the mitotic period when the spindle develops (Ris and Witt, 1981; Rieder, 1982; Godward, 1985; Rattner, 1986, 1987). A struc-

ture called the **trilaminar disk** can be demonstrated by electron microscopy. This has an inner dense layer, a less opaque middle layer, and an outer dense layer. These can be demonstrated in isolated chromosomes in a suspension medium with

---

BOX 2.1   STAINING AND BANDING OF CHROMOSOMES

The name chromosome was given these bodies because they stained heavily and easily. The original stains showed most chromosomes as mere blobs with little detail of structure except that in some, centromeric regions took little or no stain. At present, however, differential staining techniques have been developed to a high degree of sophistication and many details can be demonstrated in prophase and metaphase chromosomes.

Chromosomes can be stained for light microscopy by a great variety of dyes, ranging from the Feulgen–Schiffs reagent, which is specific for DNA, to those that react with basic proteins. Perhaps the most useful, and hence the most used stain, is Giemsa, which is actually a mixture of dyes consisting of thionine or Lauth's violet, and its methylated derivatives: azure B, azure A, azure C, and methylene blue. Most Giemsa preparations also contain eosine I bluish.

The thionine molecule itself has two free amino groups and stains chromosomes uniformly. Methylation of this molecule, 3,7-diaminophenothiazine-5-ium chloride,

gives the derivatives azure C, azure A, azure B, and methylene blue with 1, 2, 3, and 4 methyl groups, respectively. Methylene blue

METHYLENE BLUE

has been widely used as an oxidation–reduction indicator, and this property undoubtedly is important in its effectiveness as a chromosomal stain.

The thiazine moiety of these dyes

1,4 THIAZINE

interacts with the phosphate groups of DNA and side stacks along the molecule. Whether the chromosomal DNA is the only part of chromatin that reacts significantly is possible, but not proved. The eosine component of Giemsa stain is not essential for staining, but it does

THIONINE

(Continued)

BOX 2.1   STAINING AND BANDING OF CHROMOSOMES
(Continued)

enhance the patterns observed when Giemsa is employed. Depending on the mode of preparation and application, Giemsa stains give a variety of different results. Cookbook chemistry is the rule. Each stainer has his or her favorite recipe to apply to the stainee.

C-Banding

When chromosomes are fixed in 0.2N HCl and subjected to 0.07N NaOH for several minutes followed by treatment overnight with $2 \times$ SSC (0.3 M NaCl plus 0.03 M trisodium citrate) at 66°, Giemsa stains the satellite-rich centromeric heterochromatin intensely. C-banding (Fig. 1) is thus useful in identifying centromeric regions, as shown below. These chromosomes are from a cultured mouse line (C11D) in which centric fusions have occurred to form metacentrics from the acrocentrics normally present in mouse cells *in vivo*. The normal laboratory

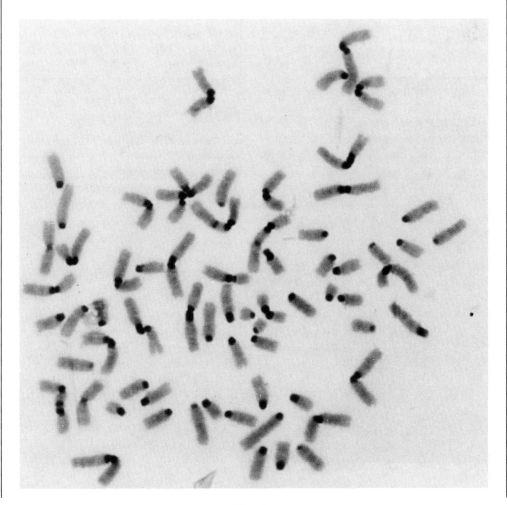

Fig. 1.

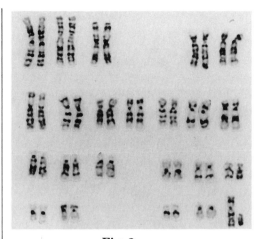

Fig. 2.

mouse genome is constituted of 20 pairs of metacentrics.

G-Banding

To produce G-bands (Fig. 2), chromosomes are first fixed in methanol-acetic acid (Carnoy's reagent) and then stained by a variety of subprocedures with Giemsa. Dark bands, the G-bands, alternating with light bands, are produced in prophase and metaphase chromosomes. The G-bands are regions of the chromosome where the DNA is presumbly free to stack with the thiazine

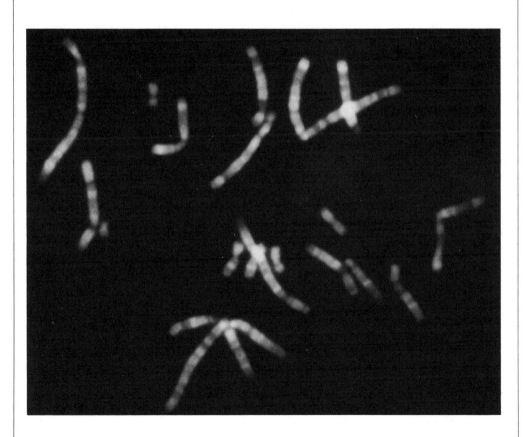

Fig. 3. *(Continued)*

BOX 2.1   STAINING AND BANDING OF CHROMOSOMES (Continued)

dyes as compared to the light bands. These chromosomes are from the cultured lymphocytes of the eminent cytogeneticist T.C. Hsu of the M.D. Anderson Hospital and Tumor Institute, Houston, Texas, and are reproduced here with his permission. The pairs are arranged so that they number from top left sequentially, left to right, to the last pair 23, the large X and smaller Y.

Q-Banding

This banding is the result of treatment of chromosomes with the drug quinacrine hydrochloride, an antimalarial. Quinacrine fluoresces strongly in the ultraviolet. AT-rich DNA reacts strongly with quinacrine and enhances its fluorescence in the UV, whereas GC-rich DNA quenches its fluorescence. Historically, Q-banding was the first of the banding procedures to be developed for mammalian chromosomes.

Hoechst-Banding

The stain Hoechst 33258 also gives an enhancement of fluorescence with AT-rich DNA and a lesser enhancement in GC-rich regions of chromosomes. The Hoechst-banded chromosomes (Fig. 3) are from cultured Chinese hamster cells.

R-Banding

Chromosomes treated after fixation with 20 mM phosphate buffer at pH 6.5 at 87°C for 10 min and then stained with Giemsa show, when viewed with phase contrast, pale G-bands and dark-stained R-bands. Also the telomeres stain well (T-banding) by this procedure. The R-bands apparently contain GC-rich DNA and are the interbands in G-banded chromosomes. Acidine orange also gives excellent R-banding.

Other Dyes

A considerable number of other dyes exist besides quinacrine that distinguish AT- and GC-rich regions of DNA. These so-called counter-stain-enhanced techniques are widely used for chromosome sorting. See Schweizer (1981) and Darzynskiewicz and Crissman (1990) for reviews and additional techniques.

---

a relatively high ionic strength, such as 75–100 mM KCl. When chromosomes are suspended in a hypotonic medium, the disk is absent and the microtubules are seen to be attached directly to the chromatin fibers (Ris and Witt, 1981). Increasing the ionic strength brings about a reversible reorganization of the trilaminar disk and concurrently an apparent reorganization of the chromatin fibers to which the microtubules (spindle fibers) are attached. They appear attached to the outer dense layer. Both this layer and the inner dense layer contain DNA. It appears that loops of chromatin fibers protrude from the inner disk layer and give rise to the outer dense layer (Fig. 2.4). This shows the chromatin 30-nm fiber from the body of the chromosome extending to the outer kinetochore plate. The trilaminar disk structure, and structures related to it, are by no means confined to the mammals. Trilaminar structures are generally found in all animals as well as in some lower eukaryotes, both plant and animal (Godward, 1985). In some cases they are clearly associated with the nuclear membrane, with the consequence that the interphase chromosomes appear attached to that membrane.

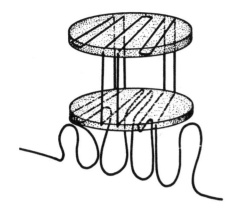

Fig. 2.4. *A model for the mammalian trilaminar disc. Reproduced from Rattner (1986), with permission of the publisher.*

3. Kinetochores and spindle fibers

Spindle fibers of the mitotic spindle consist of microtubules and associated proteins (reviewed by Olmsted, 1986). The principal protein is **tubulin**, a polypeptide of 50,000 D that occurs in two related isomeric forms, α-tubulin and β-tubulin. These two form dimers of 100,000 D that connect together with other proteins to form the spindle fibers. Two types of fibers arise with the spindle: the **polar fibers** and the **centromeric fibers**, which are attached to the kinetochores and extend toward the poles (Fig. 2.5). During metaphase tubulin is incorporated into microtubules at

the kinetochores and the microtubules play an important if not the most important, role in chromosome movement during the anaphase (Mitchison et al., 1986; Mitchison, 1988; Koshland et al., 1988). Hence kinetochores are more than passive entities. They are definitely involved in the organization of the spindle apparatus along with the centromeres of the cell (reviewed by Brinkley, 1985). It appears that the anaphase movement of the chromosomes toward poles of the dividing cell is powered and regulated by microtubule depolymerization. Depolymerization involves the loss of subunits from the microtubules at their kinetochore ends, thus shortening them and moving the chromosomes to which they are attached polewards. The kinetochore remains attached to the microtubule as it depymerizes (Koshland et al., 1988). In many organisms, including yeast, the centromeres of each chromosome are attached to a single spindle fiber via a chromatin fiber. (See review by Kubai, 1975.) In the mammals, and other organisms with more complex centromeres, many fibers arise from each centromeric region, where they are all presumably attached to kinetochores.

Kinetochores are composed of DNA, protein, and probably RNA. Several centromeric polypeptides identified as CENP-A, CENP-B, and CENP-C, with approximate molecular weights of 18 kD, 80 kD,

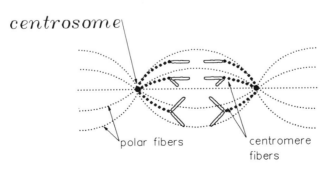

Fig. 2.5. *The spindle apparatus showing the centromeric (heavy lines) and the polar fibers (thin lines).*

and 140 kD, respectively, have been isolated and characterized in a variety of mammals (Earnshaw and Rothfield, 1985a; Valdivia and Brinkley, 1985). One of these, CENP-C, has been shown to bind specifically with the alphoid DNA of human centric heterochromatin (Masumoto et al., 1989). CENP-B also binds to a specific alphoid satellite sequence with a similarity to the helix-loop-helix family of proteins (Sullivan and Glass, 1991). How all these components are organized to form a functional centromere is not clear, but it is clear that at least some of the protein components bind to specific centric heterochromatic DNA sequences. Zinkowski et al.(1991) have proposed a model based upon tandemly arranged repetitive subunits each of which is a complex of protein and DNA (Fig. 2.6). Zinkowski et al. (1991) propose that the microtubule binding sites (MTs) consisting of the repeated protein–DNA units assembled in tandem on the 30-nm fiber constitute the functional kinetochore. In the G_2 phase it is hypothesized that in the centromeric regions the MT-binding segments of the sister chromatids are separated by linker segments

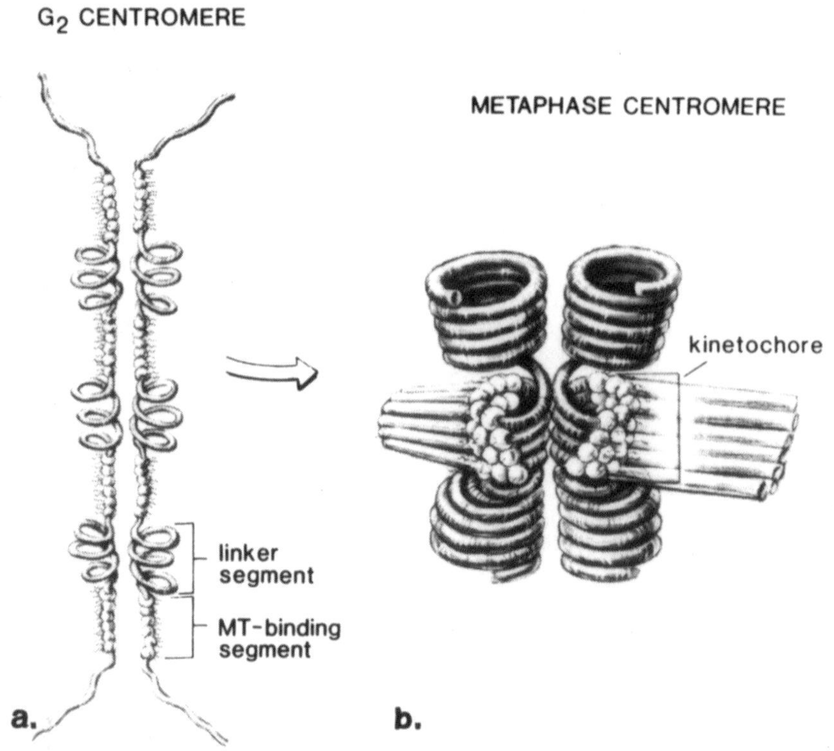

Fig. 2.6. *Model for kinetochore structure as envisioned by Zinkowski et al. (1991).* **a.** *The centromeric region of a fully replicated mammalian chromosome in G2. Each chromatid is shown with a 30-nm fiber extending continuously through its length. Distributed along these fibers are the proposed kinetochore repeated units that are capable of binding tubulin and microtubules (i.e., spindle fibers). These microtubule (MT)-binding segment stretches are interrupted by 30-nm linker segments free of MTs. In metaphase chromosomes the centromeric regions are compacted, as shown in (**b**), to form the kinetochores. It is at this level that the trilaminar structure in Figure 2.5 is exhibited. From Zinkowski et al. (1991), with permission of the publisher.*

of 30-nm fiber free of the subunits (Fig. 2.6a). In the prophase compacted kinetochores are formed that become bound to microtubules which then become the spindle fibers at metaphase (Fig. 2.6b).

II. TELOMERES ARE THE TIPS OF CHROMOSOMES WITH IMPORTANT FUNCTIONS

When a monocentric chromosome breaks at some point along its length, several different kinds of events can follow (Fig. 2.7). The broken ends can rejoin and an intact chromosome be reconstituted (Fig 2.7a). If they do not rejoin, the acentric fragment will be lost and the centric one maintained with a deficiency that will probably lead to cell death (Fig. 2.7b). Finally, if two nonhomologous chromosomes break simultaneously, a translocation can result (Fig. 2.7c and d).

In general these events occur only when otherwise intact chromosomes break. Why do they not occur with intact chromosomes by their joining at their tips? It has been long believed that they do not because chromosomes have telomeres at their tips that prevent this from occurring (Muller, 1938).

It is now apparent that telomeres are specialized structures that, like centromeres, have a unique structure distinguished from that of the rest of the chromosome. But unlike centromeres there is little to distinguish telomeres cytologically from the nontelomeric parts of the chromosome, other than that telomeric chromatin is visibly heterochromatic in some organisms. In addition to providing for a "nonsticky" end to a chromosome in contrast to a "sticky," broken end, telomeres have a structure amenable to replication of the chromosomal DNA as discussed on page 128.

A. Structure of Telomeres

A great deal has been learned about the structure of animal and plant telomeres

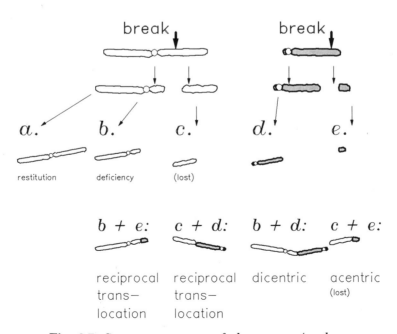

Fig. 2.7. *Some consequences of chromosome breaks.*

by analyzing the chromosomes of certain Protozoa of the Class Ciliata. Ciliates are characterized by the possession of two kinds of nuclei: (a) generative or germ line, the **micronuclei**; and (b) non-germ-line, the **macronuclei**. Micronuclei are active during meiosis, and during conjugation, when fertilization by the fusion of two haploid nuclei occurs. From this fusion nucleus new micronuclei are formed by mitosis (Fig. 2.8). Macronuclei arise from micronuclei and are unique to the Ciliata. They are the active nuclei during the vegetative period of the life cycle. Their DNA is transcribed and produces large amounts of RNA. On the other hand, the micronuclei have the role of preserving the genetic material and engaging in the sexual process. The transition from micronucleus to macronucleus may be accompanied by a loss of up to 95% of the micronuclear genome. This diminution in chromatin is discussed at length in Chapter 5. Micronuclei are not totally essential, since clones of ciliates have been established that are amicronucleate and multiply indefinitely by binary fission.

The macronucleus undergoes fission during the asexual vegetative phase (Fig. 2.8b′) but does not persist during the sexual process associated with conjugation. The macronuclei of the conjugants disappear during this process, and new macronuclei regenerate from the new diploid micronuclei of the ex-conjugants (Fig. 2.8f). Macronuclei are appropriately described as somatic nuclei, even though ciliates are considered to be single-celled organisms. (Actually it has been proposed that the ciliates are degenerate Metazoans whose macronuclei are the remains of their multicellular past.)

The ciliate *Tetrahymena thermophila*, and its amicronucleate syngen derivative, *T. pyriformis*, have been used extensively in genetic, physiological, and biochemical studies that have yielded a considerable amount of information about telomeric structure. The micronucleus of *thermophila* contains five pairs of chromosomes. However, its macronucleus has about 200–300 chromatin units prior to replication. These exist as linear subchromosomal subunits with lengths ranging from 21 kb to 1,500 kb. They are sometimes referred to as minichromosomes and, being acentric, do not undergo mitosis at the time of binary fission. The process is therefore referred to as **amitosis**; despite the apparent random nature of the distribution of the genetic material, the fission products end up with sufficient DNA of the proper type to maintain functioning daughter cells until the next division. The clones arising after series of binary fissions may show the results of segregation, since heterozygous mother cells may produce clones that are homozygous (reviewed by Gorovsky, 1980; Blackburn and Karrer, 1986).

The macronuclear DNA derived from the micronuclear genome consists of many acentric chromosomal units each of which has a functional telomere at its ends. In all ciliates that have been studied the telomeric DNA ends are inverted repeats consisting of tandem repeats of a simple sequence that varies very little from one ciliate to another. As shown in Table 2.2, for the five ciliates for which sequences have been determined, the repeat is either T_2G_4 or T_4G_2 (Gottschling and Cech, 1984; Herrick et al., 1985).

The protomonad protozoan *Trypanosoma brucei* has a telomeric repeat similar to that found in the ciliates except that an adenine is present to give the sequence 5′-TTAGGG (Table 2.2). Surprisingly the same sequence is also found in the telomeres of the vertebrates. The flowering plants Arabidopsis and maize have a closely related sequence, TTAGGGG.

The telomeric repeats of *Saccharomyces* are somewhat different from the ciliate (Table 2.2); nonetheless *Tetrahymena* re-

sexual   *asexual*

a.   conjugation of two cells   *a'.*   single cell

b.   meiosis of micronucleus   *b'.*   binary fission

c.   in each cell, mitotic division of a haploid micro-nucleus   disintegration of macronuclei

c'.   mitotic distribution of all DNA material

d.   fertilization: fusion of different haploid micro-nuclei

e.   new diploid micro-nuclei undergo mitosis   old macronuclear material eliminated

f.   each new macronucleus from a micronucleus

*exconjugants
with identical
genotypes*

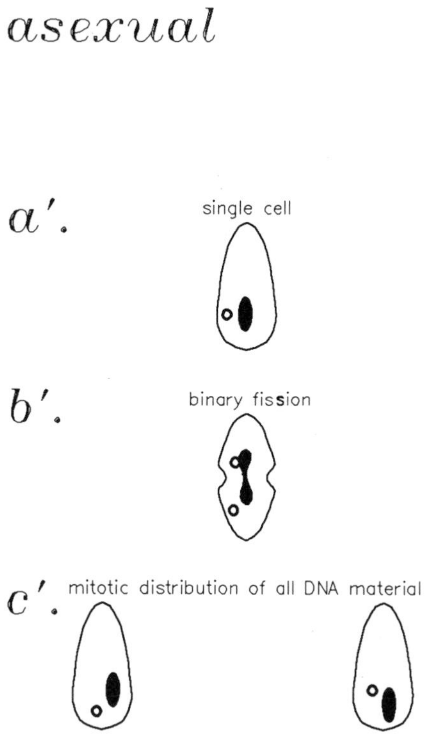

Fig. 2.8. *Sexual and asexual cycles of a ciliate protozoan. In this example, the cells have a macronucleus and a single micronucleus. When the cell begins the sexual phase, it generates four haploid nuclei by meiosis of its single micronucleus. It conjugates with another similar cell, but of different mating type (**a**). All but one of the haploid nuclei in each conjugant disintegrates and the macronucleus in each begins to break down (**b**). The two remaining haploid nuclei fuse (**c**). The macronuclei disappear completely and the fusion (2n) nucleus divides mitotically (**d**). Each conjugant receives a 2n micronucleus (**e**). The exconjugant micronuclei divide mitotically (**f**). The exconjugants separate and one of the micronuclei in each becomes a macronucleus (**a'**). The asexual phase consists of a binary fission (**b'**) in which the macronucleus divides mitotically and the micronucleus does so mitotically. Two complete daughter cells are thus produced (**c'**).*

TABLE 2.2. Telomeric Sequences from a Variety of Eukaryotes (Sequences Read From Left to Right, $5' \rightarrow 3'$)

Eukaryote	Sequence
Ciliate Protozoa	
Tetrahymena	TTGGGG
Glancoma	TTGGGG
Paramecium	TTGGGG
Oxytricha	TTTTGGGG
Stylonychia	TTTTGGGG
Protomonad Protozoa	
Trypanosoma	TTAGGG
Slime molds	
Dictyostelium	AG_{1-8}
Physarum	T_nAGGG
Fungi	
Saccharomyces	$(TG)_{1-3}TG_{2-3}$
Schizosaccharomyces	$T_{1-2}ACA_{0-1}C_{0-1}G_{1-6}$
Higher plants	
Arabidopsis	TTAGGGG
Maize	TTAGGGG
Higher animals	
Vertebrates	TTAGGG

Data from Blackburn and Szostak (1984); Blackburn et al. (1983); Moyzis et al. (1988); Emery and Werner (1981); Johnson (1980); Szostak (1984); Meyne et al. (1990).

peats function in yeast cells. By inserting fragments from the ends of *Tetrahymena* chromosomes with the T_2G_4 repeat into a circular plasmid capable of replicating autonomously in yeast cells, a linear plasmid has been formed with *Tetrahymena* termini at both ends, as shown in Figure 2.9. This linear vector was successfully cloned in yeast cells, showing that *Tetrahymena* telomeres can replicate in yeast as well as in their native cells. If one of the *Tetrahymena* termini is removed, the plasmid is not retained, but if it is modified by replacing one of the *Tetrahymena* ends with randomly selected yeast telomeric fragments, the triple hybrid vectors clone in yeast cells with no difficulty. It seems reasonable to conclude from these results that yeast telomeres, as well as being able to replicate autonomously, are quite similar

if not identical for all 17 yeast chromosomes. Also *Tetrahymena* telomeres can function in place of the yeast telomeres in yeast cells, and allow for the replication of linear DNA.

The chromosomes of the Protista and fungi are too small to provide satisfactory material for cytological analysis with a light microscope. But the higher eukaryotes do show visible differences at the tips of their chromosomes. In some, but not all, the telomeric regions are heterochromatic (Fig. 2.10), and even form "knobs" as in a maize chromosome and the salamander *Trituris*. Some species of rye, particularly the cultivated rye *Secale cereale*, have large blocks of heterochromatin associated with their telomeres (Jones and Flavell, 1983). In *S. cereale* this heterochromatin consists of repeated DNA sequences of several different kinds that have been isolated and characterized by size. These occupy a considerable proportion of the total genome in several of the species of *Secale*. It is of some interest that all five species of *Secale* have quite different proportions of three of these sequences. The 120-bp sequence is common to all and is also present in the related wheat species *Triticum dicoccum*. This repeat unit and the 480-bp unit are also found interstitially, but the 610-bp unit when present is exclusively telomeric. However, it should be noted that the method for detecting these repeated sequences requires that at least 500 copies per genome be present.

A tandemly repeated sequence from the onion, *Allium cepa* (Liliacae), constitutes about 4% of its genome and hybridizes almost exclusively to the telomeric regions of some but not all of its chromosomes (Barnes et al., 1985). Three clones derived from isolated samples of the repeat show a degree of heterogeneity of base sequence. But all three have sequences similar enough to make it evident that they are related by a common origin.

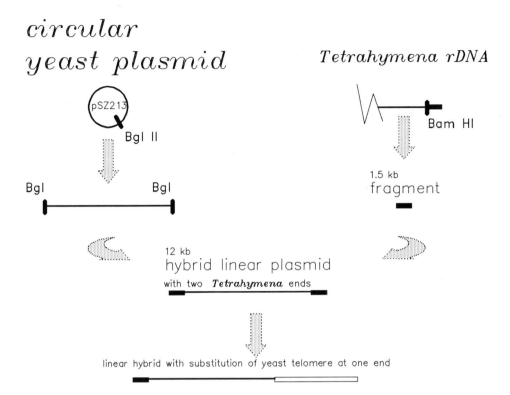

Fig. 2.9. *Telomeric Tetrahymena repeat T_2G_4 inserted into a plasmid that replicates as a linear DNA in yeast cells. The circular plasmid pSZ213 is cut with the endonuclease BglII. A linear DNA with overhanging ends A/GATCT is produced. Telomeric ends of Tetrahymena macronuclear chromatin units (minichromosomes) are cut with the endonuclease BamHl, and the fragments are inserted at the ends of the linear plasmid DNA. Data from Szostak and Blackburn (1982).*

The telomeres of *D. melanogaster* have been probed with a 12-kb fragment of *melanogaster* DNA originally cloned with a λ-phage (Young et al., 1983). This fragment, designated λ-T-A, hybridizes readily to the telomeres of all the polytene chromosomes as well as to the α-heterochromatin of the centromeric regions (Fig. 2.11). T-A is part of a complex set of repetitious DNA sequences called the TDNA (Traverse and Pardue, 1989). These sequences are confined to the chromocenter and telomeric regions of *Drosophila* chromosomes. Box 2.2 describes and explains the hybridization technique with labeled probes. The salivary gland nuclei of *D. melanogaster* contain polytene chromosomes that are described on p. 103. The homologues are tightly synapsed and fused at their centromeres to form a **chromocenter** (Fig. 2.11). The heterochromatin of this centric region is clearly of two types, α and β, as demonstrated by differential staining and DNA composition. The α-heterochromatin is composed of highly repeated simple DNA sequences, while the β-heterochromatin is much more complex and more highly polytenized than the α. λ-T-A also contains a sequence partly homologous to one cloned by Rubin (1978) that

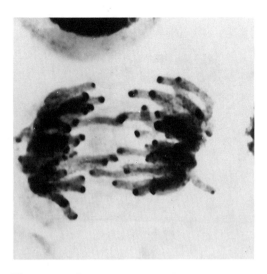

Fig. 2.10. *Onion root tip cells at anaphase showing heterochromatic telomeric regions. Reproduced from Stack and Clark (1974), with permission of the publisher.*

hybridizes to *melanogaster* telomeres. This fragment does not have C + A repeats. The evidence indicates that telomeric sequences of *Drosophila*, like those of rye, are much more complicated than what is found in the lower eukaryotes. In addition, it is possible that *Drosophila* polytene chromosome telomeres contain a high proportion of Z-DNA (Arndt-Jovin et al., 1983).

The telomeric regions of all eukaryotic chromosomes yield fragments that have highly repetitious sequences (Meyne et al., 1990). One group of related sequences $T_{2-4}G_{3-4}$ or T_2AG_{3-4} listed in Table 2.2 seems to be of wide occurrence. It is entirely possible that these sequences, all terminating in G_{3-4} at the 3' end, are of importance in the replication of the DNA of chromosomes. This is considered in detail in the next chapter (p. 128 et. seq.).

BOX 2.2 LABELING CHROMOSOMES WITH SPECIFIC PROBES

This is a powerful technique that allows one to identify the locus of a specific sequence of DNA on a chromosome. The basic procedure involves four steps:

1. Identification and isolation of a specific sequence that is then incorporated in a vector and cloned in a host bacterium or eukaryotic cell.
2. Release of the sequence from the vector by a restriction endonuclease.
3. Labeling of the sequence with either radioactive or fluorescence-labeled nucleotides.
4. Hybridization of the probe after disassociation (denaturation) with partially denatured prophase or metaphase chromosomes.

Steps 1 and 2 have already been described in Box 1.2. Here we describe the procedures involved in steps 3 and 4.

The labeling procedure used most often involves a process called **nick translation** of DNA by DNA-dependent DNA polymerases. As described in Chapter 3, all polymerases synthesize a DNA single strand complementary to the template in the 5' → 3' direction (see Fig. 3.6). The polymerase generally used is *E. coli* DNA polymerase I. Like the DNA polymerases from other organisms, *E. coli* polymerase I has exonuclease activity. It can remove nucleotides from the 3' end and the 5' end of a duplex DNA strand as shown in Fig. 1. It thus has both 3' → 5' → 3' exonuclease activity. A modified form of the polymerase known as the Klenow fragment lacks the 5' → 3' activity.

Nick translation involves the combined activities of both the 5' → 3' polymerase and 5' → 3' exonuclease

3′ → 5′ exonuclease activity

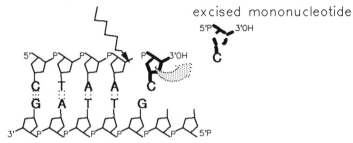

excised mononucleotide

5′ → 3′ exonuclease action

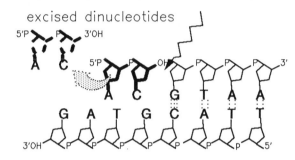

excised dinucleotides

Fig. 1.

a.   duplex DNA

b.   DNase I creates
nick with free 3′ OH

c.   5′ → 3′ exonuclease
removes nucleotides
from the 5′ side of nick

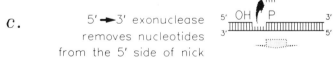

d.   excised nucleotides
replaced by polymerase
with labeled nucleotides.
exonuclease removes unlabled nucleotides

Fig. 2.

(Continued)

BOX 2.2   LABELING CHROMOSOMES WITH SPECIFIC PROBES *(Continued)*

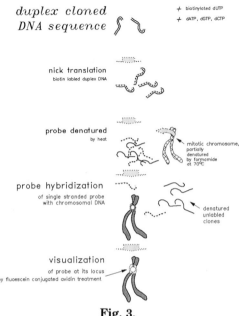

duplex cloned DNA sequence

+ biotinylated dUTP
+ dATP, dGTP, dCTP

nick translation
biotin labled duplex DNA

probe denatured
by heat

mitotic chromosome, partially denatured by formamide at 70°C

probe hybridization
of single stranded probe with chromosomal DNA

denatured unlabled clones

visualization
of probe at its locus by fluoescein conjugated avidin treatment

Fig. 3.

activities of polymerase I. When a "nick" is created by DNase I as shown in Fig. 2 in a duplex DNA, a free 3′ OH is produced. In the presence of the polymerase, one to several nucleotides are removed from the 5′ side of the nick by the polymerase 5′ → 3′ exonuclease activity. The excised nucleotides are then replaced by labeled nucleotides incorporated by the polymerase 5′ → 3′ activity. The result is a translocation of the nick and a uniform labeling of the synthesized DNA strand.

The label may either be ^{32}P, ^{3}H, or ^{14}C or a nonradioactive one that fluoresces under the near ultraviolet. Figure 3 illustrates the use of fluorescent label that is now widely applied to identifying loci on chromosomes. The cloned sequences are incubated with d-ATP, d-GTP, d-CTP, and a biotinylated d-UTP [5-(N-bio-tinyl-E-aminocaproyl-3-aminoallyl) deoxyuridinetriphosphate]. This compound can replace d-TTP in the nick translation process. The biotinylated DNA is then denatured by heat to the single-strand state and incubated with prophase or metaphase chromosomes partially denatured in the presence of formamide (HCONH$_2$) at 70°C. Loci at which hybridization of biotinylated sequences with complementary sequences occur in the chromosomes can then be identified by a biotin-binding protein, such as avidin, that has been coupled with a fluorescing dye.

If the probe has been labeled with a radioactive isotope, essentially the same procedure is used to hybridize it with chromosomal DNA, but the locus is visualized by autoradiography, as illustrated.

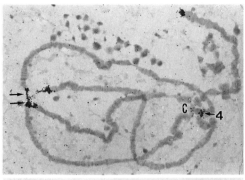

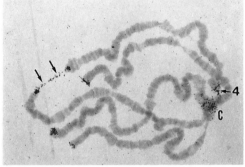

Fig. 2.11. *Salivary gland polytene chromosomes showing sites of hybridization with tritium-labeled T-A. C, Chromocenter. Chromosome 4 is heavily labeled, as are the tips of the other chromosomes. Reproduced from Young et al. (1983), with permission of the publisher.*

B. Healing of Broken Ends

In maize dicentric chromosomes, formed as the result of an inversion described on page 200, or as the result of a transposition as described on page 191, may break at the time of formation of an anaphase bridge during cell division. This can lead to what is called the breakage–fusion–bridge cycle, which is described by Mc-Clintock (1951). In this cycle (considered in detail in Chapter 5), the broken ends of the chromosomes or chromatids fuse to form rings and dicentrics, because the ends are sticky. However, under certain conditions, depending on the nature of the tissue cells, the broken ends heal and their subsequent behavior resembles that of a normal, nonbroken end. This healing re-

sults in the formation of functional telomeres.

Another kind of healing occurs in the coccid insects. All coccids have diffuse centromeres, and breaking the chromosomes with X-rays does not result in the formation of acentrics, since every broken fragment has centromeric activity. Under certain conditions the broken ends in the somatic cells heal and the fragments segregate normally. But in spermatocytes undergoing meiosis broken ends fuse (Hughes-Schrader and Ris, 1941). The similarity to the situation in maize is striking in the sense that cellular environment seems to be important for telomere formation, if we define a telomere as a nonsticky end without regard to the nature of its DNA sequences.

In the latter part of the last century Boveri discovered in the nematodes of the family Ascaridae a peculiar fragmentation of the chromosomes of those nuclei destined to pass into the somatic line of cells. This results in what Boveri (1899) called **chromatin diminution**. This process is illustrated in Figure 2.12 for *Parascaris univalens* (formerly *Ascaris megalocephala*), which has a diploid number of two. The first division of the zygote is a normal mitosis, but the second division of one of two daughter cells results in two cells with about 30 small chromatin fragments each of which subsequently acts as a separate chromosome with healed ends. Each also has a centromere, because ascarids have polycentric chromosomes. A considerable portion of the chromatin present in the zygote is cast out in these cells to become the somatic line. The eliminated chromatin is all, or nearly all, heterochromatin (Goday and Pimpinelli, 1984). Only one cell retains all the chromatin and this becomes the precursor of the germ line as shown in Figures 2.12 and 5.48. This phenomenon resembles that which occurs in the formation of the macronuclei in the ciliates (Yao et al., 1987;

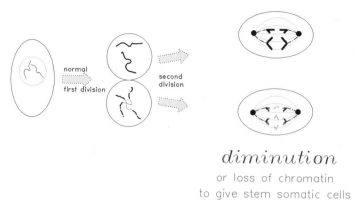

Fig. 2.12. *Chromatin diminution in Parascaris univalens. See text.*

Austerberry et al., 1984). In these Proto-zoa the residual chromatin bodies also have healed ends. Diminution with the formation of new telomeres also occurs in certain insects in the Orders Coleoptera and Diptera, and in the copepods of the Class Crustacea. The healing of broken ends of the chromosomes of *D. melanogaster* has been associated with the *HeT* family of repeats. Broken ends of chromosomes that have become healed have acquired *HeT* DNA in contrast to those ends that have not healed (Biessmann et al., 1990).

III. SPECIFIC DNA SEQUENCES ARE SPATIALLY ORGANIZED IN CHROMOSOMES

An understanding of chromosomes and genome structure and function is contingent upon a knowledge of the DNA base sequence organization of chromosomes. In this section this topic is introduced with the objective of first presenting a global view to orient the reader. Various additional phases of the subject are elaborated upon, and emphasized later in this chapter and in those that follow.

A. Distinguishing the Different Kinds of Sequences in Genomes and Their Locations

1. Reassociation kinetics analysis

The basics of the methodology for the analysis of DNA by following the reassociation of small complementary segments of single-stranded DNA was introduced in Chapter 1. Reassociation kinetics is a powerful tool for the analysis of DNA, for it enables one to separate DNA segments with repeated base sequences from those that are not repeated but are unique. Actually the resolving power for the separation of the two types is at best about 10 repeats. Identical repeats that occur with a frequency of less than 10 are not distinguishable for the most part from those that occur only once per genome.

The fact that in a mole of single-stranded DNA repeated complementary sequences are at higher concentration than unique segments, and hence have a lower $C_0t_{1/2}$ value, enables one to stop reassociation and capture the reassociated repetitive fraction by passing the DNA sample through a hydroxyapetite column, or by hydrolysis of the remaining single-stranded segments with enzymes that do not attack duplex DNA. One can also fine-tune the procedure by simply removing reassociated strands at earlier and earlier periods after the onset of reassociation, thus separating sequences with different frequencies of repeats.

2. Analytical centrifugation

Short segments of native double-stranded DNA suspended in a molar solution of a cesium salt will upon centrifugation at high gravities come to an equilibrium in the cesium salt gradient formed during centrifugation (Fig. 2.13). The equilibrium point will be at the level in the centrifuge tube where the buoyant density of the DNA sample corresponds to the density level in the cesium gradient. The buoyant density of a DNA sample is dependent on a number of factors such as the GC content of the sample, and the degree of methylation of the cytosine bases. Frequently more than one band is formed. Besides a major band containing the bulk of the DNA, minor bands of greater or lesser buoyant density will often be generated (Fig. 2.13). These are called satellite bands and their DNA is referred to as **satellite DNA**. Satellite DNA is almost always found to be highly repetitive, and it has formed the source

for the early studies on repetitive DNA. It is important to understand, however, that not all repetitive DNA components show up as satellites after centrifugation. Also the treatment of the DNA prior or during centrifugation will determine the number of satellites produced from a given sample of DNA.

3. Analysis with radioactively labeled or fluorescence-labeled DNA probes

Segments of single-stranded DNA isolated from repetitive DNA components can be used to identify the location of these repetitive arrays on chromosomes. The technique is described in Box 2.2.

B. Partitioning the DNA of a Genome

Table 2.3 gives an example of one way human DNA can be partitioned into five components. The data in Table 2.3 are derived from a variety of sources by a number of means such as reassociation kinetics, centrifugation, and others to be described in this and succeeding chapters. Here the components themselves are defined to give an overview of what holds for the vertebrate animals in particular,

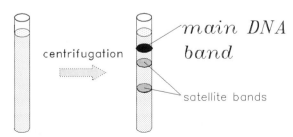

Fig. 2.13. *Satellite DNA demonstrated by centrifugation. See text.*

TABLE 2.3. Partitioning DNA in the Human Genome

	Type of DNA	% of total	Copy no.	Functions and locations on chromosomes
I.	Transcribed and translated	2	Unique	Exon parts of genes in euchromatin transcribed into mRNA and translated into polypeptides
II.	Transcribed but not translated	10	Unique and repetitive	Intron parts of genes and the DNA transcribed to rRNA, tRNA, etc.
III.	Not transcribed	5	Unique	Regulatory sequences adjacent to transcribed regions (e.g., promoter and enhancer regions)
IV.	Not transcribed	58	Unique	Unknown, scattered
V.	Not usually transcribed	25	Repetitive	Unknown functions, but sequenced; some scattered throughout genome, some confined to centromeric and telomeric regions as long tandem arrays

About 42% of the genome is more or less identified, and 58% is unique and unidentified with respect to structure.

although the same general picture is probably to be found in all higher eukaryotes.

Type I DNA (described in Table 2.3) is the unique portion constituted of "structural"-type genes encoding polypeptides. One current estimate is that this component makes up only about 2% of the total genome, if the nontranslated intron parts are excluded. However, the estimate may be too low, it may be as high as 10% as discussed on page 379.

Type II DNA is identified as that which is transcribed into identifiable forms of RNA, such as the intron parts of genes and the known forms of RNA other than mRNA that function in protein synthesis and certain aspects of regulation in the cell. Part of this component is repetitive,

particularly the DNA transcribed into rRNA as discussed later in this chapter. Type III DNA is not transcribed but is identified as that part adjacent to genes in the type I DNA that regulates their transcription. These Type III sequences are discussed in detail in Chapter 6.

Type IV DNA is unique and scattered. Its function is unknown. It is almost certainly important, however, since about 58% of the genome falls into this category.

Type V DNA constitutes about 25% of the human genome. It is the repetitive sequence DNA not identified with type II functions and occupies a relatively high fraction of the genomes of all higher eukaryotes, ranging up to 50% in some and less than 25% in others. This kind of DNA is recognized by the kinetic studies described in Chapter 1 and by using methods described in the following sections.

The partioning of the various DNA sequence types in plant chromosomes presents a picture similar to that found for animal genomes. A kinetic analysis of the DNA of 12 species of plants in the genus Lathyrus shows that on average about 40% falls in the unique category, and 50% in the middle repetitive category (Narayan, 1991). In plants as well as in animals the unique category of DNA remains relatively constant in amount among different species. It is the middle repetitive fraction that has the greatest ranges in amounts.

C. Heterochromatin Versus Euchromatin

The term **heterochromatin** was originally introduced by Heitz (1928) as a result of his studies of moss chromosomes. He sought to differentiate between pycnotic, or densely staining, regions of chromosomes and those regions that stained less darkly or not at all. He called the pycnotic regions heterochromatin. It is now known that the intense staining properties of heterochromatin in interphase is a result of an increased concentration of DNA caused by a more extensive condensation of the chromatin than the rest of the chromosome, the **euchromatin**. However, condensation and inactivation are not necessarily completely equivalent conditions of DNA.

Regions of chromosomes that take stains strongly are characteristically associated with the centromeres, telomeres, nuclear organizer regions, or NORs, as discussed in the following section, and entire arms or whole chromosomes as in some sex chromosomes. Some of these heterochromatic regions seem to be always condensed while others are only transiently so. Brown (1966) called "permanently"

heterochromatic regions **constitutive** and the transient ones **facultative**. However, this differentiation was not meant to imply that constitutive heterochromatin is always genetically inert, whereas facultative is not. It is convenient to retain these two terms, since there do appear to be permanently pycnotic regions as well as those that are cyclically pycnotic. Even so, some constitutive heterochromatin may occur in alternative states, condensed and decondensed, depending on the cell type.

Studies centered on the chromosomes of *Drosophila sp.* especially *D. hydei*, have resulted in some illuminating insights into heterochromatin versus euchromatin (Hennig, 1986). In *D. hydei* the entire Y chromosome of the male and one arm of the metacentric X are heterochromatic in the somatic cells. Also heterochromatin is present in the centromeric regions of the five autosomes of the genome. These heterochromatic regions would, by the usual standards, be considered constitutive and rich in undispersed repetitive DNA. However, the Y chromosome has a relatively small amount of repetitive DNA, and in neuroblast cells it is not entirely heterochromatic (Hennig, 1985). In *D. melanogaster* and *D. hydei* the Y chromosomes carry genes necessary for male fertility. Those regions of their Y chromosomes in which the fertility genes are located are clearly less condensed than the residual parts of the chromosomes. In contrast to the circumstances of the somatic cells, the male germ cells have essentially no heterochromatin in any of the chromosomes. The Y chromosome is not only decondensed, but it has been shown to be active in RNA synthesis in the primary spermatocytes (Hennig, 1985).

The important points raised by the example given above for *Drosophila* are that

• what may appear to be constitutive heterochromatin in one cell type may not in another, and

• the phases of the meiotic and mitotic cycles may show differential condensation of the chromatin.

Different species may exhibit quite different patterns of condensation. For example, in contrast to *Drosophila*, the water flea *Cyclops* has heterochromatin present in most stages of the germ cells, but not in the somatic cells of some species (Beerman, 1984). In the rodent *Microtus agrestis* thymus cells cultured *in vitro* show decondensation of the heterochromatic regions of the chromosomes upon treatment with thyroid hormone (Schneider et al., 1973). And in the orchid *Spiranthes sinensis*, the amount of heterochromatin per chromosome varies with the environment. In a cold climate only 2 of the 15 chromosomes are euchromatic, the rest being heavily heterochromatic. In a warm environment 6 of the 15 are euchromatic.

These examples make it quite clear that no definite line separates the two forms of heterochromatin or even euchromatin from heterochromatin. The notion of constitutive heterochromatin is a somewhat vague concept. It has heuristic value, but needs further study to characterize it definitively. With respect to the early previously held position that heterochromatic regions are genetically inert, it can now be stated that this is not true at least for *D. melanogaster* (Hilliker et al., 1980). Genes with various functions have been mapped in the heterochromatin of the X and Y chromosomes 2 and 3 of *melanogaster* (Gatti and Pimpinelli, 1983; Pimpinelli et al., 1985; Marchand and Holm, 1988a,b).

In addition to the genetic elements found in the centromeric regions of *Drosophila* chromosomes, it has also been reported that the centric heterochromatin of human chromosome 2 has within it the pepsinogen genes, *PGA3, PGA4*, and *PGA5* (Taggart et al., 1985). These appear to be present in varying numbers as determined by the use of cDNA probes containing a nu-

cleotide sequence corresponding to a part of the exon portion of the *PGA* gene.

D. Middle Repetitive Sequences Are Either Dispersed or Undispersed in Chromosomes

It is characteristic of the nonfungal eukaryotics that their chromosomes have significant amounts of specific types of middle repetitive DNA segments dispersed primarily in the euchromatic regions, and other types arranged in tandem, or undispersed, primarily in the constitutive heterochromatic regions.

1. Dispersed repetitive sequences

The regions of the chromosomes bearing unique sequences have middle repetitive DNA sequences interspersed among the nonrepetitive sequences. This has been shown for the frog *Xenopus* (Davidson et al., 1973), Syrian hamster (Moyzis et al., 1981a), and human genomes (Schmid and Deininger, 1975; Moyzis et al., 1988), among others, and is probably true for the higher eukaryotes in general.

This interspersion has been demonstrated by reassociating trace amounts of radiolabeled DNA single-strand fragments of various lengths with short, 0.2-kb, unlabeled "driver" DNA in severalfold excess, to a C_0t value at which only repetitive DNA will have reassociated. If repetitive DNA sequences are interspersed with the unique, the amount of reassociated double-stranded DNA with labeled tracer bound to hydroxyapetite should increase with increased tracer-labeled fragment length, assuming that the repetitive and nonrepetitive sequences are covalently bound to each other. Figure 2.14 illustrates the results obtained with human DNA. It shows clearly that the amount of radiolabeled DNA (R_L) bound to hydroxyapetite increases with an increase in labeled fragment size over the range of

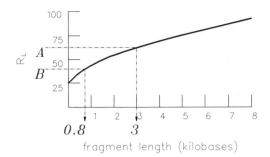

Fig. 2.14. *Relation between amount of reassociated duplex radiolabeled DNA (RL) consisting of repetitive and unique segments and the length of the fragments. See text.*

0.2–8kb. If, on the other hand, the unique and repetitive sequence components were sequestered from one another so that the unique sequences are covalently bound to one another, the kind of relationship shown in Figure 2.14 should not be expected; the reassociation DNA samples collected at a C_0t value at which **only** repetitive DNA should reassociate will contain only tracer repetitive DNA, and the effect of fragment length should be not striking, but minimal, as in Figure 2.14. It has been shown mathematically that a model based on the assumption that at least some of the repetitive sequences are randomly dispersed yields a curve that fits the experimental

data shown in Figure 2.14 (Moyzis et al., 1981a, 1989).

The "average" distance between interspersed repetitive sequences in human DNA can be estimated from Figure 2.14 to be about 3 kb. This is close to the distance of 3.4 kb reported by Deininger and Schmid (1976), which was determined by an electron microscope analysis of duplex DNA. The average length of an interspersed repeat is estimated at about 0.8 kb from the curve in Figure 2.14. This is based on the assumption that 25% of human DNA is repetitive, as indicated by the intersection of the curve with the ordinate, and about 7.5% of this repetitive DNA is assumed to be noninterspersed long tandem arrays.

A number of types of middle repetitive DNA have been identified in primate DNAs. They are grouped into two major families, the *Alu* and the *L1* (*Kpn*), which are also referred to as the *SINES* and *LINES* families, respectively. In addition, other repeat types occur such as stretches of (GT:AC)$_n$ (Rich et al., 1984) and *THE* and *VNTR* sequences. But the *Alu* and *L1* families are the dominant ones; they make up about $\frac{1}{3}$ of the total repetitive DNA. Table 2.4 lists all of these types and gives the frequencies of their occurrence and ap-

TABLE 2.4. Major Human Repetitive DNA Families

	Repeat	Length	Copies per genome	Fraction of genome mass
I	*Alu* (*SINES*)	0.3 kb	500,000	0.05
II	*Kpn* (*LINES*)	7–8 kb	50,000	0.04 (0.035–0.135)
III	α-SAT	0.17 kb	>225,000	(0.02–0.05)
IV	SAT I	0.04 kb	?	(0.02)
V	SAT II	0.05 kb	?	(0.02)
VI	SAT III	0.05 kb	?	(0.02)
VII	*THE*	3.4 kb	4,000	0.004
VIII	(GT:AC)	0.04 kb	100,000	0.0013
IX	(TTAGGG)	0.006 kb	30,000	0.00007
Total				~0.26
VNTRs, etc. Total repetitive DNA				~0.25

Possibly 1,000 families are left to be discovered.

proximate distribution in the human genome, as well as consensus sequences derived from other primate DNA data. What is most significant about their distribution in chromosomes is that they occur at extremely low to undetectable frequency in the centric heterochromatin and telomeric regions. Furthermore, the *L1* family repeats dominate in the Giemsa (G) and quinacrine (Q) bands, whereas the *Alu* family members dominate in the reverse-stained (R) bands (Korenberg and Rykowski, 1988). The G/Q bands are relatively rich in adenine and thymine (Therman, 1986) and the R bands in contrast are relatively rich in guanine and cytosine. Other differences noted between the two types of bands is that the DNA in G/Q regions replicates late in the S period of the cell cycle, and the R regions early in the S phase. Since the R regions appear to be those in which most active genes are located, it follows that the *Alu* repeats are mostly associated with active genes while the *L1* are less so (Kuhn and Therman, 1979).

2. *Undispersed repetitive sequences*

The undispersed repetitive sequences of DNA are mainly localized in the heterochromatic regions of the chromosomes, in contrast to the dispersed sequences such as *Alu* and *L1*. They are frequently referred to as satellite DNAs, since they were first discovered and described as satellite bands upon centrifugation in density gradients. However, they do not always have buoyant densities different from the main mass of the genomic DNA, and the satellite terminology is now an anachronism in a sense. In addition, a satellite band does not generally contain only one kind of repeated sequence but is frequently a collection of different sequences. The bands resulting from centrifugation are perhaps best described as classical satellites and designated with roman numerals (I, II,

. . . , N), whereas the "pure" sequences within them will be described in this discussion by arabic numerals $(1, 2 \ldots, n)$ or by Greek lowercase letters relating them to the classical satellites from which they are derived. To introduce what is at best a very complex subject, a brief description of satellites is given below.

Human centromeric DNA consists of tandem arrays of α-satellite (Willard and Waye, 1987), β-satellite (Waye and Willard, 1989), and the three other satellite classes designated as I, II, and III (Prosser et al., 1986).

Undispersed satellite DNAs are large tandem arrays of repeated sequences of various lengths. The human classical satellites I, II, and III have been extensively analyzed after isolation by centrifugation (Corneo et al., 1971); each satellite contains a number of different sequences, but Prosser et al. (1986) have shown that each contains a major component consisting of a single family of simple repeated sequences. These "pure" repeats are designated satellites 1, 2, and 3. Analysis of these repetitive components by first digesting them with restriction endonucleases and then subjecting the digests to electrophoresis on polyacrylamide gel results in "ladder" patterns such as those shown in Figure 2.15. Ladder patterns are expected when covalently linked repetitive sequences are subjected to digestion by endonucleases that recognize specific sequences in the chain of repeats. The ladder pattern results because some of these recognized sequences are altered by mutation, with the result that DNA segments of various lengths are produced. The separation of the gels is a function of the molecular weight of the segments, not their base sequences—hence the ladder effect.

The sequences of the DNAs isolated by electrophoresis can be determined by methods similar to those discussed in Chapter 1 and a **consensus** sequence derived. A consensus sequence is a kind of

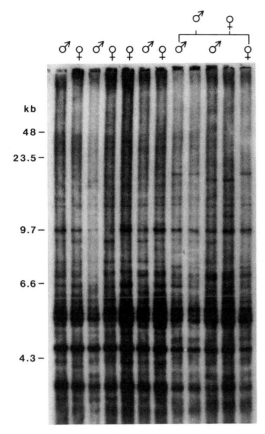

TABLE 2.5. Human Sample Sequence Isolates From Satellites I, II, and III

Satellite 1	Rsa I ladder
	Highly (A + T)-rich repeat
	units, A(17 bp) and B(25 bp)
	in arrays -A-B-A-B-A . . . n
Satellite 2	Frequent Hinf1 and TaqI sites,
	poorly conserved 5-bp
	repeat, GGAAT
Satellite 3	Hinf1 5-bp ladder,
	CAACCCGA^A_CT (GGAAT)$_n$

Data from Prosser et al. (1986).

Fig. 2.15. *Ladder-like electrophoretic pattern obtained from human satellite III DNA obtained from a number of different individuals including a family of five. Note that a polymorphism was segregating in the family. DNA was cut with restriction enzyme Sau 3Al. Courtesy of J. Longmire, Los Alamos National Laboratory.*

average derived from a series of sequences that may differ from one another by only a few bases but are otherwise obviously rated. Table 2.5 gives the results of the analysis of the three satellites by Prosser et al. (1986). Three different restriction enzymes were used in the analysis and the data clearly indicate that the satellite 2 and 3 sequences have some relationship, but satellite 1 is quite different from both of them. The highly diverged pentamer of satellite 2, GGAAT, bears striking resemblance to the sequences in the core of the yeast centromere in which variations on the theme GGAAT such as CGAAT, TGAAT, TAAAT, and so forth, occur. The same is also true of satellite 3. As for satellite 1, which is rich in A and T repeats, the same is true for element II of the yeast centromere core described in Section 1.B. Eukaryotic centromeres have cores with conserved DNA sequences (Grady et al., 1992). This is not unexpected, since centromeres from whatever source serve the same basic function.

All vertebrate animals whose DNA has been analyzed for repetitive sequences have one or more that appear as satellites (reviewed by Miklos, 1985). Bovine thymus DNA has eight detectable satellites that constitute a total of 27% of the nuclear DNA. A major one of these, referred to as satellite III, is 4% of the total genomic DNA. Since the average vertebrate genome has about 3×10^9 bp, this means that one type of repetitive sequence has 1.2×10^8 bp. This is about the DNA content of a single cattle chromosome. Hence one is dealing here with one kind of sequence that makes up a considerable portion of the cattle genome.

When satellite III is digested with the restriction enzymes Bsp and EcoRII, repeating segments about 2,350 bp long are produced. These segments can be further digested with other restriction enzymes, principally, Sau3AI, BclI, PvuII, and AluI. Sau3AI and BclI cleave to produce segments about 11 bp long or multiples thereof. AluI has cleavage points all through the 2,350-bp repeat to produce fragments that are multiples averaging about 22.3 bp.

The sequences of the 2,350 satellite fragments obtained by SauAI digestion were determined from material cloned in plasmids grown in *E. coli*. It is readily apparent on comparing sequences that they consist of 23-bp repeats along with 11- to 12-bp repeats. Representative sequences are shown in Figure 2.16a along with a

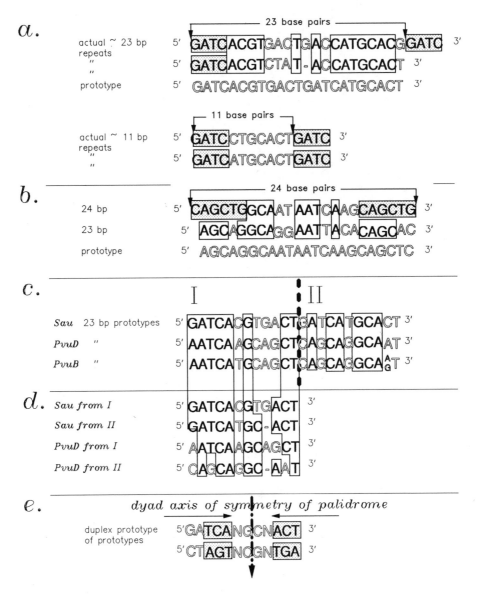

Fig. 2.16. *Comparisons of the repeated sequences obtained by restriction enzyme cutting of the 2,350-bp calf thymus repeat. See text. Data from Pech et al. (1979a, b).*

23-bp "prototype." The fragments from Pvu have also been sequenced. Representative samples and prototypes are shown in Figure 2.16b. The prototypes of the Sau and Pvu segments are also compared in Figure 2.16c. There are not only obvious similarities between the Sau and Pvu segments, but the prototypes appear to be dyads. If the 11- 12-bp segments are lined up as in Figure 2.16d, a prototype 12-bp sequence can be derived as in Figure 2.16e. Furthermore, this has a dyad structure with palindromic properties. By dissecting the satellite sequence in this fashion it becomes evident that the long 2,350-bp repeat unit is in turn made up of smaller repeat units. A basic 12-bp unit, 5'GATCANGCNACT3', exists which may be called the theme. Repeated variations on this theme arranged in tandem constitute the 1.706 satellite. Furthermore, the Sau prototype contains inverted repeats as well as true palindromes, as shown in Figure 2.17. This repetition of small units to build larger repeating units is characteristic of satellite DNAs in general. An additional example drawn from among many is the rat satellite I. It has a 370-bp repeat unit that can be shown to consist of four 92- to 93-bp subunits that have a high degree of homology, as shown in Figure 2.18 (Pech et al., 1979a).

The highly repeated centromeric sequence GGAAT$_n$ is interspersed in human chromosomes with the consensus sequence CATCATCGA(A/G)T as shown in Table 2.7. The (GGAAT)$_n$ repeat is highly conserved. It is present not only in human satellites II and III, but in other mammals and in the chromocenter of *Drosophila* polytene chromosomes, and it is similar to the III element of the centromere core of *Saccharomyces*, as shown in Figure 2.3. Thus it seems possible that the centromeres of all eukaryotes have similarities in their molecular structure (Grady et al., 1992).

Sequence similarities that are probably homologies exist among different groups of vertebrates. For example, a 170- to 173-bp sequence, found originally in the α-satellite of the African green monkey, is also

Sau prototype

5' GATCACGTGACTGATCATGCACT 3'

TCACGT   TGCACT

inverted repeat palindromes

Fig. 2.17. *The prototype of the 2350-bp repeat from calf thymus DNA is constituted of a basic 12-bp unit that contains inverted repeats and palindromes.*

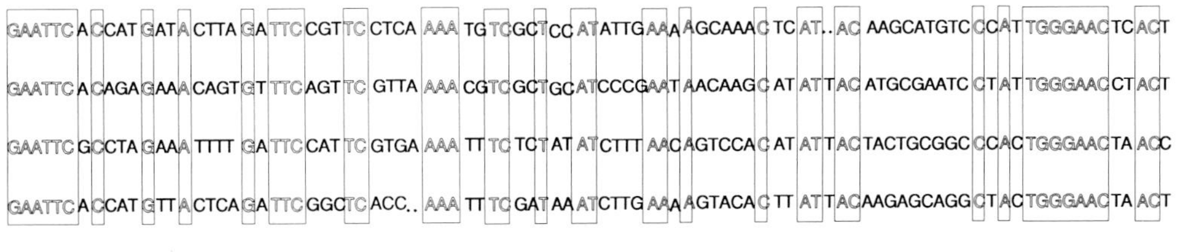

GAATTCACCATGATACTTAGATTCCGTTCCTCAAAATGTCGCTCCATATTGAAAAGCAAAGTCAT..ACAAGCATGTCCCATTGGGAACTCACT

GAATTCACAGAGAAACAGTGTTTCAGTTCGTTAAAACGTCGCTGCATCCCGAATAACAAGCATATTTACATGCGAATCCTATTGGGAACCTACT

GAATTCGCCTAGAAATTTTGATTCCATTCGTGAAAATTTCTCTATATCTTTAACAGTCCACATATTTACTACTGCGGCCCACTGGGAACTAACC

GAATTCACCATGTTACTCAGATTCGGCTCACC..AAATTTCGATAAATCTTGAAAAGTACACTTATTTACAAGAGCAGGCTACTGGGAACTAACT

Fig. 2.18. *Sequence of the 370-bp repeat unit of satellite I DNA showing that it consists of four 92- to 93-bp subunits. Data from Pech et al. (1979a).*

found with variations in other primates including humans (Singer, 1982). This family of sequences, generally referred to as the α-satellite or alphoid DNA, has a readily identifiable consensus (Meyne et al., 1989), but shows significant variation in sequence among different members of the same species as well as among different species of the Order Primates. The α-satellite itself may compose as much as 25% of the total genome in some species.

It is found in the centromeric heterochromatin of most, if not all, chromosomes, but each chromosome may have a distinct variation on the consensus theme, and the different chromosomes of a genome may be individualized by their patterns of sequence variation (Jabs and Persico, 1987). Figure 2.19a shows the distribution of an α-satellite type in the human chromosome set. Specific DNA probes need not hybridize only with centric heterochromatin. They

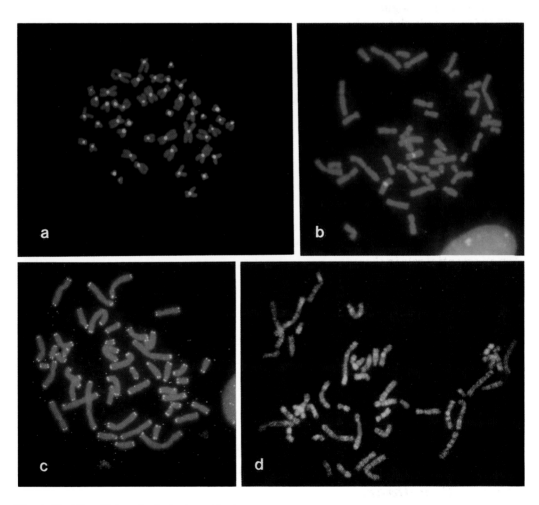

Fig. 2.19. *Identifying the loci of specific DNA sequences on chromosomes.* **a.** *α-satellite DNA at centromeric regions of all human chromosomes.* **b.** *Sequence specific for the centromeric region of human chromosome 9.* **c.** *The telomeric sequence TTAGGG on human chromo-* *somes.* **d.** *Alu distributions on a human chromosome set. The yellow fluorescence identifies Alu; the red areas identify centromeric regions in which Alu appears to be absent. Courtesy of J. Meyne, Los Alamos National Laboratory.*

may also hybridize at specific areas of chromosomes outside both the centric and the telomeric areas of specific chromosomes (Devine et al., 1985).

Sequence similarities are found not only within the same order of mammals, but in different orders. In Figure 2.20 segments of certain satellites of mouse, rat, human, calf, and monkey are depicted to show that similar sequences are apparently present in the representatives of three orders, Rodentia, Artiodactyla, and Primates. This dispersion of homologies is probably not confined to a single class such as the Mammalia. It has already been noted that the hexanucleotide repeat TTAGGG is found in all vertebrate animals and even among certain protozoa (Table 2.2). It is located in the telomeric regions of 91 species of vertebrates ranging from bony fish to humans (Meyne et al., 1990). This repeat and related ones seem to be universal among the eukaryotes as noted on p. 80. The presence of the $5'(G)_n3'$ nucleotide sequence, where n may range up to eight in some of the repeats, and the apparent constancy of some of the repeats such as TTAGGG in animals, are a good indication that they play an essential role in chromosome function. However, some repeats, particularly in the centromeric heterochromatic regions, may vary tremendously in quantity. This is particularly true of the nondispersed repeats characteristic of heterochromatin. For example, the HS satellite of the kangaroo rat *Dipodomys*

ordii contains TTAGGG as its principal constituent, and it comprises 30% of the genomic DNA. The presence of this satellite in DNA is a close relative, *D. deserti*, is barely detectable (Hatch and Mazrimas, 1974). There is a further degree of specificity in the distribution and arrangement of the various forms of satellite DNA on the different chromosomes of a genome. In the human chromosome complement, different classes of highly redundant DNA can be shown to be specific for the short arms of the acrocentric chromosomes 13, 14, 15, 21, and 22 and the long arm of the Y, and around the centromeres of the metacentrics 3, 4, 9, 16, and 17 (Willard, 1985; Moyzis et al., 1987; Waye and Willard, 1989). It is possible that each chromosome in the human genome has a centromeric sequence unique to it. Figure 2.19b shows that a specific sequence is to be found in the centric region of human chromosome 9, for example. Some centric sequences are common to all chromosomes (Fig. 2.19a), or a certain set of chromosomes. There appears to be a specific DNA sequence associated with the parameres of chromosome 9 (Mitchell et al., 1986). Parameres are small bodies associated with chromosome 9 at pachytene that appear on none of the other human chromosomes (Sumner, 1986). But again it must be recognized that a sequence, particularly if it is a short one, may be present but in such low numbers that it is not recognized with the standard procedures.

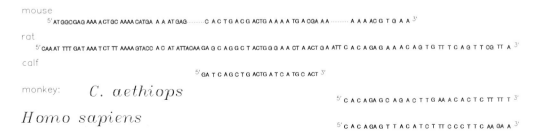

Fig. 2.20. *Sequence similarities evident for DNA from mouse, rat, calf, human, and monkey. Data from Pech et al. (1979b).*

A number of *Drosophila* species, including *D. melanogaster*, have been the principal invertebrates investigated for the repetitive DNA content of their genomes. *D. melanogaster* has a significant amount of its chromatin in the form of heterochromatin. Figure 2.21 illustrates an estimate that has been made (Miklos, 1985). Approximately 8.8×10^7 bp of DNA is estimated to be in the heterochromatin regions. This constitutes about 42% of the total DNA of the genome. Table 2.6 lists some of the satellite DNAs characterized for *D. virilis* and *D. melanogaster*. The total content of these sequences per genome accounts for most of the repetitious DNA, and there are obvious similarities in sequences between the two species. For example, compare the 1.672 satellite of *melanogaster* with the Ic satellite of *virilis*. Indeed there are similarities between sequences of the satellite 1.688 of *melanogaster* and the bovine satellite III Sau 3A and Pvu II regions (Table 2.7). Significant similarities are also found with African green monkey, human, and rat sequences. Whether these are homologies or just chance similarities may never be known, but they do indicate that a certain sequence may have been conserved because it has been selected for.

One of the more remarkable examples of the confinement of specific types of repetitive DNA to a single chromosome of a genome has been described for *D. nasutoides*. In this species with $n = 4$ chromosomes about 60% of the DNA is repetitive and the bulk of this repetitive component is confined to one chromosome, chromosome 4 (Cordeiro-Stone and Lee, 1976; Wheeler and Altenberg, 1977; Lee, 1981). Four satellite DNAs have been demonstrated in buoyant density centrifugation and all four appear to be confined to chromosome 4 (Wheeler et al., 1978). Labeled DNA from the four satellites hybridized detectably only with the DNA of this chromosome when in metaphase. The other autosomes as well as the X and Y do have heterochromatin as expected, but it apparently consists of repetitive sequences not related to the satellite sequences on the large metacentric. These may, of course, be present on the other chromosomes but not detectable by hybridization techniques.

The satellite-bearing chromosome 4 or *nasutoides* is an **isochromosome**, so-called because both arms are apparently identical. This kind of chromosome presumably arises during the cell cycle when the two sister chromatids of an acrocentric or telocentric chromosome remain fused at their centromeres, thus forming a metacentric that replicates as a single chromosome in subsequent cell divisions. Chromosome 4 is not entirely heterochromatic. It does appear to have euchromatic regions, as demonstrated by differential staining with Giemsa, quinacrine, and Hoechst 33258. (See Box 2.1.) What is further remarkable about it is that the satellite DNAs are dispersed and scattered along the arms and not detectable in the centromeric regions.

One other invertebrate group besides *Drosophila* has had its repetitive DNA in-

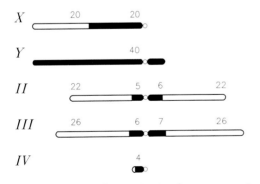

Fig. 2.21. *Heterochromatin in the genome of Drosophila melanogaster. The numbers represent the number of megabases in the heterochromatic (black) regions as opposed to the euchromatic regions. Data from Miklos (1985).*

TABLE 2.6. Satellite DNAs Isolated From *D. Melanogaster* and *D. Virilis*

Bouyant density (g/cc)	Nucleotide sequences of predominant repeats	Percentage of total genomic DNA
	D. melanogaster	
1.672	60% 5' AATAT; 40% 5' AATATAT	4.2
1.686	80% 5' AATAACATAG	3.3
1.688	5' TTTCC + others	4.6
1.705	90% 5' AAGAG; 10% AAGAGAG	4.7
1.697	Probably ribosomal genes	4.5
1.690	Rich in GC dinucleotides?	3.0
	Total	24.3
	D. virilis	
1.664 (III)	ACAAATT	8.
1.680 (II)	ATAAACT	9
1.685 (I)	ACAAACT	23
1.686 (Ic)	AATATAG	0.1
	Total	40.1

Data for *melanogaster* from Peacock et al. (1973) and Brutlag et al. (1978); for *virilis* from Gall and Atherton (1974) and Mullins and Blumenfeld (1979). Reviewed by Brutlag (1980).

TABLE 2.7. Comparison of Two Repetitive Satellite Sequences Found in *Drosophila Melanogaster* and Bovine DNA

D. melanogaster	TTGTCTGAATATGGAATGTCAT
Bovine Sau	TTGGCTGATCATGCAATGATCAT
D. melanogaster	CAGCTTTGCGAGGTATGACATTCCA
Bovine PvuD	CAGCTGAGC AGGCAGGA ATTACA

Data from Brutlag (1980).

vestigated extensively. Satellites of various species in the arthropod Class Crustacea, principally crabs in the Order Decopoda, have been analyzed. Two species, *Cancer borealis* and *C. irroratus*, have satellites that are composed 90%–97% of repeating AT units (Skinner, 1972). In the case of *C. borealis* this satellite constituted 30% of its total genome, or ~500 × 10⁶ bp. Another crab, the Bermuda land crab, *Gecarcinus lakeralis*, has a satellite that is (G + C)-rich and accounts for 3% of the total genomic DNA. When treated with restriction enzymes, this satellite DNA can be shown to have a monomer repeat unit of 2.07 kb. It occurs as a cluster of 10 or more tandem repeats and has about 1.6×10^4 copies per genome. Similar repeats occur in other crustacean satellites. They may therefore have a common origin. The basic sequence shows considerable heterogeneity, however, so that it has not been conserved intact.

Repetitious DNA is a significant part of the genomes of higher plants, and, as in animals, many of these repeated sequences are found in long tandem arrays in the heterochromatic regions of centromeres and telomeres (Flavell, 1980). Examples of repeated sequences in telomeric heterochromatin are found in species of *Secale* (rye). Table 2.8 lists the occurrence of four different repeat family units in six species of Secale. The cultivated type *S. cereale* and its relative *S. montanum* have all four types; the others do not. Only the 120-bp repeat is present in all six species (Jones and Flavell, 1983). However, it is to be noted that copies present with less than 500 bp per genome may not be detectable by the hybridization technique used; they may therefore actually be present but at low frequencies. Although each of the repeat units listed in Table 2.8 are present in each of the telomeres of the genome ($n = 7$), two of them, the 480-bp and 120-bp units, are also distributed interstitially. The 610-bp repeat is exclusively telomeric but again it may be present at undetectable levels interstitially. It is of some interest that certain wheat species (genus *Triticum*) show the presence of the 120-bp sequence both telomerically and to a less extent interstitially following hybridization with a labeled 120-bp probe from *Secale* (Jones and Flavell, 1983). As in animals, therefore, similar repeat sequences may be widespread in different taxa. Also (as noted in Table 2.2) the repeat TTAGGGG appears to be general, among at least the flowering plants, at the telomeres.

Plants, especially the flowering ones, have a wider range of repetitive DNA content per genome in related taxa than is found in animals except for the Amphibia. It is not unusual for some plants to have $\sim 10^{11}$ bp per genome. However, an angiosperm, *Arabidopsis thaliana*, a member of the mustard family, has a genomic DNA content of about 7×10^7 bp (Leutwiler and Hough-Evans, 1984). This is certainly evidence that higher eukaryotic life is possible without a large amount of repetitive DNA.

IV. NUCLEOLAR CHROMOSOMAL REGIONS

All eukaryotes appear to have the structural genic elements for the synthesis of ribosomal RNAs. These occur as tandem DNA repeats clustered at specific chromosomal sites called the **nucleolar organizer regions** (**NOR**) (reviewed by Babu and Verma, 1985). NORs generally appear as secondary constrictions in mitotic chromosomes, but in interphase chromosomes spherical structures called **nucleoli** develop at the NORs (Fig. 2.22). It is within the nucleoli that the ribosomal RNAs are synthesized, as described in Chapter 6.

Not all chromosomes of a genome will bear secondary constrictions with NOR function. Only the five pairs of the human acrocentric chromosomes, 13, 14, 15, 21,

TABLE 2.8. Repeated DNA Sequences in the Telomeric Heterochromatin of *Secale* and *Triticum*, Given as Percentage of Total DNA per Genome

	480 bp	610 bp	630 bp	120 bp
S. cereale	6	2.7	0.6	2.5
S. montanum	2.5	0.5	0.16	2.5
S. vavilovi	3.5	—	0.02	2.5
S. africanum	—	0.04	—	2.5
S. silvestri	—	—	—	2.5
T. diococcum	—	—	—	1.5

Data from Jones and Flavel (1983).

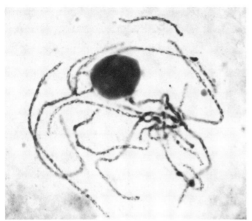

Fig. 2.22. *Nucleolus in the nucleus of a maize cell.*

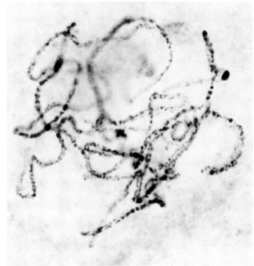

Fig. 2.23. *Chromomeres of maize chromosomes. The very dark regions are "knobs" and not chromomeres.*

and 22, bear NORs on their short arms. Nor do all NORs produce nucleoli: In humans usually only five to eight of the 10 NORs bear active nucleoli (Babu and Verma, 1985). As in most organisms, human nucleoli can form a single large and common nucleolus in the interphase and prophase. Thus the actual number of NORs per cell must be determined by the use of stains specific for the NORs rather than by counting nucleoli. In *Drosophila melanogaster* the NORs are confined to the X and Y chromosomes at the locus of the bobbed (*bb*) gene.

V. CHROMOMERES AND OTHER MORPHOLOGICAL MARKERS

A. Chromomeres

When chromosomes first become visible under the light microscope in mitotic or meiotic prophase, they are still in a partly extended condition and generally exhibit chromomeres. These chromomeres may be visible without staining under the phase contrast microscope, but they are generally best observed after staining as bands or beads (Fig. 2.23). The significance of

chromomeres is not entirely understood, but that they are important structural elements is well-established. They are arranged in a specific and consistent pattern along each chromosome at certain stages of the cell cycle and may serve to identify one chromosome from another in the same genome under both the light and electron microscope (Bahr and Larsen, 1974; Comings, 1978). As the chromosomes become more condensed going into anaphase from metaphase, the chromomeric structure becomes less and less evident, although the use of stains continues to show that a difference of chromatin structure persists linearly along metaphase chromosomes. Metaphase and anaphase chromosomes show little or no chromomeric structure without the use of stains except at the centromeric region, where there is a primary constriction, or at points where there may be secondary constrictions and where "knobs" are present (see Fig. 2.23).

Chromomeres, whether manifested morphologically as density of chromatin by electron microscopy (Bahr and Larsen,

1974) or as bands made visible by staining (Comings, 1978; Haapala, 1984a), are distinct structural elements of chromosomes above the 30-nm fiber level. As the chromosomes contract a compaction of the 30-nm solenoid apparently occurs differentially along its length. Superimposed upon this folding or compaction of fibers is the coiling of the chromosomes, which occurs mainly between early prophase and metaphase and is coincident with the separation of sister chromatids (Haapala, 1984b; Zatsepina et al., 1983).

The banding of chromosomes with the various Giemsa staining techniques (described in Box 2.1) is by far the best present method of demonstrating differentiation of structure along the length of chromosomes (Jhanwar and Chaganti, 1981, Jhanwar et al., 1982). Yunis (1981) has demonstrated about 2,000 light and dark bands in the human genome by means of trypsinization and Giemsa staining of mid-prophase chromosomes. According to Comings (1978) the Giemsa dark bands are regions where the chromatin DNA is free to stack with the thiazine dyes of the Giemsa dye mixture. The light Giemsa interband regions are either regions with less DNA because of less compaction, or those in which the DNA phosphate groups are covered, or both. The general opinion seems to be that the histones of chromatin are not important in banding, but this is by no means proven in all cases (Burkholder and Duczek, 1982). Since the thiazine dyes appear to interact with the DNA phosphate groups, it is logical to assume that the state of the DNA is important. However, histones can also be phosphorylated, and it is possible that they too can play a role in the formation of banding patterns.

B. Other Markers

Some chromosomes have **satellites** that are segments separated from the main body of the chromosome by a constriction (Fig.

2.24). The constrictions are stable and inherited in a regular fashion. They are particularly common among plants. The short, usually heterochromatic, arms of acrocentric chromosomes are sometimes referred to as pseudosatellites.

In some plants, such as maize, one or more of the chromosomes in the genome may have **knobs** (Fig. 2.24). These are not chromomeres, but regions of heterochromatin with repeated sequences. Their position and size within different maize lines are constant and they can serve as markers.

VI. ORGANIZATION OF GENES ON CHROMOSOMES

When work with species of *Drosophila* other than *melanogaster* was begun in the 1930s, it was found that a great deal of chromosomal homology existed within the genus (Sturtevant and Novtski, 1941; Patterson and Stone, 1952). Not only were the same identifiable gene loci linked, but the banding patterns of the polytene chro-

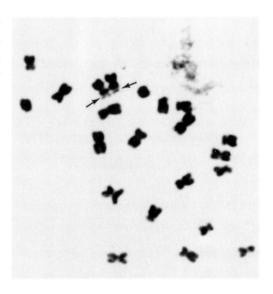

Fig. 2.24. *Plant chromosome satellites. Courtesy of Bryan Kindiger.*

mosomes indicated homologies of whole chromosome arms. Since then a similar **conservation of linkage** and **banding patterns** have been found among the mammals and some other vertebrates. This is a topic we consider at length in Chapter 9. Here we wish only to state that **genes are not randomly distributed on chromosomes**. Certain groups of genes stay linked for hundreds of millions of years of evolutionary change in phenotype. Whether this is just a happenstance or whether they stay together because of selection pressure is still open to question. But the fact remains that a certain degree of organization of genes is to be found among all eukaryotes for which there are sufficient linkage data. These linked genes may or may not appear to have functional affinities.

Linked genes that are functionally related form groups known variously as **gene complexes**, **multigene complexes**, **gene clusters**, and **gene families**. Examples have been found in all eukaryotes that we have extensive linkage data for, and it appears likely that they are in general characteristic of eukaryotic genomes. They have apparently arisen from duplication of primordial, ancestral genes (p. 381). The higher one goes on the eukaryotic phylogenetic tree, the more complex and abundant these groups seem to become. Not only are sequences of DNA with known functions distributed in certain patterns on the chromosomes, but many sequences without identifiable functions are also distributed in definite patterns, while others are randomly distributed.

VII. SPECIAL AND HETERODOX FORMS OF CHROMOSOMES

There is no such thing as a standard chromosome. The usual paradigm is a single molecule of double-stranded DNA associated with histone and other proteins and bearing special regions, such as centromeres, telomeres, and chromomeres. This model suffices for most chromosomes, at least in the G1 phase of the cycle. There are deviations, however, from this model that deserve special mention because they do help to understand certain aspects of the structure, behavior, and function of chromosomes in general.

A. Polytene Chromosomes

The usual chromosome life cycle is one in which the chromosome enters the S phase from G1 by beginning to replicate to form sister chromatids that generally enter into the G2 phase and the prophase while still held together at their centromeric regions. The chromatids separate and become known as chromosomes in late metaphase and anaphase. If the DNA content of the G1 chromosome is designated as 1C, then at G2 it is 2C and becomes 1C again at anaphase. **Polytene chromosomes** are the result of a duplication of the DNA without a following mitosis, so that the C-values are higher than 2C and will be $C \times 2^n$, where n equals the number of replications if the replications follow the normal pattern and the chromatids do not separate (see Nagl, 1978, for review).

Polytene chromosomes are found in the macronuclear anlage of some ciliate Protozoa, such as *Stylonchia*, *Euplotes*, and *Oxytricha*, and in certain tissues of the Metazoa and Embryophyta listed in Table 2.9. The familiar polytene chromosomes of *Drosophila melanogaster* are illustrated in Figure 2.25. They, like dipteran polytene chromosomes in general, are actually paired homologues, and each homologue presumably has 1,024 chromatids (2^{10}) so that the pair is 2,048 C. In general the homologous polytene chromosomes in plants do not show somatic pairing, as found in the Diptera, and the bands are not as prominent (Nagl, 1978). The stained bands of polytene chromosomes are generally considered to be chromomeres within which the concentration of DNA is greater

TABLE 2.9. Examples of Plants and Animals with Polytene Chromosomes

Tissue	Species
Plants	
Antipodal cells	*Papaver rhoeas* (poppy)
Synergids	*Allium ursinium* (onion)
Suspensor cells	*Phaseolus* sp. (beans)
Endosperm	*Allium ursinium*
Animals	
Macronuclei of some ciliates	
Salivary glands of Diptera; the silkworm *Bombyx,* and a number of *Collembola* sp.	
Trophoblasts of rodents	

et al., 1981), while another has reported the bands to be the richest in this form of DNA (Arndt-Jovin et al., 1983). It is probable that these differences in results are related to the methods of preparation of the chromosomes prior to treating them with Z-DNA antibodies. It has also been reported that the telomeric regions of these chromosomes are rich in Z-DNA (Arndt-Jovin et al., 1983). This is not unexpected, since the base sequences in the telomeric DNA are conducive to the formation of Z-DNA.

The polytene chromosomes of the Diptera show transitory changes in specific sections (Pavan and Da Cunha, 1969). These changes are considered to be related to specific synthetic activities of the chromosomes, such as transcription to form mRNA and translation to produce specific

than the interband regions. The interband regions of *melanogaster* DNA have been reported to be exceptionally rich in Z-DNA by one group of investigators (Nordheim

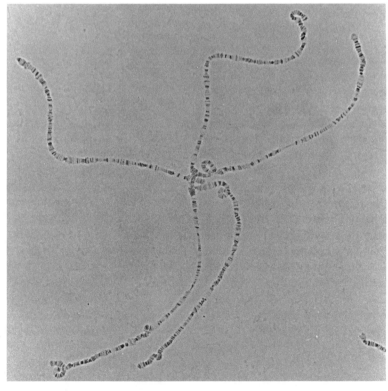

Fig. 2.25. *Polytene chromosomes in salivary gland cells of* Drosophila melanogaster. *Courtesy of Jon Lim, University of Wisconsin-Eau Claire.*

proteins. These regions generally appear as "puffs," and the extremely large ones seen in the chironomid Diptera are called Balbiani rings. Morphologically, puffs have two characteristics worth noting. First, the chromatin material of the puff region unfolds to form large loops of fibrils down to about the diameter of the 10-nm fibers from the 30-nm chromatin solenoid; second, these regions are characterized by high concentrations of RNA particles (Meyerwitz et al., 1985). These puffs are found not only in Diptera but also in certain Protozoa, such as *Stylonchia*, and plants, such as *Phaseolus*.

B. B-Chromosomes

B-chromosomes are a heterogeneous class of chromosomes also sometimes referred to as **accessory** or **supernumerary** chromosomes. They occur in some plants and animals and are usually, but not always, heterochromatic and telocentric. What sets them apart from other chromosomes in the genome they occupy is that they seem to have little, if any, effect on the phenotype; they vary in ploidy from one cell type to another; they may occur in some individuals of the same species, but not others; they may segregate normally in mitosis and meiosis but not always.

B-chromosomes have been most studied in maize (Carlson, 1978). The B-chromosome is about half the length of the shortest chromosome in the maize genome and contains both euchromatin and heterochromatin, although the heterochromatic region is exceptionally large for maize chromosomes. In some strains of maize, chromosome 10 appears abnormal in its chromomeric pattern and contains an extra amount of chromatin in its left arm. This segment has a large heterochromatic knob and a euchromatic region. B-type chromosomes are by no means confined to maize. They are found, for example, in a number of different species of other plants.

They are not uncommon in animals. The Class Orthoptera in particular seems to have an abundance of species with B-chromosomes. The vertebrate animals, however, have only a few species in which these supernumeraries have been reported. One mammal, the raccoon dog, *Nyctereutes procyonoides*, has two to four B-chromosomes per diploid set (Wurster-Hill et al., 1988).

Various hypotheses have been advanced for the existence of this class of chromosomes, ranging from their having some effect on fecundity of the individuals carrying them, to their being parasitic remains of chromosomes that have lost their function. Both may be true to a certain extent, of course. Their existence does show, in any case, that DNA with no apparent significant function can be maintained indefinitely in a population.

Sex chromosomes in animals that are XO were originally called accessory chromosomes. That appellation no longer is used, and is found only in the older literature.

C. Chromosomes of the Protista

The Protista comprise a heterogeneous group of eukaryotic organisms that have few things other than their unicellularity in common. Some are photosynthetic, some are not, and some are between being facultative photosynthesizers and heterotrophs. With all this diversity one might expect to find different kinds of chromosome types, but on the whole this is not true. For the most part their chromosomes are chromatin constituted of histones and DNA. However, in many cases no centromeres are evident, at least with the light microscope, and the nuclear membrane does not break down during cell division (in these cases it is called **amitosis**) and microtubules are not formed.

The most aberrant group with respect to chromosome structure is the Dinoflagellata (reviewed by Dodge, 1985). It has already been mentioned that no proteins of the typical eukaryotic histone type are present in the dinoflagellate chromosomes. A basic protein is present but it does not seem to be closely related to the histone, since it contains both cysteine and aromatic amino acid residues (Rizzo and Nooden, 1974). In addition, the DNA of dinoflagellate chromosomes is remarkable because a good deal of its thymidine is replaced by 5-hydroxy-methyl-uracil. In one species, *Prorocentrum micans*, 63% of the expected thymidine is replaced by 5-hydroxy-methyl-uridylate (Herzog and Soyer, 1982). Added to these unexpected findings, X-ray microanalysis has revealed that the chromosomes contain high levels of iron, nickel, copper, zinc, and calcium (Sigee and Kearns, 1982). It is believed that the function of the ions is to stabilize the chromosome structure by forming ionic bridges between the nucleic acid and protein. When isolated chromosomes are incubated in EDTA, the chromosomes collapse and their fibrillar substructure dissipates. Electron microscope and chemical studies have led Herzog et al. (1984) to propose the structure for dinoflagellate DNA shown in Figure 2.26.

Originally it was thought that dinoflagellate chromosomes had no centromeres. But studies with the electron microscope show that in some species the chromosomes are attached to the nuclear membrane by disk-like structures and microtubules. Thus it does appear that

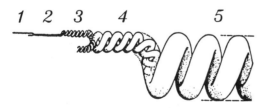

Fig. 2.26. *Proposed representation of the helical compaction of dinoflagellate chromosomal DNA. Reproduced from Herzog et al. (1984), with permission of the publisher.*

centromere-like structures do occur (Spector, 1984) and spindle fibers are involved during karyokinesis (Triemer and Fritz, 1984).

Only a small percentage of the Protista have been studied cytologically, and it is quite possible that there are other peculiar deviations from the "standard." In any case it is to be appreciated that eukaryotes do exist that function genetically without the presence of the otherwise almost universally occurring histones in their chromosomes.

D. Lampbrush Chromosomes

Lampbrush chromosomes occur in the primary oocyte and spermatocyte nuclei of some Metazoa. The lampbrush state is phase-specific and reversible. The largest and most developed lampbrush chromosomes are found in the urodele amphibians but they occur in many other animals both vertebrate and invertebrate as well (see Callan, 1986, for review). However, they are not ubiquitous in their occurrence. Like the polytene chromosomes of the Diptera these chromosomes are extremely large, but they are not polytene. They exist as bivalents and are restricted to the diplotene stage of the first meiotic prophase. In the salamander *Triturus viridescens* they may reach a diameter of about 500 μ and a length of 400–1,000 μ. Figure 2.27 is a composite diagram of a section of such a chromosome. Essentially the structure is dominated by a series of chromomeres of different sizes and shapes. The bivalents consist of two homologues presumably held together by chiasmata. Each homologue consists of a pair of closely united sister chromatids, and the chromomeres are formed by the looping of the bivalent pairs, as shown in Figures 2.27 and 2.28 (Callan, 1963). As these chromosomes proceed toward metaphase, the loops disappear and the typical compact metaphase chromosome structure appears.

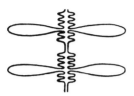

Fig. 2.27. *Generalized depiction of a lamp-brush chromosome structure showing the characteristic loops.*

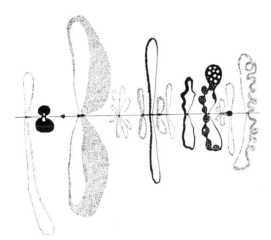

Fig. 2.28. *Loops in lampbrush chromosomes. See text. Reproduced from Callan (1955), with permission of the publisher.*

The loops of the lampbrush stage chromosomes are heavily endowed with RNA, but their structural framework is the DNA to which the RNA is attached, since the structure is resistant to both proteases and RNAses but not to DNAses. The RNA matrix is Feulgin-negative and readily becomes radioactive in *Triturus* newts that have been injected subcutaneously with [^{3}H]uridine (Gall and Callan, 1962). Transcription occurs in the loops (Miller and Hamkalo, 1972). These chromosomes are undoubtedly important elements in the functions that operate in the amphibian egg.

The pattern of loops is constant within a species or subspecies but may be quite different between closely related species or subspecies (Callan and Lloyd, 1960). Figure 2.29 shows the difference in loop patterns for two subspecies of *Triturus cristatus*. Hybrids between the subspecies have loop patterns characteristic of both parents, so the patterns are clearly inherited.

E. Sex Chromosomes

In those eukaryotes with identifiable opposite sexes there generally are a pair of chromosomes identified as sex chromosomes. In most of them, whether plant or animal, the male is the heterogametic sex which simply means that it has heteromorphic, partially homologous sex chromosomes, the X and Y. These pair in the first prophase of spermatogenesis and therefore two kinds of gametes, the X and the Y are produced, whereas the homogametic sex, the female, has two X chromosomes and yields only one kind of egg at oogenesis. Some plants and animals have female heterogamety. In these the sex chromosomes are designated as the W and Z, the W being analogous to the Y of the male heterogametic species. Little is known about the W and Z chromosomes compared to the Y–X type and we confine our discussion here to the latter (see Bull, 1983, for review). Furthermore, we consider here primarily the morphological differences between the X and Y, rather than their activities in sex determination.

Some male heterogametic species are XO in type, lacking a Y entirely. These are found mainly in invertebrates. In the vertebrates, however, an XY pattern is the general rule. In those species with extreme heteromorphism the Y differs from the X in size and shape, in identifiable gene content, and in amount of heterochromatin. The Y is usually smaller than the X, has none or very few of the gene activities identified with the X, and is mostly composed of heterochromatin. The X chromosome, on the other hand, is more like an autosome in function and appearance. However, since it exists in the monosomic state in the heterogametic sex, as

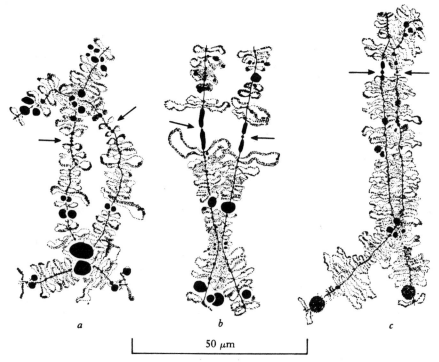

a   b   c

|_____ 50 μm _____|

Fig. 2.29. *Hybrid lampbrush chromosomes. See text. Reproduced from Callan and Lloyd (1960), with permission of the publisher.*

compared with the disomic state in the homogametic sex, **dosage compensation**, which is discussed in Chapter 5, is apparently engaged to equalize the gene dosage effects in the different sexes. One of these types of compensation involves the facultative heterochromatization of one of the Xs in the XX female mammal. Both W, X, Y, and Z functions and structures are discussed further in subsequent chapters.

F. Germ Line Chromosomes

The chromatin content of germ line cells of the eukaryotes above the level of the fungi can vary greatly from what is found in the somatic cells after gametogenesis and the early phases of development. We have already mentioned the diminution of DNA content in the ciliate protozoa and the ascarid nematodes. Analogous kinds

of quantitative changes also occur in the higher animals, including the vertebrates. Examples of this phenomenon are discussed further in Chapter 5. For the most part the chromatin eliminated is heterochromatin (Hennig, 1986).

The formation of lampbrush chromosomes has been described only in the meiocytes of animals and mostly in oocytes, although the Y chromosomes in *Drosophila* spermatocytes assumes configurations similar to the lampbrush type (Hennig, 1985; Callan, 1986). No examples seem to have been described in somatic cells.

The topology of DNA from somatic, sperm, and spermatocyte cells of a number of amphibians has been analyzed to determine the relative degrees of relaxation versus supercoiling (Risely et al., 1986). The interesting observation is made that negatively supercoiled DNA is found in cells that have a full complement of histones, but not in those in which histones are replaced by protamines. In many animals the sperm proteins are all, or predominantly, protamines rather than histones. This seems to affect the topological form of the DNA so that it is more relaxed than that found in somatic cells, in which the proteins are invariably histones. Whether nucleosomes are also found in protamine-rich animal sperm chromatin is unclear.

Many other differences between somatic and germ line chromatin are known to exist and some of these, such as genomic rearrangements in the immune system genes and the amplification of ribosomal and other DNA sequences, are discussed at some length in Chapter 5. Transposition of DNA (discussed in Chapter 5) may occur in either germ or somatic cells.

A general conclusion that can be made from a study of these heterodox chromosomes is that chromosomes are flexible. They assume the states necessary for carrying out their functions in the different parts of the soma. Again we see that the old idea of the absolute constancy of chromosome structure and function is a myth. However, the identity of a species must still be maintained by the conservation of a specific body of information.

VIII. SUMMARY

1. The centromeres of chromosomes are generally visible in monocentric chromosomes as constrictions. Ordinarily they are also regions in which the flanking DNA is highly compacted or condensed and described as visibly pycnotic in the interphase nucleus. During metaphase centromeres are the attachment points for spindle fibers, which, during the anaphase, function to move the chromosomes toward the poles of the spindle structure of the dividing cell. The actual site of the spindle fiber attachment is believed to be an aggregation of protein(s), called the ki-

netochore, imbedded in the centromeric region. Centromeres without kinetochores or with defective kinetochores do not function in karyokinesis.

2. Chromosomes of some eukaryotes may have several centromeres, in which case they are described as polycentric or holocentric. Or they may have diffuse centromeres, in which case every point on the chromosomes shows spindle fiber attachment. Plants and animals with multiple or diffuse centromeres are in the minority, however. Most have monocentric chromosomes.

3. Centromeric chromatin contains unique sequences of DNA usually not found in other parts of the chromosomes. In the higher eukaryotes in particular it is called heterochromatin and contains middle repetitive, undispersed sequences of DNA, such as $(GGAAT)_n$. This type of DNA may constitute a significant portion of the genomic DNA, ranging from 20% to 40% of the total.

4. The tips of chromosomes are called telomeres. These regions, like centromeres, contain specific DNA sequences that are generally moderately repetitive and different from those characteristic of the rest of the chromosomes. The taxonomically most widely distributed telomeric sequences consist of about six (such as $TTAGGG)_n$ to ten nucleotides with a series of guanines at the 3'OH end.

5. The repetitive sequences of DNA present in the genomes of all eukaryotes are not distributed randomly. There are two general types: (a). those that are concentrated in the centromeric and telomeric regions, and called undispersed; and (b) those dispersed in the euchromatic regions of the chromosomes.

6. The DNA of the human genome can be partitioned into five component categories. Two of these, representing about 83% of the total, have largely unknown functions; 58% of this fraction consists of unique sequences. The remainder is con-

stituted of the moderately repetitive sequences of either a dispersed or an undispersed type.

7. The dispersed moderately repetitive sequences of the human genome are interspersed with the nonrepetitive unique sequences. These repeats fall into two major families, the *Alu* and *L1*. These two families constitute about one-third of the total repetitive DNA. They are not detected by present methods in either the centromeric or telomeric regions. They appear to be randomly distributed in the euchromatin with average intervening spaces of about 3 kb.

8. The undispersed moderately repetitive sequences are, with few exceptions, found in the centromeric and telomeric re-

gions, where they constitute heterochromatin. The centromeric heterochromatin repeats are of a number of different types, of which some, but not all, are described as satellites, since they have different buoyant densities from the bulk of the genomic DNA.

9. The undispersed repeats vary in length from a few nucleotides to several thousand. But nearly all of them are in turn constituted of tandemly arranged shorter repeats that over time have been modified by mutation but still show substantial similarity in sequence.

10. The nucleolar organizer regions of chromosomes (NORs) occupy specific loci on specific chromosomes. Nucleoli develop at the NORs and within them the rRNAs

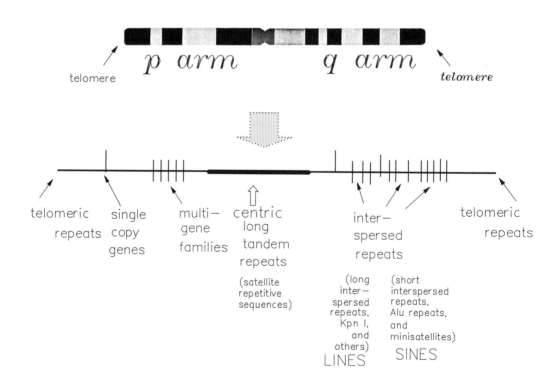

Fig. 2.30. *Summary of chromosome organization at the molecular level.*

are transcribed. Nucleoli also secrete per-ichromonucleolin, which is involved in the condensation and decondensation of chromosomes during the mitotic cycle.

11. In addition to the morphologically distinct centromeres and telomeres of chromosomes, other markers such as chromomeres exist that occupy specific loci on specific chromosomes. Also banding patterns can be developed with stains that are specific and allow for identifying the different chromosomes of the genome. Figure 2.30 summarizes organization at the molecular level.

12. Figure 2.30 attempts to summarize some of this organization. Chromosomes are not only highly organized, but each member of a genome is demonstratively different, not only because it bears different genes, but because it may bear certain specific repetitive sequences. In addition chromosomes are flexible insofar as they can assume a state necessary for performing their mission in specific cells at specific stages of the hosts life cycle.

3

Chromosomes and the Cell Cycle

Despite the baffling complex and seemingly erratic character of biological things, and their change at the touch of analysis—characteristics which have made them such a happy hunting ground for obscurantists—it is obvious that the whole congeries of variable processes of each kind of organism tends to go in a succession of great cycles, or generations, or even alternations of generations in turn made up of smaller cell cycles and that at the end of every greatest cycle something very like the starting-point is reached again.

H.J. Muller, 1947

The replication of chromosomes is one of the most important phenomena in the living world. The replication and repair of DNA are the basic processes involved in reproduction and the maintenance of genetic stability. Both processes occur during what is commonly referred to as the **cell cycle** or very often the **mitotic cycle**. This cycle not only involves the replication of cells, but also includes those periods in the lives of cells when they carry out their special functions in metabolism and act together to form and maintain an organism. For this reason the term cell cycle is preferable to mitotic cycle. The specific processes of replication of DNA and chromosomes are key elements in the cell cycle. But all the processes that constitute the cell cycle are interrelated. Therefore, we introduce the discussion of replication by first considering the cell cycle *in toto*.

I. THE EUKARYOTIC CELL CYCLE HAS MANY FACTORS CONTROLLING IT

The cell cycle, as it occurs in proliferating somatic cells, consists of four relatively distinct phases: **M**, the mitotic phase proper, the **G1 phase**, the **S phase**, and the **G2 phase**. The latter three phases collectively constitute the **interphase**—at one time called the resting phase, which it definitely is not. Resting cells are dead cells. Figure 3.1 diagrams the cycle and indicates the phases.

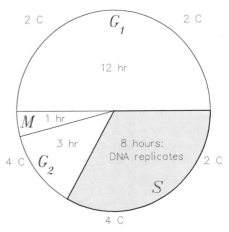

Fig. 3.1. *The cell cycle using the course of the 24-h cycle of a cell in culture as an example. Generally the G1 phase is the longest and the M phase the shortest in duration. Many cell types may spend long periods in the G1 phase and, in fact, never get out of it; these may be fully differentiated cells like nerve cells. These types of cells are said by some authors to be in a G0 phase. The shaded portion is the interphase, which includes the S phase, during which scheduled DNA replication occurs.*

Figure 3.2 depicts the results obtained by flow cytometry microfluorimetry (FMF) for a population of cells in culture. (Flow cytometry and FMF is explained in Box 3.1.) When FMF is used with a population of cells in suspension that are unsynchronized with respect to the cell cycle, the cells can be partitioned or sorted according to their DNA content. If the cells are stained beforehand with fluorescent dyes specific for binding DNA, the intensity of the signal is proportional to the DNA content per cell. Thus the pattern displayed in Figure 3.2 shows that an unsynchronized population of cells contains those with (1) a minimal amount of DNA (G1 peak); (2) an intermediate amount (S); and (3) double the minimal amount (G2 + M).

In the G1 the chromosomes are considered to be **unineme**, meaning that each has a single duplex DNA molecule running its length. The unineme status is des-

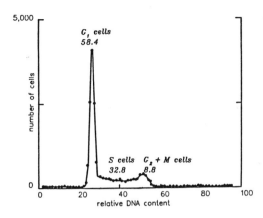

Fig. 3.2. *Distribution of DNA content of unsynchronized cells in population as determined by flow cytometry. The DNA of the cells has been stained with a fluorescent dye. See text. Courtesy of H.A. Crissman, Los Alamos National Laboratory.*

ignated as **C** to indicate that DNA content. In a diploid cell this amount of DNA would be 2C and is so designated in Figure 3.1. In the S phase the DNA is replicated and at the end of S the G2 phase starts with a 2C DNA content per genome, or 4C per cell, since each chromosome consists of two unineme **sister chromatids**. The end of G2 begins with the prophase of M. G2 and M cell types are not distinguishable with FMF.

This description of the cell cycle is a generally accepted paradigm supported primarily by observations on mammalian cells in culture. Actually this is a variation on a more basic type of cycle, the S–M cycle, which is found in rapidly dividing animal embryonic cells. It entails only an S and M phase with no detectable G1 and G2 phases (Cross et al., 1989).

BOX 3.1   FLOW CYTOMETRY AND CELL SORTING

Flow cytometry is a system in which about 4,000 cells per second flow single file through a chamber and are analyzed for particular properties such as the amount of DNA per cell. As shown in Figure 1, the entrance chamber of the flow instrument has a laser beam entering it that focuses on each cell as it passes through the beam. A piezoelectric transducer vibrates a crystal tuned to about 40,000 Hz and breaks the emerging stream of physiological saline in which the cells are suspended into uniform droplets at a rate of about 40,000 droplets per second. Thus about one of ten droplets contains a cell. The cells are stained with a fluorescent dye that is excited by the laser beam, and the intensity of the fluorescence of each cell is measured to determine the amount of stained material, such as DNA, per cell. This information is fed into a computer and the output is plotted as shown in Figure 3.2. The cells can also be sorted according to their DNA or other content. As the droplets pass between the electrodes, they can be charged either positively or negatively, depending on their cell contents, and separated from the uncharged stream. In this way 2N cells can be separated from 4N cells, for example, as they pass through the deflection plates. This basic procedure is also used to sort chromosomes of a genome by using stains that identify and sort specific chromosomes.

(Continued)

BOX 3.1   FLOW CYTOMETRY AND CELL SORTING
(Continued)

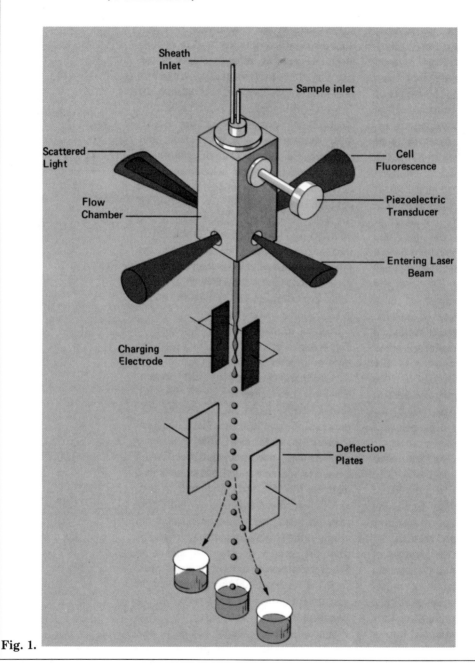

Fig. 1.

Whatever the complexity of the cycle it is regulated at steps called checkpoints by Hartwell and Weinert (1989). A number of these have been described in yeast, in which temperature-sensitive mutants are used to establish the specific point at which the cycle is arrested (Hartwell et al., 1974). (Temperature-sensitive mutants do not grow at "nonpermissive" temperatures that are above or below the optimal "permissive" one, unless a nutrient not needed at the permissive temperature is added. In the case of cycle-temperature-sensitive mutants these will complete the cycle at the permissive but will be blocked at the nonpermissive temperatures.) A general example of a checkpoint is that in all eukaryotes the arrest of DNA synthesis in the S phase by mutation of the genes encoding replication enzymes, or by damaging the DNA by irradiation, prevents cells from entering the M phase.

A. Mitosis, or Karyokinesis

The mitotic process can be divided into two stages: (1) the establishment of the **mitotic apparatus** (Mazia, 1961, 1987; Inoué, 1981), and (2) the equal distribution of chromosomes to the opposite poles of the cell by this apparatus. A highly diagrammatic representation of the mitotic apparatus in an animal cell is depicted in Figure 2.5. The formation of the apparatus begins at different periods in different cells, but in general it begins with the appearance of **centers** (or **central bodies** or **centrosomes**, as they are variously called), which polarize the cell by defining poles toward which the chromosomes move in the anaphase of M. These centers may have no visible structure to begin with or there may be **centrioles** associated with them as in most animals (Fig. 3.3). Centrioles are organelles that are involved in tubulin assembly into spindle microtubule fibers of the polar type, p. 75. Except in certain Protista and fungi the centrioles are out-

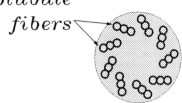

Fig. 3.3. *Schematic representation of a centriole in cross section, as found in higher animal cells. Centrioles are cylindrical and are about 150 nm in diameter and 300–500 nm in length. They occur in pairs within the centrosomal region. The internal structure is composed of nine groups of triplet microtubules. Centrioles are considered to be self-replicating by some observers, and some evidence exists that they contain DNA. Their role in mitosis is not understood.*

side the nucleus, and as the cell begins to enter the M phase they start to separate and the spindle fibers radiate from them. The mitotic apparatus centers in higher plants are anastrel and do not show centrioles.

Typically mitosis consists of four key events (reviewed by McIntosh and Koonce, 1989):

1. Prophase, the condensation of interphase chromatids
2. Prometaphase, the further condensation and arrangement of the sister chromatid pairs on the metaphase plate
3. Anaphase, the separation of the sister chromatids to the opposite poles of the cell
4. Telophase, the formation of two daughter nuclei and their entrance into the interphase.

The onset of mitosis is accompanied by the presence of an active **maturation-promoting factor (MPF)**. This is an activity, or checkpoint, that is involved in the G2–M transition (Kishimoto et al., 1984; reviewed by Cross et al., 1989). MPF isolated and purified from frog eggs consists of two protein subunits: one 34-kD and the other 45-kD (Lohka et al., 1988). The 34-kD subunit has an amino acid sequence similar to the protein product, p34^{cdc2} encoded by the $cdc2^+$ gene of *Schizosaccharomyces pombe* (Lee and Nurse, 1988). Both have a serine–threonine protein kinase activity that phosphorylates histone H1, among other proteins. This protein kinase, generally designated as p34, has been found in all eukaryotes examined. The genes encoding them are probably homologous, since the human gene $cdc2^+$ can complement, or replace the $cdc2^+$ gene of *S. pombe*. The **protein kinases** involved in the regulation of the cell cycle constitute a family of related proteins encoded by members of a gene family. A CDC protein kinase, p58, that is structurally related to p34 has been identified in humans. Abnormally elevated expression of this kinase causes abnormalities in the mitotic phase of the cycle in hamster CHO cells in culture (Bunnell et al., 1990).

The 45-kD component of the MPF of *Xenopus* has been identified as a **B-type cyclin** (Gautier et al., 1990). Cyclin, which is encoded by the *cdc13* gene of *S. pombe*, is synthesized in the late interphase and complexes with p34 to form pre-MPF, which becomes active when the p34 subunit is dephosphorylated (Dorée et al., 1989). The gene *cdc25* of *S. pombe* encodes an activity that is required for the dephosphorylation of p34, and is thus another factor in triggering mitosis. A homologous gene, *string*, has been identified in *Drosophila melanogaster* (O'Farrell et al., 1989). Immediately after active MPF triggers the onset of mitosis, cyclin activity is degraded, the nuclear membrane begins to break down, chromosome condensation leading to the level of metaphase compaction begins, and the spindle apparatus is organized (Umek et al., 1989). An **A-type cyclin** has also been identified, but its role in relation to the B-type is not clear.

The beginning of **prophase** is not well defined, but cytologically it is the first visible sign that the mitotic process has begun (Fig. 3.4). When the G2 chromosomes condense enough to become visible, prophase has begun. Early prophase is a propitious time to observe chromosome bands made with stains. The early prophase genomes may show thousands of bands, in contrast to metaphase genomes, which may show fewer than 100 (Yunis and Prakash, 1982). The formation of poles by the separation of centers may be one of the first or the last events in prophase, but in any case their formation is ordinarily accompanied by dissolution of the nuclear membrane and the disappearance of the nucleoli. The end of prophase is signaled by the beginning of chromosome movement in relation to the poles. This marks the beginning of **metaphase**, sometimes called prometaphase or metakinesis.

The movement of chromosomes at the end of prophase into their final metaphase positions on the **metaphase plate** constitutes the metaphase. The kinetochores play a role in this maneuver. Acentric chromosomes are incapable of organized movement. The growth of spindle fibers and attachment to the pole centers occurs in prophase, and by the time the metaphase plate is formed the complete mitotic apparatus is ready to operate in the separation of the sister chromatids. This mitotic apparatus can be isolated by proper procedures from sea urchin eggs and grasshopper spermatocytes, so it has a well-defined coherent structure (Mazia, 1987).

A nuclear mitotic apparatus protein with specific antigenic properties has been identified in eukaryotic cells (Pettijohn et al., 1984; Pettijohn and Price, 1988). This protein, identified by the acronym **NuMA**, is confined to the nucleus in interphase cells. But when cells enter M, NuMA accumulates at the poles of the mitotic spindle, where it remains through anaphase. NuMA is a nonhistone protein. When iso-

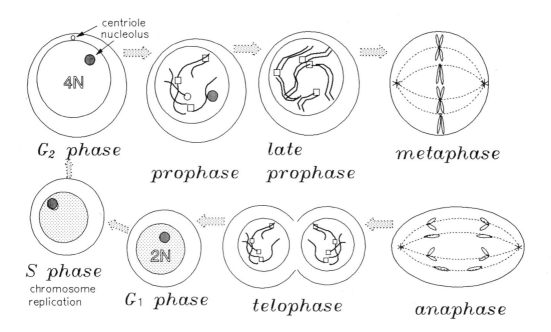

Fig. 3.4. *Diagrammatic representation of the mitotic process. An animal cell is used as an example showing the division of the chromosomes. This cell has the chromosome number 2N = 4 before chromosome replication. The late prophase shows chromosomes consisting of pairs of sister chromatids that separate at the beginning of anaphase or the end of metaphase.*

lated from human cells, it appears to be a phosphoprotein with an Mr of about 250 kD.

Sister chromatids separate at anaphase from their positions on the metaphase plate. The centromeric regions are associated up to this stage and their separation signals the beginnings of **anaphase**. As discussed on p. 75, the movement of the chromatids, now called chromosomes, toward the poles is apparently the result of the shortening of the fibers attached to the kinetochores by loss of subunits of tubulin rather than by contracting like rubber bands.

The end of anaphase is the beginning of **telophase**. In this phase the chromosomes become decondensed and elongated, the nuclear membrane is reformed, and the NuMA protein disperses throughout the nucleus as the spindle centers assume the interphase condition. The separation of sister chromatids is called segregation by the geneticist. It is a precise process that results in daughter cells with equal numbers and kinds of chromosomes in each daughter cell. Failure of this separation is called nondisjunction, a process that generally results in inviable daughter cells as discussed on p. 272 and in Figure 5.55.

B. Cytokinesis

Cytokinesis is the division of the cytoplasm, in distinction from **karyokinesis**, the division of the nuclear contents. It may occur either during mitosis or after.

Karyokinesis can occur repeatedly, as in the plasmodia of slime molds, with no cytokinesis to form large multinucleate structures, or syncytia, without cellular partitions. Fungi like *Aspergillus* and *Neurospora* have cell walls in their hyphae, but these cells may have a variable number of nuclei, since the walls have holes in them through which the nuclei flow back and forth through the hyphae. In addition, if cytokinesis occurs, the division of cytoplasm need not be equal, as in the case of budding yeast cells or the formation of polar bodies in the oogenesis of animal egg cells.

C. The Interphase

The **interphase** period of the cell cycle is an active one consisting of G1, S, and G2. Figure 3.2 shows the distribution of cells in a population with respect to their DNA content, and makes it clear that the shift from 2C to 4C occurs in the interphase. In addition to the determination of the DNA contents of cells by flow cytometry, RNA and protein can also be measured. The RNA and protein contents of Chinese hamster ovary (CHO) cells in culture have been determined by using dyes specific for RNA and protein (Crissman,

1985; Crissman et al., 1985). Table 3.1 gives results obtained for exponentially growing CHO cells, and for cells arrested at G2 by isoleucine deprivation. The values given for RNA and protein content are relative and on a per cell basis. It is evident from the data that RNA and protein synthesis appears to occur continuously during the interphase of the cell cycle. Furthermore, the ratios of RNA to protein remain remarkably constant, indicating that transcription in the nucleus and translation in the cytoplasm go hand in hand, at least during the interphase. The RNA/DNA ratio is also fairly constant, indicating that transcription is not slowed by DNA replication.

During the G1 phase the preparations for the S, or replicative, phase are begun. A quiescent phase into which the G1 may shift is generally designated as the G0 phase. Nonproliferating cells are generally in this phase (reviewed by Pardee, 1989). Once the shift from G0 to G1 occurs, cells become more metabolically active. Proteins of various kinds such as kinases, new hnRNAs, and enzymes involved in DNA synthesis appear. The stage is set for replication of DNA in S phase and the processes of transcription.

TABLE 3.1. RNA and Protein Content and RNA–DNA and RNA–Protein Ratios for G1, Middle S, and G2 + M Populations of CHO Cells From Exponentially Growing and Isoleucine-Deprived Cultures

Cell cycle phase	RNA content/cell	Protein content/cell	RNA–DNA	RNA–protein
		Exponentially growing cells		
G1	57.3	67.4	44.0	35.6
Middle S	82.7	94.7	46.2	36.9
G2 + M	100.7	115.1	39.3	36.8
		Noncycling (isoleucine-deprived) cells		
G1	51.3	59.1	40.3	35.4
Middle S	72.3	83.6	40.6	36.4
G2 + M	103.8	119.2	38.9	37.1

Results are expressed in arbitrary units of mean fluorescence intensity.
Data from Crissman et al. (1985).

The transition from G1 to S involves triggering by an MPF identified in the mouse with p34-like and cyclin-like proteins that are distantly related to the A- and B-type cyclins (Matsushima et al., 1991). A similar control mechanism for entrance into S phase from G1 has been identified in *Saccharomyces cerevisiae*, which also produces cyclin-like proteins (Surana et al., 1991; Ghiara et al., 1991). It is clear that cyclin synthesis and degradation is a key element in the oscillation of dividing cells from interphase to mitotic phase to interphase.

II. THE BASIC MECHANISMS OF DNA REPLICATION ARE SIMILAR FOR ALL LIFE FORMS

The synthesis or replication of DNA in the cell nucleus during the S phase of the cell cycle requires the presence of many different proteins that act together as an organized **replication complex**. Much of what we presently understand about the workings of this complex comes from studies with phage, animal viruses, and *Escherichia coli*. Work with these prokaryotes and eukaryotic cells has made it clear that the basic aspects of replication are common to all forms of life.

The replication of DNA is always **semiconservative** in the sense that a new double helix consists of an old strand and one new strand complementary to it. Therefore each new double strand is properly described as being semiconserved. The basic processes of DNA replication are the following:

● separation of the two complementary chains of the DNA, and
● synthesis of new chains at what is called the **replication fork** by the incorporation of the 5'triphosphate deoxynucleotides (dTPNs) of adenine, thymine, cytosine, and guanine in the presence of DNA-directed **DNA polymerases**.

As shown in Figure 3.5, synthesis occurs by the formation of a phosphodiester bond between the 3'hydroxyl group at the growing end of the DNA **primer** chain, and the 5'phosphate group of the incorporating triphosphonucleotide. The primer chain grows in the $5' \rightarrow 3'$ direction using the complementary chain as the **template**. The distinction between the primer and template functions is an important one to be borne in mind when considering DNA replication; the primer is the growing chain, the template is the complementary "static" directing chain.

A. DNA Polymerases

The DNA polymerases involved in DNA synthesis **always**, so far as known, operate to attach 5'phosphate groups to 3'OH groups. Hence a free 3'OH group is necessary for chain growth to proceed in the $5' \rightarrow 3'$ direction. The overall reaction catalyzed by DNA polymerase is

template DNA $(dMPN)_n$ + dTPN

$$DNA (dMNP)_{n+1} + PPi$$

in which M = monophosphate, T = triphosphate, and N = any one of the four nucleotides. Mg^{2+} is required as a cofactor for this reaction by all DNA polymerases at a concentration of about 10^{-3} M *in vitro*.

B. Initiation of Synthesis

Notwithstanding the fact that DNA polymerases operate only in the $5' \rightarrow 3'$ direction, the two strands of the replicating DNA double helix are synthesized simultaneously. This occurs as shown in Figure 3.6 with a primer chain, designated the **leading strand**, growing on one chain leading into the replication fork and the other primer, designated the **lagging strand**, growing on the other chain in the direction **away** from the replication fork. Replication is continuous for the leading

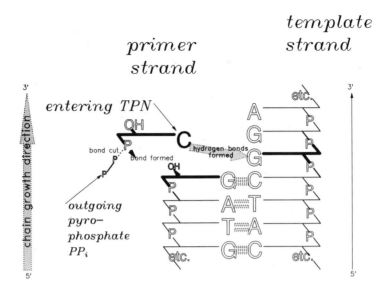

Fig. 3.5. *Synthesis of DNA in the presence of DNA polymerase and the tri-phosphonucleotides. Note the difference in identity of primer and template strands. Strand growth is on the primer strand in the 5′ → 3′ direction, using the complementary 3′ → 5′ strand as the template.*

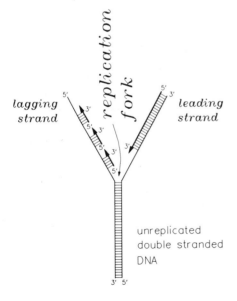

Fig. 3.6. *Simultaneous DNA synthesis on both the leading and the lagging strands.*

strand and discontinuous for the lagging strand. This combination of continuous replication in one arm of the fork and discontinuous replication in the other is referred to *in toto* as **semidiscontinuous** replication, and probably occurs in all prokaryotes and eukaryotes. Either strand of the DNA may support continuous or discontinuous replication.

The initiation of discontinuous DNA synthesis depends upon the activity of an RNA polymerase to transcribe short RNA chains that form RNA–DNA hybrid duplexes (Fig. 3.7). The RNA polymerase also operates in the 5′ → 3′ direction, but unlike the DNA polymerase it does not require a 3′OH primer end. A DNA template is all that is necessary. **RNA priming**, (illustrated in Figure 3.7) has been demonstrated in plasmids, viruses, bacteria, and eukaryotes so it is probable that it is

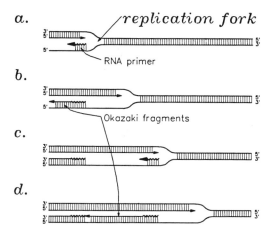

Fig. 3.7. *RNA primer and discontinuous DNA replication.*

of general occurrence. Newly discontinuously synthesized DNA consists of fragments that are 0.1–2 kb in length. These fragments, which are shorter in eukaryotes, can be removed from the replication fork region experimentally and are referred to as **Okazaki fragments**.

The role of RNA polymerase in the initiation of DNA synthesis by transcription of a short segment of DNA was discovered in connection with the formation of Okazaki fragments in discontinuous synthesis. It has now become evident that **transcriptional activation** of this kind by the polymerase may also be involved in the initiation of synthesis of the continuous strands (Campbell, 1986). After transcriptional activation, a nucleoprotein complex is formed by the binding of an initiator protein to a DNA sequence, called the **replicator** (Fuller et al., 1984; Dodson et al., 1985). The replicator is presumably part or all of the DNA sequence that is transcribed by the polymerase (Fig. 3.7). A binding protein of this type has been identified in *E. coli*, phage, animal viruses, and eukaryotes.

Replication of circular DNA molecules such as are found among the prokaryotic life forms can presumably occur by the rolling circle method illustrated in Figure 3.8. By this method the DNA is replicated unchanged each generation. However, linear DNA molecules such as are found in eukaryotic chromosomes are subject to the problem of gaps occurring at the 5′ ends (Fig. 3.9).

C. The Unwinding of the Double Helix

In order for single strands to be made available for synthesis of new DNA, the duplex must unwind at the region of the fork. The unwinding occurs in the presence of **DNA-dependent ATPases** (called **helicases**) that break the hydrogen bonding of the complementary base pairs and drive the unwinding of the double helix. These have been described extensively for

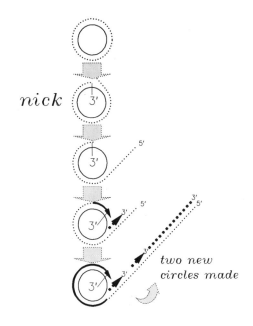

Fig. 3.8. *The rolling circle method of DNA replication proposed for organisms with circular DNA. This type of replication may occur in eukaryotes that synthesize extrachromosomal DNA.*

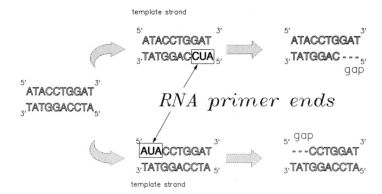

Fig. 3.9. *Synthesis of new strands of duplex, linear DNA will result in gaps at the 5′ ends, where RNA primers are situated and later released.*

E. coli, which has at least three helicases operating on the lagging strand and other active proteins on the leading strand. A similar process must occur in the eukaryotes, too, but their helicases are not as well understood as those in prokaryotes. Two have been described, one in *Lilium* and the other in human cells, but little is known about them other than that they are 130 kD and 110 kD long.

Control of the unwinding of the strands instituted by helicases is important in DNA replication. If unwinding is by rotation at a free end, an obvious problem is the flailing of unwinding arms. This is not a problem in forms with circular genophores or chromosomes, because there are no free ends, and its occurrence in eukaryotes with linear DNA is highly improbable because telomeric ends are probably hairpins. But even if the hairpins are cut, the length of most chromosomes is such that simple unwinding would create great disorder in the interphase nucleus during the S phase. If, in some way, the ends of the molecule are prevented from rotating and strand separation proceeds by internal unwinding, positive supercoiling should occur. We discuss this very important question further, in connection with chromosome replication in the next section.

III. REPLICATION OF CHROMOSOMAL DNA

Although the fundamentals of DNA replication in eukaryotes are the same as in the prokaryotes, important differences exist. Chromosomal DNA is linear in contrast to that of most of the prokaryotic genophores; chromosomal DNA molecules, especially in the nonfungal eukaryotes, are frequently longer than prokaryotic genophores by a factor of 10 or more; chromosomal DNA is complexed with large amounts of protein to form a highly organized chromatin.

A. Replication of Yeast Chromosomes

The yeast *Saccharomyces cerevisiae* has been a powerful tool in the analysis of eukaryotic replication. Its life cycle is well understood and easily manipulated, making it possible to analyze the replicative processes during the various phases of the cycle (review by Campbell, 1986; Newlon, 1988). The yeast genome consists of 17 small, linear chromosomes having 250–2,000 kb. The total genomic DNA content, exclusive of the mitochondria and plasmids, that yeast cells carry is about 1.35×10^7 bp. This is only about 3.2 times greater than the DNA content of an *E. coli* genophore.

During S phase of the eukaryotic cell cycle replication begins at many points along each chromosome instead of at one or two sites as in prokaryotes. At these points **replication bubbles** form that are generally referred to as **replicons** (Fig. 3.10). They are units of replication defined as DNA replicated from a single origin. The bubbles are formed by the unwinding and separation of the strands of the duplex; upon the onset of replication within the bubbles, many bubbles are formed per chromosome, the distances between their centers averaging about 36 kb. Therefore about 380 replication origins are formed per yeast genome. Replication is bidirectional and proceeds at a rate of about 6–18 kb per minute (Fig. 3.10). As each replicon grows it meets its neighbors on each side, and fusions occur presumably by ligase action. The completion of the process results in two chromatids. The separation of strands to form the bubbles will, of necessity, cause a strain on the wound part of the DNA on either side in the direction of positive supercoiling. As depicted in Figure 1.16b, the partial sep-

aration of two strands of negatively supercoiled circular DNA can result in relaxation. Carried beyond the point of relaxation, it will result in further change in the linkage number, L, and positive supercoiling. It is apparent that topoisomerase I is involved in the relaxation of the strains involved. These kinds of strains are expected to occur with linear DNA. If the DNA is negatively supercoiled on a nucleosome, as described on page 56, then separation of strands to accommodate the origins of replication forks will relieve negative strain. However, continued unwinding can also produce positive supercoiling; this can be relieved by single-strand nicking by topoisomerase I (TOPOI) (Fig. 3.11), in which a complementary strand "passes through" the break caused by the nick. Theoretically the strain could also be relieved by topoisomerase II (TOPOII), which cleaves **both** strands and allows for strand rotation.

The yeast genes *TOP1* and *TOP2* encode TOPOI and TOPOII, respectively. Temperature-sensitive mutants with inactive TOPOI do not exhibit aberrant phenotypes, but the inactivation of *TOP2* causes a block in mitosis at the time of disengagement of the sister chromatids (Goto and Wang, 1985). However, the double mutants *TOP1⁻, TOP2⁻* grow poorly at permissive temperatures and are defective in DNA replication (Brill et al., 1987).

The DNA sequences involved in the origins of replication are designated as ARSs (autonomous replicating sequences). These contain a 10- to 11-bp AT-rich consensus sequence (A/T)TTTAT (A/G)TTT(A/T) that is also proposed to be the recognition site for a protein analogous to the T antigens of SV40 (Umek et al., 1989). The number of ARSs per cell corresponds closely to the approximate number of replicons formed during the S phase. Furthermore, ARS segments incorporated into plasmids control the rep-

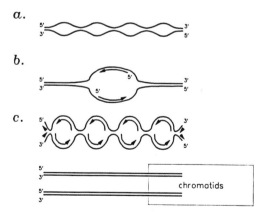

Fig. 3.10. *Replicons formed by the unwinding of chromosomal DNA coincidentally with the beginnings of DNA synthesis at replication forks (b). See text.*

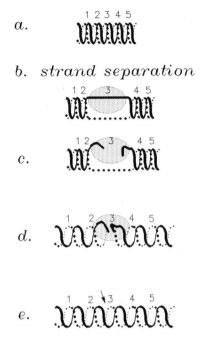

a.

b. strand separation

c.

d.

e.

Fig. 3.11. *Relief of strain due to positive supercoiling by the action of topoisomerase I.* **a.** *a segment of relaxed linear DNA.* **b.** *Partial unwinding of DNA to cause increase of the linking number and thus positive supercoiling on either side of the bubble.* **c.** *Nicking of the DNA on the left side of the bubble.* **d.** *Passing of a loop of the complementary strand through the nick.* **e.** *Ligation of the nick and relief of the strain by restoration of relaxed breakage numbers. The point of ligation is indicated by the arrow. This process can be imagined to continue as the strands continue to unwind.*

lication of the plasmids; replication is initiated at the ARS. ARSs are probably protein-binding sites and contain regions flanking the AT-rich sequences that are also important for autonomous replication. At least one of these flanking sites promotes the unwinding of the double strands necessary for the synthesis of DNA (Umek and Kowalski, 1988).

The initiation of DNA synthesis requires the action of the products of at least 11 genes. Altogether at least 19 genes are known to interrupt DNA synthesis and abort the cell cycle in yeast when they mutate to an inactive condition.

B. Eukaryotic DNA Polymerases

Three general types of DNA polymerizing enzymes exist in eukaryotes: (1) those that use DNA only as a template (these are the true DNA replicative polymerases); (2) the **reverse transcriptases,** which use RNA as template; and (3) the **terminal transferases,** which add nucleotides on the terminal 3′OH of a DNA strand independently of a template. Here we consider only those polymerases that are dependent on a DNA template. At least five kinds of these are now known to exist in eukaryotic cells; they are generally designated as Pola, b, c, d, and e (or **Polα, β, γ, δ,** and **ε**) (reviewed by Wang, 1991).

TABLE 3.2. **Eukaryotic DNA Polymerases That Use DNA as a Template**

Polymerase	Unit or subunit	Location	Inhibitor[a]	Prevalence
Alpha	165 kD	Nucleus	A, E, B	All eukaryotes
	70 kD			
	55–60 kD			
	48–49 kD			
Beta	40 kD	Nucleus	D	Vertebrates Only?
Gamma	140 kD	Mitochondria	E, D	All
Delta	125 kD	Nucleus	A, E, C	All?
	48 kD			
Epsilon	255 kD	Nucleus	A, C	All?

[a]Inhibitors: A, aphidicolin; E, N-ethylmaleimide; B, butylphenyl-dGTP; D, ddNTP; C, carbonyldiphosphate.

Table 3.2 gives some descriptive data about these five, all of which are found in vertebrates. Polβ may only be confined to the vertebrates, but the remaining four are apparently present in all eukaryotes, including the fungi.

Polα is a **polymerase/primase** heterotetramer complex consisting of the four subunits listed in Table 3.2. It is so named because it contains primase subunits of ~58 kD and ~48 kD that are tightly bound in the complex but can be separated and shown to be structurally distinct from the catalytically active 165-kD polymerase subunit (Plevani et al., 1985, 1988). The primase activity catalyzes the synthesis of the RNA oligomers that are necessary for DNA synthesis. None of the other polymerases have this activity. Polα is strongly inhibited by **aphidicolin**. The aphidicolin sensitivity is associated with the 165-kD subunit, which also has the polymerase activity. Primase activity is not inhibited by aphidicolin. Human Polα is encoded by a gene on Xp22.1 → p 21.3.

Polδ and Polε are also inhibited by aphidicolin and they, along with Polα, are the basic eukaryotic nuclear DNA replicative enzymes. Polδ and Polε have different structures (Table 3.2) and are functionally different, since δ is inhibited by N-ethylmaleimide, while ε is not. Also δ is dependent on a 36-kD protein known as PCNA (proliferating cell nuclear antigen) for its activity in the replication process, while ε is not.

Polα and Polδ have been implicated as acting jointly in replication with Polα primase activity, starting on a lagging strand with the replicative process and then switching over to Polδ action on a leading strand, as shown in Figure 3.12 (Kelly, 1988; Tsurimoto and Stillman, 1991). The exact role of the Polε in the replicative process is not clear. It has been shown to be active in repair of ultraviolet-induced DNA lesions. Both it and Polδ have 3'–5' exonuclease activity, indicating that they

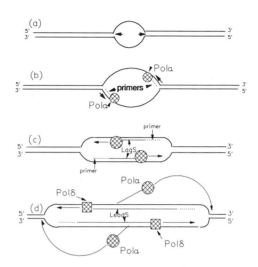

Fig. 3.12. *Proposed mechanism for the synthesis of eukaryotic DNA involving both Polα and Polδ, in which the former, having the primase activity, acts on the lagging strand and the latter on the leading strand. A molecule of Polα synthesizes a limited new stretch of DNA, which may be thought of as an Okazaki fragment, and then moves back to the replication fork to start synthesis anew. Its place is taken by leading strand synthesizer Polδ. LagS, lagging strand; LeadS, leading strand. Based on the model proposed by Tsurimoto and Stillman (1991).*

have proofreading capabilities as described on page 145.

Polβ appears to be a repair polymerase acting in unscheduled synthesis (Hubscher et al., 1981). It has gap-filling properties, indicating a role in short-patch repair of DNA. Since it does not appear to exist in invertebrates, this role probably is taken over by α, δ, and ε in these forms. A cloned sequence of human DNA encodes a polypeptide of about 42 kD with Polβ activity and has been localized on chromosome 8 in the region 8p11 → p12 (Chang et al., 1982).

Polγ is active in the replication of mitochondrial DNA. Hence, like Polβ, it is not involved in the synthesis of nuclear DNA during the scheduled synthesis of the S phase. Like δ and ε it has 3′–5′ exonuclease activity.

The function of the polymerases in DNA synthesis in all life forms is certainly among the most important in the biosphere. Therefore it is not surprising that elements of them have been found to be highly conserved. Human Polα has conserved regions of sequences of yeast Polα, various animal viruses, T4 DNA polymerase, and polymerases of prokaryotes (Wang, 1991). A conserved 125-kD polypeptide has been isolated from cell extracts of *E. coli, Ustilago maydis, Drosophila melanogaster*, rat neurons, calf thymus, human fibroblasts, and HeLa cells that possess DNA chain elongation activity (Hubscher et al., 1981). Polypeptides from the different sources have the same properties, but the amino acid sequences have not been determined. They may or may not be related to the holoenzyme subunits described in Table 3.2 with Polα activity.

C. Replication of Linear Duplexes Is Dependent on Telomeric Sequences

The termination of DNA synthesis at the ends of chromosomes without loss of important functional sequences has been a long-standing problem seeking a solution. Recent revelations concerning the structure and function of telomeres are beginning to lead to an understanding of the termination process, which is apparently similar in all eukaryotes. It is pointed out in Chapter 2, page 78, that the telomeres of chromosomes have repeated sequences with a high degree of similarity in the various eukaryotic taxa, ranging from unicellular to the complex multicellular; As shown in Table 2.2, the repeats are relatively short oligonucleotides consisting of six to 11 deoxynucleotides (reviewed by Zakian, 1989; Zakian et al., 1990;

Blackburn, 1990). The most obvious relationship among them is that they all end with guanines at the 3′ end. Thus a consensus repeat sequence may be written as $5'T_{0-4}/A_{0-1}/G_{1-8}3'$. Other than the fungi and slime molds, the animals and plants have closely related repeats of the type $5'T_{2-4}/A_{0-1}/G_{3-4}3'$, and terminal duplexes should be of the form:

$$5'-GGG3'$$
$$\cdots$$
$$\cdots$$
$$3'-CCC5'$$

with the G strand running away from the centromere 5′ → 3′, while the CCC strand runs 5′ → 3′ toward the centromere. All vertebrates have the terminal repeat 5′TTAGGG (Meyne et al., 1990).

A telomere terminal transferase discovered first in *Tetrahymena* cell-free extracts by Greider and Blackburn (1985) elongates the G-rich strand of telomeres in the absence of a DNA template by synthesizing the *Tetrahymean*-specific sequence d(TTGGGG). This telomere terminal transferase has been given the name **telomerase** and is shown to be able to recognize single-stranded oligonucleotides of telomeres ending in -G3′- in a variety of eukaryotes including other ciliates (Zahler and Prescott, 1988, 1989), HeLa cells (Morin, 1989), and fungi (Zakian et al., 1990). The telomerase is a ribonucleoprotein (RNP) enzyme. Its RNA moiety for the ciliate *Euplotes crassus* has the sequence 5′CAAAACCCCAAA3′. This RNA sequence is the template for the synthesis of the terminal 5′TTTTGGGG3′ of *Euplotes* telomeres (Shippen-Lentz and Blackburn, 1990). In this sense the telomerase is a kind of reverse transcriptase, since RNA codes for a DNA sequence. It is hypothesized that the terminal single-strand 5′TTTTGGGG3′ can fold upon itself by non-Watson-Crick guanine-to-guanine base pairs forming a hairpin, as shown in Figure 3.13 (Henderson et al., 1987). As a result, a free 3′OH G is present, making it possible for the DNA polymerase to function in the 5′ → 3′ direction. The ubiquitous presence of apparently free single strands of -GGGG3′ ends among all eukaryotes analyzed makes it highly probable that the problem of replication at the end of a single strand of linear chromosomal DNA is essentially solved.

D. DNA Replication in Higher Animal Cells Has a Definite Tempo and Mode

1. Number and utilization of replicons

Most prokaryotes, viruses, and plasmids have one origin of replication and therefore a single replicon. Eukaryotes in

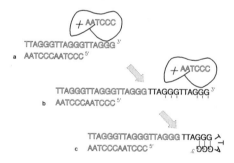

Fig. 3.13. *Proposed model for replication of a linear duplex DNA molecule such as those found in eukaryotic chromosomes. The telomeric region has the repeat TTAGGG, which can be extended by the telomerase RNP.* **a.** *The telomerase attaches to the 3′ end of the extended T_2AG_3 end and adds the ribonucleotide complementary to the T_2AG_3 repeat.* **b.** *The extension of the ribonucleotide A_2TC_3 acts as template for the further extension of the T_2AG_3 repeat. The telomerase can continue removing distally from the centromere while adding more T_2AG_3 repeats.* **c.** *Movement can stop and a hairpin structure can form by bonding between guanines. This forms a 3-OH end of a primer, which then allows for duplex formation by DNA polymerase action that can continue into the non-T_2AG_3 repeat region toward the centromere.*

contrast have large numbers of replicons to facilitate the timely replication of their very much larger genomes. An *E. coli* genophore consisting of about 4.2×10^6 base pairs of DNA replicates in about 20 min under optimal conditions. The rate of synthesis at the two growing forks is about 1,700 base pairs/s per fork. The rate at which forks grow in eukaryotic cells is considerably slower. Human cells in culture have a fork growth rate of about 100–150 base pairs/s and it is approximately the same for most other eukaryotes.

The length of the S phase of the cell cycle varies considerably from one organism to another and from one developmental stage or tissue to another as well. In *Drosophila*, for example, the S phase of embryonic cells is completed in about 4 min but in *Drosophila* cells in culture it takes about 600 min. Fork growth is about the same in both kinds of cells (Blumenthal et al., 1974). Probably all possible origins of replication are utilized in embryonic cells to allow rapid synthesis of the 1.8×10^6 base pairs of DNA in the chromosomes, while in other tissues and stages that grow more slowly only a small percentage of the origins are utilized. Generally, mammalian cells in culture take about 14–24 h to complete a cell cycle (Fig. 3.1).

2. The nucleus in S phase

The state of affairs in the interphase nucleus during the period of chromosome replication is one that we consider in some detail in Chapter 7. At this point, however, it should be recognized that a good deal of what goes on in the S phase in the so-called **nuclear matrix** is presently hidden from us. This matrix, which can be described as the skeletal framework of the nucleus, can be isolated from nuclear fractions free of cytoplasm. If actively replicating cells are subjected to labeling with ^{3}H-thymidine, it can be shown that the newly synthesized DNA is tightly attached to the matrix with a multienzyme complex called a **replicase** consisting of at least six enzymes involved in DNA synthesis (Reddy and Pardee, 1980). Polα and topoisomerase II are strongly bound within the nuclear matrix as major protein components in *Drosophila* cells (Berrios et al., 1985). These findings about the matrix strongly indicate that a complex machine exists within the S phase nucleus for the synthesis of new chromatids.

3. Regulation of replication

The regulation of replication in eukaryotes is poorly understood compared to prokaryotes. In both cases one of the keys to understanding this regulation is the initiation of replicon formation at the **origin** of replication. In the eukaryotes there is one origin per replicon. As we have already indicated, there is evidence, at least in yeast, that specific sequences with origin functions (**ARS**s) are present in about 400 places in the yeast genome (Stinchcomb et al., 1981). This number is concordant with the number of replicons formed in the S period of the yeast cell cycle. The ARS nucleotide sequences are by no means identical, but they all do seem to have an 11-bp "consensus" sequence. In view of the fact that not all parts of a chromosome are replicated at the same time, this finding of some nonhomology of ARS sequences is understandable. An ARS might possess a "master" sequence as the actual initiation point, but sequences associated with it might respond to signals that determine the time of initiation of the master origin sequence. As noted in the previous section, good evidence exists in *Drosophila* that the number of replicons utilized in embryonic cells is much greater than in cells in culture; thus response to signals to initiate replicon formation may be different in different tissues of the higher eukaryotes.

One of the facts about the replication of eukaryotic chromosomes is that replicons originate at different times along the lengths of the chromosomes. One of the first observations made regarding the timing of the onset of replication at different periods of the S phase was that the heterochromatic regions of chromosomes of higher eukaryotes replicate later than the euchromatic regions (Lima de Faria and Jaworska, 1968). This is not the case for yeast, however. Replications of centromeric DNA occurs early in the S phase in this fungus (McCarrol and Fangman, 1988). The difference in the observed replication time of centromeric DNA between higher and lower eukaryotes may well reside in the fact that the former tend to have heterochromatic centric regions, whereas the latter have little or no centric heterochromatin.

Three phases of chromosomal replication have been delineated as reference points: early, middle, and late. A decrease in replication activity occurs between the early and late phases (Schmidt, 1980). This relaxation of activity is called the 3C pause and constitutes the middle phase (Holmquist et al., 1982). When euchromatin becomes inactivated by condensation, as in the formation of Barr bodies in XX female mammals, the inactivated chromosome replicates later than the active, uncondensed homologue. Thus facultative heterochromatin can assume the status of constitutive heterochromatin with respect to timing of replication (Nicklas and Jaqua, 1965).

Two general types of genes are delineated in the genomes of multicellular animals and plants. They are categorized as **housekeeping** and **tissue-specific**. The housekeeping types encode proteins that are actively transcribed in all cell types whatever their degree of specialized function. For example, genes that encode enzymes active in glycolysis and the Krebs citric acid cycle are housekeeping genes. Tissue-specific genes, as the name implies, are transcribed only in certain specific cells. For example, the genes encoding α- and β-globins are transcribed in erythroblasts, and those that encode immunoglobulins are active in the white cell components of blood, but not in erythroblasts.

In the vertebrate animals, at least, many housekeeping genes have been reported to replicate early in S phase along with certain families of moderately repetitive sequences interspersed among them, whereas tissue-specific genes replicate later (Goldman et al., 1984; Holmquist, 1987). As a general rule this may be under certain conditions, but the situation is far more complex than envisioned at first. Hatton et al. (1988) and co-workers have shown by the use of a number of different mouse cell lines in culture that the time of replication of a gene is dependent on factors other than its role in housekeeping or tissue specificity. Figure 3.14 illustrates some of their results diagrammatically. It is

evidence that cells that are expected to make globins such as the erythroleukemia cells have the two globin genes replicating early (Fig. 3.14a). Plasmacytoma cells are expected to make Ig components and their genes replicate early (Fig. 3.14b). α-Fetoprotein and albumin are normally made in the liver, and their genes are replicated early in the hepatoma cells (Fig. 3.14c). Cells are expected to make Tcβ of the T-cell receptors, and this gene is replicated early in the T-cell leukemia line (Fig.3.14d).

From these data one can draw the conclusion that transformed mouse cells derived from different tissues replicate the genes that transcribe their expected normal products very early in S. This is true for both housekeeping and tissue-specific genes. Actively transcribed genes usually replicate early in S. It is also evident from Figure 3.14 that nontranscribed, unexpressed genes can replicate at any inter-

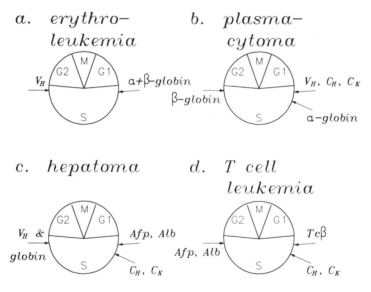

Fig. 3.14. *Times of replication of some housekeeping (Afp and Alb) and some tissue-specific genes (α- and β-globin, V_H, C_H, C_K, Tcβ) in four different mouse cell lines. Afp, α-fetoprotein; Alb, albumin; V_H, immunoglobulin heavy chain variable region; C_H, heavy chain constant region; C_K, constant region of the kappa light chain locus; Tcβ, constant region of the β chain of the T-cell receptor. For further details see text. Data from Hatton et al., 1988b.*

val of S, but generally later than the transcribed genes. Other results reported by Hatton et al. (1988) show that housekeeping genes such as *Dhfr* (dehydrofolate reductase), *H2A* (histone H2A) and *MT* (metallothionein), which are expected to transcribe actively in all cell types, can also replicate late in S, depending on the cell type. It has also been found that members of multigene families that occupy contiguous loci replicate coordinately. While these experimental results obtained with cells in culture are quite convincing in themselves, it must be recognized that replication control in the cells of an intact animal may be different than in isolated cells that are transformed, and hence not exact replicas of the parental cells from which they were derived.

In addition to early and late replication of identifiable structural genes, it has also been shown that certain families of interspersed repetitive sequences replicate in the early (S_E) phase and other different families of repetitive sequences in the late (S_L) phase (Goldman et al., 1984). There also is evidence pointing to the possibility that a gene's position on a chromosome determines the time of its replication (Calza et al., 1984). The mouse β-cell plasmocytoma line, MPC11, has two copies of the $C\gamma2\beta$ gene, a member of the Ig family. The normal position of this gene is on mouse chromosome 12. The $C\gamma2\beta$ copy of the gene on 12 is actively transcribed and replicates very early in the S phase. But the other copy is involved in a translocation that places it next to oncogene *c-myc* on chromosome 15. It replicates very late in S. This gives some credence to the possibility that the position of a gene within a genome is not a random matter, a topic we consider further in Chapter 7.

The early- and late-replicating regions of the genome are clearly delineated in the mammalian prophase and metaphase chromosomes of cells in culture. The S_E phase occurs primarily in the Giemsa light (R) bands and the S_L primarily in the dark (G) bands (Schmidt, 1980; Holmquist et al., 1982; Holmquist, 1989). However, there is not an absolute correlation between band regions and time of replication, since both types of bands replicate later on the inactive X chromosomes of female mammals.

A chromosome band contains a very large amount of DNA, and therefore multiple origins of replication must be contained within it. Since the origins within a band begin replication either early in light bands or late in dark bands, some control signals must exist to coordinate these events. The control mechanism could be similar to that which makes all replicons on the inactive X chromosome late-replicating. There is evidence that this control may involve methylation of cytidine bases (Gregory et al., 1985; Cooney et al., 1988).

The initiation of replication in the formation of replicons is obviously an important aspect of its regulation. We have already mentioned the ARS sequences of yeast earlier in this section and on p. 125. Two DNA fragments have been isolated from a human DNA library that, when introduced into yeast cells by transformation, function as autonomously replicating sequences. The sequences of these two fragments, **ARS1** and **ARS2**, have been determined and are shown to have a number of homologous regions as well as a definite homology to the yeast ARS consensus sequence.

It is not at all clear, however, that ARS-like sequences are important in initiation of replication in all eukaryotes. The DNA of SV40 virus injected into unfertilized eggs of *Xenopus laevis* is replicated semiconservatively. SV40 DNA lacking the origin of replication can also act as templates for multiple rounds of replication. Therefore it has been concluded that a specific ARS-type sequence is not required to act as an origin of replication in the *Xenopus* egg (Harland and Laskey, 1980). This may be true for "foreign DNA" injected into *Xen-*

opus eggs, but it is not necessarily true for *Xenopus* chromosomes *per se.* Chromosomes replicate within nuclei, and cytoplasmic synthesis in eggs may be an artifact that tells us little about the synthesis *in vivo.*

It has been shown that the slime mold *Physarum polycephalum* has its rDNA organized in such a way that each of the repeats encoding the 26s, 5.8s, and 19s rRNA is separated from its neighbor repeat by a ~20-kb sequence that is not transcribed, but contains four putative replication origins (PROs), as shown in Figure 3.15. Not all four PROs are active simultaneously, however. For a typical rDNA stretch only one origin is active per replication cycle (Vogt and Braun, 1977). The nontranscribed regions between the genes containing the PROs have their cytosines heavily methylated, but the transcribed cDNA regions do not (Cooney et al., 1984). When replication starts, one of the PROs becomes hypomethylated, and it has been proposed that this is the active replication origin, whereas the remaining three within the nontranscribed spacer remain hypermethylated and inactive (Cooney et al., 1988). Methylation appears to act by shutting down replication origins.

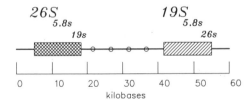

Fig. 3.15. *Map of a region of a Physarum chromosome showing two rDNA transcribing regions (cross-hatched) separated by a central nontranscribed spacer containing four symmetrically arranged replication origins (indicated by circles). Each Physarum nucleus contains a few hundred rDNA molecules with about 60 kb. This constitutes about 2% of the genome. After Cooney et al. (1988).*

It would follow from this that a change from early to late replication, discussed in the previous section, may be controlled by early hypermethylation to be followed by demethylation and the onset of late replication.

There is some evidence that the initiation of replication of chromosomal DNA during the S phase is at the point of attachment of DNA loops to the nuclear matrix with the possible involvement of the nuclear membrane (Cook and Lang, 1984). This possibility is treated more fully in Chapter 7 in connection with the discussion of the nuclear matrix. If the matrix is involved, it may indicate that there are indeed specific sequences of origin that recognize specific signals or "receptors" in the matrix and/or membrane.

4. *Replication in meiocytes*

Practically all the studies on DNA replication in higher eukaryotes have been done with somatic cells in culture. In somatic cells replication is completed by the end of the S phase. This is not necessarily true in meiocyte–germ line cells about to enter into meiosis. In *Lilium* the microsporocytes do not complete DNA replication until zygotene of the first meiotic prophase (Hotta et al., 1966). During this period of chromosome pairing 0.1%–0.2% of the chromosomal DNA previously unreplicated is synthesized in segments about 4–5 kb in length on all chromosomes (Hotta and Stern, 1984). This DNA is transcribed into a poly(A)$^+$RNA called **zygRNA**, which is detectable only in meiotic cells starting in leptotene and is absent by midpachytene. It apparently is of general occurrence in meiocytes, since chromatin of mouse spermatocytes treated with S1 nuclease releases a zygRNA that hybridizes with *Lilium* zygDNA (Hotta et al., 1985). The appearance and disappearance of zygDNA in mouse first-prophase spermatocytes is concordant with that found in *Lilium*. This matter is treated further in Chapter 4 in connection with the meiotic process.

IV. REPLICATION AND ASSEMBLY OF CHROMATIN

In the preceding sections we have considered the replication of DNA only, and with good reason, since it and not the protein of the chromosome is the genetic material. However, chromosomes are made of chromatin, and the protein component of chromatin is involved in important ways in the replication of the total chromosome as well as its function.

A. The Histones and Replication

During the S phase of the cell cycle the synthesis of DNA in the nucleus and histone proteins in the cytoplasm occurs coordinately (reviewed by Osley, 1991). Cells contain a number of free histone pools (Bonner et al., 1988); "fast" and "slow" pools for H4, H3, H2A, and H2B have been detected. Both pool types are present in all phases of the cell cycle, but the fast pool is linked to DNA replication. When this pool is depleted (as, for example, by inhibiting protein synthesis), the rate of DNA synthesis declines. For cells in the G1 phase, in which there is no DNA synthesis, histone synthesis proceeds at a low level (Wu and Bonner, 1981). Those histones that are synthesized are incorporated into nucleosomes (Wu et al., 1983; Jackson and Chalkley, 1985). For the most part the histone-encoding genes are turned off except during the S phase.

These separate histone pools appear to be associated with "histone carriers"— proteins that complex to histones and transport them from cytoplasm to nucleus. Two kinds of complexes have been isolated from *Xenopus* oocytes. One contains H3 and H4 bound to acidic proteins (Kleinschmidt et al., 1985) and the other

complex contains H2A and H2B bound to nucleoplasmin (Dillworth, 1987). These two complexes are required for nucleosome assembly in *Xenopus* egg extracts.

The participation of histone H1 in chromosome replication has been studied in cells in culture. Histone synthesis occurs in the G1 and S phases of mammalian cells in culture (Groppi and Caffino, 1980), but the rate is not necessarily constant. Results with hamster CHO cells show that the rate of H1 synthesis starts to increase in the G1 phase, and as the cells enter S phase there is a 22-fold increase in synthesis. In contrast, the rate of H1° synthesis remains relatively constant throughout the G1 and S phases (D'Anna et al., 1985a). H1 and H1° possess many structural features in common and presumably bind to DNA in a somewhat similar fashion, but they have differences in sequences in their central regions, in their content in chromatin, and in the rate of their synthesis, at least in cultured cells (D'Anna and Prentice, 1983; D'Anna and Tobey, 1984). This difference in rate of synthesis of the two forms of H1 was observed in synchronized populations of CHO cells that were pulse-labeled at different times with [^3H]lysine and the cells were not perturbed by blocking DNA synthesis at any point. Therefore it may be that the differences are natural for these cells. But at the same time it must be recognized that the relation between what happens in cells in culture, and in the intact organism in different tissues, is quite ill-defined in this case. It is known that H1° occurs in some tissues, but not others.

Synchronization of CHO cells by isoleucine deprivation, followed by blocking in early S phase by the use of hydroxyurea or 5-fluorodeoxyuridine, which are inhibitors of nucleotide biosynthesis, or by aphidicolin, which specifically inhibits DNA polymerase α activity, results in the loss of H1 from the cells and changes in the structure of the chromatin within the cells (D'Anna and Prentice, 1983; D'Anna and Tobey, 1984). In addition, although these inhibitors allow cells to enter the S phase, they retard replicon elongation and the turnover of H1 (D'Anna et al., 1985b). Thus there would appear to be a rather complex relationship between H1 synthesis and DNA synthesis.

A mutant strain of Chinese hamster V79 lung cells designated *ts14* is temperature-sensitive and ceases protein biosynthesis within 2h after a shift to the nonpermissive temperature of 39°C. Cells incubated at this temperature initially complete one round of DNA replication before ceasing growth, but during this round of replication they synthesize significantly less histone protein than the wild-type cells under similar conditions (Roufa, 1978). Hence, histone synthesis is probably not essential for DNA replication, if sufficient histones are already present to sustain it. After a period of deprivation, however, insufficient histones almost certainly will cause a cessation of chromatin synthesis. Of course a complete cessation of all protein biosynthesis, as in this case, would be expected to cause all cellular functions, including DNA synthesis, to grind to a halt. The replication of eukaryotic DNA is clearly associated with and dependent upon histone availability in the cell. The synthesis of both must occur in a coordinated fashion for chromosomes to be replicated *in vivo*.

B. Organization of Chromatin

The presence of histones is necessary not only for chromosome replication but also for the organization of the higher-order structure of chromatin up to and above the 10-nm level. Changes in chromatin structure occur preferentially in the early-replicating replicons (D'Anna and Tobey, 1984). A model chromatin assembly system has been devised in the presence of all five histones and poly (dA-dT)

synthetic DNA (Stein and Bina, 1984). An approximation of a short 10-nm fiber was obtained with this system. A further step was taken by Simpson et al. (1985), who prepared a series of DNA sequences from the sea urchin *Lytechinus variegatus* that span the nucleosome repeat lengths for intact chromatin. Fragments ranging from 172 bp to 207 bp repeated three to about 50 times were cloned in a plasmid. Each repeat contained the region that is involved in forming a positioned histone core for a nucleosome on a 5rRNA gene. The fragments were associated with chicken erythrocyte core histones, and nucleosomes were formed on the DNA fragments at nonrandom positions.

During DNA replication nucleosomes are formed and integrated into the newly formed DNA strands behind the replication fork (Svaren and Chalkley, 1990). The preexisting nucleosomes come apart to form (H3H4)2 tetramers and (H2AH2B) dimers, which recombine with newly synthesized histones. Thus the new nucleosomes consist of a mixture of old and new.

The topological configurations assumed by chromosomes during S phase in order to contain the simultaneous processes of DNA replication, the assembly of histones and other proteins such as TOPOII with the nascent DNA strands, and the transcription of active genes are far from being understood. A model of a chromosome considered to have a "fractal" structure has been proposed by Takahashi (1989). He considers that a chromosome has but one continuous duplex DNA fiber with radial loops that make up what are essentially minichromosomes. These represent replicons and are repeated units or fractals of replication, structure, and function.

Studies made with SV40 indicate that newly synthesized DNA accumulates as intertwined catenated dimers (Sundin and Varshafsky, 1981). SV40 has a genophore with nucleosomes and therefore some similarity to eukaryotic chromosomes. Intercatenation of sister chromatids during S phase is also a possibility (Murray and Szostak, 1985). The precise nature of most of the events occurring in the cell cycle between the beginning of S phase and the segregation of chromatids at the end of metaphase remain clouded in deep mystery. Understanding these processes is one of the major problems in present day biology. They form the basis for the understanding of the reproduction of eukaryotic systems and the things, such as cancer, that can go wrong with them.

V. ENDOREDUPLICATION AND ENDOMITOSIS

Among the eukaryotes a number of phenomena occur as a part of the chromosome replication process that are described variously as **endoduplication, endomitosis, endopolyploidy, endoredupli-**

TABLE 3.3. Some Examples of C Values per Nucleus Among Eukaryotes

Phylum, class, or order	Species	Cell type	C value	n
Sarcodina	*Cibicides lotatulus* (foraminifera)	"Somatic"	30	—
Cnidaria	*Carymorpha palma* (jellyfish)	Endoderm	16	4
Nematoda	*Ascaris suum*	Esophageal gland	262, 144	18
	Panagrellus silusiae	Gut	10	—
Annelida	*Ophryotrocha puerilis*	Nurse	256	8
Mollusca	*Aplysia californica*	Giant neurons	80–200	—
	Helix pomatia	Salivary gland	64	6
	Triodopsis divesta	Neuron	32	5
Insecta	*Dermestes maculatus*	Testes wall	64	6
	Chironomus thummi	Malpighian tubules	16, 384	14
		Salivary gland	8, 192	13
	Drosophila melanogaster	Salivary gland	2,048	11
	Bombyx mori	Silk gland	524, 288	19
Aves	*Maleagris gallapavo* (turkey)	Heart muscle	8	3
Mammalia	*Mus musculus*	Trophoblast	512–1,024	9–10
		Liver	16	4
	Rattus rattus	Trophoblast	1,024	10
		Trophoblast	4,096	12
	Oryctolagus cunniculus (rabbit)	Trophoblast	256	8
	Homo sapiens	Myocardium	16	4
		Auricular tissue	96	—
		Bone marrow Megakaryocytes	128	7
Angiospermae	*Phaseolus coccineus*	Suspensor	8,192	13
	P. vulgaris	Suspensor	2,048	11

Data from Nagl (1978).

cation, and **polyteny** (see page 103). These are the results of modifications of the usual mitotic cycle. Generally elementary biology texts state that all somatic cells of individual plants and animals contain the same number of chromosomes, implying that all nuclei have the same content of DNA 1C in haploids and 2C in diploids. This is not true. Although **most** somatic nuclei in a eukaryote may have a 2C complement of DNA when in G2, in just about every eukaryote ranging from Protista to higher plants and animals there are cells that deviate in the G1 from the accepted norm for the species (Nagl, 1978). These cells may have DNA contents ranging from 4C to 524,288C, although in the latter case not all sequences may be replicated. The latter value is found in the silk gland nuclei of *Bombyx mori*, in which 19 replications must occur to give the value of 524,288C ($= 2^{19}C$) (Gage, 1974). The insects have some of the highest C values, but high values are also found in practically every other group of plants and animals. Table 3.3 gives some representative examples in both animal and plant groups. Only two examples are given for plants, but practically all flowering plants have cells with high C values. For a more complete list of C values in plants see Nagl (1978). In all examples given in Table 3.3 the C values were determined by actual DNA measurements rather than by counting the number of DNA strands.

These observed increases in the DNA C values of nuclei are the result of a manifold replication of chromosomal DNA in the S phase, rather than single replication, as in the typical cell cycle in which daughter cells receive the same DNA content as the mother cell. The chromatids formed as a result of **endoreduplication** may separate and remain in the same nu-

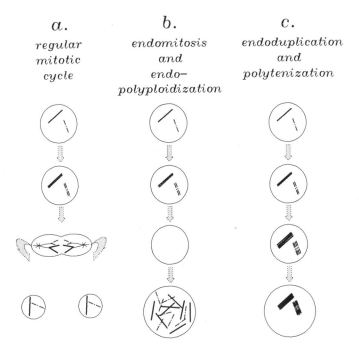

a.
regular
mitotic
cycle

b.
endomitosis
and
endo-
polyploidization

c.
endoduplication
and
polytenization

Fig. 3.16. *Endomitosis and endoduplication—compared with the normal mitotic cycle (**a**)—result in endopolyploidization (**b**) and in polytenization (**c**), respectively.*

cleus without a following karyokinesis, in which case the cells become **endopolyploid** and the chromosomes can be counted under the microscope (Fig. 3.16). This form of somatic polyploidization is frequently described as the result of **endomitosis** and is quite common among the insects. If the chromatids do not separate, "giant" **polytene** chromosomes will result, as we have described in Chapter 2. This endoduplication results in polytenization (Fig. 3.16c). Polytene chromosomes are most commonly formed among the Diptera, in which they occur in the salivary glands of the larvae, and certain other tissues depending on the species; they are also found among other insects, such as *Bombyx*, in the ciliate Protozoa, and in the tropho-

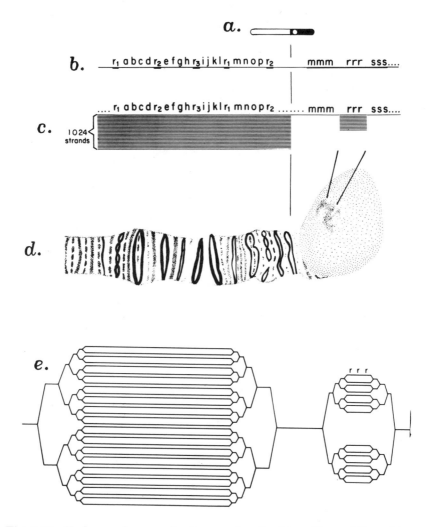

Fig. 3.17. *Under- and overreplication as visualized in a Drosophila polytene chromosome (a) and (b) nonpolytene X chromosome of D. hydei; (c) uniform replication of the euchromatic segment of the X (r_1 to r_2); slight underreplication of heterochromatic region (rrr) containing rDNA; (d) nonreplication of centric heterochromatin (mmm) and other DNA at terminal knob (sss); (e) attempts to show how the unreplicated and overreplicated regions may be related. From Laird (1973), with permission of the publisher.*

blasts of mammals. In general, angiosperms have them in the ovarian and other tissue cells of their flowers, but other cell types, such as root cells, can also have polyteny (Nagl, 1978). In addition, endomitotic polyploidy is common in angiosperms and bryophytes. Curiously it does not appear to occur in ferns and gymnosperms.

Figure 3.16 diagrams the results of two types of endo-cycles: endomitosis with polyploidization and endoreduplication with polytenization, in contrast to the true mitotic cycle. During the endo-cycle heterochromatic and other repetitive DNA appear to be replicated in step with endoreduplication in *Bombyx mori* (Gage, 1974), in the trophoblast of the mouse (Barlow and Sherman, 1974), in the giant nuclei of the plants *Vicia faba* (Millerd and Whitfeld, 1973), and in *Gossypium* (cotton) (Walbot and Dure, 1976). However, both underreplication and overreplication of specific DNA sequences also occur. Endocycle is a convenient term that was introduced by Nagl (1978) to signify the repeated replication of DNA within the nuclear envelope without the formation of a spindle and the following anaphase at the end of each DNA replication cycle. As a rule, endopolyploid nuclei and polytene chromosomes never complete an entire mitotic cycle; they are a somatic dead end.

A. DNA Regional Underreplication

The polytene chromosomes of *Drosophila* result from an unequal endoreduplication; the heterochromatic and telomeric regions are grossly underreplicated and appear to fit the model diagrammed in Figure 3.17 (Laird et al., 1974). There is evidence that most of the heterochromatin does not replicate at all during the formation of these chromosomes in *Drosophila* (Rudkin and Tartof, 1973; Laird, 1973). In addition to heterochromatic DNA, the rDNA sequences are also underreplicated in the polytene chromosomes of *Drosophila* (Spear, 1977). Also those regions of the *Drosophila* chromosomes that show localized constrictions that have been described as intercalary heterochromatin are underreplicated (Lamb and Laird, 1987).

The polyploid tissues of *Drosophila virilis* have been studied by Endow and Gall (1975) by determining the satellite content of their DNA with buoyant density centrifugation. Table 3.4 gives the results obtained with eight different tissues in the whole newly emerged adult females. Brain and imaginal disk tissue of *Drosophila* larvae are constituted of diploid cells, and DNA from these cells of *D. virilis* show a

TABLE 3.4. Satellite DNA From Tissues of Adult *Drosophila virilis*

Tissue	% of total genome		
	Main peak	Satellites I + II	Satellite III
Brain	53.0	38.4	9.4
Midgut	73.5	20.4	6.1
Hindgut	65.3	27.5	7.3
Malpighian tubules	100.0		
Salivary glands	76.6	20.3	3.1
Thoracic muscles	64.6	32.5	2.9
Ovaries	51.9	39.6	8.6
Pupal ovaries	46.0	43.5	10.6
Whole adult	62.2	33.1	4.7

Data from Endow and Gall (1975).

main band of DNA at 1.700 g/cc and three satellites: I = 1.692 g/cc, II = 1.688 g/cc and III = 1.671 g/cc, which represent 25%, 8%, and 8% total DNA, respectively. It is quite evident from the data that the DNAs from midgut, hindgut, malpighian tubules, salivary glands, and thoracic muscles have low satellite contents. This is particularly true of the malpighian tubules, which have highly polytenized chromosomes. No satellite DNA whatever was detected in these cells and very low amounts in salivary gland cells. Ovarian tissues from adults and pupae had high concentrations, however. It is apparent that all cells that are known to have polytene chromosomes or endopolyploidy, the satellite DNAs, and hence presumably heterochromatin, are underreplicated relative to the euchromatin. The cytological observations are thus confirmed by the molecular ones. The satellite DNAs of embryonic tissues of *D. virilis* have been analyzed and compared with those found in the adult by Blumenfeld and Forrest (1972). Embryonic cells are diploid, whereas in the adult not all are, as pointed out above. The main peak DNA found by these workers was 60% for embryonic DNA and 72% for the whole adult. The difference could be ascribed to the expected higher concentrations of satellite DNA in the diploid embryonic tissues, and these values are therefore consistent with the results of Endow and Gall (1975).

A detailed study has been made of the ovarian nurse cells of *D. melanogaster*. The chromosomes of these cells undergo endoreduplication and endomitosis with the result that the cells become endopolyploid (Painter and Reindorp, 1939). Polytene chromosomes are not seen in these cells after they mature, and the increase in DNA accompanying the endopolyploidy is not the result of a regular doubling (Hammond and Laird, 1985a). This condition is not at all unique to *Drosophila*. As shown in Table 3.3, C values resulting from reg-

ular doubling of all, or most, chromosomal DNA are not always found. Note, for example, the C values for the foraminiferan, the nematode *Panegrellus*, the mollusc *Aplysia*, and human auricular tissue.

Hammond and Laird (1985b) incubated whole ovaries with [³H]thymidine for 20 min and then stained them with Feulgen stain. The nurse and follicle cells of the ovaries that had not replicated during the period of incubation, as shown by their not taking up [³H]thymidine, fell into eight classes as measured by their DNA content with Feulgen staining. The data obtained show that the observed DNA contents of the nuclei do not fall into classes predicted by full replication. Rather they show that at least after the fifth doubling about 25% of the DNA does not replicate. To determine whether there was underreplication of specific sequences rather than random underreplication, probes were prepared for 28S + 18S RNA, 5S RNA, and histone genes, satellite sequences, a telomeric sequence, and a *copia*-type sequence described on p. 223. These radiolabeled probes were then hybridized *in situ* to test for the amount of hybridization within the squashed nuclei of nurse cells after appropriate treatment to denature the DNA of the chromosomes. The resulting measurement of hybridization for each of the different probes showed that the satellite and histone sequences underreplicate during nurse cell growth whereas 5S rDNA genes replicate fully with the rest of the DNA. A slight underreplication of the 28S + 18S rDNA and telomeric sequences was noted but may not be significant.

In addition to the general conclusion that satellite and certain other kinds of repetitive DNA in *Drosophila* cells are underreplicated during endoreduplication, it is also evidence that replication of a given satellite DNA is independent of replication of other satellites. Compare, for example, the values for thoracic nuclei with midgut nuclei data in Table 3.4. Furthermore, in pupal ovarian tissue the satellite DNAs seem to be overreplicated. Findings such as these and those of Hammond and Laird (1985b) cast considerable doubt on the often stated hypothesis that satellite DNA is an unimportant constituent of chromosomes.

B. DNA Regional Overreplication

In general, the puffs that develop in polytene chromosomes do not have a disproportionate replication of DNA. Rather, they are associated with an uncoiling of DNA and synthesis of RNA. Hence, they are appropriately called RNA puffs. However, in the dipteran family Sciaridae the polytene chromosomes develop both RNA and DNA puffs. DNA puffs are the result of regional DNA overreplication (Pavan and daCunha, 1969). In the Sciarid *Rhynchosciara angelae*, polytene chromosomes can be observed in synchronously developing larvae over a period of many days. This makes possible a detailed study of the course of development of puffs. The puffs last from a few hours to 6–7 days and each appears at a specific time of larval development. Puffs appear in all tissues that have polytene chromosomes, but each tissue has its own specific array of puffs. The DNA puffs have not only a larger amount of DNA than the surrounding chromosomal parts, but also a large amount of RNA, which may be synthesized along with the replicating DNA or following the completion of DNA synthesis.

Another member of the Sciarid family, *Hybosciara fragilis*, not only develops DNA puffs in its polytene chromosomes, but the puffed bands, some of which may have the characteristics of nucleoli, reach a large size and release extrachromosomal DNA in the form of micronuclei (daCunha et al., 1969; daCunha, 1972). These have a central core of DNA surrounded by material rich in RNA. They cease being formed as the puffs responsible for them appear. The usual interpretation of puffing, as exhibited in the salivary gland polytene chromosomes of *Drosophila*, is that extensive transcription of DNA occurs at the puff sites (Ashburner and Berendes, 1978; Ashburner, 1990). However, high levels of transcription of genes are observed in the absence of a puff and puffs are observed in the absence of transcription (Meyerwitz et al., 1985, 1987). Presently we cannot be sure just what all puffs represent.

Cells in culture frequently show spontaneously endoreduplicated chromosomes. Diplochromosomes consisting of four chromatids and quadruple chromosomes with eight chromatids are observed in metaphase preparations after treatment with colcemid or colchicine. The chromatids seem to be held together at a common centromeric region (Takanari, 1985).

VI. CHROMOSOME DAMAGE OCCURS CONTINUOUSLY IN ALL ORGANISMS

It is a fact of life that DNA is a chemical substance subject to alterations, some of which are deleterious to the individual. If these alterations were not in many cases repairable, life as we know it would long ago have disappeared from this planet (see review by Friedberg, 1985). Hence the repair of DNA is secondary in importance only to its synthesis. Damage to DNA by external agents such as radiation and chemicals, which may modify bases and cleave the DNA chain, as well as mistakes made during the synthesis process, must be repaired at least at a rate consistent with maintaining the organism's ability to survive and reproduce. Damage to the protein part of chromatin must also occur through the action of such agents as radiation and chemicals. However, very little is known about this type of damage.

Therefore our primary consideration here must relate to DNA damage and repair.

A. Sources and Types of Damage

DNA alterations can arise "spontaneously" (meaning they have no known external agent producing them) simply as a result of mistakes made during the replication process. The mispairing of bases that cause mismatches (Radman and Wagner, 1986) and the production of frame shift mutations (Ripley, 1990) are two examples of alterations that can occur in the replication process. Actually polymerases are quite accurate and repair editing is quite efficient so that replicative fidelity is high and the error frequency in newly replicated DNA is about 10^{-9} to 10^{-10} per nucleotide replicated (reviewed by Echols and Goodman, 1991). Polymerases are exquisitely accurate in placing the proper dNTPs in place during the replicative process, but they also edit as they function, so that if an incorrect dNMP is inserted at the end of a growing chain, it is generally quickly replaced by exonucleocytic enzymatic activity. Even if it is not replaced immediately, an incorrect dNMP can be corrected by insertion of the correct nucleotide in its place postreplicatively. This called **mismatch repair** (Friedberg, 1988, 1991).

In addition to polymerase errors purine and pyrimidine bases can be structurally altered chemically by tautomeric shifts and the deamination of cytosine, adenine, or guanine. Also bases can be lost and as a consequence DNA strand copying can be interrupted.

Chromosome damage that results in breakage is a common occurrence. Such damage can be the result of **clastogenic** (clastic = breaking into fragments) **factors** of either endogenous or exogenous origin. Table 3.5 describes four human inherited diseases in which there is an elevation in the frequency of detectable chromosome breakage, and susceptibility to exogenous clastogenic factors (data from Cohen and Levy, 1989). The phenotypic effects are complex and many. The factors involved in inducing the chromosomal damage are both endogenous and exogenous in the form of increased susceptibility to radiations, and other clastogenic radiomimetic chemicals such as bleomycin, an antibiotic, have an extreme clastogenic effect especially in persons with ataxia telangiectasia (Joenge, 1989).

Endogenous clastogenic factors are formed by the metabolic activities of cells. For example, normal human lymphocytes cultivated in the culture medium from cells of persons with either ataxia telangiectasia or Bloom's syndrome show a significant increase in chromosome breakage (Shaham et al., 1980; Emerit and Cerutti, 1981) The effect is striking: a six- to twelvefold increase in chromosome aberrations was detected in normal lymphocytes cultured in Bloom's syndrome culture media.

In addition to complex phenotypic effects the diseases described in Table 3.5 have complex genetic backgrounds. This is especially true of xeroderma pigmentosum probands who exhibit eight to nine complementation groups and ataxia telangiectasia probands with five to nine complementation groups. Thus more than one gene is involved in both diseases, making it evident that different mechanisms of defects are operating.

The most important sources of chromosomal DNA damage relevant to our discussion here are those that cause double-strand breaks followed by total chromosomal chromatin breaks. Ionizing radiations are a potent causitive agent of such breaks. It has been shown that the number of double-strand breaks produced intracellularly in the presence of oxygen is linearly related to the dose of radiation (reviewed by Ward, 1990). For a double-strand break to occur, both strands must break more or less simultaneously and close

TABLE 3.5. Inherited Human Disorders Accompanied by Abnormally High Chromosomal Instability

Name	Phenotypic manifestation	DNA synthesis effects
Ataxia–telangiectasia (A-T)	High frequency of chromatid breaks with consequent gaps and deletions, as well as acentrics etc.; immune dysfunctions; neurological abnormalities; elevated α-fetoprotein; extreme susceptibility to ionizing radiation and radiomimetic chemicals; predisposition to cancer at early age.	Only slight inhibition of DNA synthesis after X-radiation in contrast to normal cells. Maybe at least 8 genes giving similar phenotypes.
Bloom's syndrome (BS)	High frequency of sister chromatid exchanges (SCE) and chromosome breaks; immune dysfunctions; some afflicted show elevated susceptibility to UV and ionizing radiation; short stature and narrow face; predisposition to cancer at early age.	Abnormalities in cell cycle; possible ligase defect, but no concrete evidence for DNA repair defect.
Xeroderma pigmentosum (XP)	Chromosome instability low to normal; severe UV sensitivity with chromosome breaks; immune dysfunctions; some have neurological abnormalities; predisposition to cancer at early age, especially skin cancer.	Low unscheduled DNA synthesis ability; deficient in incision step of excision repair of induced UV lesions; at least 8 complementation groups, XP-A to XP-G, giving similar phenotypes have been identified.
Fanconi anemia	High incidence of chromosome breaks and endoreduplication; skeletal abnormalities; short-wave (254-nm) UV increases chromatid breaks; predisposition to cancer at early age.	DNA synthesis generally slow; different strain sources have quite different molecular manifestations.

Data from Cohen and Levy (1989).

together along the double helix to produce a complete chromosome break. Single-strand breaks by ionizing radiation are initiated by loss of a hydrogen atom on a deoxyribose, thus forming a radical. The radical then reacts with oxygen to form a peroxy radical and a strand break (von Sonntag, 1987).

Thus ionizing radiations presumably cause breaks in DNA strands either by direct action of the radiation energy breaking phosphodiester bonds, or indirectly from the interaction of reactive substances with DNA formed in the cell by radiation energy. Cells are mostly water and the irradiation of water results in the formation of hydrogen peroxide, hydrogen atoms, hydrated electrons, and hydroxyl radicals, all of which are putative DNA-damaging agents (Ward, 1975). Free radicals or chemical entities conducive to free radical formation are all suspect.

Ultraviolet light in the range of 260 nm is readily absorbed by DNA, and its effects on DNA have been more widely studied than any other source of DNA damage, especially in *E. coli*. In terrestrial animals and plants its effects are only to be expected on cells at the surface, since water among other cell contents is relatively opaque to UV. However, as noted in Table 3.5, the skin cells of persons with xeroderma pigmentosum are very sensitive to UV exposure even in the range above 300 nm, and tumors occur frequently among

them in sun-exposed areas of the skin and eyes (Kraemer et al., 1987).

The most prominent effect of UV irradiation of DNA is the formation of pyrimidine dimers (Fig. 3.18). DNA exposed to UV radiation is subject to the formation of four-membered cyclobutyl rings linking two adjacent pyrimidines in the same chain. This kind of lesion interrupts polymerase action and can result in cell death, via nonsense mutations and strand breaks, unless repaired.

B. Repair of Damage

Many different types of repair have been described, but they can all be classified as either repair by direct reversal of damage, or by excision of the damaged segment of DNA and its replacement by a normal nucleotide sequence. An example of repair by reversal of damage is that which occurs after UV-radiation-induced formation of pyrimidine dimers. Cells in which pyrimidine dimers have been formed can have

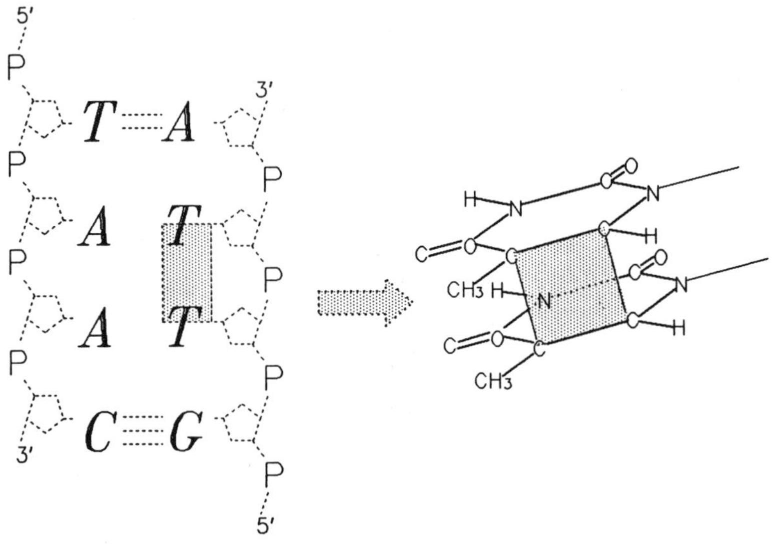

Fig. 3.18. *Formation of a pyrimidine dimer between two adjacent thymidines. The cyclobutyl ring is shown shaded.*

this damage reversed by enzymatic photoreactivation. This occurs when the UV-damaged cells are irradiated with light in the range of 300 nm to about 400 nm, with the maximum effect for yeast cells at about 370 nm. The light activates the enzyme DNA photolyase, which has been found in all eukaryotes examined, with few exceptions. The action spectrum for photolyase activation in human leukocytes is somewhat shifted toward the red with a peak about 425 nm. Other forms of reversal of damage have also been identified. One of the more important ones from the standpoint of chromosome breakage repair is the rejoining of single-strand broken ends by DNA ligase(s).

Apparently most of the damage sustained by DNA is the type that is not repaired by direct reversal, but by **excision reversal**, in which damaged sections are excised and replaced by the proper, normal sequences. Repair by excision and replacement can be by three distinct biochemical mechanisms: (1) by nucleotide excision repair, (2) by base excision repair, and (3) by postreplicative mismatch correction or repair. As in the case of direct reversal, most of the studies of excision repair mechanisms have been with prokaryotes. While similar mechanisms of repair are expected in both pro- and eukaryotes, it must be recognized that eukaryotic chromosomes are endowed with a large mass of protein organized in a very specific way with DNA as chromatin. The DNA of nucleosomes, for example, is bound to histones of the core, and the SARs at the base of the loops with TOPII. These relationships between DNA and proteins, and the formation of coils and loops involved in chromosome compaction as described in Chapter 1 isolate the DNA from the internal environment of the cell to a considerable degree. Since excision repair involves first enzymatic incision followed by an excision and then by enzymatic replacement, the damaged DNA must be accessible to the enzymes. Of course this problem may be alleviated if the repair enzymes are part and parcel of the chromatin, and this may well be the case.

Figure 3.19 illustrates how coordinated incisions, excisions, and replacements can be visualized by using replacement of pyrimidine dimers as an example. One route of repair is for endonucleases to make nicks on either side of the dimer, the incised sequence to be excised, and repair to proceed by unscheduled DNA synthesis via DNA polymerase activity. Alternatively a single nick followed by strand displacement can also occur. In eukaryotes the most active repair polymerase appears to be POLβ, but POLα and POLδ cannot be eliminated as contributors under some circumstances.

These are but two of the many examples of DNA excision repair that have been studied and described in the literature. The replacement of damaged or lost bases has also been demonstrated experimentally. As might be expected, the evidence indicates that the most likely damage to occur in eukaryotic DNA is in the linkers between the nucleosomes. It would be in these nuclease-sensitive regions that chromosomes would most likely break, and be repaired.

C. Repair of Damage to DNA Involves the Action of Many Genes

Most of the studies on genes involved in excision repair have been done with *E. coli* (reviewed by Van Houten, 1990). Among the eukaryotes *S. cerevisiae* has been the most studied, and the work with it has begun to provide us with a yeast model for developing some understanding of repair genes in eukaryotes in general (reviewed by Friedberg, 1988, 1991).

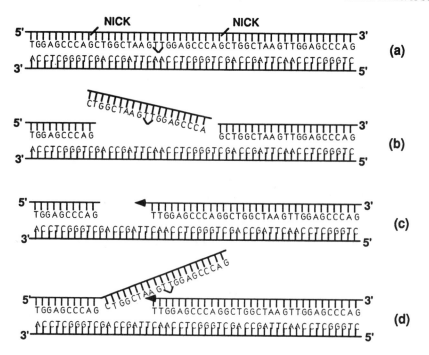

Fig. 3.19. *Incision* (**a**), *excision* (**b**), *and repair* (**c**) *at a pyrimidine dimer site.* **d.** *A modification called strand displacement.*

At least ten or eleven genes are known to be involved in nucleotide excision repair (NER) in yeast. These have been identified among the mutants that are abnormally sensitive to UV radiation. Five of these, designated *RAD1, RAD2, RAD3, RAD4,* and *RAD10,* encode protein products that appear to be the primary agents involved in the recognition of DNA damage and its repair. *RAD1* and *RAD10* are also involved in recombination (Schiestl and Prakash, 1988; 1990). Five other genes: *RAD7, RAD14, RAD16, RAD23,* and *RAD24,* when they mutate, confer a deficient rather than a defective NER. Their role in the repair process is not understood. Altogether 27 genes have products that are involved to one degree or another in DNA repair in yeast.

The *RAD3* gene has been sequenced, and its encoded protein has been purified and identified as a DNA-dependent ATPase/

DNA helicase involved in the unwinding of duplex DNA (Sung et al., 1987; Horosh et al., 1989). It has been suggested that this helicase activity is necessary for accurate DNA synthesis and also for postreplicative mismatch correction, as well as NER. The product of *RAD3* may participate in several different complexes, one of which is a necessary helicase in normal synthesis and others of which are involved in repair proper (Sung et al., 1987).

Some of the yeast repair genes have apparent homologues identified in mammalian genomes. *RAD3* has homology with an *RCG-2* gene in the Chinese hamster and the *ERCC2* gene in humans. *ERRC2* is the gene associated with the XP-D complementation group of XP, as shown in Table 3.5. Another human NER gene, *ERCC3,* encodes a protein that also may be a helicase similar to that encoded by *RAD3.* But this gene is not homologous to

RAD3, since it is associated with the XP-B complementation group of XP. *ERCC3* is on chromosome 2q21 and *ERCC2* on 19q13.2. These apparent homologous relationships are only a few of the many that probably exist in the eukaryote group. We deal here with an ancient group of genes and their products. The mechanics of repair may well be as ancient as the mechanics of synthesis of DNA. Furthermore many, if not most or all, of the enzymes involved in repair are also involved in synthesis or functions associated with synthesis.

VII. SUMMARY

1. The cell cycle consists of four distinct phases: G1, S, G2, and M. The DNA content of the nucleus of a diploid cell is 2C during the G1 phase. During the S phase it becomes 4C and remains so through the G2 phase. During the M (mitotic) phase it returns to the 2C state in the resulting daughter cells.

2. In the mitotic phase the chromosomes compact and become visible in the prophase. The mitotic spindle or apparatus forms, the nuclear membrane dissipates, and metaphase begins with the attachment of the spindle fibers to the kinetochores in the centromeric regions of the chromatids. Chromatids separate in the anaphase, which ends with the beginnings of telophase, the reformation of the nuclear membrane, and the decondensation of the chromosomes. The G1 phase then begins.

3. The interphase of the cell cycle consists of the G1, S, and G2 phases. Chromosomes replicate in the S phase, which is also called the period of "scheduled" synthesis. Active transcription occurs during the interphase.

4. The number of genes involved in the maintenance and regulation of the eukaryotic life cycle, and hence chromosome replication and segregation, may well be in the hundreds. The determination of their function and expression at specific times in the cycle is an important aspect of the study of causes relating to the onset and progress of cancer.

5. Protein kinase(s) is an important regulator of the cell cycle. One of them, identified as p34 and encoded by *cdc2$^+$*-type genes, is found in yeast and all other eukaryotes studied. Protein kinases are encoded by proto-oncogenes. They phosphorylate side chains of tyrosine, serine, and threonine residues in proteins. The one identified as p34 (which may be a family), together with members of the cyclin family, operates at two points in the cell cycle: the transitions from G1 to S and from S to M. The phosphorylation and dephosphorylation of p34 and the degradation of cyclin, once these transitions are initiated, are key elements in the oscillation of the cell cycle from interphase to mitotic phase to interphase in actively dividing cells.

6. Replication of DNA is semiconservative. Synthesis of chromosomal DNA is catalyzed by DNA polymerases in the 5′ → 3′ direction. The growing, or primer, strand must have a free 3′OH end and copies from the template. Synthesis of DNA among eukaryotes occurs on both strands of duplex DNA simultaneously. It is necessary for an RNA primer to be present that is excised and for the gap to be filled in as synthesis proceeds. Synthesis of the "leading" strand involves the 3′ strand as template. "Lagging strand" synthesis occurs intermittently on the 5′ → 3′ template strand.

7. The mechanism of DNA synthesis has a basic similarity among all cellular organisms, both prokaryotic and eukaryotic. This similitude extends even to the phages and viruses that infect prokaryotes and eukaryotes alike.

8. The basic mechanisms of DNA replication are the following:

a. initiation of synthesis at a replication fork created by the separation of the strands of a duplex to form a replication bubble;

b. unwinding of the helix as replication proceeds at the fork; and

c. synthesis of new strands on the leading and lagging strands.

9. The replication of linear chromosomal DNA molecules such as those found in all eukaryotes is apparently made possible without loss of DNA segments at every replication cycle by the presence of specific repeats with a series of guanines at the telomeres. The non-Watson-Crick type of bonding of terminal guanines produces free $3'OH$ ends for continuation of $5' \rightarrow 3'$ synthesis at the ends.

10. The immense lengths of DNA molecules in a eukaryotic chromosome require many origins of replication to form replicons in order for their replication to occur at a rate constant with the rate of cell division.

11. The replication of chromosomal DNA does not occur at the same time and rate in different parts of chromosomes. Some genes replicate early, some later in the S phase. The time of replication also varies with the state of differentiation of the cell, at least for some genes.

12. A relationship exists between the degree of methylation of the cytidines of the chromosomal DNA and the mutation of replication, at least in some eukaryotes.

13. The replication of chromosomal DNA is dependent upon the availability of histones in the cell in S phase.

14. DNA replication in some eukaryotic cells may occur without mitosis, in which case the nuclei can have very high C values ranging from 4C to as high as 524,288C. If the chromatids separate, the result is called endopolyploidy; if they do not, the result is polyteny.

15. Polytene chromosomes may show regional underreplication and also overreplication of DNA.

16. Chromosome damage probably occurs continuously in all cells.

17. The types of damage are varied, ranging from the mispairing of bases, which results from mistakes made by polymerases during replication, to chromosome breakage.

18. Factors involved in the production of damage are also varied and many. Endogenous factors may be the result of the production of metabolic mutagens and clastogenic factors within the cells themselves. Exogenous factors are intrusive radiations such as ionizing radiations, which can cause chromosome breakage, or ultraviolet light, which can cause the formation of pyrimidine dimers and can lead to gene mutations and strand breakage.

19. Life survives despite gene damage because repair processes are normally active in all cells. The types of repair can be classified as repair by direct reversal of damage and by excision of a damaged segment of DNA followed by its replacement. Many genes are involved in the repair process by encoding products that recognize damage, that can excise damage, and that repair damage. The DNA polymerases themselves are involved in many of these activities.

4 Genetic Recombination

From the fact of sexual reproduction, which brings together equal amounts of paternal and maternal germ-plasm at each fertilization, I inferred not only the composition of the germ-plasm out of the number of units, the ancestral plasms, but also the necessity of a reduction of the germ-plasm each time to one half of its bulk, as well as the reduction in the number of ancestral plasms within it.

A. Weismann, 1892

During the nineteenth century, when the groundwork was laid that led up to the development of our present understanding of cell structure and function, little attention was paid to the sources of variation in natural populations even though the existence of such variation was obvious to all. Darwin's theory of natural selection had as its basic tenet that variants in populations of organisms supplied the material upon which natural selection operated. But neither Darwin nor his contemporaries had any idea about the basic nature and sources of these inherited genotypic variations. It was not until after 1900, when it was realized that genes mutate and that chromosomes and their parts reassort in various ways, that an understanding of the origins and perpetuation of variants began to evolve. It is now known that the basic material source of organismic variation is mutation of the DNA. However, superimposed upon changes in DNA structure is one other process the importance of which is often ignored. This is the **recombination** of genetic material—the reassortment of genetic material already present.

It has been asserted that if all new mutations were to cease in any sexually reproducing population, such as the human, the level of variation within the population would continue essentially the same for hundreds of generations. Because of recombination, the number of different genotypes produced from a cross between two eukaryotes, human or otherwise, is essentially infinite. Thus to understand the role of chromosomes in the maintenance of variation, it is necessary to have some understanding of the recombinatorial processes in which they engage.

Genetic recombination is the exchange of corresponding chromosome portions between genomes (or partial genomes); a ploidy level above haploid at the outset must exist for at least those parts involved in the exchange. It may involve exchange at low frequency between matching chromosomes, or even short duplicated segments in somatic tissue, but it is a vitally important part of the regular process of reduction to haploidy at meiosis in eukaryotes.

Recombination in eukaryotes takes several different forms, including, impor-

tantly, (1) the *independent assortment* at meiosis of genes carried by the various different nonhomologous (heterologous) chromosomes of eukaryote genomes; (2) *crossover reciprocal exchange* of corresponding parts of homologous chromosomes, a process apparently essential to normal chromosome distribution during meiosis in many organisms by way of chiasma formation, and (3) *conversion*, which involves the substitution of a short portion of a chromosome for its counterpart in a homologue without reciprocal exchange. The mechanisms of these processes are understood in varying degree. In this chapter we outline current understanding of these forms of recombination as well as the context in which they occur, and also point out unsolved problems, some of which are colossal.

Independent assortment was first described by Mendel (1866). It was the basis for his so-called Second Law. The dihybrid crosses between heterozygotes he made resulted in the familiar 9:3:3:1 ratio. Since he did not have knowledge of chromosomes, he could go no further than stating that in crosses involving two pairs of marked genes, the contrasting characteristics assorted independently.

In 1905 Bateson, Saunders, and Punnett described a case of **linkage** in the sweet pea. A cross between plants with purple flowers and long pollen grains to plants with red flowers and round pollen grains gave F_1 progeny all purple and round, thus establishing dominance of purple and round. However, the F_2 progeny ratio did not obey Mendel's second law. Instead, a ratio of 11 purple long: 1 purple round: 1 red long: 3 red round was obtained. Purple and long appeared to be "linked," as did red and round. The explanation for this linkage began to be developed by T. H. Morgan as a manifestation of genes being linked on the same chromosome (Morgan, 1911; Morgan and Lynch, 1912). Morgan's findings with *Dro-*

sophila melanogaster led to the two hypotheses, soon to be verified: (1) that linked genes can be separated by crossing-over, or reciprocal exchange, between homologous chromosomes, and (2) that the number of sets of linked genes should correspond to the number of chromosomes in a genome. Sturtevant (1913a, b) next showed by using crossover data from crosses with *Drosophila* that genes are linked in linear order on the chromosomes; this led to the construction of genetic maps as described in Chapter 8.

Conversion was first described as gene conversion by Winkler (1930). The phenomenon, which results in the unequal recovery of marker alleles in the offspring from a single meiosis, was generally ignored by geneticists until Lindegren (1953) clearly demonstrated its occurrence in *Saccharomyces*. A heterozygous yeast cell, *Aa*, should, upon undergoing meiosis, produce a one-to-one ratio of *A* and *a* haploids for each ascus. But in organisms such as yeast and other ascomycetes, in which all four products of a single meiotic event are kept together in ascus sacs, *Aa* zygotes may produce haploids in the ratios of $3A:1a$ or $1A:3a$, as well as the expected 1:1 ratio. The 3:1 and the 1:3 ratios are evidence of conversion. Since conversion at a single locus usually occurs at a very low frequency of about 0.1%–0.4%, it would not be recognized if the meiotic products were simply a random sample, as is usually the case.

I. MEIOSIS AND RECOMBINATION INVOLVE COMPLEX CHROMOSOME MANEUVERS AND STRUCTURES THAT ARE LITTLE UNDERSTOOD

Meiotic recombination has been considered to represent the crucial component of sexual reproduction, the element capable of conferring potential adaptive value to changing environment through enhanced variability. A variety of models (to be outlined) have been proposed to account for the evolution of meiosis itself. Much of the intricate cellular machinery utilized at meiosis is essentially the same throughout the eukaryote kingdoms with few exceptions. This is commonly taken as evidence that the processes and structures involved arose early and have been strongly conserved.

Meiosis consists of the two divisions that typically precede gamete or spore formation in eukaryotes (Fig. 4.1). The second division is superficially similar to a mitotic division, but the first division entails chromosomal maneuvers that lead to the reduction in chromosome number from diploid to haploid level, so that each gamete or spore at the end of meiosis contains one of every kind of chromosome. The occasional errors that occur generally lead to nonviable or defective offspring. One round of chromosome replication precedes the first division, and the second division follows the first without further chromosome replication. Gamete nuclear fusion at fertilization eventually restores the chromosome number to diploid level at a subsequent stage in the life cycle, which, depending on the organism, occurs immediately following meiosis (diplonts), immediately preceding meiosis (haplonts), or at some other period between meiosis and fertilization (diplohaplonts). Meiosis and fertilization are the counterbalancing ingredients of sexual reproduction.

Evidence exists that the process of DNA synthesis at premeiotic S phase differs from that at premitotic S in several ways. Fewer sites of initiation seem to be utilized, so the synthetic period required is prolonged (Callan, 1972). In addition, DNA synthesis of short segments scattered throughout the genome at least in some organisms seems to be delayed until prophase of the first meiotic division (reviewed by Stern

Meiosis

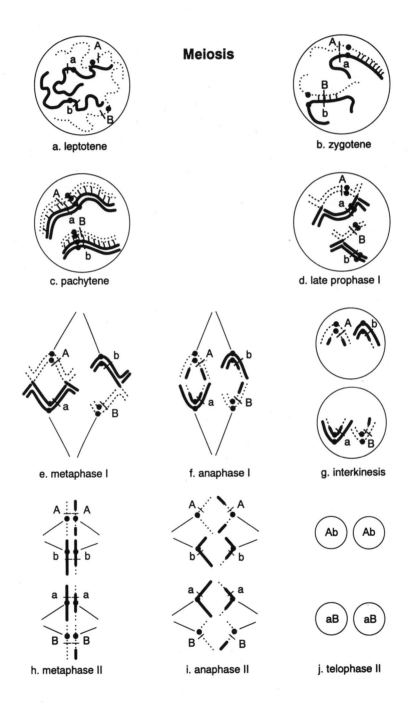

a. leptotene

b. zygotene

c. pachytene

d. late prophase I

e. metaphase I

f. anaphase I

g. interkinesis

h. metaphase II

i. anaphase II

j. telophase II

and Hotta, 1987). It has been suggested that such delayed DNA synthesis may function importantly in the subsequent meiotic processes.

A. Prophase I and the Synaptonemal Complex

In early prophase of the first meiotic division partially condensed sister chromatids generally become at least segmentally and often entirely bound to each other by a proteinaceous structure called the **axial core**. This axial core is visible with electron microscopy (EM) resolution in favorable material. It is destined to become a fully extended lateral element of a structure virtually universal in early prophase meiocytes, known as the **synaptonemal complex** (SC) (reviewed by Westergaard and von Wettstein, 1972; Wettstein et al., 1984; Gillies, 1984; Loidl, 1991). At **synapsis**, each chromosome approaches and pairs with its matching mate or homologue along its length. When corresponding parts of homologues reach a proximity of about 300 nm at this stage, the central proteinaceous component of the SC is formed between the two lateral elements, binding them together. When synapsis is complete, the SC generally extends from end to end of each of the pairs of associated homologues, which are now called **bivalents**. The tripartite completed SC consisting of two lateral elements and a central element can be seen with EM resolution at this stage (Fig. 4.2). There is remarkable structural similarity among the SCs of all the widely divergent organisms in which it has been studied, although details are somewhat variable. Some fluctuations in structural detail occur in the course of stage progression and localized differentiations have been observed along the length of completed SC (reviewed by Dresser, 1987).

Reciprocal recombination via crossing-over is initiated probably at (or just before) the early stages of installation of the SC central component, or during the residence of the completed SC structure, or both. Conversion is probably also initiated during this time frame, and these two forms of recombination are thought to share some component mechanisms, although they probably do not occur concurrently. Both are probably completed during the presence of the SC. Special structures called **recombination nodules** (RNs) (Fig. 4.2a) can be seen during the synaptic stages with EM resolution, often apparently spanning

---

Fig. 4.1. *Diagrammatic representation of meiosis I* (**a–f**) *and meiosis II* (**h–j**) *(omitting prophase II). Two homologous pairs of chromosomes are shown that differ in total length, centromere position, and genes, with one pair carrying gene A/a and the other gene B/b. In each case the chromosome carrying the dominant allele is indicated by dotted lines, and the chromosome carrying the recessive allele is indicated by solid lines. Centromeres are shown as closed circles. The synaptonemal complex (SC) is indicated by crosshatch lines. At leptotene and zygotene* (**a, b**) *chromosomes are seen as single lines although they are known to be double (already replicated for the most part), because their doubleness cannot be visualized. Note especially the following points: Homologous chromosomes are held together by the SC during zygotene and pachytene, and by chiasmata (the result of crossovers) following pachytene until anaphase I. Centromeres are not located on the equatorial plate at metaphase I; instead centromeres are already located in poleward positions at this stage, and chiasmata are located on the equatorial plate, constraining the separation of homologues at this stage. At telophase II if, as shown here, no chiasma has been located between the genes under consideration and their respective centromeres, from any single meiosis only two of the four possible genotypes with respect to the two genes will occur. The other two possible genotypes would result from a similar meiosis in which one of the two pairs of associated homologues had the opposite orientation on the metaphase I plate. That those two kinds of orientation occur at random is the source of independent assortment, with the four possible products occurring in equals number from the whole population of cells undergoing meiosis: 1/4 Ab, 1/4 aB, 1/4 AB, 1/4 ab.*

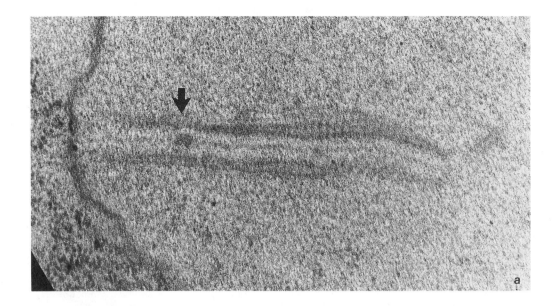

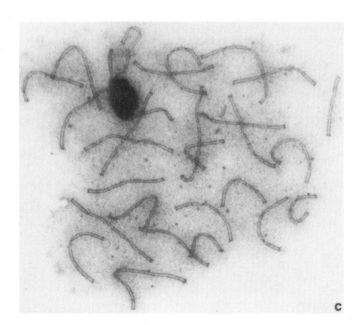

Fig. 4.2. *Electron micrographs of synaptonemal complex structures.* **a.** *Thin section of SC from Sordaria humana. Lateral elements, central element and a recommendation nodule (RN) (arrow) are visible. Reproduced from Zickler and Sage (1981) with permission of the publisher.* **b.** *Thin section of SC from a Bombyx mori 2n male showing central region, lateral elements, and associated chromatin. (Courtesy S.W. Rasmussen.)* **c.** *Silver-stained spread SCs from a Bombyx spermatocyte showing entire bivalent complement. Lateral elements are darkly stained, central elements are not visible, and chromatin is diffuse. (Courtesy S. W. Rasmussen).*

the central region of the SC, at positions thought to be potential sites for crossing-over and/or conversion (Carpenter, 1979; Bojko, 1989). The numbers and positions of these RNs at later synaptic stages have been generally reported to correspond to the numbers and positions of **crossovers**, which can be visualized as **chiasmata** at still later meiotic prophase in material with relatively large chromosomes. At the end of the synaptic period the SC seems to be generally degraded and progressively lost from its position between homologues, and the homologues condense drastically but remain bound to each other at chiasma positions throughout the remainder of prophase I. This chiasma-related binding together of homologues apparently serves to promote correct **disjunction** at anaphase.

Although the SC is most commonly found to associate homologous chromosome regions, it is also found occasionally between nonhomologous segments. The pattern of the pairing found is consistent with the notion that at synaptic stages the SC will be formed between any two chromosome segments that become aligned at a distance as close as 300 nm, and that such alignment is most efficiently achieved by homologous parts, so that these are the portions most frequently associated. Although some limited nonhomologous synapsis has been found at early synaptic stages, particularly in relatively short "foldback" configurations (Fig. 4.3), the preponderance of nonhomologous synapsis seems to occur late and between parts that have failed to synapse homologously. However, a startling phenomenon has been reported in some mammals such that chromosome parts that were synapsed homologously at early meiotic stages later seem to change their pairing mode to nonhomologous. This is called **synaptic adjustment**. Probably the clearest examples have been illustrated in mouse spermatocytes in chromosome regions heterozygous for an inversion (Moses et al., 1982). At first, homologously synapsed loop configurations (Fig. 5.15) are found in the heterozygous inversion region, but at later stages no loops can be found, so the paired chromosomes have the appearance of homozygous normal sequence bivalents. At these later stages, then, synapsis must be nonhomologous in the inverted region, as a result of an active alteration from apparently initial homologous synapsis. This synaptic adjustment phenomenon is not understood and seems to be absent in maize stocks heterozygous for an inversion that have been intensively studied in this respect (Anderson et al., 1988).

Meiotic crossing-over and conversion are presumed to occur at the synaptic stage, since close association of matching parts is essential to both these processes.. They in fact probably require complementary DNA base pairing for segments of substantial length. Conceivably such base pairing could take place at any stage during which matching parts are closely associated, including the entire time period from first association in early **zygotene** until synapsis is reversed at the end of **pachytene** (see Fig. 4.1). Commitment to crossing-over may not occur until after synapsis is complete at the start of pachytene, or it may be initiated during zygotene. Detailed studies (Henderson, 1988) report that crossover frequency can be experimentally altered at four different stages during the progress of meiosis: at premeiotic S phase, at pairing stage, during synaptic completion, and during very early pachytene. The strongest effect was found at pairing initiation, and the least effect was found at very early pachytene. There is good reason to believe, as will be described later, that the crossover process normally requires a series of steps. Very early steps in preparation for crossover events may occur during the interphase preceding meiosis or even in earlier cell generations. Other steps are thought to

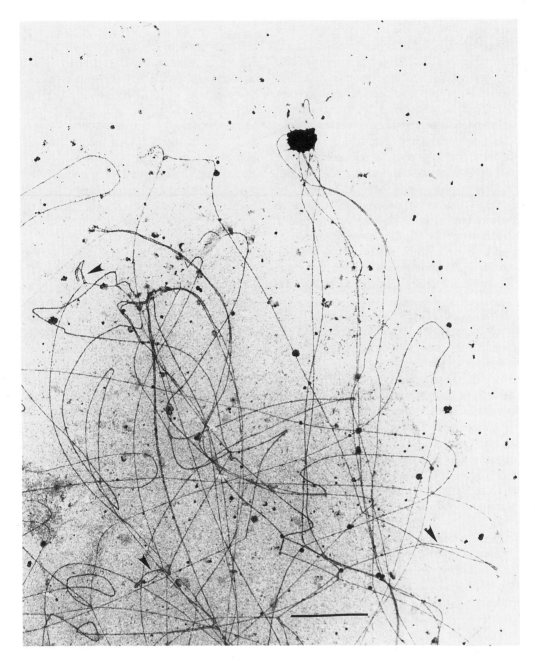

Fig. 4.3. *Electron micrograph of silver-stained axial cores at zygotene in a Tradescantia ohiensis microsporocyte. Homologous synapsis has been initiated at a few locations; nonhomologous, short-interval foldback synapsis can be seen here at three locations marked by arrowheads. (Courtesy Clare A. Hasenkampf.)*

occur in sequence with meiotic progression. Precondition events in fact include homologue pairing. Studies of meiotic mutant effects, especially in *Drosophila* and yeast (Baker et al., 1976; Carpenter, 1984), have helped to establish the sequence. The resolution of certain intermediates may occur in alternative directions, one leading to crossing-over, the other to conversion only, which does not yield chiasmata. Another late step apparently involves repair of mismatched bases, and in some configurations this can give rise to increased recombination. Final commitment to crossing-over is not complete as long as further change is possible, although potential sites for crossing-over may be established earlier.

Plausible fine-structure models of the crossover event are presented on p. 163 *et seq.*, but at this point, an overview of the understanding of the process is important, since it is often erroneously diagrammed. In effect, breaks occur at corresponding positions in nonsister chromatids and the broken ends are rejoined as indicated in Figure 4.4a. During this process, with the exception of the limited region surrounding the breaks, sister chromatids within each homologue appear to remain tightly associated with each other. Indeed they are thought to be strongly associated along their length by the lateral element of the SC. Therefore, the commonly presented diagram of Figure 4.4b is erroneous, as a moment's reflection indicates, for it includes no breaks and reunions. So far it has not been possible to visualize crossovers at pachytene stage, even with EM resolution, although they are thought to be present by the end of pachytene or earlier. They are visualized readily, especially in organisms with large chromosomes, during the remainder of meiotic prophase I, and at metaphase I, as chiasmata. Photomicrographs of chiasmata are presented in Figure 4.5.

Chiasma

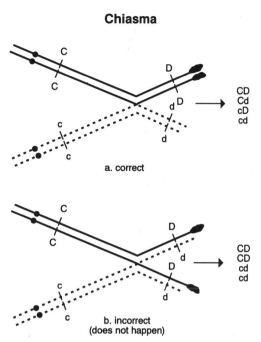

Fig. 4.4. *Correct diagram* (**a**) *and incorrect diagram* (**b**) *of a chiasma. Configurations as seen at late prophase I of meiosis in heteromorphic homologues and the products that would result at the end of the meiosis. One homologue is indicated by solid lines, the other by dotted lines. One homologue carries dominant alleles (C, D) the other recessive alleles (c, d). Centromeres are indicated by closed circles. A heterochromatic region present on the solid-line homologue but absent from the other is indicated by a large dark protruding region at the right end. Sister chromatids are consistently adjacent in the correct version, but separated distal to the crossover in the incorrect version. If the latter kind of separation occurred, it could be visualized in material heterozygous for such heterochromatic regions.*

B. Metaphase I

At the end of prophase I, the nuclear membrane system generally seems to disintegrate, and the bivalents (pairs of associated homologues) become positioned on the plate of the metaphase I spindle, in such a way that pairs of sister chromatid centromeres are not only oriented

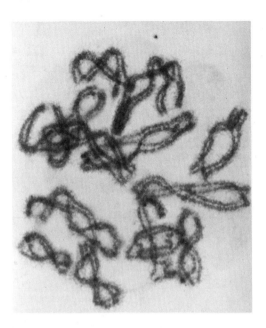

Fig. 4.5. *Photomicrograph of bivalents of Oed-ipina uniformis at late prophase I of meiosis in an acetic-orcein squash. Points of association of homologous chromosomes generally repre-sent chiasmata. (Courtesy James Kezer.)*

to opposite poles but have approached these poles as much as the existing chiasmate associations permit (Fig. 4.1). The meta-phase I resting position of each bivalent on the spindle apparently represents a compromise resolution of forces directed to opposite poles (Nicklas, 1988, Nicklas and Kubai, 1985). For each bivalent the poleward orientation of the components of each homologue seems to be determined by the course of development of poleward tensions without regard to parental source, such that commitment to an independent assortment of chromosomes occurs at this stage. The outcome is exchange or recom-bination of the sequences of maternal and paternal alleles located on different chro-mosomes, so that gametes produced can have different combinations of alleles from those of the two parental contributions. A large part of the total meiotic recombi-nation thus results from this independent assortment. Consider, for example, the fact that if each pair of homologues in a human diploid set bears at least one gene with two different alleles, then $2^{23} = 8.4 \times 10^6$ different haploid gametes are possible.

C. Anaphase I

At anaphase I, chiasmate associations are broken or released in some unknown way, and pairs of chromatids, termed dyads, with sister centromeres associated and leading, proceed to opposite poles of the spindle. It is important to note that if chiasmate associations had not been es-tablished earlier and maintained until this stage, bivalents would have separated to single chromosome univalents at the end of the synaptic period. In normal material this happens only rarely. When univa-lents occur, they usually are randomly distributed at anaphase I, or they lag and separate equationally, leading to irregu-larity of chromosome distribution at the second meiotic division. In either case, production of aneuploid gametes with missing or extra chromosomes is a notable consequence. In most organisms that have been studied in this respect, at least one chiasma per bivalent is usually formed, even in relatively short bivalents, so uni-valent formation does not occur. Also, al-though exceptions have been noted as in the male of *Drosophila*, an apparent chiasma usually represents at least one crossover event. Separating pairs of dyads then typically carry reciprocal exchange chromatids so that distal to the position of a single crossover exchange, nonsister chromatids travel together, resulting in equational separation at anaphase I for such chromosome regions. This can be ob-served in cases of heteromorphic homo-logues (Fig. 4.6), where there is hetero-zygosity for cytological markers, and an opportunity therefore exists for cytologi-cal estimation of crossover frequency.

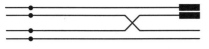

a. heteromorphic homologues with proximal chiasma

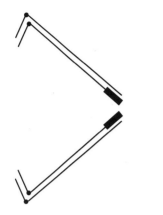

b. anaphase I equational separation of heteromorphic dyads

Fig. 4.6. a. *Diagram of a chiasma proximal to a terminal heteromorphic region in a pair of homologues at late prophase I of meiosis.* **b.** *The resulting equational separation of the heterochromatic region sister chromatids at anaphase I.*

D. Telophase I and Division II

At telophase I, apparently complete nuclear membranes are generally reestablished, and chromosomes decondense somewhat; cytokinesis usually occurs to generate two separate cells. Each of the nuclei then normally progresses through prophase II recondensation of chromosomes, and the dyads in each nucleus become briefly oriented on the metaphase II plate, followed by anaphase II chromatid distribution to opposite poles and the final formation of four separate cells, each with the haploid chromosome number and a DNA content of 1C.

E. Oogenesis

In the meiosis of most female animals (Fig. 4.7) there is unequal division of the cytoplasm at the end of meiosis I such that most of the cytoplasm is incorporated into one of the daughter cells; the other becomes the first **polar body**, which commonly disintegrates without going through meiosis II. The meiosis I product with most of the cytoplasm progresses through meiosis II, followed again by uneven cytokinesis. The meiosis II daughter cell with most of the cytoplasm becomes the ovum, or functional female gamete, while the other daughter cell, known as the second polar body, generally disintegrates. Thus, female meiosis in animals usually produces only a single functional gamete. In plants, mitotic divisions at the haploid level after meiosis generally form gametophytes of varying complexity, and gametes are finally produced following a mitotic division.

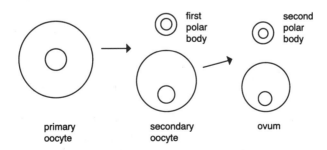

first polar body

second polar body

primary oocyte   secondary oocyte   ovum

Fig. 4.7. *Diagram indicating unequal division of the cytoplasm at meiosis I and II in animals.*

F. Variations on the General Meiotic Theme

Exceptional chromosome behavior has been observed at meiosis in a variety of organisms, and these unorthodox forms occasionally provide sources of insight into the more usual forms of behavior (Wise, 1988). An exceptional behavior occurs in spermatocytes of the Dipteran Sciara, in which one complete set of chromosomes (the set contributed by the male parent) is eliminated and only the maternally contributed set of chromosomes is passed on to the next generation. It has been reported that the two genomes occupy separate and different parts of the nucleus at meiotic prophase I (Kubai, 1982); their separation is maintained, and homologues do not pair. Another example of unusual behavior occurs in *Drosophila melanogaster* spermatocytes, in which a synaptonemal complex is not formed between the homologues, and crossing-over does not normally occur, although homologues are loosely paired as they probably are in all cells of *D. melanogaster*. In the absence of normal chiasmate association resulting from crossing-over, another mechanism, not well understood, serves to maintain homologue association until correct anaphase I disjunction has been established. Cytologically, the associations superficially resembles crossover chiasmata. In a number of other unrelated organisms, crossing-over does not normally occur in one sex or the other, but a special mechanism appears to have evolved in each case that ensures correct anaphase I disjunction. In the silkworm (*Bombyx mori*), for example, homologues pair in the oocytes and an SC is formed between them, but crossing-over does not normally occur. In this case, the disjunctive mechanism involves exceptional maintenance of the SC into metaphase I (Rasmussen, 1977). In *Drosophila melanogaster* females and also probably in yeast (*Saccharomyces cerevis-*

iae), although crossing-over usually occurs at least once per bivalent, it appears that there is a backup disjunctive system that is not well understood. This mysterious system tends somehow to mediate the distribution of members of pairs of chromosomes of similar size and shape to opposite poles at anaphase I when they have not been associated by a chiasma (reviewed by Grell, 1976; see also Dawson et al., 1986). Such chromosomes, which have not been involved in crossing-over, are referred to collectively as the distributive pool in these cases. Their distributions are probably mediated by spindle interactions in which segregating elements are bonded by interchromosomal microtubules, and small chromosomes may form spindle associations earlier than large ones (reviewed by Carpenter, 1991).

When multiple crossovers occur within bivalents, they usually do so very much less frequently in short regions than random expectation predicts, an effect that tends to disappear over longer regions. This phenomenon is entitled positive crossover (or chiasma) interference and has the apparent practical effect of preventing close positioning of adjacent crossovers. This interference appears to be lacking in species of several fungi (e.g., *Aspergillus* and *Schizosaccharomyces*), but otherwise it seems to be virtually universal in eukaryotes.

II. MOLECULAR MODELS OF CROSSING-OVER AND CONVERSION ARE FITTED TO GENETIC DATA

The molecular mechanisms of crossing-over and conversion are unknown. Working models (reviewed by Orr-Weaver and Szostak, 1985) were developed only after phenomena seemingly in defiance of Mendel's law of equal segregation were noted and appreciated, following **tetrad anal-**

ysis of meiotic products from certain fungi. Fungi such as *Neurospora crassa* maintain the four products of each meiosis in a bag called an ascus. Some of these fungi (*Neurospora* included) also keep the four products in an order such that sister cells of the second meiotic division are adjacent in a column, and a normal mitotic division follows meiosis, retaining the ordered arrangement so that a column of eight ascospores results. Adjacent pairs of ascospores are sister cells of the mitotic division that followed meiosis and are therefore normally expected to be genetically identical. Overall, then, from a heterozygous meiocyte ($+$/m) a 4$+$:4m segregation is expected according to Mendel's laws. However, with a frequency typically of the order of about 1/1,000, aberrant asci were found with respect to segregation of a specific locus, although other heterozygous loci in the same ascus could be seen to follow normal Mendelian expectations. Therefore the linear order of ascospores within the ascus had been maintained. The aberrant asci contained ratios of 6$+$:2m, 2$+$:6m, 5$+$:3m, 3$+$:5m or 4$+$:4m with an aberrant order as shown in Figure 4.8. The 6$+$:2m and 2$+$:6m asci were called **gene conversion** asci, and the 5$+$:3m and 3$+$:5m asci were called half-conversion asci. The 5$+$:3m, 3$+$:5m, and 4$+$:4m aberrant asci contained pairs of sister spores that differed genetically and therefore showed genetic segregation following the mitotic division. This was called **postmeiotic segregation**, a very puzzling phenomenon when first observed. The frequencies of aberrant asci, while low, were too high to be explained as mutation events, and new alleles were not found. However, asci that were aberrant for a specific locus were found to be reciprocally recombinant for closely flanking linked heterozygous alleles about half the time (Fig. 4.9). In other words the production of an ascus that was aberrant for a specific locus was found to be frequently associated with a cross-

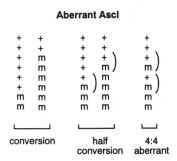

Fig. 4.8. *Spore contents of ordered linear asci showing various types of aberrant segregation from a $+$/m heterozygote following meiosis and a subsequent mitotic division. Products are kept in order such that adjacent pairs of spores are sister cells of the mitotic division, and products from sister cells from the second meiotic division are together within a hemisphere. Bracketed pairs show postmeiotic segregation. Some other sequences of spores not shown here are also possible for each of the categories.*

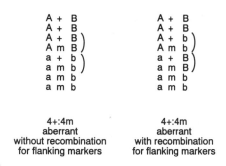

Fig. 4.9. *Asci with 4:4 aberrant segregation with and without recombination of flanking markers from a meiocyte that was A$+$B/amb. Brackets mark spore pairs showing postmeiotic segregation. Note that in the ascus with reciprocal combination for flanking markers, spores in the pairs with postmeiotic segregation are also the spores that are recombinant for the flanking markers. This is consistently the case and provides evidence that crossing-over and aberrant segregation are physically associated.*

over event. This relationship provided clues for the development of models for the molecular mechanism of crossing-over. It was realized that postmeiotic segregation would be an expected result of the formation without correction of hybrid or **hetero-duplex DNA** at the time of a prospective crossover event. Nonsister chromatids in a meiocyte heterozygous for a pair of alleles ($+/m$) should differ by at least one base pair. If corresponding portions of complementary DNA strands from two such chromatids were hydrogen-bonded, this would constitute a heteroduplex with mismatched bases at the site of heterozygosity but complementary base pairs elsewhere. The two strands of such a heteroduplex would be finally segregated to separate cells after the DNA replication for the mitotic division that follows meiosis. Without correction of the mismatched base pairs such events give rise to postmeiotic segregation that is scorable in the ascus of organisms that produce ordered octad asci such as *Neurospora* or scorable as sectored colonies produced directly by meiotic products in yeast.

It was realized that various DNA strand breaks and reunions, in conjunction with heteroduplex formation and the options of repair or no repair of mismatched bases, all together could account for gene conversion, crossing-over, and postmeiotic segregation, depending in each case on the details of the sequence of events. Modes of repair of mismatched bases are illustrated in Figure 4.10. Three molecular models designed to account for either gene conversion or crossing-over are illustrated in Figures 4.11, 4.12, and 4.13. These are the **Holliday model**, the **Meselson–Radding model**, and the **double-strand break repair model**. With a process called **isomerization** (Fig. 4.14), crossing-over with recombination of flanking markers results. Without it there may be simple gene conversion, half-conversion, or $4+:4m$ aberrant segregation, depending on whether there is repair or no repair of mismatched bases and on the direction of repair in heteroduplex DNA. One of the possible outcomes produces no visible manifestation of these events. This occurs

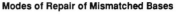

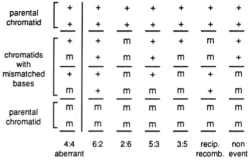

Fig. 4.10. *Modes of repair of mismatched bases showing the various categories of outcome. Each $\pm$ or $\underline{m}$ represents bases for wild-type or mutant, respectively. Adjacent pairs of bases in a column represent chromatids; top and bottom chromatids in each column represent the parental chromatid not involved in the recombination, whereas the central two chromatids in each column were involved in heteroduplex formation. The left-most column shows 4:4 aberrant segregation; without repair of mismatched bases, each strand of the heteroduplex in this case is simply used as a template in the replication following meiosis. In the other columns various repairs occurred before replication, and each strand was promoted to a full DNA molecule by the replication. For example, in the column with 6:2 at the bottom, in both involved chromatids a patch of bases including the mutant base or bases was removed, and the remaining complementary patch of bases on the wild-type strand was used as template for repair so that each of these two chromatids became wild-type. Happenings in the other columns can be readily deduced. Note that the right-most two columns show (a) the outcome indistinguishable from the outcome of a crossover event located anywhere between the $+/m$ locus and the centromere, and (b) the outcome expected if nothing had happened to affect segregation at the $+/m$ locus.*

when there is no crossing-over and mismatched base correction restores the original genotype. In the case of the double-strand break repair model, if a marker is located within a double-strand gap, it is converted by a double-strand transfer in the absence of heteroduplex formation. For

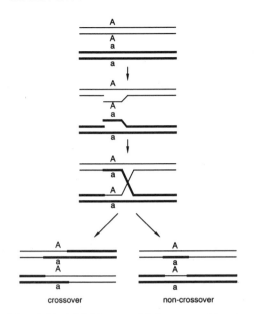

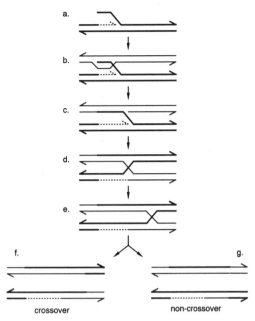

Fig. 4.11. *Holliday model for recombination. Strands of the same polarity are nicked at homologous sites and are then exchanged to produce symmetric heteroduplex DNA. The crossed strand, or Holliday junction, can be resolved either with or without exchange of flanking markers, depending on which two strands are broken and recombined. If the symmetric heteroduplex produced by strand exchange is not repaired, an aberrant 4A:4a segregation results. After Orr-Weaver and Szostak (1985).*

Fig. 4.12. *Meselson–Radding model.* **a.** *Recombination is initiated by a single-strand nick that serves as a primer for DNA repair synthesis. This displaces a single strand that can then pair with a homologous region on the other chromatid.* **b.** *The resulting D-loop is degraded, and the asymmetric heteroduplex DNA is enlarged by DNA synthesis on the donor chromatid coupled with degradation on the recipient duplex.* **c.** *Branch migration and ligation of the nicks produces a Holliday junction that can be isomerized.* **d.** *Symmetric heteroduplex DNA can be formed by branch migration of the Holliday junction.* **e.** *Resolution can yield either the crossover (**f**) or the noncrossover (**g**) configuration, depending on which two strands are broken and recombined. After Orr-Weaver and Szostak (1985).*

this portion, then, there is no postmeiotic segregation. The Holliday model was the earliest of the three to be proposed; the other two represent refinements to accommodate ever-increasing genetic information (reviewed by Orr-Weaver and Szostak, 1985). It is important to realize that the model diagrams (Figs. 4.11, 4.12, and 4.13) do not show chiasmata, where sister chromatids are closely associated. Only the two chromatids involved in the recombination processes are diagrammed, and sister chromatids that are not involved are omitted. In Figure 4.15 all four chromatids are diagrammed for bivalents destined for aberrant segregation with and without reciprocal crossing-over on the

basis of the Holliday model. Similar adjustments can be added to the other models as well.

Heteroduplex DNA may be formed on two chromatids, in which case it is called symmetric heteroduplex, or on only one chromatid, called asymmetric heteroduplex. Closely linked genetic markers are often gene-converted together, an outcome called co-conversion. If conversions

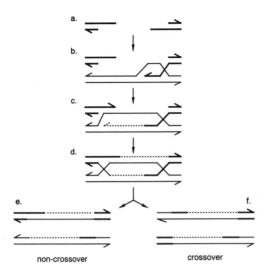

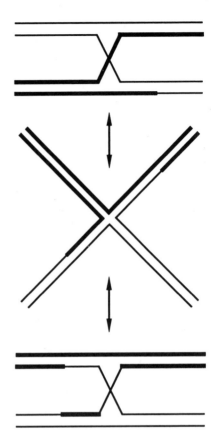

Fig. 4.13. *Double-strand break repair model.* **a.** *A double-strand cut is made in one duplex, and a gap flanked by 3' single strands is formed by the action of exonucleases.* **b.** *One 3' end invades a homologous duplex, displacing a D-loop.* **c.** *The D-loop is enlarged by repair synthesis until the other 3' end can anneal to complementary single-stranded sequences.* **d.** *Repair synthesis from the second 3' end completes the process of gap repair, and branch migration results in the formation of two Holliday junctions. Resolution of the two junctions by cutting either inner or outer strands leads to two possible noncrossover* **(e)** *and two possible crossover* **(f)** *configurations. In the illustrated resolutions, the right-hand junction was resolved by cutting the inner, crossed strands. After Orr-Weaver and Szostak (1985).*

Fig. 4.14. *Isomerization of a Holliday junction. The Holliday junction can isomerize through a symmetrical intermediate without bond breakage. Therefore, resolution can occur by cutting either the originally crossed strands or the noncrossed strands. After Orr-Weaver and Szostak (1985).*

to wild-type, and to variant form, are equal in frequency, the situation is described as **parity; disparity** implies unequal frequency of conversion in the two directions. Either condition may prevail. The frequency of aberrant segregations varies with different genes and even with different sites in the same gene, which may also vary with respect to the forms of aberrant segregations that are predominant. Frequencies of gene conversion at sites within a gene also tend to vary with the position of the sites, often but not necessarily with

frequencies higher at one end of the gene and decreasing toward the other end. This effect is called **polarity**. It is thought to reflect the existence of fixed sites for initiation of recombination in such a way that probability of inclusion of a site in recombination events declines with increasing distance from the initiation position. The frequency of association of crossing-over with aberrant segregation varies widely. For example, crossover frequencies associated with conversions of different alleles of the *arg 4* locus in yeast have been found

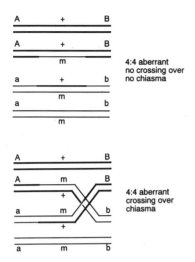

Fig. 4.15. *Bivalents destined to have 4:4 aberrant segregation on the basis of the Holliday model. In the upper bivalent the Holliday structure was resolved in such a way that there is no crossover or chiasma formation. In the lower bivalent, isomerization of the Holliday structure occurred so that breaks and reunions gave rise to crossing-over and chiasma formation. The inclusion of the two uninvolved parental chromatids allows visualization of the chiasma. Note that at both ends of these bivalents sister chromatids are together; they are bound together in each homologue by a lateral element of the SC, an association typical of sister chromatids at pachytene.*

to range from 18% to 66% (Fogel et al., 1979). Crossover frequencies at some loci also vary with the type of aberrant segregation. Where crossing-over is associated with aberrant segregation in the case of 4:4 aberrant segregation (Fig. 4.9), it was readily deduced that the two chromatids involved in postmeiotic segregation were also the two reciprocally recombinant chromatids of the crossover event (Kitani et al., 1962). This information, where the nature of events is most visible, provides strong support for the inference of physical association of crossover and aberrant segregation events. It is also of interest that conversion events for adjacent loci associated with a crossover show chiasma interference, but such conversion events without an associated crossover apparently do not (Mortimer and Fogel, 1974). In other words, crossover events do not ordinarily occur in each of two adjacent loci, but conversion events may be so positioned.

In yeast informative experiments on the stage of commitment to recombination have been possible because meiosis can be induced by exposure to sporulation media where meiosis occurs, and the return to vegetative media after various intervals in sporulation media yields varying results. After prolonged exposure to sporulation media, diploid cells are found to be committed to meiosis completion and spore formation before resuming vegetative growth. It has been learned that with shorter periods in sporulation media the commitment to meiotic recombination occurs simultaneously with or shortly after commitment to the premeiotic DNA synthesis. Interestingly, although crossing-over and chiasma formation seem to be essential to the completion of a normal reductional first meiotic division, commitment to recombination is separable from commitment to meiosis I, which appears to be a subsequent event. In yeast it is therefore possible to obtain meiotic levels of recombination in the absence of a reduction division, by the appropriate media manipulations.

Fungal genetic studies have yielded a great deal of insight into the processes of recombination. The question naturally arises as to how generally applicable inferences from such studies are to higher eukaryote systems, for which the tools of tetrad analysis are not available. Similar though less detailed findings, however, have resulted from studies with *Drosophila*, in which the techniques of half-tetrad analysis allow the recovery of two copies of one chromosome arm attached to the same centromere following sexual reproduction (Chovnick et al., 1970). Locations of crossover junctions (in cases where reciprocal exchange products were recovered) and also locations of probable gene conversion tracts (from base mismatch repair) have recently been determined by DNA sequencing techniques and denaturing gel electrophoresis (Curtis et al., 1989). Approximately half of the reciprocal crossover events were found to have such conversion tracts at the crossover junction, as expected in these studies, if there is no bias in the direction of conversion, and half were converted to the undetectable parental form, and all crossovers were associated with the generation of heteroduplex DNA.

Enzymes associated with recombination in eukaryotes are largely unknown (reviewed by Orr-Weaver and Szostak, 1985). Probably the best characterized is the *rec 1* normal gene product of *Ustilago maydis*, the maize smut fungus. The activities of this enzyme seem to be similar to those of the *E. coli rec A* protein, which catalyzes homologous DNA strand transfer with a requirement of ATP presence and of a polarity in the direction of strand transfer (Kmiec and Holloman, 1984). The *rec 1* protein, for example, catalyzes the pairing of a single strand circle with the 5′ end of a linear duplex, displacing the 3′ end of the duplex. This directionality is opposite that of the *rec A* protein. The formation of paranemic joints, in which the single-stranded DNA pairs but does not interwind with its complement in the duplex molecule, seems to be coupled to the formation of left-handed Z-DNA. In *rec 1* mutant strains, both recombination and some pathways of repair of DNA damage are blocked. A variety of evidence is associated with apparently relevant enzymes that are less well known. Deficiency for an endoexonuclease present in wild-type cells accompanies *RAD52* mutants in yeast. This enzyme shows cross-reactivity with a *Neurospora crassa* single-strand DNA endoexonuclease. The yeast nuclease is induced during meiosis with maximum activity at the time of commitment to recombination. The purified protein in yeast is too large to be encoded by the *RAD52* gene, and therefore may instead be under the regulation control of *RAD52*. A cell-free extract from yeast has been produced that catalyzes both gene conversion and reciprocal recombination between circular and linear substrates (Symington et al., 1983). This provides a valuable tool for analysis.

III. BIOCHEMICAL STUDIES OF MEIOSIS HAVE BEEN INITIATED AND PROVIDE IMPORTANT CLUES

Most of the investigation of meiosis at the molecular level has been accomplished by H. Stern and Y. Hotta and their collaborators, who for the most part have worked with *Lilium* and to a lesser extent with mouse (reviewed by Stern and Hotta, 1987). They have focused on chromosome replication, synapsis, and recombination, with some consideration of the consequences for disjunction. They have noted that extrapolation to other organisms may be risky but that apparently similar processes occur generally in eukaryotes at meiosis and have in fact been required for most sexual reproduction since its beginnings. Since basic mechanisms tend to be evolutionarily conserved, it is argued that similar mechanisms found both in lily and mouse may well have widespread prevalence. What follows in this section, except where otherwise noted, is the work of Stern, Hotta, and collaborators (reviewed by Stern and Hotta, 1987).

A. Premeiotic S Phase and Zygotene

When meiotic cells were compared with mitotic or somatic cells in general, specific proteins and enzymes were found to be present exclusively in meiotic prophase cells, as though their production for this stage had been turned on, and then turned off again with further progression into later meiotic stages. Several striking features of meiosis have been noted at the premeiotic DNA synthesis stage. The duration of the S phase in cells destined to undergo meiosis is typically much longer than that of the S phase in somatic mitotic cells (Callan, 1972). But an especially remarkable feature of the premeiotic DNA synthesis in *Lilium* is that it seems to be incomplete. About 0.1%–0.2% of the genome is apparently not replicated until zygotene. Such delayed replication appears to be restricted to segments 4–10 kb in length, which are broadly distributed throughout the genome. The DNA of these sequences (called **zygDNA**) was found to have low copy number, but it is possible that a small fraction of repeated sequences is normally present within them. These very late-replicating sequences are thought to remain in duplex configuration until zygotene. Further investigation has suggested that a specific lipoprotein (called leptotene or **L protein**) serves as a suppressor of zygDNA replication during the S phase. This protein is present during late premeiotic S phase, and its artificial removal stimulates zygDNA synthesis. The L protein is reported to bind strongly to purified zygDNA and the binding site specificity may be limited to about 1% of that zygDNA. Replication and transcription of the zygDNA seem to occur during zygotene where the L protein again seems to play an important role. It has also been inferred that the zygDNA itself functions not only in chromosome synapsis but also in meiotic chromosome disjunction. The replication of the zygDNA that occurs during zygotene still seems to be incomplete at segment ends. This is inferred from the fact that discontinuities remain that cannot be repaired by ligase only but require polymerase action. It is suggested that

these short unreplicated segments, which are finally closed after pachytene before meiosis I is complete, may function in the maintenance of sister chromatid cohesiveness until correct meiosis I disjunction has been established. Transcription of the zygDNA to **zygRNA** apparently occurs during homologue synapsis. zygRNA is found in rapidly increasing quantities during late leptotene, reaches a maximum at middle zygotene, and then falls sharply again. Its function is therefore thought to be complete by middle pachytene, and since it is found only in small quantities in the case of special hybrid material where synapsis is reduced and abnormal, transcription of zygRNA indeed seems to be coordinated with the synaptic process. Its function is not understood, however. Evidence that the zygDNA sequences have been evolutionarily conserved has emerged from the finding of cross hybridization under moderate stringency conditions between zygDNA and zygRNA sequences in lily and mouse. An additional function of the L protein has been inferred as noted above. The L protein apparently nicks one strand of the DNA when it is bound to zygDNA in the presence of ATP. It is posited that this may serve to facilitate the formation of a single-strand tail that may hybridize to a complementary single-strand tail from the homologous chromosome to yield a heteroduplex. Such a heteroduplex would necessarily be dissociated with replication of the zygDNA in which it resides. It has been further speculated that during the brief interval of existence of the heteroduplex, an SC segment formation is initiated to serve the function of stabilizing homologue alignment. Thus an important function of the zygDNA may be to promote homologue alignment during synapsis.

B. Pachytene

Another set of studies by Stern, Hotta, and collaborators, which focused on the pachytene stage, at which synapsis is complete, has provided evidence for the existence of special repair replication of specific DNA sites at this stage (reviewed by Stern and Hotta, 1987). Pachytene DNA synthesis is judged to be repair synthesis because it is relatively insensitive to inhibition by hydroxyurea and follows nicking of both DNA strands (Hotta and Stern, 1974). Such repair DNA synthesis has been assumed to be related directly to the crossover process. It has been found to be severely reduced or absent in material in which synapsis is reduced or absent, as in certain hybrids. The DNA sequences involved seem to constitute a moderately repetitive fraction, with evidence of evolutionary conservation as noted for zygDNA. These sequences were found in segments of 150–300 bp positioned at opposite ends of an internal DNA segment of 800–3,000 bp. The short segments are referred to as **PsnDNA**, and each set of the two flanking segments together with their internal segment is called PDNA. All or most of the PsnDNA regions seem to be nicked at pachytene by an endonuclease, and these sequences seem to be positioned in chromatin that has been modified to provide accessibility to the endonuclease. In addition, nicking does not occur in the absence of synapsis. The modification of the chromatin appears to be mediated by a small nuclear RNA, called **PsnRNA**, that is homologous to the PsnDNA, together with a nonhistone protein, called Psnprotein, that in fact binds to the PsnRNA. These components are found only during zygotene and pachytene. Apparently the usual chromosomal histones in the PsnDNA regions are replaced by Psnprotein and PsnRNA, beginning during zygotene in the presence of synapsis, and the function of these is to render the PsnDNA accessible to the endonuclease. It has been proposed that the special chromosomal sites serve as potential initiating sites for crossing-over. Such potential

sites are acknowledged to be far more numerous than the number of crossovers eventually formed, and it is reasoned that a large excess of potential sites somehow assures that each pair of homologues will probably become bound together by at least one chiasma. In this way, correct disjunction will tend to prevail later in meiosis.

A group of proteins thought to be relevant to recombination has been found to be present only during leptotene, zygotene, and pachytene (reviewed by Stern and Hotta, 1987). They generally reach maximum abundance during the period from late zygotene to early pachytene. Three such proteins are an endonuclease, a DNA-unwinding protein, and a protein that catalyzes DNA reassociation. The endonuclease produces single-strand nicks in double-stranded DNA but only in DNA in chromatin that has been modified as described above by PsnRNA and Psnprotein, in the presence of synapsis. The nicks are then apparently extended to gaps by an unknown enzyme. The DNA-unwinding protein (called U-protein) unwinds about 400–500 bp from a nick, and this constitutes adequate length for heteroduplex formation. The reassociation protein (called R-protein) binds strongly to single-stranded DNA and mediates double-stranded association. Unlike the other two proteins, the abundant presence of the R-protein seems to require the existence of synapsis.

It is emphasized that the assumption that crossovers are initiated at PsnDNA sites is based on a presumed coincidence of occurrence of crossing-over and PsnDNA repair synthesis, together with the supposition that crossovers require some DNA repair synthesis (although most models call for only very small amounts at crossover and conversion sites). Also, both PsnDNA sites and crossovers appear to be restricted to euchromatin and to synapsed chromosomes. Other plausible explanations for these coincidences have not been ruled out, such as the possibility that repair is necessitated by SC metabolism, the effects of which would be expected to be much more frequent in euchromatin than in heterochromatin, where, SC association per unit chromosome length is minimal because of condensation. But the conjectures present an attractive model. So far, crucial biochemical information on the nature of crossover site determination is lacking.

IV. MAJOR MEIOTIC MECHANISMS REMAIN UNEXPLAINED

The mechanisms of at least three vital steps in the meiotic process have not been resolved: homologue pairing, determination of sites of crossover commitment, and maintenance of chiasmate association until anaphase disjunction has been established. The molecular mechanisms that have been proposed for crossing-over and conversion seem in principle to be at least superficially consistent with a number of observations, although their accuracy remains to be established.

A. Homologue Pairing

The problem of homologue pairing remains a particularly perplexing puzzle. In a number of organisms it appears that homologues are not nearer each other than random distribution predicts in most cells at the onset of meiosis (Walters, 1970; John, 1976). Exceptions include Diptera, in which homologues seem to align at least loosely at zygotic stage and maintain or reestablish this alignment at each mitotic prophase. In the Diptera, then, as well as in organisms with zygotic meiosis, homologue alignment must begin to occur soon after fusion of the nuclei contributed by the two gametes. Probably the homologue pairing process has been most directly observed in *Coprinus*, a fungus with zygotic

meiosis, although observations have so far been limited to fixed and stained material at defined intervals following syngamy (Lu and Raju, 1970). *Coprinus* has fairly large chromosomes that appear to condense progressively during the pairing of homologues. At first, the two input genomes are separately contained in two distinct clusters of chromosomes. These clusters seem to loosen and commingle; then homologues apparently become progressively aligned and synapsed from initiation points usually located, at first, distally but not terminally (Fig. 4.16). Phase microscopy images of living material with synapsis in progress in *Coprinus* are unfortunately not currently feasible. In most organisms chromosomes are very elongate and generally rather decondensed at early synaptic stages, so that bright-field light microscope images are not clear. From EM observations of serial reconstructions of thin sections, and recently especially from EM observation of whole-mount spread preparations of SCs, it has been learned that in a number of organisms synapsis is initiated at multiple points within homologue pairs. In widely divergent organisms such as *Allium* (Albini and Jones, 1987) and *Psilotum* (Anderson and Stack, 1988) SC lateral elements at zygotene appear to be connected at a number of points by filaments usually carrying a nodule indistinguishable in appearance from an RN (Fig. 4.17). Synapsis very likely proceeds to completion from at least some of such points. It is not known whether crossovers occur at some or none of these sites, but known crossover frequencies are much lower than the frequencies of these sites, and it has been suggested that abortive recombination events might tend to be located at such positions (Powers and Smithies, 1986; Smithies and Powers, 1986; Carpenter, 1987). It has been suggested that the first steps in the recombination process occur at this time and are reserved without leaving a lasting recombinant ef-

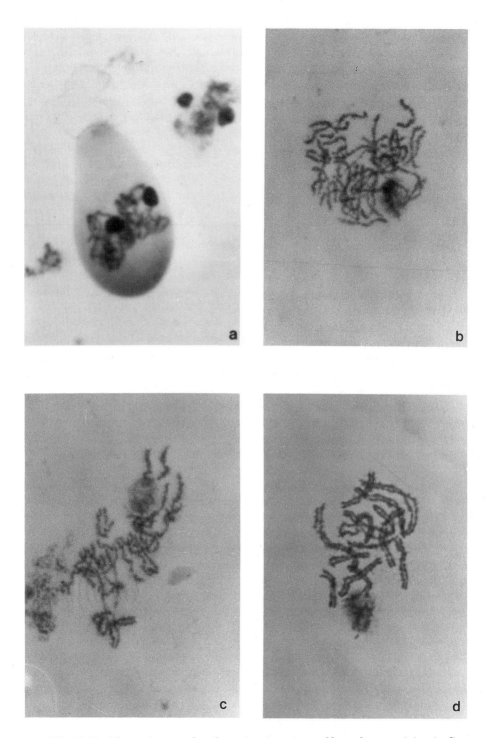

Fig. 4.16. *Photomicrographs of progressive stages of homologue pairing in Co-prinus lagopus. Reproduced from Lu and Raju (1970), with permission of the publisher.*

fect, or they lead to conversion. It is not known whether specific sites for initial association have been evolutionarily differentiated and fixed and therefore represent the positions of DNA sequences specialized for pairing, but limited evidence from other kinds of studies suggests that such sequences exist (Hawley, 1980). In addition, it has been proposed that chromosome segments in which meiotic DNA synthesis is delayed until synaptic stage may represent sites of synaptic initiation established by complementary base pairing (Stern and Hotta, 1987). Another recently proposed scheme envisions four-strand parallel association of guanine-rich DNA sequences, involving a strand from the DNA molecules of each of the four chromatids (Sen and Gilbert, 1988). These four strands would necessarily have the same polarity, and guanines could be bonded together by unconventional Hoogsteen base pairing in such a way that complexes to complementary strands could be simultaneously maintained. Such associations are apparently formed between oligonucleotides at physiological salt concentrations, and it is suggested that homologue pairing in register would be facilitated by this process. Once matching regions to be associated have met with contact, or very close juxtaposition (about 300 nm), a variety of forms of association might conceivably serve the binding function for initiation of synapsis. The more problematical aspect of meiotic homologue pairing is the achievement of the initial positioning of homologous segments in close juxtaposition to allow alignment for synapsis. Telomeres are commonly but not necessarily found to be associated with the nuclear envelope at interphase and confined to an area that varies with the organism from a small proportion to about half of the total area. However, even the smallest area of telomere location (found with "bouquet" formation) leaves large distances in physical

chemical terms between homologous telomeres; when synapsis begins, it does not appear to be initiated at telomeres. Also ring chromosomes, which are devoid of telomeres, as well as heterozygously translocated internal chromosome segments, seem to synapse efficiently. There are, in fact, a number of cases where telomere–nuclear envelope association can be ruled out as a principal source of initial homologue alignment. In addition, it seems unlikely that simple diffusion could generally stir the chromosomes of higher eukaryotes so efficiently that matching parts would reliably meet by chance at early meiotic prophase for initiation of synapsis. Such chromosomes are often relatively immense and extended. Conspicuous heterochromatic regions are not required for homologue pairing; chromosome segments devoid of such heterochromatic regions regularly achieve homologous pairing, but shorter repetitious DNA sequences cannot presently be ruled out as possible specialized pairing centers, when these are brought into close proximity. Similarly, centromeres are not essential to homologue alignment, since heterozygously translocated internal segments without centromeres seem to pair with good regularity. On the other hand, pairing may normally be facilitated to some extent by remnant chromosome arm positioning imposed at anaphase of the last mitotic division and therefore indirectly by centromere positions. Chromosomes usually move very little with respect to each other during interphase; therefore matching chromosome parts tend to occupy matching latitudes at entry into prophase because of the common polar region location of their centromeres. Relative chromosome arm length has in fact figured prominently in a recently proposed model for homologue pairing (Bennett, 1982). From observations of relative distances between centromeres in somatic mitotic metaphase configurations, it has

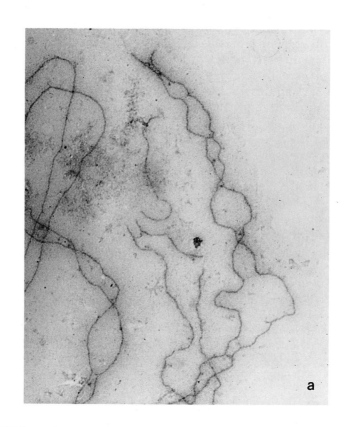

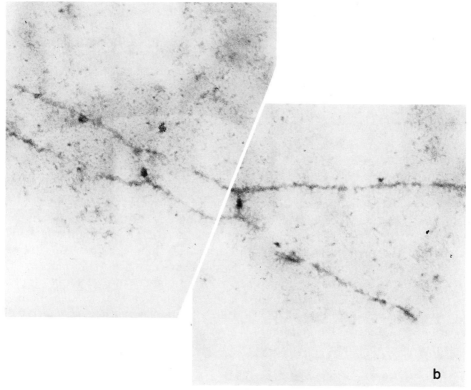

been concluded that the heterologous chromosomes of a genome tend to be positioned in a specific sequence such that long arms of most similar length and short arms of most similar length are adjacent. Although the validity of the statistics on which this conclusion is based has been questioned (Callow, 1984), the attractive suggestion is that at meiosis, homologue pairing is accomplished by appression of the two similar genome sequences. If the chromosome contributions from the two parents are each arranged in the same order in a chain, merely bringing the two chains together accomplishes pairing. But if chromosome arm length plays an important role in homologue pairing, it might be expected that selection pressure would have favored the development of distinctly different arm lengths within genomes to promote error-free pairing; instead species with normal meiosis can have genomes with chromosome arms of very similar length. Also, heterozygous rearrangements that substantially alter relative chromosome arm lengths within a genome do not generally upset homologue pairing at meiosis. If relative positioning of chromosomes within the premeiotic nucleus plays a prominent role in homologue pairing, it is unlikely that the positioning is determined solely by relative arm lengths. Other mechanisms must be sought. Arguably, it is conceivable that although diffusion of chromosomes at the synapsis stage is probably inadequate to provide for haphazard meeting of matching specialized pairing centers in order to initiate synapsis of homologues at several sites dependably, these may be as yet undiscovered mechanisms that somehow serve to stir chromosomes within the nucleus with respect to each other and accomplish this result. Direct observations of living material at this stage are generally frustrated by the usual decondensed state of the chromosomes, but there are reports of signs of chromosome motion (Parvinen and Söderström, 1976). Other scattered reports describe instances of apparent homologue pairing at mitotic divisions immediately preceding meiosis, and it has been suggested that motions of chromosomes at these mitotic divisions might provide opportunities for the strategic establishment of homologue associations at a few specialized locations in advance of meiosis, setting the stage for synapsis.

B. Chiasma Maintenance

Another unsolved but important problem within the meiotic system is concerned with the mechanism of chiasma maintenance during the period between the end of the pachytene stage and the establishment of an orientation of homologous centromeres to opposite poles at metaphase I to provide for normal disjunction. The crossover process *per se* does not produce a structural basis for such chiasma maintenance. In fact there is evidence that separate genetic control of crossing-over and chiasma maintenance exist. Mutants have been studied in maize, yeast, and possibly *Drosophila* that seem to have normal crossover function but impaired chiasma maintenance function (Maguire, 1978; Rockmill and Roeder, 1988). In some organisms the positions of chiasmata seem to shift progressively toward chromosome ends during the course of prophase I and metaphase I, perhaps as a result of condensation. This apparent shifting, termed terminalization, has been shown in some cases to be an observational artifact, and two or more adjacent chiasmata may in fact be seen as one following condensation. But of overriding importance to normal

Fig. 4.17. *Electron micrographs of axial cores with points of apparent association marked by the presence of recombination nodule-like granules.* **a.** *Allium cepa. (Courtesy Albini and Jones, 1987.)* **b.** *Psilotum nudum. Reproduced from Anderson and Stack (1988), with permission of the publisher.*

disjunction is the fact that at least one chiasma per bivalent persists until bivalent orientation on the metaphase I plate.

Darlington hypothesized in 1932 that chiasmata are probably maintained by some process of sister chromatid association, a suggestion that has been widely ignored. Actually, twisting of homologues about each other and relational coiling of sister chromatids are commonly absent, although they have been thought to impose topological constraints to homologue separation during prophase I following crossing-over and disintegration of the SC. The general mode of proposed functioning of sister chromatid cohesiveness as a chiasma binder as illustrated in Figure 4.18 along with possible complications imposed by the presence of various forms of double-crossover events. Several different mechanisms for sister chromatid cohesiveness at meiotic prophase have been suggested. For example, it is conceivable that the chromatids are associated by hydrogen bonds of the DNA helix at positions not yet replicated, for it has been reported that unreplicated gaps may still be present at late prophase I (reviewed by Stern and Hotta, 1987); the potential strength of overall hydrogen bonding that might be available from such a source cannot be presently estimated. Another suggestion is that in the course of normal DNA replication, catenations may be routinely formed at the sites where replication forks meet in such a way that sister chromatids remain associated at these positions until the catenations are resolved, possibly by the function of a topoisomerase activity (Murray and Szostak, 1985). A variety of circumstantial evidence (Maguire 1988b; Rockmill and Roeder, 1988) and some direct observations (Moens and Church, 1979) tend to implicate a conjectured late function of the SC in the provision for persistent sister chromatid cohesiveness before the SC itself disintegrates. An additional possible mechanism for sister

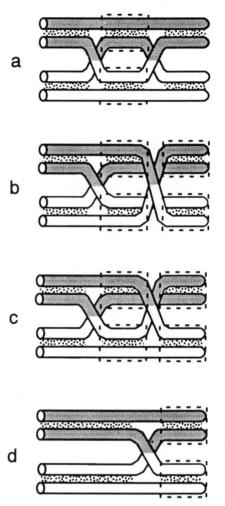

Fig. 4.18. *Diagrammatic representation of bivalents with two-strand double crossovers* (**a**), *four-strand double crossovers* (**b**), *three-strand double crossovers* (**c**), *and a single crossover* (**d**). *In each case one homologue is shaded, centromeres are represented by circles at the left ends, sister chromatid cohesiveness is represented by flecking between sister chromatids, and critical regions for chiasma maintenance are enclosed by dashed lines. After Maguire (1980).*

chromatid cohesiveness has been emphasized by studies of mitotic chromosomes in cultured cells in which apparent associations of sister chromatids were observed at mitotic prophase and metaphase at heterochromatic regions (Lica et al., 1986). While some apparent sister chromatid cohesiveness is a common feature of mitotic prophase and metaphase chromosomes, the special demands of meiosis seem to call for enhanced cohesiveness (reviewed by Maguire, 1990).

C. Establishment of Crossover Sites

An especially puzzling aspect of meiosis is concerned with the mechanism by which crossover sites are established. As stated earlier, where two or more crossover sites are found within chromosome arms, they are typically positioned nonrandomly, that is, farther apart than random expectation predicts. In contrast, conversion events can be closely positioned with respect to each other. This is surprising in view of the common belief that the two kinds of events may represent differing forms of resolution of DNA heteroduplex formation, both of which require intimate homologous pairing and in fact probably complementary base pairing at the DNA level as in Holliday structures. Current evidence is inconclusive as to whether every Holliday-type structure can be resolved either way at some stage following its formation, or each has a distinct destiny in this respect from its start. It is not known whether initiation of Holliday-type structures is mainly restricted to a narrow stage or can occur with equal facility at any phase while matching parts are closely associated. The role of the SC is obscure. Widespread belief favors the view that the structure is essential to meiotic-level crossing-over and that it probably functions by holding homologues in register in close juxtaposition. However, this view has recently been questioned on the basis of observation of "ectopic" crossovers in yeast, in some cases approaching normal meiotic levels, between dispersed repeated copies of DNA sequences that share as little as 2.2 kb of homology (Jinks-Robertson and Petes, 1986; Lichten et al., 1987). In these cases the heterologous chromosomes containing the dispersed sequences were thought to be normally synapsed with their respective homologues during pachytene so that the ectopic crossing-over most likely happened in the absence of SC accommodation. It is inferred that recombination nodules (RNs) (Fig. 4.2) function in various aspects of recombination (Carpenter, 1979), but it is still unclear whether RNs associated with conversion-only events differ from those associated with reciprocal recombination, whether the same kind of RNs are involved with both kinds of recombination, or (perhaps most likely) whether RNs at early stages of their existence may tend to be associated with conversion-only events while persistent, possibly modified, RNs are associated with reciprocal recombination. It is not known whether there is turnover of RNs such that some exist transiently and are lost without association with any form of recombination, but their observed high number at early stages in some cases (Stack and Anderson, 1986) suggests that this is so. In any case by middle to late pachytene, RN number and location have been found to correlate with chiasma number and distribution observed at later prophase stages in a number of organisms. In organisms for which very clear preparations are available, RNs have been observed during the stage of synaptic initiation and extension, often apparently at or near sites of synaptic initiation (Stack and Anderson, 1986). Many of the RNs seen at this stage were apparently precipitously lost by early to middle pachytene so that there was then a number of RNs that corresponded to the number of chiasmata found at later prophase. It was suggested that, as a rule, the RNs present earliest in a region may tend to have the highest probability of being associated with a crossover. Then, to account for crossover interference, it was proposed that the first successful reciprocal recombinant event (crossover) causes a signal to be transmitted in either direction along the SC, releasing in its path RNs that have not yet established crossover intermediates (Stack and Anderson, 1986). This scheme is consistent with the fact that only small numbers of crossovers per bivalent are usually observed, despite the fact that most bivalents contain at least one crossover. If it is additionally supposed that those RNs in the path of the signal that have established preliminary crossover intermediates by that time are limited to mediation of conversion-only events, more nearly random distribution of conversion-only events might be expected. Other models devised to account for crossover and conversion distribution along chromosomes call for completed SCs, extending from end to end of bivalents, to serve somehow as conduits for transmission of overall sensing and regulation of the location of multiple sites on chromosome pairs (Carpenter, 1987). More work is needed. Two organisms stand out as seeming to lack crossover interference. These are *Schizosaccharomyces pombe* (Olsen et al., 1978; Snow, 1979) and *Aspergillus nidulans* (Egel-Mitani et al., 1982). Curiously, no SC has ever been found in either of these organisms, although it has been searched for. It has even been suggested that fully formed SC structure may tend to prohibit crossover site initiation so that it might be extension of SCs that serves as a transmitted signal to restrict the distribution of crossover sites (Maguire, 1988a).

V. MITOTIC RECOMBINATION IS GENERALLY RARE BUT HAS KEY EFFECTS AND SOME DIFFERING MECHANISMS

Mitotic recombination occurs spontaneously at frequencies much lower than meiotic recombination. It includes both crossover and conversion events. In yeast, in which it has been intensively studied (reviewed by Roeder and Stewart, 1988), it is 100–1,000 times less frequent than meiotic recombination and occurs predominantly by gene conversion (Roman, 1956). But mitotic recombination can be increased to meiotic levels by such agents as ionizing radiation, ultraviolet light, and chemical mutagens. Spontaneous mitotic reciprocal exchange occurs as frequently in *Drosophila melanogaster* males as in females, although meiotic exchange is normally absent in males. Such mitotic exchange, although rare, is generally more frequent in *Drosophila* of both sexes than in most higher eukaryotes, probably because there is apparently at least loose homologue pairing in all somatic cells. Important meiotic machinery, such as the SC, is typically absent at mitosis, but at least short homologous sequences must pair intimately, perhaps by chance, for recombination to occur.

The consequences of mitotic recombination, even at low frequencies, are important. A repair process for ultraviolet light-induced DNA damage, for example, depends upon a recombination system. Recombination provides perhaps the only efficient repair mechanism for double-strand breaks. It can also give rise to homozygosity and therefore to expression of a recessive allele previously present heterozygously—an effect that may actually contribute to carcinogenesis in higher eukaryotes when there is heterozygosity for cancer-inducing recessive alleles. When it occurs between dispersed repeated sequences (such as transposable elements) located in nonhomologous positions, reciprocal translocations and inversions can result. On the other hand when it occurs between tandemly repeated sequences, duplications or deletions may be formed, and the elimination of new mutations can occur. In multigene families it may contribute to maintenance of sequence homogeneity. Gene conversion may provide a mechanism of somatic hypermutation of mammalian immunoglobulin genes (Maizels, 1989) as well as targeted diversification of specific genes in the course of developmental regulation in a variety of organisms. Mating-type switch in yeast (Haber et al., 1980; Klar et al., 1980) and diversification of trypanosome variant surface glycoprotein genes (Reynaud et al., 1987; Thompson and Neimann, 1987) are likely examples of this effect. Gene conversion probably also has played an important role in the evolution of the mammalian genome (Slighton et al., 1980). For example, the origin of a large region (1.5 kbp) of homology shared by the two fetal γ-globin genes is probably best explained as conversion-mediated.

While meiotic recombination seems to be initiated after DNA replication in yeast, spontaneous mitotic recombination is usually initiated in G1 phase prior to DNA synthesis (Esposito, 1978). However, experiments have indicated that repair recombination induced by DNA damage can be initiated either before or after DNA synthesis (Fabre et al., 1984). Models for mitotic recombination at the two-strand and four-strand stages are illustrated in Figures 4.19 and 4.20.

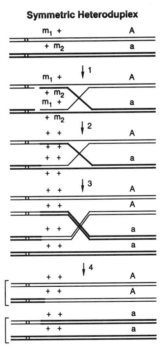

Fig. 4.19. *Mitotic recombination at the two-strand G1 stage. A sectored, prototrophic colony can be produced from a G1 conversion event mediated by either asymmetric or symmetric heteroduplex DNA. Nine classes of marker segregation are possible, only two of which are shown here. A sectored prototrophic colony containing only one of the two alleles is diagnostic of a G1 event.* **1.** *Heteroduplex DNA is formed by homologous strand exchange.* **2.** *Mismatch repair of the heteroduplex DNA (in the case of symmetric heteroduplex, only one of the chromatids need be corrected to wild-type to result in a sectored, prototrophic spore).* **3.** *DNA replication through the crossover resolves the recombination event.* **4.** *Chromatids 1 and 3 segregate from 2 and 4 to produce the sectored colony. Symbols above and below a chromatid refer to the two strands of the chromatid. After Orr-Weaver and Szostak (1985).*

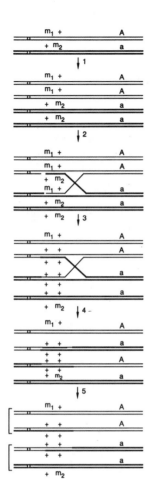

Fig. 4.20. *Mitotic recombination at the four-strand G2 stage. A sectored, prototrophic colony can be produced from a G2 event only if symmetric heteroduplex DNA is formed and corrected to wild-type. This type of an event results in a colony that retains both of the mutant alleles. 1. Replication of the diploid DNA, producing four chromatids. 2. Strand exchange to form heteroduplex DNA on each of the two recombining chromatids. 3. Mismatch repair of the heteroduplex DNA to wild-type. 4. Resolution with a reciprocal exchange of the flanking marker A. 5. Segregation of chromatids 1 and 3 from 2 and 4 to produce a prototrophic colony, sectored for marker A. Where symbols appear both above and below a chromatid, they refer to the two strands of the chromatid. After Orr-Weaver and Szostak (1985).*

It is not known to what extent mitotic and meiotic recombination share common pathways, but some differences prevail (reviewed by Orr-Weaver and Szostak 1985). Experiments suggest that symmetric heteroduplex is more commonly formed in spontaneous mitotic than in meiotic recombination (Golin and Esposito, 1981). Certain yeast mutants have different effects on mitotic and meiotic recombination. For example, one such mutant, *spoI*, abolishes meiotic recombination but has only small effects on mitotic recombination in specific intervals, while another, *RAD50*, reduces meiotic recombination but increases mitotic recombination. Polarity seems to be absent in mitotic recombination possibly as a result of formation of long heteroduplex regions; or possibly initiation sites used in meiotic recombination are not recognized at mitosis. Evidence suggests that there may be two pathways for mitotic recombination, one for gene conversion and some crossovers, and another for reciprocal crossing-over. Some mitotic crossovers seem to be independent of the gene conversion mechanism. Analysis utilizing various recombination assays and the finding and use of more mutants with specific effects at mitosis and meiosis should prove revealing.

VI. THE EVOLUTION OF MEIOSIS MUST HAVE OCCURRED EARLY AND RAPIDLY AFTER THE ORIGIN OF EUKARYOTE DIPLOIDY

Sexual reproduction provides the potential advantage of more rapid adaptation than is deemed feasible with asexual reproduction, through variation produced by recombination. Beneficial mutations may be incorporated while deleterious mutations are eliminated. But adaptive combinations of alleles produced by recombination can also be lost by recombination, and it is frequently questioned whether natural selection, by favoring advantageous combinations where they exist, could actually favor those organisms capable of producing them enough to balance the costs of sexual reproduction. Since two parents must be maintained to reproductive stage for sexual reproduction, the number of offspring per parent is halved with sexual reproduction. Since selection for a *potential* advantage is difficult to envision, other, more instant possible advantages of sexual reproduction have been sought. For example, under some conditions, the larger cells imposed by diploidy could be desirable as long as the possibility of shifting efficiently to the smaller cells of haploidy is also available (Margulis et al., 1985). Thus cell and nuclear fusion to yield diploidy, seen to represent a primitive form of fertilization—followed, perhaps a number of cell generations later, by reduction to haploidy—could achieve this purpose.

Some fungi, such as *Aspergillus*, have the capabilities for such a cycle, called a parasexual cycle (as well as a conventional fertilization and meiosis) (Käfer, 1961). The parasexual cycle (Fig. 4.21), during which haploidization is achieved by a largely hit-or-miss series of mitotic nondisjunction events, can be visualized as representing a remnant primitive form.

The chromosome behavior in some flagellates has also been envisioned as representing various stages in the evolution of meiosis (Cleveland, 1947), although cell fusion is not apparently a part of the life cycle. Actually, chromosome behavior in these flagellates probably bears only superficial resemblance to meiosis. In some of these flagellates chromosomes seem to exist permanently in a kind of double state, so that in all cell divisions, following replication of the already double elements, pairs of dyad-like structures separate at anaphase. This kind of division has been equated to a first meiotic division, but what

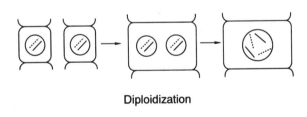

Diploidization

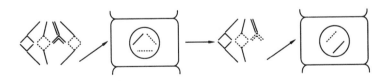

Haploidization

Fig. 4.21. *The parasexual cycle. Diploidization includes cell and nuclear fusion. A number of mitotic divisions at the diploid level may then ensue before the sequence of nondisjunction typical of haploidization occurs. Shown here at haploidization are early anaphase of two mitotic divisions in sequence with the results of each. A chromosome complement of only two pairs is shown here for simplicity.*

was called pairs of synapsed homologues seem instead to result simply from replication in place of structures already double. In other flagellates, following the kind of division just described, a second kind of division can occur without replication, in effect reducing the doubleness to the single state. This kind of division has been equated to a meiotic second division. But the special doubleness reported in these flagellate chromosomes could not have arisen in the first place by fusion of separate nuclei. Instead it apparently arose by replication of the chromosome material without the usual subsequent anaphase separation, thus leaving the products of replication at least partially associated with each other. This apparent double state has been equated both to diploidy and to the polyteny of giant chromosomes, although it was perceived to be due to change in the centriole–chromosome duplication schedule.

An essential element and major hurdle in the evolution of meiosis, however, is the development of a mechanism for the pairing of matching chromosomes initially contributed from separate nuclei; this remains a major unsolved problem, although such mechanisms must have arisen very early in the evolution of meiosis, in ways that would seem irrelevant to the special behavior of the chromosomes of the flagellates. Those unorthodox organisms may not be primitive, but instead specialized in their own idiosyncratic fashion. Their alternation between single- and double-chromosome states may indeed provide for alternation between large and small cell forms and in fact may do so far more simply and directly than would be possible by means of meiosis and fertilization.

Other probable sources of immediate advantage of diploidy and meiosis have been proposed. It has been noted that double-strand DNA damage requires diploidy (or presence of a normal sister chromatid)

for repair and that both conversion and crossing-over may be byproducts of the double-strand repair process (Bernstein, et al., 1988). Repair provides obvious immediate advantage. Conversion can produce homozygosis, and in most organisms crossovers, which yield chiasmata, furnish a reliable system for the orderly distribution of perfect sets of chromosomes to gametes at meiosis after the pairing of homologues. Double-strand repair processes may have arisen first in prokaryotes, along with the essential ingredients of crossing-over, in such a way that they were carried into the progenitors of eukaryotes.

An imaginative scheme for the evolution of sex in eukaryotes that depends upon transposable elements has been proposed (reviewed by Hickey and Rose, 1988). It suggests that at the initial stage of the evolution of meiosis, parasitic DNA sequences promoted genomic fusion to produce diploids from initially separate haploid cells. Such fusion is seen as having been advantageous to these DNA sequences because it provided for their intergenomic spread. Although these parasitic DNA sequences would be expected to have had some negative effect on the host cells, it is calculated that moderate levels of negative selection would not have prevented the spread of these elements in a population. But once diploids had been produced, further spread presumably depended upon the evolution of primitive meiotic mechanisms to increase the number of element-carrying cells. Thus the evolution of syngamy and meiosis are seen to have served the interests of the parasitic DNA at first, and then to have provided the benefits of double-strand DNA repair to host cells during the diploid phase, and advantages of potentially adaptive variation to the host population as well.

Eukaryote evolution involved development of linear chromosomes and mitotic spindle apparatus following installation of the nuclear membrane system (Cavalier-Smith, 1988). While diploidy can result either from disturbance of the mitotic cell cycle, which yields doubling of the chromosome number with homozygosity, or from the fusion of distinct cell nuclei from separate sources with possible heterozygosity, homologous recombination gives rise to variability only where there is heterozygosity. Simple chromosome doubling may have been irrelevant to the evolution of sexual reproduction. Once primitive diploidy with heterozygosity had been produced, restoration of haploidy appears to have been a much more complex proposition. One possible route is the rather haphazard haploidization process of the parasexual cycle typical of *Aspergillus*. The other route, virtually universally present in eukaryotes, including *Aspergillus*, is meiosis. With a primitive meiosis in an organism with more than one pair of linear chromosomes, segregation and recombination by way of independent assortment would occur even in the absence of crossing-over. But in the absence of chiasmata some mechanism for the regular disjunction of homologues would be required, as in the male of *Drosophila*. Where diploidy existed without a regular mechanism for the restoration of haploidy, it is doubtful that it would have persisted very long in recognizable form. Mutations and chromosome rearrangements would be expected to modify the chromosomes so that matching pairs (homologues) would no longer exist. Instead there would be a new haploidy at the doubled chromosome number. With the exception of some parthenogenetic forms, which have probably arisen relatively recently, diploidy in the absence of regular sexual reproduction seems to be a rare circumstance. With the diploidy produced by syngamy and karyogamy, homologue pairing with SC formation and provision for chiasma formation and maintenance for correct chromosome disjunction must have arisen

and then prevailed widely (Maguire, 1992). These features involve complex functions with structures that are virtually universal. Common origin is generally inferred.

An advantage for diploidy could have been provided by complementation of preexisting harmful mutations if diploidy was produced by fusion of cells from differing sources. But since it is doubtful that true diploidy would have been long maintained in the absence of sexual reproduction, once it was achieved as a part of the life cycle, meiosis may have served as a kind of guardian of diploidy. In fact, the origin of meiosis probably predated the evolution of diplontic multicellular organisms with large genomes, and it may have facilitated their evolution by serving in the guardian of diploidy capacity, so that the complementation advantage prevailed. This advantage, in balance with other considerations, is more likely to have been important in the long term for more complex diplontic organisms with large mutable genomes (Charlesworth, 1991; Kondrashov and Crow, 1991; Perrot et al., 1991).

VII. SUMMARY

1. Recombination, often considered the essence of sexual reproduction, provides the potential advantage of relatively rapid evolution because of the resulting variation available for selection. In eukaryotes it occurs frequently at meiosis and infrequently but importantly at mitosis.

2. At meiosis recombination takes three main forms: (1) mendelian-independent assortment, which shuffles parental input combinations of alleles into new output combinations in gametes; (2) reciprocal exchange between homologous chromosomes by way of crossing-over, which reassorts linked alleles; (3) gene conversion, which involves the substitution of a short portion of a chromosome for its counter-

part in a homologue, without reciprocal exchange, and results in unequal recovery from a heterozygous parent of marker alleles in a single meiosis.

3. At mitosis both reciprocal crossover recombination and gene conversion are known to occur spontaneously but infrequently in most cases. Important and varied consequences nevertheless follow mitotic recombination.

4. Mendelian-independent assortment is the outcome of nearly ubiquitous meiotic mechanisms. Crossing-over and conversion are not thoroughly understood, but molecular models have been proposed that yield results mainly consistent with genetic data. Tetrad analysis, which is possible in many fungi, has figured promi-

nently in the development of these models. Postmeiotic segregation has furnished important clues, since it implies DNA heteroduplex formation. Intimate homologue pairing at least for short matching segments is thought to be required for both crossing-over and conversion.

5. Chiasmata, which apparently result from crossing-over together with sister chromatid cohesiveness, serve the important role of maintaining homologue associations until correct disjunction is established at meiotic metaphase I.

6. Major meiotic mechanisms remain unexplained. These include homologue pairing, chiasma maintenance, and crossover site establishment.

5 Variations in Chromatin Organization and Amount

The genetic situation in an interbreeding species with a finely divided population structure provides a virtually infinite field of potential variability which permits ready adaption on the part of the species to fluctuating environmental conditions, ready phylogenetic adaption to secular changes in conditions, effective exploration by trial and error processes of possibilities of breaking through the restraining pressures even under static conditions, and the rapid exploration by similar means of any major ecologic opportunity.

Sewall Wright, 1949

Chromatin is both stable and unstable. If we were to take an Olympian view, we would assert that the Grand Design maintaining the state of the biosphere can be conceived as an equilibrium between the opposing processes of alteration and repair of DNA. Both processes are, and have been, responsible for the variety of living things now on earth. If the equilibrium should shift too far in the direction of alterations, life would become untenable because of the cost of maintaining large numbers of individuals with the low fitness inevitably created by the alterations. If, on the other hand, it were to shift too far to the stable condition with no changes, evolution would cease, and changing ecological conditions would eventuate in drastic effects on organisms unable to adapt to the new conditions.

Genomic alterations among the Eukaryota are considered here to be the result of seven types of events: (1) recombination; (2) gene or point mutations; (3) DNA duplication of genic or "nongenic" nucleotide sequences, leading to reiteration of these sequences; (4) changes in ploidy of a euploid or aneuploid type; (5) deletion and diminution of chromatin; (6) gross rearrangements of chromatin categorized as inversions or translocations, respectively; and (7) transpositions of mobile elements of DNA. In this chapter we consider types (3) (4) (5) (6) and (7) while bearing in mind that the distinctions among all seven types are not always clearly defined.

I. CHROMOSOME BREAKS AND THEIR CONSEQUENCES

Breaks in chromosomes can lead to a variety of aberrations, some of which are rearrangements of chromatin within chromosomes and some between nonhomologous chromosomes.

A. Types of Aberrations

Chromosome breaks are generally repaired by rejoining of the broken ends, the chromosomes being left intact. Alternatively the broken ends may join with different other broken ends within the same chromosome that has two breaks, or with broken ends in other nonhomologous chromosomes.

Breaks that are not rejoined can presumably occur at any stage of the cell cycle, but in general they will be recognized only in the mitotic and meiotic stages, when the chromosomes are most visible. A chromosome with one break can either be reconstituted to its original state by the rejoining of the broken ends, or give rise to a deficient centric and an acentric fragment, if rejoining does not occur. In the latter case the cell containing the break will in the following divisions lead to aneuploid cells, and they will probably die as a result of the deficiencies.

1. Inversions and translocations

Two general types of rearrangements occur: inversions and translocations. These rearrangements are defined here as *gross*, since they are visible with the light microscope. They can result from breaks in whole chromosomes at the 2C stage. Many different kinds of aberrations of this type have been recognized by observing the metaphase and anaphase stages of cell division. Some of them are diagrammed in Figures 5.1 and 5.2. Two breaks in the same chromosome can produce an **inversion**, if the excised fragment is reinserted after being rotated 180°, and proper polarity of the reinserted DNA is maintained with the rest of the DNA (Fig. 5.1). Note that the inverted fragment can include the centromere, in which case it is termed a **pericentric** inversion. If the centromere is not included, it is called a **paracentric** inversion. A pericentric inversion can change the position of the centromere and cause a metacentric to become a submetacentric and vice versa.

Three breaks in a single chromosome can result in an intrachromosomal trans-position (Fig. 5.2a). The transposed segment may or may not be inverted. Single breaks in each of two homologues can result in a wide variety of aberrations, some of which are illustrated in Figure 2.6 and Figure 5.2b–d. A reciprocal translocation is formed by the fusions of broken ends from two nonhomologues as shown in Figure 5.2b, which shows two types of expected results: a **balanced** one, in which each translocate has a centromere, and an **unbalanced** one, in which one translocate is a dicentric and the other an acentric. If two breaks occur in one chromosome, the released segment may intercalate within a nonhomologue with one break and thus become transposed (Fig. 5.2c). Thus interchromosomal transpositions as well as intrachromosomal transpositions

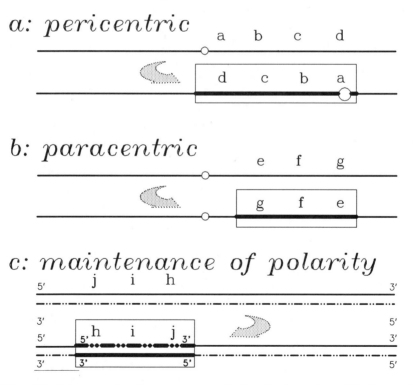

Fig. 5.1. *Inversion of a chromosome.* **a.** *Pericentric inversion with change in the position of the centromere.* **b.** *Paracentric inversion.* **c.** *Inversion of a segment of DNA requires exchanges of strands to maintain polarity.*

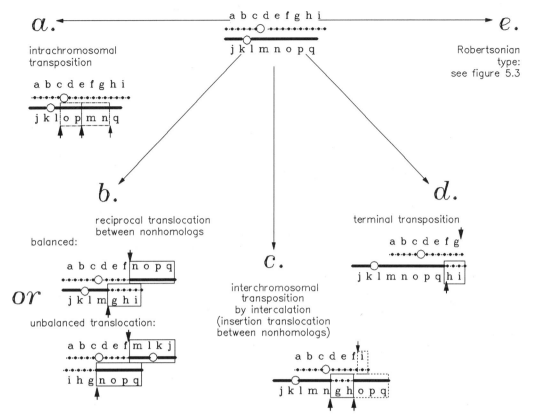

Fig. 5.2. *Some examples of observed types of translocations.* **a.** *Intercalation of a segment of chromosome to another position in the same chromosome.* **b.** *Reciprocal translocation between nonhomologues. Either a balanced pair of functional chromosomes is formed on fusion of broken ends, or a pair with one dicentric and* one acentric will be found, which will lead to an unbalanced aneuploid condition in the next division. **c.** *Intercalation of a segment from one chromosome into a nonhomologue.* **d.** *Like (c) but the segment is transposed to the end of the receiving chromosome.* **e.** *A Robertsonian event.*

may occur. Terminal transpositions of the type illustrated in Figure 5.2d may also occur. It is generally assumed that the telomeric end of the acceptor chromosome is deleted, thus forming a "sticky" end. But this may not always be the case, since intercalary TTAGGG sequences have been detected in some chromosomes that may represent traces of old fusions involving telomeres.

A commonly observed type of translocation results from a so-called **Robertsonian event** or type in which two acro-centric chromosomes fuse at their centric ends (Robertson, 1916) (Fig. 5.3). The association by fusion at the centric ends of two acrocentric rods can result in the formation of a metacentric, and their dissociation can result in two acrocentrics. Hence a decrease or increase of the number of chromosomes can be brought about by a fusion or separation of two chromosomal elements or **arms** (White, 1969, 1973; John and Freeman, 1975).

Much comment and debate has gone on for many years among cytogeneticists about

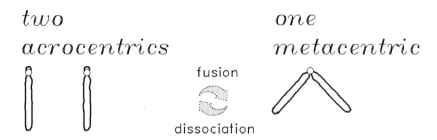

Fig. 5.3. *Robertsonian type of translocation in which two acrocentrics fuse to form a single metacentric. The opposite event, called dissociation, results in two acrocentrics from a metacentric.*

the mechanisms of fusion and dissociation. Figure 5.4 represents an exaggerated version of the possible modes of fusion. Of the 36 possibilities, 13 will result in a presumably viable metacentric, provided the loss of heterochromatin is not deleterious to the chromosome's functioning. Dicentrics, half-centrics and 1.5 centrics may also be viable. Only the acentrics are expected to be inviable. Figure 5.5 illustrates three possible modes of dissociation of a metacentric. Figure 5.5a shows a simple fission of the single centromere. Of course, the metacentric may

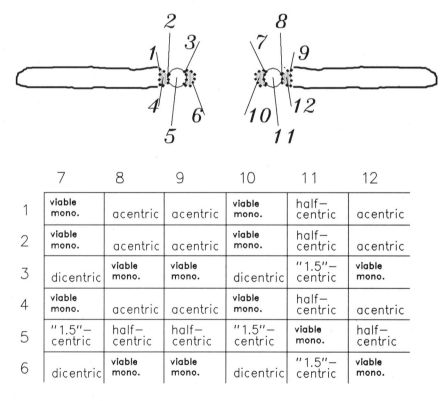

	7	8	9	10	11	12
1	viable mono.	acentric	acentric	viable mono.	half–centric	acentric
2	viable mono.	acentric	acentric	viable mono.	half–centric	acentric
3	dicentric	viable mono.	viable mono.	dicentric	"1.5"–centric	viable mono.
4	viable mono.	acentric	acentric	viable mono.	half–centric	acentric
5	"1.5"–centric	half–centric	half–centric	"1.5"–centric	viable mono.	half–centric
6	dicentric	viable mono.	viable mono.	dicentric	"1.5"–centric	viable mono.

Fig. 5.4. *Various possible modes of centric region fusion of two acrocentrics (see text).*

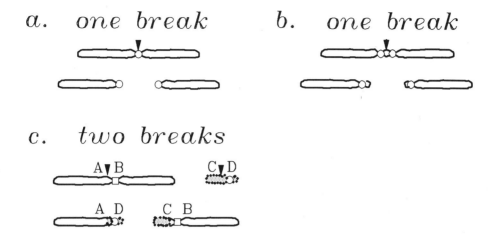

Fig. 5.5. *Dissociation of a metacentric* (**a–c**). *See text.*

be a functional dicentric as proposed in Figure 5.4, in which case a scission could simply separate the two (Fig. 5.5b). A third possibility is the involvement of another chromosome donating a centromere, as shown in Figure 5.5c. This would not result in a change in chromosome number, but a metacentric could disappear and hence the karyotype could be changed. White (1973) describes a large number of possibilities, most of which fit into the three categories described above.

Breaks occurring at the 4C stage can, on reunion, produce the types of aberrations illustrated in Figures 5.1 and 5.2 if both chromatids are broken at identical loci. Breakage of just one of a chromatid sister pair can, however, produce either intrachromosomal or interchromosomal aberrations. Figure 5.6 illustrates possibilities of interchanges between nonhomologues.

One of the results of breakage of two chromatids in the same arm of a chromosome followed by reunion of the ends attached to the centromere can result in the **breakage–fusion–bridge cycle** illustrated in Figure 5.7. This phenomenon occurs in maize and involves dicentric chromosomes broken in anaphase bridges (McClintock, 1951). It can also occur in other plants and some animals. Two kinds of the cycle have been described: the chromatid and chromosome types. Both arise from the formation of dicentric chromosomes or chromatids. Figure 5.7 diagrams the one type of cycle and the consequences. Chromosome 9 of maize bears a number of marker genes, namely (in this case), bronze (*bz1*), waxy (*wx1*), colored (*C1*), and shrunken (*sh1*), that are used to follow the events described in Figure 5.7. A transposable factor, *Ds* (dissociation), to be considered at length later in this chapter (p. 211), is also involved. Each of the chromosomes has a *Ds* factor located at the sites indicated by the asterisks, and as a result breakage may occur at these sites. This cycle may occur either at meiosis or mitosis. Meiosis results in aberrant gametes so far as chromosome 9 is concerned, but the cycle may continue in mitotically dividing cells.

2. *Deletions and duplications*

The addition and the loss of segments of chromosomes, ordinarily called **duplication** or **deletion**, respectively, by the geneticist, can be detected by cytological examination and also by their phenotypic

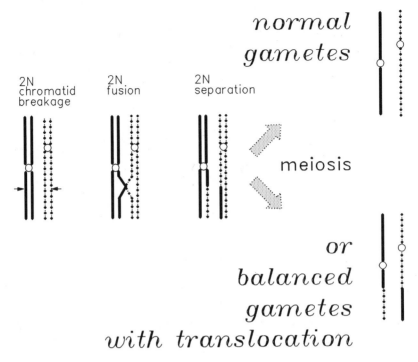

Fig. 5.6. *One kind of a result expected from chromatid exchanges between nonhomologues. Reciprocal translocations can be formed, which may be balanced.*

effects. The term deletion was originally coined to describe an intercalary loss of chromatin while deficiency described the loss of a terminal segment. Here we shall use the terms deficiency and deletion as synonyms.

Figure 5.8 illustrates duplications in the salivary gland polytene chromosomes of *Drosophila melanogaster*. These duplications are contiguous in tandem, but duplications may not always be contiguous, or in tandem. The several types of duplications that are known to occur may be interchromosomal or intrachromosomal. Interchromosomal duplications can result from the insertion of a contiguous duplicate segment internally in a nonhomologue (Fig. 5.9a), or at a different part of the chromosome (Fig. 5.9b). These resemble the translocation described in Figure 5.9, but the difference is that in a true

translocation the segment is moved from the original site, and not its duplicate. Intrachromosomal duplications are contained within the chromosome of their origin (Fig. 5.9a–e). They may also be extrachromosomal (Fig. 5.9f).

Several mechanisms for the formation of duplications have been proposed. Contiguous duplications of the type illustrated in Figure 5.9a are generally considered to be the result of unequal crossovers that produce a duplication in one chromatid and a deficiency in the other. They may occur either between homologous nonsister chromatids or between sister chromatids during the meiotic or mitotic phases of the cell cycle. As noted in Figure 5.10 crossing-over within the inverted segment in an inversion heterozygote can also result in duplication as well as deficiencies. Generally, but not al-

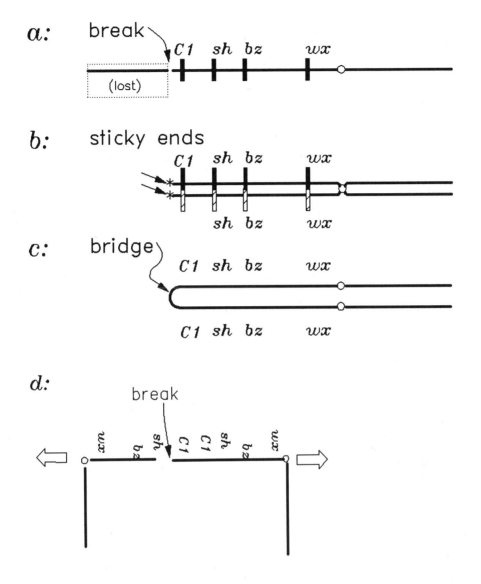

Fig. 5.7. *Breakage–fusion–bridge event involving maize chromsome 9.* **a.** *A break in the chromosome occurs at point indicated by the arrow; the acentric piece is lost.* **b.** *Chromatids are formed.* **c.** *Broken ends fuse to form a dicentric.* **d.** *Dicentric fused arms break at anaphase.* **e.** *Broken remainders are either deficient or have duplications.*

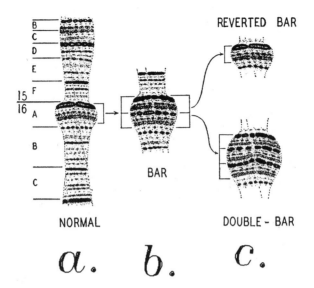

REVERTED BAR
B
C
D
E
15 F
16 A
B
C
BAR
NORMAL
DOUBLE - BAR
a.
b.
c.

Fig. 5.8. *Duplication in polytene X chromosome of Drosophila melanogaster. See text.* (**b**) *shows duplication of region 16A; double-bar in* (**c**) *shows triplication.*

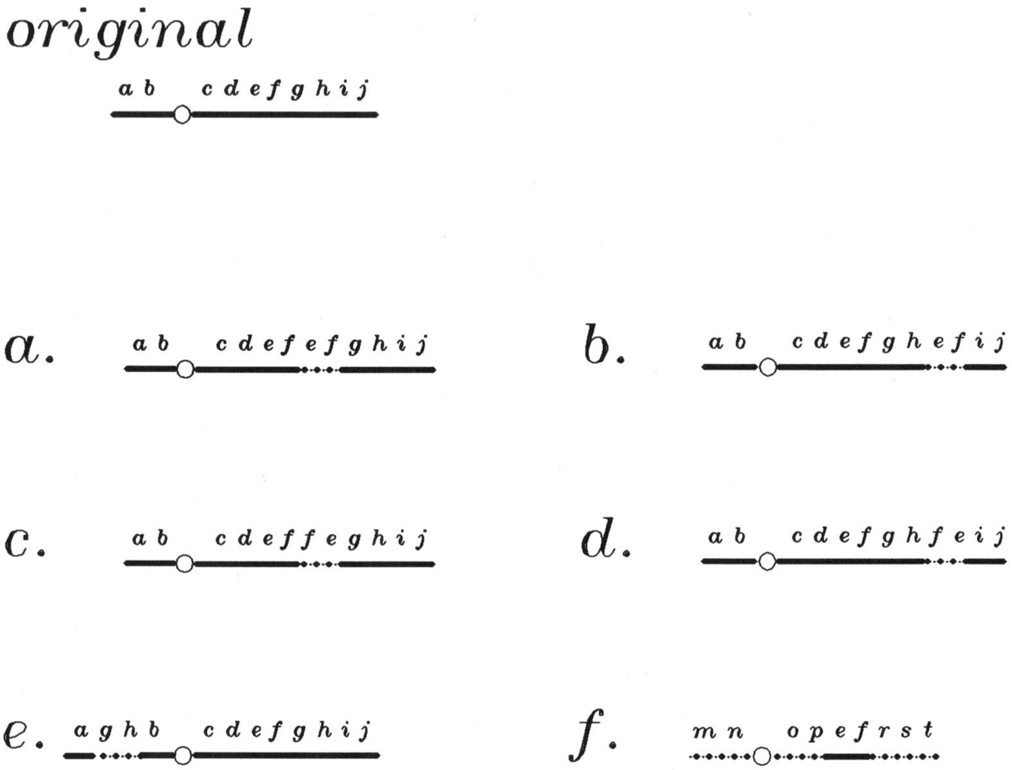

original
a b c d e f g h i j
a.
a b c d e f e f g h i j
b.
a b c d e f g h e f i j
c.
a b c d e f f e g h i j
d.
a b c d e f g h f e i j
e.
a g h b c d e f g h i j
f.
m n o p e f r s t

Fig. 5.9. *An assortment of different types of duplications known to occur.* **a–f**. *See text.*

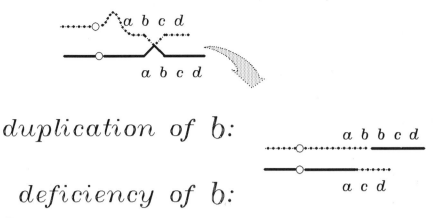

Fig. 5.10. *Duplication and deficiency resulting from unequal crossing-over between homologous chromsomes.*

ways, the products of these crossovers result in inviable zygotes after fertilization. Also, transpositions can result in duplications of segments of DNA, particularly through the action of reverse transcriptase, as discussed on p. 220.

Visible deficiencies occur in the salivary gland chromosomes of *D. melanogaster*. Deficiencies are generally lethal when homozygous and, if they are large enough, when heterozygous. Plants tolerate deletions of chromatin better than animals, but in either case the loss of functional genetic material is considered to be a retrogressive rather than a progressive alteration. On the other hand the formation of duplicate segments of DNA is generally considered to be one of the more important factors in the evolution of genomes, as discussed in Chapter 10.

B. Some Factors Involved in the Formation of Rearrangements

In the above discussion we have considered that aberrations, whether of the chromosome or chromatid type, result from breakage and reunion—the breaks being lesions that result in discontinuities. Two broken ends must then find one another to form a union. One alternative hypoth-

esis posits that the primary lesions are not breaks but local regions of instability in the chromosomes (Evans, 1962; Rieger, 1966). These "instabilities" lead to "exchange initiations" that eventuate in complete or incomplete exchange by interactions between the unstable regions. The exchanges depend on contact between the chromosomes and can be considered to be analogous to crossing-over. Ectopic pairing between nonhomologues could, for example, produce translocations of the reciprocal variety. Whatever the mechanism, the important point is that different types of gross rearrangements occur and can be maintained in a population.

The frequency of formation of aberrations can be increased experimentally by a variety of means. The induction of chromatin breaks by ionizing radiation and chemical agents has been extensively investigated (Evans, 1962, 1974, 1983). Chromosomes also break spontaneously in the absence of extrinsic agents. Intrinsic factors must be involved in these breaks. Important events leading to rearrangements must occur in the interphase nucleus, which can be called a black box. We can expect that chromosome replication and repair are involved, but how they are involved is a question yet to be answered.

It has been proposed that the break-points in human structural rearrange-ments are located primarily within the G-light or reverse (R) bands; much lesser numbers are located in the G-dark or G/Q bands. Nakagome et al. (1983) found by their own observations and a review of the literature that of 148 cases in point, 71 breaks definitely occurred in the light bands and an additional 55 probably also did. Of the remaining 22 cases, 9 positive and 11 probables occurred in R bands and only 2 seemingly occurred at the interface of dark and light bands. Whether these data reflect the actual incidence of breaks or the efficiency of repair at certain parts of the chromatin cannot be presently de-termined. Both may be involved. The tak-ing up of the thiazine component of the Giemsa stain delineates the dark bands and undoubtedly reflects the condition of the chromatin in the regions where fewer rejoined breakpoints are observed.

Logic leads one to the provisional con-jecture that those breaks that occur in re-gions of chromatin containing repetitious DNA should be those regions where re-joining of broken ends is most likely. This hypothesis presupposes that repetitious DNA segments contain more sequences in common than unique DNA, and hence re-lated complementary strands should rec-ognize one another and engage in forming heteroduplexes later to be restored to com-plete homology of complementarity by the appropriate repair enzymes. Repetitious DNA segments have been identified in the X chromosomes of *D. melanogaster* by Rudkin and Tartof (1973), and Lee (1975) found their distribution to be correlated with the loci of breaks determined by Kaufman (1946). Lee's analysis shows that the frequency of breaks involved in ab-errations of the X is greatest where rep-etitious segments have been identified (Fig. 5.11).

A unique approach to the analysis of the mechanism of rejoining of chromosome

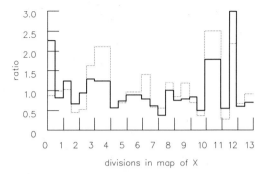

Fig. 5.11. *Histogram showing correlation be-tween known chromosome breaks in the X chro-mosome of D. melanogaster and the regions of repetitious DNA in the euchromatin. The bro-ken line indicates regions of breaks and the solid line regions of repetitious DNA. Data from Lee (1975).*

broken ends has been made by Natarajan and Obe (1984). Using endonucleases to treat permeabilized Chinese hamster and mouse fibroblast cells in culture, they demonstrated that those enzymes that in-duce blunt double-stranded breaks (EcoRV and PvuII) were more effective in induc-ing chromosomal aberrations than those that induce breaks with cohesive or "sticky" ends (BamHI, HindIII, and MboI). These results would seem to be in contradiction to those obtained by Lee (1975). However, it should be recognized that cells have ac-tive exonucleases that nibble away at one strand of a broken end and can therefore convert blunt ends to cohesive ends. Fur-thermore, it was found that endonuclease treatment preferentially induced aberra-tions near centromeres. This is precisely the part of the chromosome with the heav-iest concentration of repetitious DNA in mammals. It has also been reported that a high frequency of DNA rearrangements in a mouse cell line is associated with the centromeric satellite DNA fraction (But-ner and Lo, 1986).

Fragile sites have been identified in every human chromosome except chro-mosome 21 up to the year 1987 (Sutherland and Mattei, 1987). These are specific sites on the chromosomes, frequently expressed as nonstaining regions, that are inherited. They appear as gaps that show fragility under *in vitro* conditions when the chromosomes are exposed to specific chemical agents or conditions of tissue culture. Fragile sites are classified as either rare or common. The rare sites occur only in certain kindreds and range in incidence from one in 40 to one in several thousand individuals. The common sites are present in all chromosomes except 21 and are a structural property of human chromosomes probably common to all members of the human population. Rare sites have been identified on chromosomes 2, 6–11, 16, 17, 19, 20, 22, and the X. All sites, whether rare or common, appear to be regions of late-replicating DNA (Laird et al., 1987).

Fragile sites have been identified in other mammals, including mouse, rat, and cow. They are probably of general occurrence at least among the mammals.

Rare fragile sites are classified on the basis of the chemical compounds that induce them in cultured cells. Three types have been described: folate-sensitive, distamysin-inducible, and BrdU-requiring. Among the common sites there are also three types: aphidicolin, 5-azacytidine-, and BrdU-inducible sites (Berger et al., 1985).

A specific rare fragile site near the distal end of the human X chromosome (Xq27) is associated with the **fragile X syndrome** (Sutherland, 1982; Brown et al., 1988). Affected X chromosomes are found in about one in 2,500 live births. Fragile X carriers are frequently but not always mentally deficient, and they generally show phenotypic as well as behavioral defects. Fragile X is second only to Down's syndrome as a chromosomal defect associated with mental retardation, especially in males. Eighty percent of the male carriers show mental retardation and other phenotypic defects, but only about 30% of the

female carriers are mentally impaired and generally much less so than males. Some autosomal fragile sites also appear to be associated with mental retardation.

A number of inherited human diseases are associated with an abnormally high incidence of chromosome breaks. This chromosomal structural instability is characteristic of Bloom's syndrome, ataxia telangiectasia, xeroderma pigmentosum, and Fanconi's anemia. These are rare, recessive, autosomally inherited syndromes described in Table 3.5 (reviewed by German, 1983). The break sites on the chromosomes of patients with these diseases are not random. In Bloom's syndrome, for example, it is reported that most breaks occur preferentially in the centric regions (Kuhn and Therman, 1979). This correlates with the observation that most breaks occur in the late-replicating regions of some chromosomes (Lindgren, 1981). It is also of interest that Bloom's syndrome cells show an eight-fold increase in mutation rate as measured by the frequency of resistance to 6-thioguanidine (Vijayalaxmi et al., 1983).

II. GROSS REARRANGEMENTS IN NATURAL POPULATIONS

Rearrangements of chromatin of the gross kind have been found in all species of eukaryotes, ranging from fungi to the highly differentiated multicellular forms of plants and animals. Spontaneous inversions and translocations have been variously reported as occurring at rates of 10^{-4} to 10^{-3} per gamete per generation in natural populations (Lande, 1979). They must therefore be considered to be much less common than gene mutations, which have been estimated as occurring at the rate of one per 2–20 gametes per generation in *Drosophila* (Muller, 1950). The usually accepted mutation rate per gene is 10^{-5} to 10^{-6}. The break rate in the interphase stage may be as high as the mutation rate, however, and the repair of breaks more efficient than repair of point mutations.

A. Inversions

Because many thousands of skilled human karyotypers are busy daily all over the world examining the chromosomal configurations in cells from human fetuses, both normal and aborted, as well as newborns and adults, a great deal of information about the occurrence of inversions in the human population has been garnered. Both pericentric and paracentric inversions occur in 1%–2% of the population, and in most cases those that persist in the population have no noticeable effect on the phenotype (Moorehead, 1976; Voiculescu et al., 1986). Pericentric inversions have been most frequently observed in chromosome 9 (Vine et al., 1976). But they have been identified in nearly all others too. The breakpoints are generally nonrandom. Paracentric inversions are less frequently observed than pericentric in humans. This may simply be because they are less easy to detect, since the position of the centromere is not changed in paracentrics. In addition it is less probable for two breaks to occur on one arm of a chromosome than on different arms. Paracentric inversions have been catalogued in humans for chromosomes 16, 7, 3, 5, 1, 12 and 14.

Many different species of *Drosophila* have been investigated for the occurrence of gross rearrangements in both laboratory and natural populations. The polytene chromosomes of *Drosophila* species make them particularly valuable subjects for studying the incidence of rearrangements among related species. There are over a thousand species in the genus, so there is no lack of material. As illustrated in Figure 5.12, homozygous inversions can generally be detected readily. Figure 5.13 shows the same inversion in a heterozygote. The numbers in the diagram represent the regions of the normal X and inverted X as shown in Figure 5.12. Note the inversion loop that enables the homologous regions to pair. In many cases overlapping inversions and inversions within inversions can be detected in salivary gland preparations. Patterson and Stone (1952) have described the karyotypes of 215 species in the family Drosophilidea. More recently Carson (1983) reported on his analysis of the banding patterns of polytene chromosomes of 103 species of Hawaiian *Drosophila*. Both sets of data confirm that numerous inversions have occurred during the evolution of these many species from common ancestors. The Hawaiian *Drosophila* species all have a six-chromosome genome, and among these 103 species 213 paracentric inversions, or about two per species, have been identified. In many cases the inversions are polymorphic, that is, some individuals of the species have the inversion, others do not. Pericentric inversions also occur in different species throughout the genus, and like paracentric inversions some have been "fixed," that is, they are not polymorphic but present in all members of the species (Wharton, 1943).

Inversion polymorphisms are also extensively distributed in the Australian grasshopper, *Morata scurra* (White, 1957) as well as in other orthopteran genera, and in other dipteran families beside the Drosphilidae. (See White, 1973, for extensive review.) The malaria vector *Anopheles stephani* has autosomal inversions in both natural and laboratory populations (Sakai and Mahmaod, 1985).

Pericentric inversions have been identified in just about every mammal that has been analyzed cytologically. Mice (Capanna, 1985), rats (Yosida and Amano, 1965), and primates, including *Homo*, have been shown to have inversions within their respective populations. Comparisons of the human genome with the other members

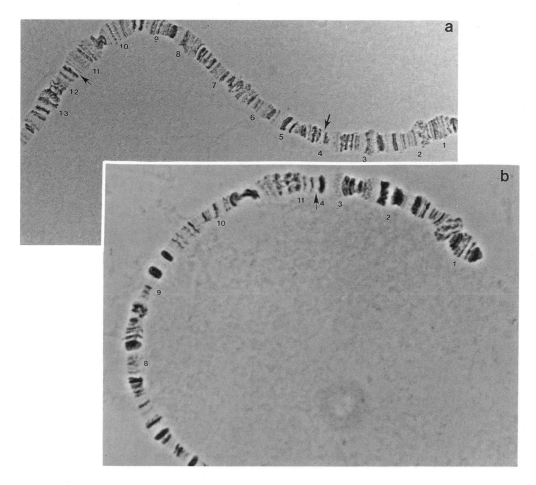

Fig. 5.12. *Polytene X chromosomes of D. melanogaster showing the distal end of the normal or standard X (a) and part of the homozygous inverted segment, In(1)delta 49 (b). As shown in (a) the two breaks were in region 4 and between 11 and 12. As shown in (b) the 4 region is united with 11. The other reunion is not shown in (b), but see Figure 5.13. Courtesy of Jon K. Lim of the University of Wisconsin–Eau Claire.*

of the anthropoid ape group reveal that inversions are quite common. Figure 5.14 compares the genomes of man, chimpanzee, gorilla, and orangutan. The banding patterns of the four species are similar and definitely show that the chromosomes within the anthropoid group have undergone inversions. Chromosomes 4, 5, 9, 12, 15, and 16 of man and chimpanzee clearly differ by pericentric inversions. Chromosome 7 of the chimpanzee and gorilla differ by a paracentric inversion in the long arm. Actually every chromosome except the X shows evidence of gross rearrangements (Yunis and Prakash, 1982). Inversions are also widely distributed among plants, and many are also polymorphic. Two genera in particular, *Paeonia* and *Paris*, have been reported to have members with high frequencies of inversions. In certain populations of *Paris quadrifolia* every member is heterozygous for one or more inversions, and these are distributed over every chromosome arm in the gen-

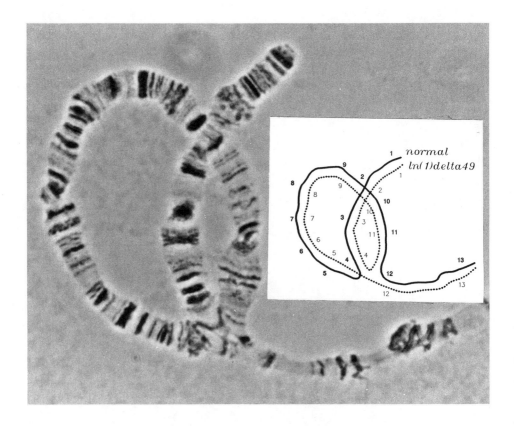

Fig. 5.13. *A heterozygous inversion polytene chromosome configuration involving the normal X and the inversion In(1)delta 49 illustrated in Figure 5.12. Courtesy of Jon K. Lim of the University of Wisconsin–Eau Claire.*

ome (Geitler, 1938). As in animals, every higher plant species has inversions in its population. (See Stebbins, 1950, for extensive additional data.) Inversions have also been found in the fungi; numerous inversions have been detected in *Neurospora crassa* (Perkins and Barry, 1977) and *Aspergillus* (Kafer, 1977), for example.

B. Meiotic Consequences of Inversions and Translocations

When a meiocyte with a chromosome pair heterozygous for an inversion enters

into prophase I of meiosis, problems arise. First, the chromosome containing the inversion must be synapsed to its homologue, which has a segment of its genes lined up in inverted order. This is generally overcome by the formation of an inversion loop, illustrated by the photograph of an inversion heterozygote in a polytene pair of homologues in *Drosophila* (Fig. 5.13) and the diagram in Figure 5.15. (If the inversion is very short, a loop may not be formed, and the homologous inverted regions may simply not synapse.) The next problem arises if a crossover occurs between the chromatids within the inversion loop. The results expected for

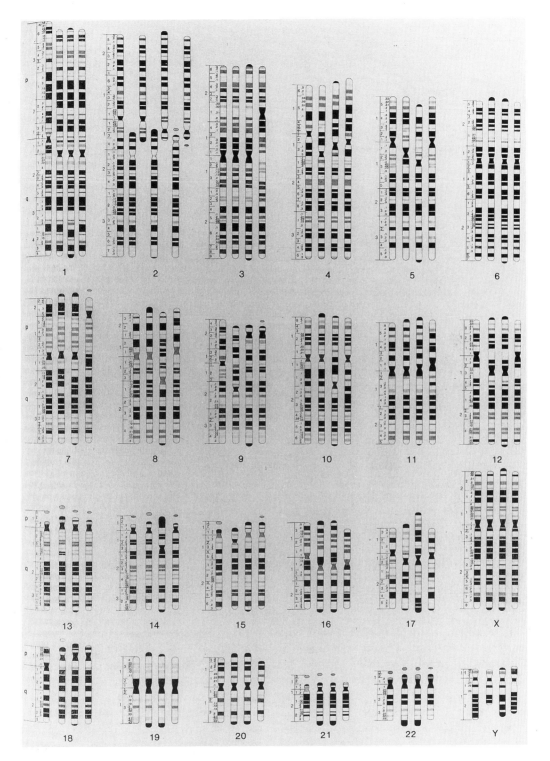

Fig. 5.14. *Comparison of the banded late prophase chromosomes of man, chimpanzee, gorilla, and orangutan (reading from left to right). Reproduced from Yunis and Prakash (1982), with permission of the publisher.*

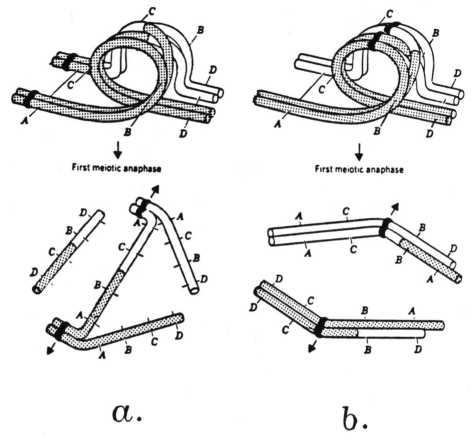

First meiotic anaphase   First meiotic anaphase

$a.$ $b.$

Fig. 5.15. *Results of crossing-over inside inversion loops in inversion heterozygotes. The results are different for paracentric versus pericentric inversions if a crossover occurs inside the inversion loop. In this diagram the standard arrangement is A B C D and the inversions are A C B D.* **a.** *The products of a paracentric inversion are an acentric (which is lost) and two deficient chromosomes resulting from breakage of a dicentric, or a dicentric, which does not get into a gamete. Only the two chromatids that do not participate in the crossover are complete.* **b.** *The crossover products from a pericentric inversion have duplications and deficiencies. Gametes carrying them will result in aneuploid zygotes that will produce defective embryos, unless the deficiencies are minor.*

single crossovers involving paracentric and pericentric inversions, respectively, are illustrated in panels a and b of Figure 5.15. Note that aberrant gametes will be produced. The products of a paracentric are an acentric and a dicentric plus two normal chromatids not involved in the crossover. On the other hand, the crossover products of a pericentric will be monocentrics, two of which will have duplications and deficiencies.

Since one expects inversions to arise in meiocytes of the germ line and thus form inversion heterozygotes, selection against them should occur until they become homozygous. In that case, selection should be insignificant, unless the new gene order itself is deleterious. This matter is an important one and is considered at length starting on p. 424.

Translocations are just as common in natural populations as inversions. Not only do they occur as polymorphisms within species, but it is clear that species derived from common ancestors have many times become endowed with different karyotypes that have resulted from translocations during the course of their evolutionary divergence. One need look no further than the prophase chromosomes and the anthropoid apes and man (Fig. 5.14) to see such evidence. Especially noteworthy is the clear relationship between the human chromosome 2 and the ape acrocentric or submetacentric chromosomes lined up alongside it in Figure 5.14. Either human chromosome 2 was derived from the ape ancestral chromosomes by fusions in the centric region as Robertsonian-type events, or the ape chromosomes in question were derived from an ancestral subhuman chromosome 2 by dissociation. Also note in Figure 5.14 that the gorilla genome has a reciprocal translocation involving chromosomes 5 and 17, and that there is an insertion of the tip of the p arm of the orangutan 20 to the centromeric region of 8q. Some Robertsonian events may be simple telomeric fusions resulting in dicentrics that become functional monocentrics, as described in Chapter 2. This appears to be the case for human chromosome 2 (Hsu, 1973) and for the domestic sheep (Bruere et al., 1974). A fusion of two acrocentrics may also involve a reciprocal translocation with a small bit of heterochromatin with associated centromere being lost and not missed. Possible examples of this have been identified by White (1973) in sheep, rodents, and insects, and the human example given above, of course, can also be interpreted in this way.

The dissociation of a metacentric into two acrocentrics (Fig. 5.5) is more difficult to explain, but good evidence that it has occurred in natural populations is available (White, 1973). For example, the involvement of a second chromosome contributing an additional centromere and

telomere may have occurred in the grasshopper *Keyacris scurra* (White, 1957).

Reciprocal translocations have been found to be polymorphic among most plants and animals that have had extensive karyotypic analyses. Generally they are recognized in cells undergoing meiosis because a typical cruciform configuration is formed at pachytene I (Fig. 5.16c). This type of configuration is formed as a result of a heterozygous reciprocal translocation (Fig. 5.16b). The heterozygous condition is called "balanced" because all the genes present in the contributing chromosomes of the parent normal-sequence homozygote are also present in the same dosage in the translocation heterozygote. By metaphase I the pachytene arrangements of synapsed chromosomes have become the types of configurations illustrated in Figure 5.17. Four alternative modes of functionally distinct distributions can occur at anaphase I, which are described as alternate-1, alternate-2, adjacent-1, and adjacent-2.

The two alternate classes involve the distribution of alternate members of the ring of chromosomes to the same pole at anaphase I. The homologous centromeres segregate to opposite poles. The result is that at the end of meiosis I the secondary meiocytes will commonly be of two types (Fig. 5.18): a set of chromosomes with normal sequences (centromeres 2 and 3) and a balanced reciprocally translocated pair (centromeres 1 and 4). Crossovers between the breakpoints and the centromeres of the translocated chromosomes during the prophase can yield a more complicated outcome but still produce viable products. The other two classes, adjacent-1 and 2, generally give rise to secondary meiocytes with unbalanced genic contents. Their chromosomes contain deficiencies and duplications as shown in Figure 5.18. In the adjacent-1 type, adjacent members of the ring of four chromosomes with nonhomologous centromeres are dis-

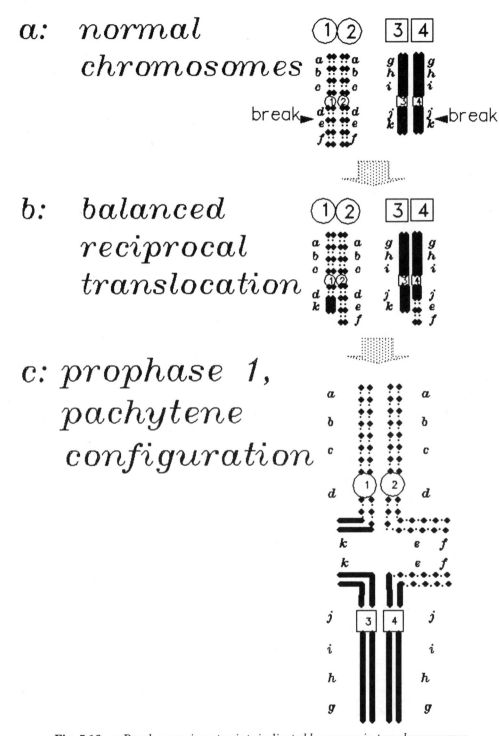

Fig. 5.16. a. *Break occurring at points indicated by arrows in two chromosomes with centromeres labeled 1 and 4 result in a balanced reciprocal translocation* (**b**). *At prophase I the chromosomes pair to produce a cross-like pachytene configuration* (**c**). *Homologous centromeres are 1 and 2 and 3 and 4.*

alternate-1 or alternate-2

homologous centromeres segregate
to opposite poles

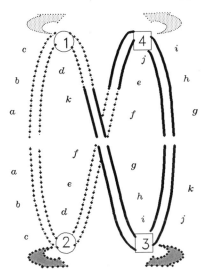 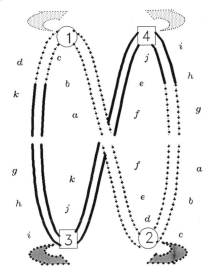

adjacent-1 or adjacent 2

non–homologous
centromeres to same
poles

homologous centromeres
to same poles

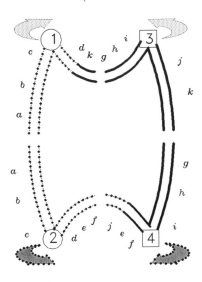 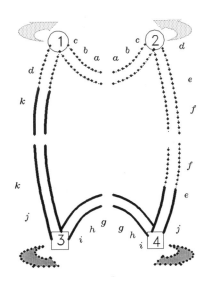

Fig. 5.17. *The four possible modes of metaphase I alignment: alternate-1, al-
ternate-2, adjacent-1, and adjacent-2.*

alternate 1st and 2nd

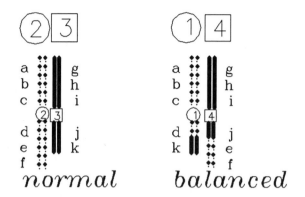

normal   *balanced*

adjacent-1   *adjacent-2*

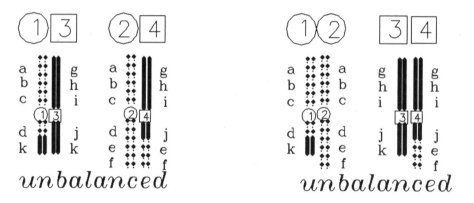

unbalanced   *unbalanced*

Fig. 5.18. *The results of meiosis I. The 2 and 3 pairs are normal; the 1 and 4 pairs bear balanced translocations; the 1 and 2 and 3 and 4 pairs are unbalanced, as are the 1 and 3 and 2 and 4.*

tributed to the same pole, while in the adjacent-2 type the adjacent members with homologous centromeres are distributed to the same pole. In either case unbalanced chromosome pairs are formed. The meiotic products with duplications and deficiencies are defective for normal function, unless the deficiencies are very small.

Translocation heterozygotes, like inversion heterozygotes, tend to produce defective gametes, leading to inviable zygotes. Hence they are expected to be selected against, especially in the case of translocation heterozygotes, because 50% sterility should result, as shown by Figure 5.18, if alternate and adjacent orientations are random. This has been found to be the case in plants simply by checking the frequency of shriveled pollen or the germination of seed produced by crosses involving heterozygotes (Burnham, 1956). Actually a wide range of types of orientation frequencies has been found, from 90% alternate for some animals and plants to almost total adjacent orientation (Rickards, 1983). In the latter case the translocations should have little chance of becoming fixed in the populations in which they arise.

If translocation heterozygotes inbreed, results such as those diagrammed in Figure 5.19 may be obtained. Standard homozygotes, translocation heterozygotes, and balanced translocation heterozygotes are expected to be viable and quite normal, but 28 of 36 possible zygote types will have deficiencies and duplications. Both deficiencies and duplications may be maintained in subsequent generations, if they are not lethal heterozygous, even after they become homozygous. Products of translocation hybrid crosses have been identified in plants such as cotton (*Gossypium hirsutum*). Menzel et al. (1986) have identified duplication and deficiency-bearing lines derived from chromosome translocation heterozygotes crossed to standard. These were identified cytologically and their phenotypes were noted. Deficiencies had a much greater effect on the phenotype than duplications, but some were nonetheless viable.

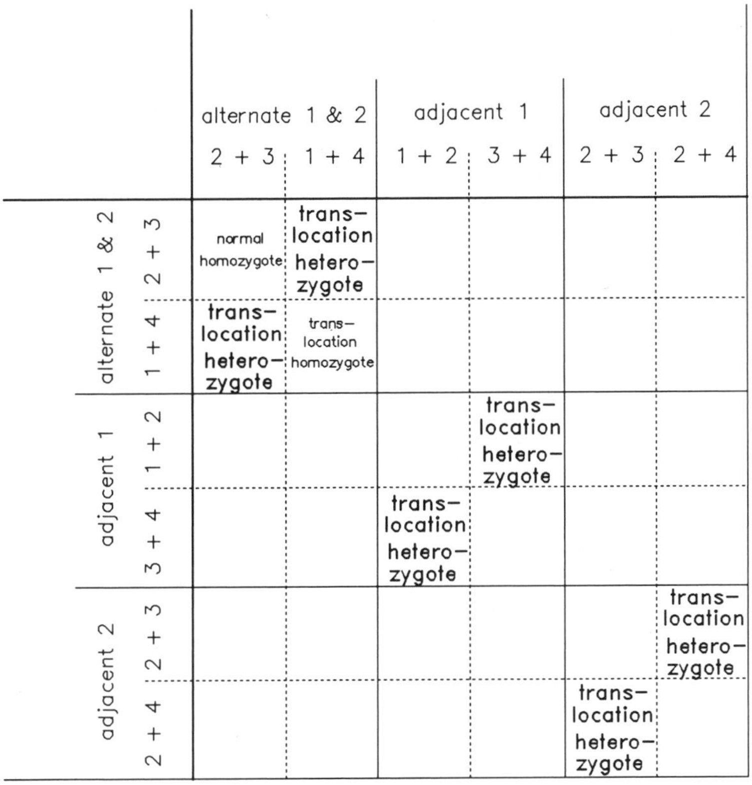

		alternate 1 & 2		adjacent 1		adjacent 2	
		2 + 3	1 + 4	1 + 2	3 + 4	2 + 3	2 + 4
alternate 1 & 2	2+3	normal homozygote	translocation heterozygote				
	1+4	translocation heterozygote	translocation homozygote				
adjacent 1	1+2				translocation heterozygote		
	3+4			translocation heterozygote			
adjacent 2	2+3						translocation heterozygote
	2+4					translocation heterozygote	

Fig. 5.19. *Results expected if translocation hybrid selfed. The balanced heterozygotes and normal homozygotes are the only ones deficient out of the 36 possibilities.*

C. Conserved Translocation Heterozygotes

One of the most remarkable examples of conserved translocation heterozygosis is to be found among plants in the genus *Oenothera*. deVries (1901) collected a series of what he called mutants of *O. lamarckiana*. It was upon the occurrence of these mutants that he based his "Mutation Theory" of evolution (deVries, 1901, 1903). This theory proposed that evolution did not proceed by a series of small, almost imperceptible changes, as proposed by Darwin, but by a series of big **saltations** (deVries, 1909). Soon after his initial proposal it became evident that genes are parts of chromosomes and most of his saltations were the result of segregation of chromosomes. It was recognized that deVries's strain of *lamarckiana* was a heterozygote of hybrid origin and that the "mutations" were for the most part new combinations of genes already present in the parent strains (Davis, 1911, 1916; reviewed by Clelend, 1972). It was through cytogenetic studies by a large number of investigators, principally Renner (1949), Shull (1928), Catchside (1936), Cleland (1932), and Emerson (1936), that it became clear that the largest number of so-called mutants arise through changes in chromosome number. One of the most important conclusions arrived at by the *Oenothera* workers was that most if not all genes in most races of *Oenothera*, including *lamarkiana*, are linked. This indicated that the "paternal" set of chromosomes separated from the "material" set at anaphase I of meiosis without independent assortment. This type of segregation was found to be the result of the formation of reciprocal translocation configurations among nonhomologous chromosomes with alternate segregation from the ring (Emerson and Sturtevant, 1931; Cleland and Blakeslee, 1931).

O. lamarckiana has a basic chromosome number of $n = 7$. At meiotic prophase I it shows a circle of 12 chromosomes plus a pair in (Fig. 5.20). This circle is the result of the pachytene configuration diagrammed in Figure 5.21 proceeding into metaphase as described in Figure 5.17. Figure 5.21 shows the pairing relationships of the six pairs of translocated chromosomes. To understand these translocation relationships better the genome of *Oenothera hookeri* has been adopted as a standard. Its genome shows no interchanges and is assumed to be the ancestral type. The seven metacentric chromosomes of *hookeri* have their arms labeled as follows: 1:2, 3:4, 5:6, 7:8, 9:10, 11:12,

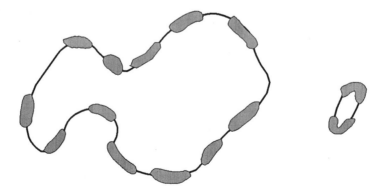

Fig. 5.20. *A diagrammatic representation of the metaphase I configuration of the 14 chromosomes of Oenothera lamarckiana. The large circle has the six pairs of translocated chromosomes. The untranslocated pair of homologues is shown at the right.*

Fig. 5.21. *Pachytene configuration of the six pairs of translocated chromosomes of Oenothera lamarckiana.*

13:14. The 1:2 metacentric of *lamarckiana* is not involved in the translocation complex; all the others are. Since the chromosomes of *hookeri*, and all other forms of *Oenothera*, are essentially equal-armed metacentrics, and the exchange breakpoints for the translocations occur near the centromeres, the configuration diagrammed in Figure 5.21 is made intelligible.

O. lamarckiana ordinarily contains two genomes of seven chromosomes each. One is designated the *velans* complex, the other the *gaudens* complex. The arm relationships for these two genomes, as compared with *hookeri*, are as follows:

hookeri   1:2 3:4   5:6 7:8   9:10 11:12 13:14

velans   1:2 3:4   5:8 7:6   9:10 11:12 13:14

gaudens   1:2 3:12 5:6 7:11 9:4   8:14 13:10

Thus the *velans* genome has a reciprocal translocation only between 5:6 and 7:8 to give 5:8 and 7:6, whereas *gaudens* has six to give 3:12, 5:12, 7:11, 9:4, 8:14, and 13:10. The result is the pairing pattern shown in Figure 5.21. It should be noted that the

numbering system used here is that of Cleland (1972). Other authors have used a different number system. The principles are the same, however, if disjunction is regularly such that alternates go to the same pole (Figs. 5.17 and 5.18), then an anaphase I pattern such as diagrammed in Figure 5.22 should be formed, and the *gaudens* and *velans* complexes separate intact. The 1:2 chromosomes segregate independently. Two kinds of gametes, gaudens plus 1:2 and *velans* plus 1:2, are produced. Therefore in the normal course of events three kinds of seed should be produced after fertilization: *gaudens.gaudens*, *velans.velans* and *gaudens.velans*. But the *gaudens* and *velans* complexes contain lethals that are different in each complex, so that they are balanced in heterozygotes and not in homozygotes. The result is that each complex remains intact and both function together only in the heterozygous state, and functionally each complex is a single chromosome. The "mutations" observed by deVries were probably due to changes in chromosome number, to occurrences of new translocations, and to crossing-over of a gene from one complex to another, or to gene mutations (Emerson and Sturtevant, 1931).

Translocation complexes of the *Oenothera* type have been identified in hybrid species of *Datura*, the jimson weed (reviewed by Avery et al., 1959) and the termite *Incitermes schwarzi* (Luykx and Syren, 1979). The translocations in this termite are found only in males and the number of them in a genome varies with the geographic location.

D. General Comments About Breaks and Gross Rearrangements

What is observed cytologically in the form of inversions and translocations in the mitotic or meiotic phases of the cell cycle is generally the result of breaks and

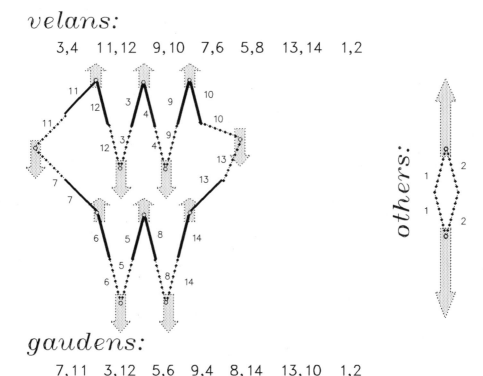

Fig. 5.22. *Alternate disjunction of translocated chromosomes of Oenothera lamarckiana.*

rejoinings in the interphase nucleus. Therefore we have essentially no good information about the frequency of breaks, their distribution among the different chromosomes of the genome, and whether some breaks repair and restitute more often than others. Thus we are innocent of a full understanding of the factors involved in the formation and survival of gross rearrangements. Those that we detect outside the interphase may be only a small sample of those that occur. What, then, can be said about the rearrangements that are detected? Probably we can state with some assurance that (1) although balanced rearrangement heterozygotes may maintain their form during mitosis, the rearrangement may be selected against and lost during meiosis; (2) those rearrangements that persist through many successive generations of sexual breeding generally appear to have no effect on the phenotype; (3) those rearrangements that do have a phenotypic effect may result in death at stages ranging from the haploid sperm or egg, or the gametophyte of an embryophytic plant, to the zygote stage or other stages prior to adulthood (hence they may never be recognized); and (4) those rearrangements that appear to have no phenotypic effect may in fact have selective value (this may not be recognized except after many generations of breeding).

III. REARRANGEMENTS BY TRANSLOCATION OF MOBILE ELEMENTS

In the 1940s McClintock began to publish a series of communications describing certain peculiar phenomena in her maize plants (McClintock, 1951). Her findings proved to be among the more important ones made in genetics in the first part of this century. They were the beginnings of the recognition that segments of DNA can move about from one part of the genome to another within and between chromosomes. This nomadic activity has somewhat the same characteristics as the insertional translocations described on p. 187 of this chapter. But it is clear that there is a phenomenon, different from translocation, that can be called **transposition**, and that results from the movement within genomes of segments of DNA called **transposable elements**. The movement of these **mobile elements** from one position to another with the involvement of an enzyme called a **transposase** frequently results in a variety of phenotypic changes not characteristic of insertional translocations. Furthermore, these elements have a specific structure and their position within the genome can affect the expression of specific genes. This phenomenon is widespread among both the Prokaryota and the Eukaryota.

Just how many different kinds of transposable elements exist among the eukaryotes is not known. At present writing at least 50 different kinds are known. Finnegan (1989) has contrived a classification scheme for 26 of the best-known ones and this is presented in Table 5.1 along with additional data taken from Weiner et al. (1986). It should be noted that some of the statements in Table 5.1 are assumptions and therefore provisional. But the table does organize what we think we know about these elements so that they can be discussed in an organized fashion pending further experimental evidence. In this classification two classes are defined on the basis of whether the elements transpose through an RNA intermediate (Class I) or directly as DNA (Class II). Each class in turn has subcategories based primarily on the nature of the terminal sequences of the elements. In this discussion we consider some examples of Class II elements first, mainly for historical reasons, as well as the fact that the transposable elements of maize and the *P* elements of *Drosophila* have received a great deal more attention than most of the other types.

A. Transposition in Plants

It is appropriate to begin the discussion of mobile elements with maize, since it was in this organism that the phenomenon was first recognized by McClintock. Comprehensive reviews of the purely genetic aspects have been made by Fincham and Sastry (1974) and Nevers et al. (1985), and a summary based on these and other voluminous commentaries is given here of the genetic findings followed by a discussion of the more recent biochemical and molecular work, which has been summarized by Doring and Starlinger (1986), Gierl et al. (1989), and Federoff (1989).

1. Cytogenetics and genetics of controlling elements in maize

McClintock (1956) coined the term **controlling element** to describe transposable elements that manifest themselves through control of the activities of known maize genes. A controlling element may inhibit the activity of a gene by becoming integrated within or close to it. It may be excised from its site of integration in either germinal or somatic tissue, in which case the gene may regain some or all its activity. The same element may transpose and reintegrate elsewhere and affect the activity of another gene.

Maize transposable elements can act either autonomously or nonautonomously. **Autonomous** elements act alone in their transpositions. **Nonautonomous** elements are controlled by other elements

TABLE 5.1. Classification of Mobile Elements According to Finnegan (1989) and Weiner et al. (1986)

Class I. Transpose by means of RNA intermediates
    A. Viral superfamily (retrovirus-like retrotransposons)
        1. These have the following characteristics:
            a. Have long direct repeats (LTRs)
            b. Encode reverse transcriptase from ORFs in DNA between LTRs
            c. Able to generate 4- to 6-bp target site duplications
            d. Have no 3′ terminal poly(A) track
            e. Are dispersed in genome
        2. Examples
            a. *Ty* (*Sacharomyces cerevisiae*)
            b. *Copia-like* (*Drosophila melanogaster*)
            c. *DIRS-1* (*Dictyostelium*)
            d. *BS1* (maize)
            e. *IAP* (rodent)
            f. *THE* (humans)
            g. *VL30* (rat and mouse)
    B. Nonviral superfamily (Nonviral retroposons)
        1. These may have the following characteristics:
            a. Have no terminal repeats
            b. Have ORFs
            c. Do not encode enzymes responsible for their transposition (passive transposition)
            d. Have 3′ terminal poly(A) tract
            e. Be dispersed in genome
        2. Examples
            a. Transcripts of RNA POLII
               F family (*Drosophila melanogaster*)
               Lines 1 family (human, ape, monkey, mouse, and rat)
               Processed pseudogenes (retropseudogenes)
            b. Transcripts of RNA POLIII
                SINES
               7SL RNA retropseudogenes
               B1 family (rodents)
                Alu family (primates)
               7SK RNA retropseudogenes
               tRNA retropseudogenes
               Polymerase unknown
               ING1/5RS1 (Trypanosomes)
Class II. Apparently transpose directly: DNA → DNA by transposase
    All have terminal inverted repeats (IRs)
    A. With short inverted repeats (SIRs)
        P; *hobo* (*D. melanogaster*)
        Ac–DA; *Spm/En* (Maize)
        Tam (*Antirrhinum majus*, Snapdragon)
        Tc1 (*Coenorhabditis elegans*)
    B. With long inverted repeats (LIRs)
        FB (foldback) (*D. melanogaster*)
        TU (*Strongylocentrotus purpuratus*)

located elsewhere in the genome. Mc-Clintock has used the term **operator** to describe the element that integrates near or within the affected gene, and the term **regulator** to describe the element acting from a distance upon the operator. These terms should not be confused with the operator and regulator functions at the *lac* operon of *E. coli*, for example. Hence it has been suggested by Fincham and Sastry

(1974) that instead of operator the term **receptor** should be used. This will be done here.

In the autonomous systems the regulator and the receptor elements are in one unit and so are described as one-element systems. Two-element, nonautonomous systems, can arise from autonomous ones by loss of the regulator. The remaining receptor can then respond to regulators elsewhere in the genome. Two families of maize transposing elements have been identified: The *Activator–Dissociation* (*Ac-Ds*) and the *Suppressor–Mutator* (*Spm*) groups of elements. Most of the phenotypic effects of these elements are recognized in the corn kernels. A diagram of a kernel is given in Figure 5.23. Maize, like other angiosperms, forms a **female gametophyte** or **embryo sac** consisting of eight haploid nuclei derived from mitotic divisions of a megaspore haploid nucleus. One of these eight becomes the egg and two move to the central part of the embryo sac and become polar nuclei. The pollen grain sends a tube into the sac and delivers two sperm nuclei; one fuses with the egg to produce the 2n zygote, and the other with the two polar nuclei to produce a 3n nucleus that by successive mitotic divisions forms the endosperm tissue. The embryo develops from the 2n zygote. Table 5.2 lists some of the maize genes whose expression in kernel tissues is modified by the elements.

Studies of the breakage–fusion–bridge cycle of maize involving the short arm of chromosome 9, as described in Figure 5.7 and on page 189, indicated that the element *Dissociation* (*Ds*) caused instability resulting in chromosome breaks when in the presence of *Activator* (*ac*). (*Ac* is also called *Modulator*, *Mp*, by some workers.) The genes Y_{g2}, C_1, Sh, Bz, and Wx are on the short arm of 9 along with *Ds* (Fig. 5.24). Some of the maize genes whose activities are affected by controlling elements are listed in Table 5.2. *Ds* has been shown to move from the locus indicated In Figure 5.24a to other positions indicated by the arrows. When moved to the C_1 locus, it causes instability of action of the C_1. A mutable allele, C_{1m-1}, has been produced at this locus by *Ds*. C_1 is a pigment gene and its recessive alleles produce no anthocyanin pigment when homozygous.

Ds is not only site-specific for chromosome breakage in the presence of *Ac*, but it can transpose only in the presence of *Ac*

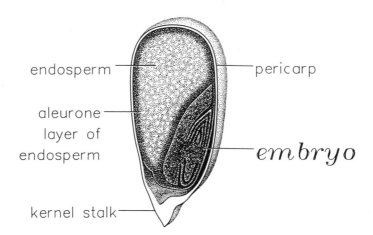

Fig. 5.23. *A section through a maize kernel showing some of its parts. The endosperm is a triploid tissue formed by fusion of a haploid pollen sperm haploid nucleus with two haploid polar nuclei in the embryo sac. The aleurone layer of the endosperm is the surface region in which anthocyanin pigment is formed. The pericarp is diploid tissue derived from the sporophyte. See text.*

TABLE 5.2. Some Maize Genes Affected by Controlling Elements

C1, R, C2, A, Bz, and *Bz2*: These are all involved in the synthesis of pigment in the aleurone of the endosperm of the maize kernel (See Fig. 7.19). The Bz gene encodes UDP-glucose:flavonoid 3-0-glucosyltransferase, which is involved in anthocyanin synthesis. The *C2* locus encodes chalcone synthase.

P: Controls pigment formation in the cob and pericarp tissue of the kernel.
Sh: Encodes endosperm sucrose synthetase. Plants homozygous for the recessive *sh* have shrunken seeds.
Wx: Encodes UDPG:glucocyltransferase. *Wx* kernels contain amylose and amylopectin while the recessive *Wx* (waxy) homozygotes have no amylase.

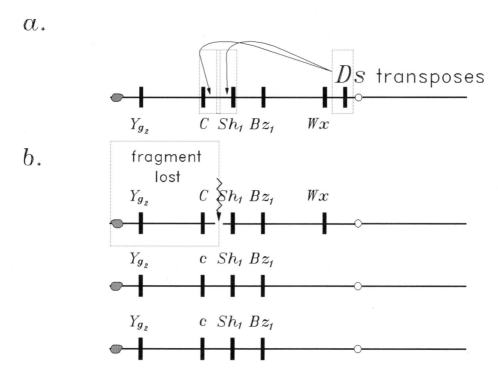

Fig. 5.24. a. *Transposition of Ds from the region between Wx and the centromere on the short arm of chromosome 9 of maize to C1, causing a mutation, or to a region between Sh and C, causing a break as shown in (b) and resulting in the loss of the C1 allele and a colorless endosperm.*

in the same cell. Thus if *Ds* causes a mutation of a gene, that mutation becomes stable if *Ac* is removed. Changes in phenotype of the endosperm of the kernel can also result from the breaks induced by *Ds*. For example, if the polar nuclei of genotype *c/c* in the embryo sac (Fig. 5.23) fuses with a pollen sperm nucleus carrying the dominant *C* allele, the endosperm will be *Ccc* and colored. But if a break occurs proximal to *C*, as shown in Figure 5.24b, the segment of the chromosome distal to the break will be acentric and lost and the endosperm will be colorless. It was events such as these that convinced McClintock that the breakage–fusion–bridge cycle was an important factor in uncovering the activities of the transposing factors.

The *Spm* element family, also called the *En/Spm* family, has both autonomous and nonautonomous elements like the *Ds-Ac* family. The nonautonomous elements are identified as *Spm* or *I* elements, and can be transposed only in the presence of an active autonomous *Spm* element. Thus they are receptor elements. The *Spm* system of transposable elements is different from and more complex than the *Ds-Ac*. Its members are autonomous and as such have a receptor component, *Rs*, a suppressor component, *Sp*, and a mutator component, *m*, as well as regulator functions. When *Spm* is present, its *Sp* component completely inhibits the functions of those mutable alleles, producing nonmutant or partially mutant phenotypes. The *m* component of *Spm* can condition a restoration of gene function in some cells, giving rise to clones of fully functional cells in a background of mutant cells.

One example of *Spm* family activity has been described by Nelson and Klein (1984). They have shown that the association of *Rs* with a pigment-forming *Bz-1* allele (Bronze-1) forms a two-element *Spm*-controlled *bz*-mutable allele identified as *bz-m13*. The normal *Bz* alleles are structural genes that encode the enzyme that catalyzes the last step in anthocyanin synthesis. The recessive alleles at this locus are nonfunctional in this respect. In the absence of *Spm*, endosperm of the genotype *bz/bz/bz-m13* has full anthocyanin production in the aleurone layer. But in the presence of *Spm*, one dose of *bz-m13* gives a variegated phenotype resulting from a high rate of change (70% or more) to stable *bz* or *Bz* derivatives. A possible mechanism suggested for these changes is the alteration of *Rs* or its transposition away from the *Bz* locus.

Schwartz (1960) early found that recessive alleles (*sh*) of Shrunken (*Sh*) on chromosome 9 and *Ds* at that locus and the protein encoded by *Sh* was missing. It was later shown that the protein was present but altered in structure and activity (Hanna and Nelson, 1976). Subsequent investigations with mutants at the *Bz* locus with *Ds* involvement extended these observations. The nonmutant alleles at the *Bz* locus encode the enzyme, uridine diphosphate glucose:flavonoid glucosyltransferase (UFGT). An analysis of several of the *bz* mutants led to the conclusion that the association of *Ds* with *Bz* can result in (1) the absence of detectable gene product, (2) an alteration of the enzyme's structure, and (3) a change in the pattern of the enzyme's production and hence expression (Dooner and Nelson, 1977a,b). These findings taken together indicate that the relationships between *Ds* and the genes with which it associates are not simple ones with a single key causation. This also became evident when it was found that UFGT production is controlled by at least three other genes besides its structural gene *Bz*. The genes *C*, *B*, and *Vp* act late in the development of the endosperm and induce high levels of UFGT. Mutations at any one of these loci result in low levels of UFGT in *Bz* strains and a reduction in pigmentation of the aleurone so that the kernels are almost colorless (Dooner and Nelson, 1979). The genes *C* and *Bz* are on the short arm of 9, but *R* and *Vp* are on 10L and 3L, respectively. Conclusions made about the action of transposable elements based only on observations of changes in phenotype must at best be considered inconclusive.

Ac and *Ds* elements have been cloned and sequenced. The *Ac* element is 4.6 kb long; it has 11-bp inverted terminal repeats (IRs) and encodes a 3.5-kb mRNA with five exons (Fig. 5.25) (Pohlman et al., 1984; Doring and Starlinger, 1986; Federoff, 1989). *Ac* structure is conserved, but not *Ds* structure. Several *Ds* elements that have been excised from mutant genes and sequenced all show structural relation-

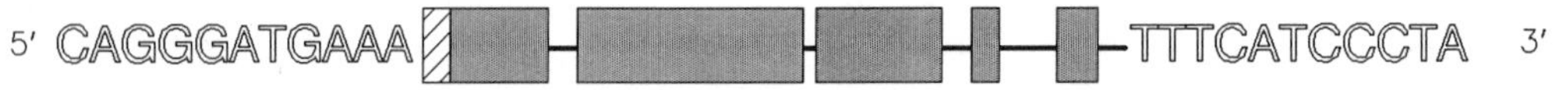

5′ CAGGGATGAAA ▨▬▬—▬▬▬▬—▬—▬—TTTCATCCCTA   3′

Fig. 5.25. *Structure of the Ac element of maize. The solid bars indicate exons and the intervening lines, introns. The inverted terminal repeats are characteristic of these maize elements. Data from Pohlman et al. (1984).*

ships to the *Ac* element, since they have the same inverted terminal repeats. They differ primarily in having internal deletions ranging from about 200 to several thousand base pairs long (Dooner et al., 1985; Federoff, 1989). They are apparently deletion derivatives of the *Ac* element and are presumably incapable of transposing independently because of these deletions. Not all *Ds* elements cause chromosome breakage. Those that do, and have been cloned, are identical in structure. They are called **double**-*Ds* because they consist of one *Ds* inserted into another.

When *Ac* inserts into a gene, it generates a short duplication of an 8-bp segment already present in the host gene before transposition. Figure 5.26 illustrates the insertion process involving the waxy gene, *Wx* (Pohlman et al., 1984). This gene encodes the enzyme uridine diphosphate glucosyltransferase, which is involved in the synthesis of amylose in the endosperm. When the *Ac* element leaves its site of integration in the host *Wx* gene, the gene may revert to the nonmutant condition, but a portion of the duplication generated remains. Transposition of *Ac* does not involve its replication independently of host chromosome replication. The element will transpose either during or after chromosome replication, giving rise

a. insertion
     site in
     Wx gene

--- TCCCGCAGCT**CATGGAGA**TGGTGGAGGA

b. Ac inserted into Wx gene

CATGGAGATAGGGATGAAA---TTTCATCCCTA CATGGAGA

c. remains of duplicated
     segments in
     Wx revetant

TCCCGCAGCT**CATGGAGA---TGGAGA**TGGTGCAGGA

Fig. 5.26. *The insertion of Ac into the waxy gene of maize.* **a.** *The boxed insertion sequence present in the Wx gene.* **b.** *The inserted Ac with only the terminal repeats and the additional insertion sequence generated by Ac.* **c.** *When Ac excises from the host gene it leaves behind some of the duplicate insertions sequence. Data from Pohlman et al. (1984).*

to one sister chromatid with and one without the element (Greenblatt, 1984). The transposing *Ac* may then insert in a new site that is generally in a hypomethylated region. *Ds* elements can transpose by the *Ac*-type mechanism, but the double-*Ds*-type transposition is frequently accompanied by chromosome rearrangements.

The *Spm* family has, like the *Ac-Ds* family, both transposition-competent and transposition-defective elements. The structure of *Spm* is diag ·ammed in Figure 5.27. It is 8.3 kb long and has terminal inverted repeats consisting of 13 nucleotides. Some similarity to the *Ac* terminal sequences exist, since both have CA at the 5′ end and a series of A's at the 3′ end. Strains of maize with an active *Spm* produce a major 2.5-kb transcript containing the exons shown in Figure 5.27 as filled boxes. The vertical arrow indicates the leader exon that is not translated but is necessary for the active functioning of *Spm*. It is also this region that is heavily methylated in those *Spm* elements that are inactive or cryptic. The long intron between this 5′ end exon and the downstream exons contains two open reading frames (ORFs; stretches of transcribed DNA without stop codons): OFR1 and OFR2. OFR1 may encode the so-called transposase, the enzyme required for the insertion of the element. The protein produced by translation of the transcript of the ex-ons at the 3′ end is designated S protein and is believed to be the *Sp* functional component of *Spm* (Gierl et al., 1988). The *dSpm* elements that have been isolated and sequenced prove to be *Spm* with deletions ranging from about 1.5 kb to almost all of *Spm* except the terminal repeats. In this sense they are analogous to the *Ds* elements.

Ac and *Spm* and their associated components can apparently insert in any gene in the maize genome. Most of the mutations that result from their actions are the result of insertions into translated exons. However, introns, untranslated exons, and 5′ flanking regions within 200 bp of the transcription start site have also been found to be involved. When *Ac* and *Spm* transpose, they usually insert on the chromosome only about 4 centimorgans (cM) away from the site from which they are excised. When these elements are inserted, they generate sequence changes of the type illustrated for the waxy gene in Figure 5.26. When *Ac* inserts, it does so at the 8-bp sequence site, CATGGAGA, at one end and generates another CATGGAGA at the other. When it is excised, it leaves behind remains of some of its duplicate sequences. The result is that even if the gene reverts to give a normal phenotype, it is still a "mutant" gene. For this reason excision of these elements provides a mechanism for producing proteins with altered

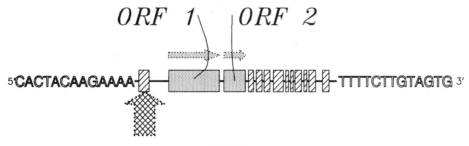

Fig. 5.27. *Structure of the Spm element of maize. See text.*

properties (Wessler, 1988). The excision of either autonomous or nonautonomous elements thus creates genetic diversity by adding nucleotides to encoding or regulatory regions. Even changes within introns can be significant, since they may change the splicing pattern of the unspliced transcript. Genes that are most likely to be invaded by these elements are most likely to be active ones, since the elements appear to fancy inserting into hypomethylated target sites. Most strains of maize do not have active *Spm* or *Ac* elements, but they do have multiple copies of sequences that are homologous to the sequences of these elements. Presumably these are maintained in an inactive, cryptic state by hypermethylation (Federoff, 1989).

2. *Transposable elements in other plants*

Transposable elements of the type identified in cultivated maize have also been found in its monocot relatives *Tripsacum* and *Teosinte*, and in the unrelated dicots *Antirrhinum majus* (snapdragon) and *Glycine max* (soybean). The transposable elements *Tam1* and *Tam2* of *Antirrhinum* show close homology to *Spm* and *Ac*, respectively (Coen and Carpenter, 1986). The *Tgm1* element from *Glycine* has inverted repeat termini almost identical to those of *Spm*. These types of elements appear to be quite ancient and constitute a family whose members are widely distributed throughout at least the flowering plants.

Transposable elements from maize have been transferred to tobacco. Baker et al. (1986) have transformed tobacco cells with plasmid vectors that contained either *Ac* or its *Ds* derivative flanked by short segments of the waxy gene of maize from which they were isolated. It was possible to show that the *Ac* element acted autonomously in tobacco and could transpose from another site in its genome. On the other hand, *Ds* did not transpose, as expected, if to-

bacco did not have an active element comparable to *Ac*.

B. Transposable Elements in *Drosophila*

Extensive studies have been done with the transposable elements of *Drosophila melanogaster*. At least six structural types have been identified: *copia-like*, *hobo*, *foldback*, *F*, *P*, and *I* (reviewed by Finnegan and Fawcett, 1986). They occupy fully two thirds of the moderately repetitive fraction of the flies' genome. This amounts to 10% of the total genome. The *P* element family has been investigated perhaps more than the other five, and we use it here as the primary exemplary case (reviewed by Engels, 1983, 1989; Rio, 1990).

1. *Drosophila P elements*

What is considered to be the primary *P* factor has been sequenced by O'Hare and Rubin (1983). It consists of 2.9 kb with a 31 bp inverted terminal perfect repeats (Fig. 5.28). Three long and one short open reading frames have been identified for the 5′ → 3′ strand. It has been proposed that these are involved in the production of *P*-specific proteins. RNAs complementary to this strand have been identified (Karess and Rubin, 1984). The ORFs apparently contribute to the encoding of a single 87-kd polypeptide with transposase activity. Besides the 2.9-kb *P* element, shorter elements have been identified. Some of these have been isolated and prove to be related to the 2.9-kb element. They all have 31-bp terminal repeats identical to those of the 2.9-kb element, but they all have internal deletions, and are nonautonomous in their activity.

P elements are not ubiquitous in populations of *D. melanogaster*. Some strains collected from the wild have them; some don't. If all members of the species had them, *P* factors would never have been

a.

$5'$ **ORF 1**   **ORF 2**   **ORF 3** $3'$

b.

$5'$ **CATGATGAAATAACATAAGGTGGTCCCGTCG** $3'$

$3'$ **GTACTACTTTATTGTATTCCACCAGGGCAGC** $5'$

Fig. 5.28. *Structure of the P element of Drosophila melanogaster.* **a.** *The element, showing the terminal repeats as black boxes and the approximate positions of the ORFs.* **b.** *Sequence of the terminal repeats.*

recognized. The reason is that crosses involving both parents with *P* elements produce normal progeny, but male flies with *P* crossed to females without *P* have strikingly abnormal progeny (Fig. 5.29). Flies without *P* are said to have *M* (maternal) cytotype in contrast to those with *P* cytotypes. *M* females crossed to *P* males produce progeny that have germ lines with elevated levels of mutation and chromosomal aberrations, and are sterile if raised at 29° because their gonads fail to develop. This is called **hybrid dysgenesis**, and does not eventuate from the other types of crosses diagrammed in Figure 5.29.

The fact that the reciprocal cross *M* males × *P* females produces normal progeny indicates a maternal effect. Apparently *M* cytoplasm lacks the repressors of transposition of *P* elements. It is these transpositions that are responsible for dysgenesis. Thus, since only *P* cytoplasm has the repressor activity, the three crosses listed in Figure 5.29b should be nondysgenic.

The major effect of *P* elements in an *M* cytotype environment is **gonadal dysgenic sterility**. Progeny of both sexes have temperature-sensitive sterility. The effect is inconsequential below 22° and increases as the temperature at which the progeny

are raised up to 29°. (Above 29° even normal flies become sterile). *P* elements are not normally active in somatic cells. Hence the effect of the *P* element is limited to the germ line and these gonadal cells die, while somatic line cells develop into adult flies. This makes it difficult to study whatever effects *P* elements may have beyond sterility. To overcome this difficulty a modified *P* element, Δ2-3, was used (Engels et al., 1987). This element has a very high level of transposase activity, but it does not itself respond to its own transposase and is therefore quite stable (Robertson et al., 1988). When present alone in the genome of the fly, it has no observable effect. Progeny arising from crosses of Δ2-3 males to *M* cytotype females are normal. But when Δ2-3 flies of either sex are crossed to a strain designated as Birmingham, a marked effect results. Birmingham flies have defective *P* elements apparently similar to the type described by O'Hare and Rubin (1983), which have neither transposase nor repressor of hybrid dysgenesis such as found in the standard *P* element strains. Despite the presence of approximately 60 of these defective *P* elements per genome Birmingham acts as an *M* cytotype. These elements are nonautonomous and hence are not mobile. In

a. dysgenic cross

M^{female} × P^{male}

progeny have temperature-
sensitive sterility
germ line cells have elevated mutation
rates including gross rearrangements

b. nondysgenic crosses

M^{female} × M^{male}
P^{female} × M^{male}
P^{female} × P^{male}

have no temperature-
sensitive sterility
progeny normal

Fig. 5.29. *The crosses of melanogaster, which produce dysgenic results* (**a**) *and normal results* (**b**).

the presence of $\Delta2$-3 they become movable and have a wide range of somatic effects. Pupal lethality is almost 100% at elevated temperatures up to 28°. Many survivors that were recovered after the larvae were raised at temperatures down to 19° show a variety of developmental anomalies such as gonadal sterility and reduced longevity. These effects are presumed to be caused by chromosomal breakage during the larval period, especially in the imaginal disks that develop into adult structure during the pupal period. The excision and transposition of the defective P elements in the presence of the transposase of the $\Delta2$-3 element is considered to be the cause of these effects (Engels et al., 1987).

Mutations can be induced by P elements by (1) the insertion of the element into the gene, (2) the excision of the element from the gene, and (3) a rearrangement with one breakpoint in a gene and one at the site of a P element within that gene (Craig,

1990). P elements have different effects depending on their sites of insertion. For example, the singed gene, sn^+, of *melanogaster* has a hot spot mutating at a rate as high as 10^{-3} during hybrid dysgenesis. This is in contrast to the *Adh* gene, which shows a mutation frequency of less than 10^{-6}. Twenty-two mutations induced at the sn locus have been analyzed for site insertion and all 22 have P inserted within the same 700-bp region. Additionally 10 of the inserts whose more precise loci have been determined had four sites within a 100-bp interval (Roiha et al., 1988).

A dysgenesis-induced allele of singed, "weak singed" (sn^w) is hypermutable. It reverts to wild-type (sn^+) or to an extreme singed phenotype (sn^e) at a frequency sometimes exceeding 50%. This hypermutability occurs only in the M cytotype, but it is stably transmitted in the P cytotype (Engels and Preston, 1984). The sn^w allele apparently has two P elements in inverted orientation. The excision of one of them gives rise to the sn^e allele, and of the other the sn^+ allele (Rioha et al., 1988).

2. The Drosophila F element, Doc

Transposing elements of *melanogaster* are possibly responsible for many of the chromosomal aberrations noted in this species. Elements other than P elements are also involved with chromosome breaks and hence rearrangements. For example, a rearrangement is associated with a dominant mutation at the Antennapedia locus (*Antp*). A heterozygous fly ($Antp/Antp^+$) has its antennae transformed into legs. *Antp* is lethal homozygous. An allele of $Antp^+$, $Antp^{73b}$, apparently arose by the insertion of one copy of an F transposon-like element, *Doc*, into the first intron of *Antp* and a second copy in the first intron of the gene *rfd* (responsible for dominant phenotype) on the same chromosome arm 3R. The resultant inversion spans the region 84B1-2 to 84D1-2 of the polytene chromosome and is cytologically detectable (Schneuwly et al., 1987a). In Figure 5.30 Panel a is a diagram of the *Antp* gene showing the site where *Doc* has been in-

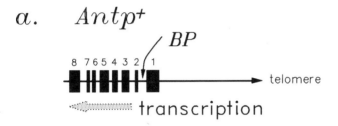

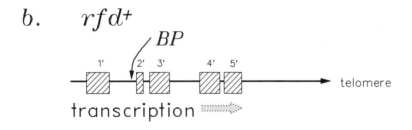

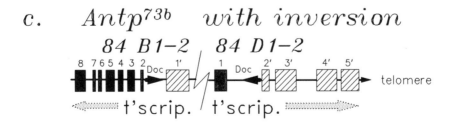

Fig. 5.30. *The conversion $Antp^+$ and $Antp^{73b}$ by insertion of two copies of the transposing element D at the sites indicated by BP.* **a–c.** *See text.*

serted by an arrow, and Panel b diagrams the *rfd* with its corresponding insertion locus for the second *Doc* element. Panel c depicts the mutant Ant^{73b} structure, showing the two *Doc* elements in opposing orientations, and also the inverted region. *Antp* and *rfd* are transcribed in opposite directions and the *Doc* elements are oriented in the same direction as transcription.

Figure 5.31 shows that the inversion could be generated by the two *Doc* elements pairing homologously to form a loop (Panel a and b). If a crossover followed within the paired *Doc* regions (Panel c) an inversion would result (Panel d). A sequence analysis of $Antp^{73b}$ has shown that the inversion depicted in Figure 30c has caused a reciprocal exchange of the promoters and first exons of *Antp* and *rfd*. The mutant phenotype produced by $Antp^{73b}$ / $Antp^+$ is presumably the result of transcription products of $Antp^{73b}$ that encode polypeptide products with altered expression owing to the coding region of *Antp* being separated from its normal control region. One transcript should have exon 1 of *Antp* and exons 2′, 3′, 4′, and 5′ of *rfd*, and the other should have exon 1′ of *rfd* and exons 2–8 of *Antp*. The dominant effect is a homeotic one in which the body plan of the fly is altered. The $Antp^+$ allele apparently dictates that the second thoracic (mesothoracic) segment develop a pair of legs, and the transformation of antennae into second thoracic legs is possibly the result of the ectopic overexpression of the *Antp*-encoded protein (Schneuwly et al., 1987b).

C. Retrovirus-Like Elements

These are members of the **viral superfamily** listed in the class I category of Table 5.1. Many members of this family encode a reverse transcriptase and an integrase or transposase. They can be defective in one or the other or both, and they depend on the other members of the superfamily present in the same genome to supply the necessary missing function. The characteristic of these elements that sets them apart from others is their undoubted close resemblance to certain known RNA retroviruses as determined by DNA sequence similarities. Their elements are therefore generally referred to as **retrotransposons** (Boeke et al., 1985; Baltimore, 1985). They differ significantly from other elements because of the involvement with **reverse transcriptase** and by the possession of direct rather than inverted terminal repeats.

The enzyme reverse transcriptase was originally discovered by Temin and Mizutani (1970) and Baltimore (1970). It has the capacity to transcribe a complementary DNA strand from an RNA template strand—hence the name reverse transcriptase. A DNA→RNA→DNA sequence is thus made possible by the action of an RNA POL followed by reverse transcriptase activity (Wintersberger and Wintersberger, 1987).

1. Ty, a retrotransposon of yeast

The *Ty* elements constitute a family in *Saccharomyces cerevisiae* (reviewed by Finnegan, 1985; Fink et al., 1986). These elements have direct terminal repeats known as δ sequences about 300 bp long. One member of the family *Ty1* is about 6.2 kb long in total length. The terminal δ sequences for different members of the *Ty* family show a considerable degree of heterogeniety. Delta sequences separated from complete *Ty* elements are called "solo δ's," and have highly variable sequences of nucleotides. They cannot transpose independently (Roeder and Fink, 1983).

Different isolates of yeast vary in numbers of complete *Ty* elements. Some have as few as 5 copies and some as many as 35. As many as 100 solo δ's may be present

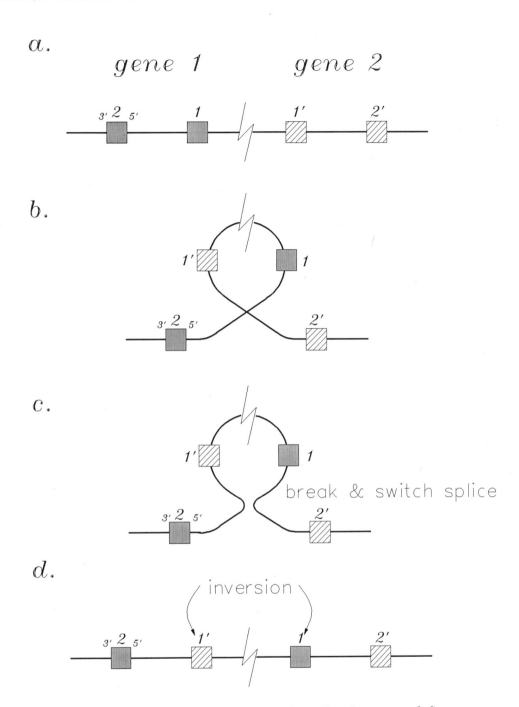

Fig. 5.31. *Model for generation of an inversion by two Doc elements.* **a–d.** *See text.*

in the genomes of some strains. Two kinds of polyadenylated RNA transcripts are found in *Ty*-infected cells: One is 5.7 kb long and the other 5.0 kb long. The 5.7-kb transcript has δ sequences at both ends. Translation of these transcripts produces a polypeptide with reverse transcriptase activity (Garfinckle et al., 1985).

Ty elements appear to have a predeliction for inserting into the regulatory regions of genes, although some few instances of insertion into the transcribed regions have been reported (Sunchen et al., 1984; Rose and Winston, 1984; reviewed by Sandmeyer et al., 1990). Insertion of *Ty* in the 5'-regulatory region may increase or decrease the degree of expression of the gene. Some *Ty* elements may in fact contain enhancer sequences (Young et al., 1982) that increase the rate of transcription.

The *Ty* elements differ from the transposing elements of *Drosophila* and plants previously discussed because they transpose through an RNA intermediate (Boeke et al., 1985; reviewed by Sandmeyer et al., 1990). The elements are indeed DNA but during transposition information in *Ty* flows from DNA to RNA to a DNA insert. As shown in Figure 5.32 the 5.7-kb RNA transcript has δ long terminal repeats (LTRs) of RNA. This transcript represents two genes, *TYA* and *TYB* (Clare and Farabaugh, 1985). *TYA* has sequence homologous to certain retroviral *gag* genes that encode the viral core proteins, and *TYB* is homologous to retroviral *pol* genes that encode proteases, reverse transcriptase, and integrase. Figure 5.32 depicts the relationships of these various entities.

One of the more interesting findings about *Ty* made by Boeke et al. (1985) is that when this element is placed adjacent to the *GAL1* promoter, its expression is under control of galactose. When the promoter is induced by galactose, the rate of *Ty* transposition is elevated. In addition the presence of *Ty* elements in the yeast genome can lead to the generation of gross rearrangements as well as solo δ ele-

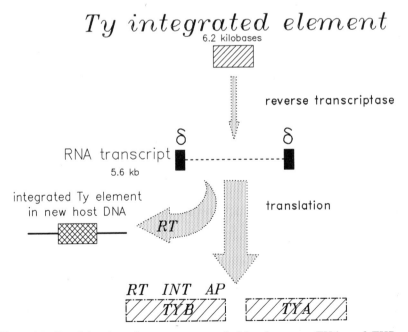

Fig. 5.32. *Participation of sequences encoded by the genes TYA and TYB of the yeast Ty element. RT, reverse transcriptase; INT, integrase; AP, protease.*

ments. Recombination between δ elements at the ends of complete Ty elements can also result in the formation of circular molecules with one δ and leave the other behind in the chromosome. This may be one way in which solo δ's are formed.

2. *Copia in D. melanogaster*

Copia was first recognized as a circular, extrachromosomal DNA molecule in cultured cells of *melanogaster* (Flavell and Ish-Horowicz, 1981). It was then found to have a sequence analogous to retroviruses (Emori et al., 1985) and to have homologous sequence counterparts inserted in the genomes of laboratory strains of *melanogaster* (Junakovic et al., 1984). *Copia*-like elements constitute a family of elements not confined to *Drosophila*, but are also found in certain Lepidoptera (Miller and Miller, 1982). One *copia* from *melanogaster* has been sequenced (Emori et al., 1985). It has 5,143 bp and terminates with 276-bp-long LTRs that are identical or very similar to the LTRs of other copias. The organization of *copia* element, *kcop88* is depicted in Figure 5.33 along with its encoded products. Two RNA transcripts have been identified: 5-kb VLP-hRNA, and a 2-kb one related to the 5-kb transcript. Both have 3' poly(A) tails and both appear to be the coding sequence for a 31-kD VLP protein that is the reverse transcriptase for the element.

Three laboratory strains of *melanogaster* have *copia* liberally distributed over the X and both arms of the second and third chromosomes (Young and Schwartz, 1981). Their loci do not appear to be completely random. Some locations on the polytene chromosomes have heavy concentrations of elements as determined by hybridization with labeled copia elements. One region of the X chromosome that appears to be especially favored is the white locus, w^+. Various mutant alleles of w^+ have resulted from the insertion of a copia retrotransposon at this locus (Bingham and Judd, 1981; Mount and Rubin, 1985). Reversions to w^+ are accompanied by the excision of copia. The copia inserted in w^+ to produce the apricot (w^a) allele is 3' to the white promoter, and is transcribed in the same direction as the gene. A partial reversion of w^a toward wild-type, designated w^aLTR, has a 276-bp insert remaining after the excision of most of copia. This residue has been sequenced and is the 3'LTR of copia (Zachar et al., 1985). Numerous other alleles of white have been analyzed and found to have copia inserts. Some of them are dominant. None are lethal homozygous—an expected result, since a homozygous deletion of white produces a fly with white eyes, that is otherwise normal.

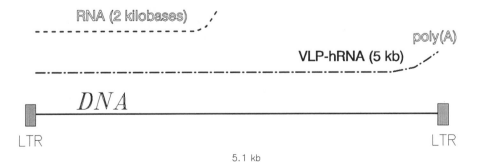

Fig. 5.33. *The structure of a copia element from Drosophila melanogaster showing the two RNA transcripts it generates. LTR, long terminal repeat.*

3. THE elements in the human genome

The human genome contains about 10,000 segments of a retrovirus-like transposable element designated *THE1* (transposon-like human element). This retrotransposon has 2.3 kb with 350 bp flanking LTRs that begin with 5'TG and end in 3'CA (Paulson et al., 1985). It apparently transcribes a 2.0-kb RNA. The LTRs are also present solo, repeated about 30,000 times. Similar elements are found in other hominoids and the Old and New World monkeys. Prosimians such as the lemur have a small number of single copies that hybridize to *THE1* probes (Deka et al., 1986). *THE* may be unique to Primates. It has not been found in rodents. Transposition of *THE* elements is believed to be through an RNA that is retrotranscribed to a circular DNA that inserts into the chromosomal DNA by crossing-over (Baltimore, 1985).

D. Nonviral Retroposons

One member of this group has been described in a previous section. The *F* element *Doc* was used as an example to illustrate the role of transposing elements in the formation of inversions, and the consequent phenotypic results. These elements have no terminal repeats and may or may not encode enzymes needed for their transposition. It is highly probable that these elements transpose via RNA, but this has yet to be shown directly. They are placed in two categories (Table 5.1), depending on which of the RNA polymerases described in Chapter 6, POLII or POLIII, they are transcribed by. In many cases they are moderately repetitious and can occupy a significant portion of nuclear DNA.

1. Nonviral elements transcribed by POLII

LINE1: One of the most interesting families of retroposons present in the gen-

omes of humans and other mammals, including the marsupials, is the *LINE1* (L1) family, a.k.a. *LINES* (long interspersed repeated sequences) and *Kpn*. Elements that have related sequences are also found in the *F* family of *Drosophila*, the *Ingi* elements of Trypanosoma, the *R2* element of *Bombyx mori*, and the *Cin-4* element of maize. *L1* members are dispersed throughout the human genome except that they are underrepresented, or absent, in the centromeric regions (Moyzis et al., 1989). Mammalian genomes have members of this family about 6.1 kb long represented by 20,000–50,000 copies. The elements have no terminal repeats, but most terminate at the 3' end with poly(A) tracts. They contain OFRs, and 6.5-kb transcripts have been identified in certain cancer cells (reviewed by Skowronski and Singer, 1986, and by Weiner et al., 1986). As shown in Table 2.4, *Kpn LINE1* elements constitute about 11% of human chromosomal DNA. The region that includes ORF sequences in primate elements has about 60% similarity with analogous regions in rodents. Untranslated regions appear to be unrelated in sequence. For this reason *L1* elements are highly heterogeneous; no two appear to be the same, but they are obviously related by the OFR sequences they have. This degree of conservation in ORF regions indicates that the putative encoding regions may have significant functions whose structures are maintained by selection pressure (Sakaki et al., 1986; Skowronski and Singer, 1986; Di Nocera and Sakaki, 1990).

No clear evidence exists that *L1* elements encode a reverse transcriptase, although ORF-2 has a sequence compatible with this possibility. Furthermore, it is clear that *L1* elements can transpose within the human genome. They have been shown to be inserted into exon 14 of the coagulation factor VII gene in two unrelated patients with hemophilia (Kazazian et al., 1988). In addition an *L1* element has been implicated in the production of somatic mutation that causes a neoplasia. A patient with a breast adenocarcinoma was found to have an *L1* inserted in the second intron of her *c-myc* proto-oncogene. Thus it is clear that these elements are not passive but can disperse in the human genome and cause disease.

Scattered throughout mammalian genomes are DNA sequences with no obvious functions, if any at all, but with obvious relationships to the sequences of functional genes in those same genomes. These sequences are called pseudogenes.

Two kinds of pseudogenes have been identified. (1) **Unprocessed pseudogenes**: These are elements usually located in a family such as the β-globin family that have introns and otherwise resemble family members, but do not function as family members. They have an accumulation of deletions and other deleterious mutations that render them nonfunctional. Presumably they were once functional but lost the function, which was taken over by a functional homologue. They are fossils. (2) **Processed pseudogenes**: These have no introns and generally no regulatory sequences and hence resemble a cDNA copy of a fully processed mRNA. They are retroposons or **retropseudogenes** with the 3′terminal poly(A) tracts of the mRNA from which they are derived by reverse transcription (Weiner et al., 1986; Wagner M, 1986).

Processed pseudogenes frequently have the entire coding region intact, but they are inactive because they lack the upstream regulatory sequences. This fact in itself points again to the essential role that these upstream regions play in gene action. If a retropseudogene inserts by chance downstream from a foreign regulatory region, it can become an active gene. This may have happened in the chicken, which has a functional, intronless calmodulin gene, as well as a functional one with introns (Stein et al., 1983). The functional *PGK-2* gene on human chromosome 19 has a functional counterpart, *PGK-1*, on the X chromosome. *PGK-1* has 11 exons and 10 introns. *PGK-2* completely lacks introns, has remnants of a poly(A) tail, and direct terminal repeats like most pseudogenes (McCarry and Thomas, 1987). It is not active in all cells, however. It is active in spermatogenesis, which *PGK-1* is not. This indicates that the two are under control of different promoter elements, as in the case of the chicken calmodulin genes. Also, retroposition of an mRNA may occur on rare occasions along with the normal promoter region. These functional retrogenes have apparently been formed in some mammals and then maintained in the germ line. The rat and mouse preproinsulin I genes are an example of this kind of incident (Soares et al., 1985). These rodents have two preproinsulin genes that are equally expressed. Sequence analysis of the two genes, I and II, lead one to the conclusion that gene I resulted from reverse transcription of gene II along with its regulatory region. This event may have occurred 35 million years ago.

Retropseudogenes are common in mammalian genomes but not in other vertebrate genomes. The chicken calmodulin gene is the sole example outside the Class Mammalia (Wagner, 1986). Most of the available data are from studies with human, mouse, and rat genomes. Table 5.3 lists the pseudogenes known in the human genome. Many of the active genes listed in Table 5.3 with their pseudogenes, with the exception of the globin genes, are one-copy housekeeping genes. The α-globin pseudogenes are intronless. Processed pseudogenes may have hundreds of copies per genome as in the case of the mouse glyceraldehyde-3-phosphate gene (Piechaczyk et al., 1984). They are not necessarily confined to one chromosome. The 14 pseudogenes of the human arginosuccinate synthetase gene are distributed over 11 chromosomes, including the X and Y. The sole functional gene is on chromosome 9.

TABLE 5.3. Genes in the Human Genome With Pseudogenes

Active genes			Pseudogenes	
Symbol	Locus	Name	Number[a]	Chromosomes
ACTB	7pter-q22	β-Actin	6	X, 5, 18, 7, 8
ACTG	17	γ-Actin	2	3, Y
ALDOA	16q	Aldolase A	2	3, 10
ASS	9q34-qter	Arginosuccinate synthetase	14	2, 6, 9, X, Y, 3, 4, 5, 7, 11, 12
CALC	11p	Calcitonin	1	11
CPP	3q	Ceruloplasmin	1	8q
CSHI	17q	Chorionic somatomammotropin	1	17
CYP21	6p21	Cytochrome P450 (steroid 21-hydroxylase)	1	6p21
DHFR	5	Dihydrofolate reductase	3	18, 6, 3, 5
FUCAI	1p35	Fucoside, α-L-1	1	2
GAPD	12p13	Glyceraldehyde-3-dehydrogenase	1(16)	1, X, 2, 4, 6, 8, 18, 7, 5
GLUD	10q23	Glutamate dehydrogenase	1	X
GPXP	3q11	Glutathione peroxidase	2	21, X
HBA	16p13.3	α-Globin	2	16P13.3
HBB	11p15.5	β-Globin	1	11p15.5
HBZ	16p13	ζ-Globin	1	16
HMG17	1p36	High-mobility group protein 17	2	?
HPRT	Xq26	Hypoxanthine phosphoribosyl transferase	4	3, 5, 11
HRAS	11p15	Harvey rat sarcoma oncogene homologue	1	X
IFNA	9p	α-Interferon	1	9p
IGHE	14q22.3	Immunoglobin IgE	2	14q32.3, 9
IGHG	14q32.	Immunoglobin IgG	1	14q32
IGKV	2p12	Immunoglobin κ variable	9	1, 15, 22
MT2	16q21-q22	Metallothionein-2	1(4)	4, 1, 18, 20
KRAS	Gp12	Kirsten rat sarcoma oncogene	1	6p12
NRAS	1p13	Neuroblastoma RAS oncogene	2	9, 22
PGK1	X13-q21	Phosphoglycerate kinase	2	X, 6
PPOL	1q41	Poly(ADP-ribose polymerase)	2	13, 14
RAFI	3p25	Murine leukemia oncogene	1	4
RNU1	1p36.1	RNA, U1, small nuclear	4	1q12-q22
RPL32	15	Ribosomal protein L32	1	6
RRM2	2p24-p24	Ribonucleotide reductase	4	1, X
TF	3q21	Transferrin	1	3
TUBB	6pter-p21	β-Tubulin	2	8, 13

[a]Numbers in parentheses indicate "like"-genes, which may or may not be pseudogenes.

2. Nonviral elements transcribed by POLIII

These retrotranscripts are 70–300 bp long and are designated SINES (short interspersed repeated sequences). The human Alu family of SINES is the most studied of the elements transcribed by POLIII. As shown in Table 2.4 the human genome has about 5×10^5 repeats of Alu constituting 5%–6% of the total mass of genomic DNA. Human Alu sequences are about 56% GC, and concentrated in the reverse (R) bands of the chromosomes (Korenberg and Rykowski, 1988). The R bands are rich in CG, but *Alu* integrates preferentially into short d(AT)-rich regions (Daniels and Deininger, 1985). R-band DNA replicates early, condenses late, and appears to have a high concentration of active genes. It does appear that most breaks that result in exchanges occur in the R bands. Whether the presence of *Alu* elements is involved in these breaks and rejoinings is not known, but if it is true that the frequency of rejoinings to form aberrations is greatest where repetitious DNA occurs, then they may be.

Alu elements are at least 50 times more underrepresented in centric heterochromatin than in the euchromatin. Although there appears to be no clear segregation of *L1* in specific bands of the human chromosomes, at the molecular level it is clear that *L1*s are segregated from one another. *Alu* elements are not inserted into *L1*s and vice versa (Moyzis et al., 1989).

Alu elements are found throughout the Order Primates, including prosimians. The consensus sequences of *Alu* elements from chimpanzee and the New World monkeys are indistinguishable from the human (Willard et al., 1987). Some gene families such as the σ-globin cluster on chromosome 16 have *Alu* elements scattered throughout (Sawada et al., 1985). The human TNF (tumor necrosis factor) genes (α and β) are surrounded by *Alu* elements upstream and downstream (Nedospasov, 1986).

Rodents have a *B1* family of repeats with a considerable degree of homology with the *Alu* repeats. *B1* elements differ from *Alu* primarily in being monomeric in contrast to the dimeric *Alu* elements. *Alu* and *B1* elements are processed 7SL pseudogenes derived originally from 7SL RNA by reverse transcription (Ullu and Tschudi, 1984). 7SL RNA is a small cytoplasmic RNA of about 290 nucleotides (Li et al., 1982) that is associated with six polypep-

tides to form a ribonucleoprotein that functions in protein translocation and protein integration into the endoplasmic reticulum membrane (Walter and Blobel, 1983). The right monomer of dimeric *Alu* is 31 bp longer than the left. Both monomers have deletions in the 7SL pattern. The transcription of *Alu* by POLIII initiates at the 5′ end of the element and terminates at the 3′A-rich end. This terminal tract has 4–50 or more AU base pairs, and is characteristic of most *SINES* elements.

Alu elements are undoubtedly very ancient and appear to have arisen from different but related progenitors (Brown, 1984; Bains, 1986; Britten et al., 1988). An analysis of the human genome indicates that it contains at least two distinct families of sequences. One is more closely related in sequence to 7SL RNA than the other, and may have arisen more recently (Jurka and Smith, 1988). No definite evidence exists for actual transposition of *Alu* in present-day mammals.

Many of the other *SINES* families are retrotransposons of tRNA or tRNA genes (Rogers, 1985, 1986). Elements with strong homologies to tRNA have been found in a number of different mammalian genomes such as rodent, primate, and artiodactyl members. They are probably also present in other vertebrates, since fish, amphibians, reptiles, and birds have highly repetitive sequences with tRNA-like structures (Endoh and Okada, 1986).

E. Origins of Transposable Elements

The Class I elements listed in Table 5.1 have in common the presumed requirement of reverse transcriptase for transposition by the DNA→RNA→DNA sequence, if and when they transpose. All cells have the RNA POLs for direct transcription, but not for the reverse. So far as known reverse transcriptase is not encoded by the endogenous DNA of the eu-

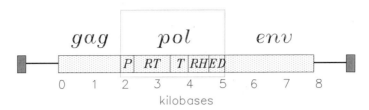

Fig. 5.34. *The basic structure of a retrovirus genophore. The black bars at the termini represent the LTRs (long terminal repeats). The gag region has genes encoding internal structural genes. The pol region has genes coding for viral enzymes: P, protease; RT, reverse tran-* *scriptase; T, connecting segment; RH, ribonuclease H; ED, endonuclease. The env region genes code for viral glycoproteins in the viral coat or envelope. Not all retroviruses are this complete and need "helpers" to complete their life cycle (see Temin, 1989).*

karyotic cell, but only by the Class I elements listed in Table 5.1.

Retroviruses are a family of RNA viruses with the consensus structure shown in Figure 5.34. They replicate through a DNA intermediate by reverse transcriptase as shown in Figure 5.35. In the replication process they use the enzymes of the host cell for going from the DNA phase to the RNA phase. So far as is known, all retrotransposons, such as the yeast *Ty* element (p. 220), have sequences that are homologous to some of the *gag* and *pol* genes of functional retroviruses. The retroposons have the *pol* genes only (reviewed by Temin, 1989). Thus both the nonvirus-like retroposons and the virus-like retrotransposons have retroviral affinities. They have probably been derived from retroviruses and may be thought of as defective retroviral remains that have lost the capacity to go into the extracellular, vegetative phase.

The evolutionary origin of the elements that require reverse transcriptase has been considered by Rogers (1985) and M. Wagner (1986). One can conceive of the possibility that the transposing elements have arisen from the retroviruses by unequal crossing-over, for example. The retroviruses do have the direct terminal repeats such as those found in the retrotransposons. Homologous pairing between regions of chromosomes bearing inserted re-

troviral DNA followed by unequal crossing-over (Fig. 5.36) could lead to the loss of segments of DNA between the direct repeats.

As for the origins of the Class II transposons one must look to sources other than the retroviruses. It is possible that they have arisen from DNA viruses, but if so, relationships between the maize elements and the *Drosophila P* elements have yet to be established. The fact that these elements encode transposases suggests their origin from virus-like DNA elements that have the means of integrating into the host genome. DNA viruses that can integrate are known.

But then again what did these and the RNA viruses arise from in the first place? A good argument can be made that RNA

Fig. 5.35. *Structure and life cycle of the HIV retrovirus particle.* **a.** *The vegetative phase showing a particle in section with its icosahedral core surrounded by an envelope derived mostly from host membranes. The viral genome consists of two identical RNA molecules within the core; gp120 is a viral envelope protein that binds to the target cell to be infected.* **b.** *1–2. The virus particle binds to and enters the host cell and loses its envelope.* **3–5.** *The free viral RNA is transcribed by reverse transcriptase, the proviral DNA enters the host cell nucleus, and the viral DNA integrates into the host chromosome to become an episomal provirus.* **6–8.** *Synthesis of new viral RNA by transcription then leads to the synthesis of new virus particles, followed by their extrusion from the cell.*

a.   structure of HIV

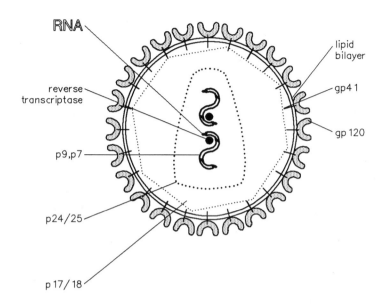

b.   retroviral life cycle

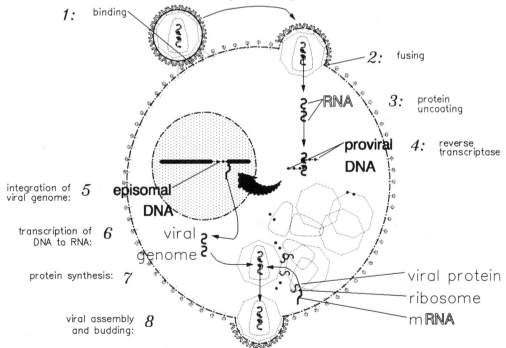

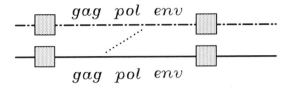

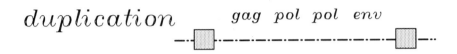

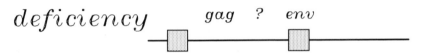

Fig. 5.36. *Possible mode of origin of a retrotransposon from a retrovirus. By unequal crossing-over between two chromosomes in a cell homozygous for a provirus, a deficient element such as the yeast Ty could be generated that has, for example, only part of the pol region.*

is the first nucleic acid to arise and DNA is secondarily derived from it. The retrovirus line may be the remnants of an early form of life that gave rise to a DNA form, and then became dependent upon the DNA form for its reproduction.

F. Retroviruses and Their Cognates

A retrovirus first described by Rous in 1911 has since been labeled the Rous chicken sarcoma virus. Its discovery initiated the study of animal virology and led to the beginnings of the understanding of the relationship of viruses to various types of cancer. The retroviruses form a huge family of related entities that infect plants and animals. They all have in common a single-stranded RNA phase and a DNA replicative intermediate formed by a reverse transcriptase that they encode. Not only is this enzyme present in all strains,

but the structure of the encoding genes has been conserved to a high degree compared with the other parts of the viruses shown in the diagram in Figure 5.34 (McClure et al., 1988).

Retroviruses can be transmitted from one carrier eukaryotic cell to another in the form of the extracellular particles as illustrated in Figure 5.35. This exogenous form of reproductive activity is accompanied by an endogenous form in which the double-stranded DNA of the virus becomes integrated into the host's chromosome, is reproduced with it, and is passed from generation to generation through the germ line. In the latter integrated form it is called a **provirus**. The provirus may produce no apparent effect, but it may also result in diseases such as leukemia and AIDS.

Retrovirally derived DNAs with the ability to induce neoplastic transformations are probably derived from genes normally present in the host cells that have important functions. These are designated *proto-oncogenes* or *c-oncs*. They are not oncogenic in their natural state. About 50 of these have been identified in the human genome. They occupy specific loci on specific chromosomes and encode proteins that perform essential functions. Table 5.4 lists proto-oncogenes and their linkages on the human chromosomes as of the year 1988. There may be many more.

Proto-oncogenes are not all unrelated in function and structure. They occur in families. For example, the SRC gene on chromosome 20 is related to ABL, FES, MOS, ERBB, FMS, ROS, YES1, RAF1, REL, and at least 10 others listed in Table 5.4. They constitute the SRC family, and all of them apparently encode protein-tyrosine or serine (threonine) kinases. RSK1, HRAS1,

RAS2, and NRAS1 constitute the RAS family, members of which regulate by conveying signals to adenylate cyclases. FOS and MYC are involved in transcriptional control through DNA-binding proteins they encode. All of these are extremely important functions having to do with the regulation of the cell cycle and cell growth, as described in Chapter 3. These genes are not confined to mammals but are found in all eukaryotes including the fungi.

The evidence is now quite convincing that during the replicative cycles certain retroviruses "capture" host cell *c-oncs* whereupon they become transduced to *v-oncs*, or *oncogenes* capable of initiating neoplasms (Varmus, 1988). Two mechanisms by which they may occur are illustrated in Figure 5.37. Invasion by either mode may cause the gene to overproduce

TABLE 5.4. Proto-Oncogenes Identified in the Human Genome

Chromosome and locus	Oncogene symbol	Chromosome and locus	Oncogene symbol
1p36	*FGR (SRC 2)*	8q24	*PVT1*
1p32	*BLYM1*	8q24	*MYC*
1p32	*MYC*	9q34	*ABL*
1p22	*NRAS1*	11P15	*HRAS1*
1q22	*SK*	11q13	*INT2*
1q24	*ARG*	11q13	*SEA*
1q31	*TRK*	11q23	*ETS-1*
2p24	*NMYC*	12q12	*RAS-2*
2p13	*REL*	12q12	*INT-1*
3p25	*RAF1*	13q12	*FRT*
4pter-p15	*RAF2*	14q21	*FOS*
4q11	*KIT*	14q32	*AKT-1*
5q33	*FMS*	15q25	*FES*
6pter-q12	*PIM1*	17q21	*ERB-B2*
6q12-p11	*RASK1*	17q12	*NGL*
6q21	*SYN*	18q21	*BCL-2*
6q22	*MYB*	18q21	*YES-1*
6q22	*ROS*	18q22	*ERV-1*
6	*YES-2*	19p13	*MEL*
7p14-p12	*ERBB*	19	*RRAS*
7p11	*ARAF2*	20q12	*SRC*
7p11	*PKS1*	21q22	*ETS2*
7q22	*MET*	21	*ERG*
8q13	*LYN*	2q12	*SIS*
8q22	*MOS*		

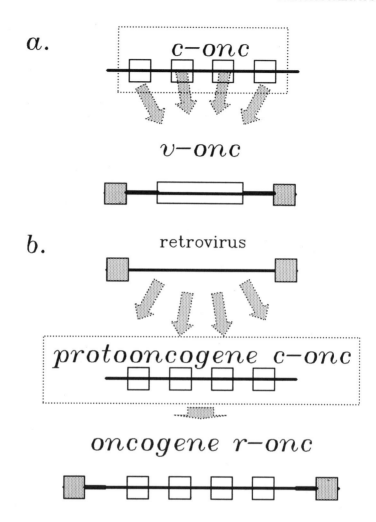

Fig. 5.37. *Two mechanisms by which retroviruses and protooncogenes interact to generate oncogenes or oncoretroviruses.* **a.** *Insertion of the exons of a protooncogene into a retrovirus by transduction.* **b.** *Insertion of retroviral elements on either side of the c-onc. The stippled blocks represent the retroviral LTRs.*

its product and cause the cell to enter the uncontrolled growth phase characteristic of cancer cells (Bishop, 1983; Varmus, 1988). The transduced *c-onc* is definitely related to the *v-onc*. Both are transcribed and the transcripts are translated into similar proteins. The close relationship between the *c-oncs* and their cognate *v-oncs* have been proved by sequence determinations. For example, *c-mos* and *v-mos* differ at only 25 positions out of 1,157 nucleotides, and their protein products by only 11 amino acids (Bishop, 1983, 1985, 1986, 1987). Also different members of *c-onc* families have homologous sequences. The SRC family members have a 45%–80% sequence similarity in nucleotide sequences compared with *c-src*. Considering their wide distribution from yeasts to mammals it is probable that these *c-onc* genes are among the most ancient in the eukaryotes, and their presence may in fact constitute a fundamental difference between eukaryotes and prokaryotes.

Besides being involved with retroviruses proto-oncogenes, when moved to positions in the genome other than their normal ones by transpositions, inversions, or translocations, may become involved with neoplasms without the apparent participation of retroviruses. This relationship between chromosomal rearrangements and cancer is discussed in Chapter 7, p. 341.

G. Significance of Transposing Elements

The widespread occurrence of transposing elements among the prokaryotes and eukaryotes raises the obvious question about their roles in the lives of the organisms they inhabit. The central question is Do they play a significant role in the functioning and evolution of those species in which they occur? It has been suggested that the repetitious DNA of both the dispersed euchromatic and the undispersed centric types is nonfunctional "junk" and "selfish" DNA, (Orgel and Crick, 1980.) One can either accept this hypothesis or form the null hypothesis to test whether the repetitious elements do have a function. As shown in Tables 2.4 and 5.1, the major repetitive elements in the euchromatic regions of the human genome are members of families of retroposons and retrotransposons, namely the *Alu, L1,* and *THE* elements. All of these have presumably arisen by retrotransposition. Repetitive elements of the same or related types are found in other primates and the rodents. The mammals may all share similar euchromatic dispersed elements. Other vertebrates have somewhat dissimilar

elements. For example, nonfunctional processed pseudogenes seem to be confined primarily to the mammals.

Not all forms of life have the same distribution and kinds of transposable elements. This may be sheer accident, or it may be significant. Phenotypic differences may in part be dictated by differences in these elements. Although the processed pseudogenes of the mammals appear for the most part to be completely inactive and neutral with respect to selection, this may not be true of the more abundant elements such as *L1*. The fact that these elements are not distributed at random in the euchromatin can be significant. *Alu* is segregated from *L1*. *Alu* is concentrated in R bands and therefore concentrated in those parts of the chromatin that has the highest apparent concentration of known genes. Many of these genes are surrounded upstream and downstream by *Alu* elements. It would be interesting to know more about the genetic elements associated with the *L1* elements. Both of these families are represented in the rodents and primates, two orders that are separated by about 60 million years from a common ancestor. Yet there is a significant degree of similarity in the sequences of the families in the different orders. Either there has been a conservation of structure maintained by selection pressure, indicating a positive role played by these elements, or the members of the different orders have recently been infected with related retroviruses. The former may well be the case, since the *L1* family has members that produce long transcripts and have long conserved reading frames. These have been conserved in both taxa, whereas the sequences in other parts of the elements have not been to the same degree (Rogers, 1985). It has been suggested that this type of apparent selection pressure is simply a reflection of the elements' capability of preserving themselves as neutral, nonfunctional entities—the "selfish gene" concept—rather than conferring an advantage upon the host (Doolittle, 1982).

It is well established that transposable elements promote genotypic instability that results in gene mutations and chromosomal aberrations in every organism that has them in active form (Nevers and Saedler, 1977; Nevers et al., 1985; McClintock, 1984). Do they have selective advantage because of this? Without mutation there would be no evolution, but with too high a rate an organism can become extinct. An equilibrium must be maintained, as stated on p. 185. The effects of P elements on fitness in $D.$ $melanogaster$ has been investigated, and preliminary evidence indicates that viability may indeed be affected positively (MacKay, 1985). There are also negative effects. Whether the beneficial effects outweigh the deleterious ones is a matter yet to be determined. It must also be recognized that sources of mutations other than those caused by transposable elements exist. Conceivably life could go on without them.

McClintock (1984) and Wessler (1988) have made an objective appraisal of the possible role of maize transposable elements in generating the necessary genetic diversity for the plant to cope with life-threatening situations. Not all maize strains exhibit active transposing elements, but most if not all do have inactive Ac and Spm elements. It has been shown that certain inactive Ac elements are hypermethylated, and reactivation of them is accompanied by demethylation (Schwartz and Dennis, 1986). It is not unreasonable to posit that a plant under stress from changes of temperature in the environment, the onset of drought, infection by a virus, and so on, may respond to the shock by its cryptic elements becoming active by demethylation, resulting in the generation of genotypic variants that enable it to survive. Evidence does exist to support the null hypothesis that trans-

posable elements may play a role in evolution, but it is presently certainly not overwhelming.

IV. SPECIALIZED SOMATIC REARRANGEMENTS OF GENIC DNA

During the first half of the twentieth century, the general consensus was that the theory of the genetic continuity of the chromosomes applied not only to germ cells, but to somatic cells as well. The "specific constancy of the chromosomes" was maintained through the many generations giving rise to the billions of cells constituting the soma of a multicellular eukaryote (Wilson, 1925). Once the specific content of the zygote nucleus was established at the time of fertilization nothing changed thereafter except for those rare instances when chromosome diminution occurred. This was in direct conflict with the theory of August Weismann (1892), who proposed that:

> Ontogeny depends on the gradual process of disintegration of the germ-plasm which splits into smaller and smaller groups of determinants in the development of each individual. . . . Finally, if we neglect possible complications, only one kind of determinant remains in each cell, viz., that which has to control that particular cell or group of cells. . . . In this cell it breaks up into its constituent biophores and gives the cell its inherited specific character. (From the translation by E. B. Wilson, 1925)

Weismann's theory is wrong, but the more correct idea that the specific constancy of chromosomes is maintained throughout embryogeny and subsequent stages is now subject to some modification. While it is true that some nuclei from highly differentiated cells preserve their totipotency as shown with transplanted frog nuclei, it is also true that some somatic cells do undergo changes in their genomes not only during development, but in the fully developed adult. Nuclei from frog erythrocytes, when transplanted into enucleated oocytes, can direct the development of that egg to the adult stage (Orr et al., 1986). This does not mean, however, that all differentiated cells are totipotent.

Two kinds of events can and do occur to alter the genomic structure in the somatic line cells derived from vertebrate fertilized eggs. First, somatic mutations may occur at any stage in ontogeny and in the adult. Such mutations can even become homozygous by mitotic crossing-over (p. 176). Chromosomal aberrations and transpositions also can occur in the soma, as previously discussed in this chapter. These changes are, in general, random and fortuitous. Second, however, programmed alterations do occur and are executed following a schedule incorporated into the genome itself (Borst and Greaves, 1987). Several types of programmed alterations such as chromosome diminution and gene amplification are described later in this chapter, and the programmed inactivation of the mammalian X in the female has been previously described (p. 280). Here we consider a number of specialized types of rearrangements that occur a response to extrinsic factors and are, in a sense, programmed.

A. The Immunoglobulin Genes

A defence mechanism widespread among the vertebrate animals is based on the production of antibody proteins in response to "foreign" antigen molecules. The immune system of mammals and some other vertebrates has two subsystems: one that recognizes circulating antigens, the **B-cell immunoglobulins (Ig)** or antibodies, and a second one, the antigen receptors called the **T cells**. Both originate from lymphocytes that arise from stem cells in the bone marrow. In mammals lymphocytes develop into T cells only in the thymus gland,

while B cells arise directly in the bone marrow. (The reason B cells are so labeled is that they were originally described in birds where they arise in the Bursa of Fabricius, a tissue type not found in mammals.)

1. Basic structure of the immunoglobulins

Immunoglobulin molecules are composed of a number of different polypeptide units as shown in Figure 5.38. They possess two identical **light (L) chains** and two identical **heavy (H) chains** held together by disulfide bonds. The amino terminal ends of both the L and H chains are highly variable in amino acid sequence and are designated V_L and V_H, respectively. These variable regions are the ones that are involved in antigen contact and their domains are the ones primarily involved in conferring antigenic specificity upon the molecules. The CL, CH_1, CH_2, and CH_3 and the **hinge (H)** domains are constant regions whose exact sequences characterize the different classes of immunoglobulins as described in the following.

The immunoglobulin system is capable of dealing with a virtually limitless number of different kinds of antigens by producing immunoglobulin molecules specific for each kind of antigen. The mammalian repertoire of immunoglobulin types is estimated to be at least one million, and may be even greater by a factor of 10. In a mammalian circulating system they are a major class of protein constituting about 20% of the plasma protein by weight.

2. The classes of immunoglobulin molecules

Five different classes of immunoglobulin molecules exist in mammals: γ, α, δ, μ, and ε. These have sequence differences in their H chains and when combined with the L chains produce IgG, IgA, IgM, IgD,

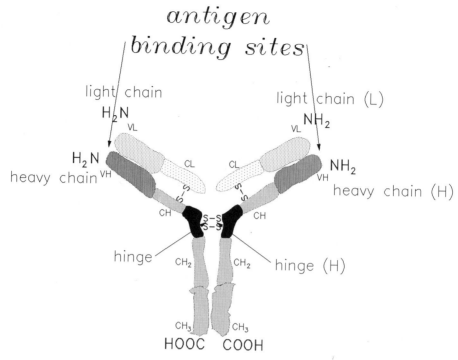

Fig. 5.38. *Model of an immunoglobulin molecule. See text.*

and IgE molecules, respectively, as shown in Table 5.5. These all have different roles to play in the immune system that we shall not discuss here, since the primary aim is to describe how immunoglobulin differences arise rather than how they function.

In addition to having five classes of H chains immunoglobulins also have two types of L chains, κ and λ. Either one of these may be associated with any class of H chain, but an immunoglobulin has one or the other, never both. As shown in Table 5.5, ten basic classes of immunoglobulins result from the various kinds of combinations. Since all immunoglobulins always have identical H and L chains, they always have two identical antigen-binding sites.

3. Genetic variations on a general theme

The V_L and V_H variable regions are about 100 amino acids long. The variability of the sequences within these domains provides the diversity of antigen-binding sites required for the production of the many different kinds of antibodies needed for recognizing antigens. Within each variable region are hypervariable segments, as diagrammed in Figure 5.39. These regions are designated as CDR_1, CDR_2, and CDR_3; the less variable regions are FR_1, FR_2, FR_3, and FR_4. The two FR_4 segments are also called the J segments, and the CDR_3 segments the D segments.

TABLE 5.5. The Different Types of Immunoglobulin Molecules

L-chain types	H-chain classes				
	γ	α	μ	δ	ε
λ	IgG	IgA	IgM	IgD	IgE
κ	IgG	IgA	IgM	IgD	IgE

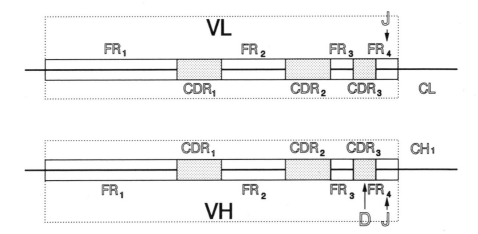

Fig. 5.39. *The variable regions of the light (V_L) and heavy (V_H) chains of an immunoglobulin.*

The immunoglobulins are encoded by part of a gene superfamily that incorporates eight multigene families and 12 single-gene representatives (Hunkepiller and Hood, 1986). Details about this superfamily are discussed in Chapter 9, p. 385. Here we are concerned with the subfamily encoding the immunoglobulin antibodies only. The DNA sequences encoding the C regions of the H chain are all on chromosome 14 at q32 in the human genome. The genes are in a cluster identified as IGH (Table 5.6). This cluster encodes the C_H chains for IgA, IgD, IgE, IgG, and IgM. Also included are the genes for the joining region, the variable region of H and pseudogenes for the H ε and λ chains. A cluster encoding the immunoglobulin κ constant and variable regions (IGK) is on 2p12. The two genes encoding the λ constant and variable regions are on chromosome 22q11. In the same region of 22 are three pseudogenes for the κ polypeptide. A similar arrangement is found in the mouse, which also has three unlinked gene clusters: the λ light chain genes on chromosome 16, the

TABLE 5.6. Loci of Genes Encoding Immunoglobulins

Chromosome	Gene symbol	Locus	Name
2	*IGKC*	p12	Kappa constant region
	IGKV	p12	Kappa variable region
	IGKDE	p12	Kappa deleting element
14	*IGHC*	q32.3	H constant regions for α, γ, δ, μ, and ε
	IGHPs	q32.3	Pseudo γ and ε genes
	IGHDY1	q32.3	H variable region
	IGHJ	q32.3	Joining region
22	*IGKVPs*		Kappa pseudogenes
	IGLC	q11.1-11.2	Lambda constant region
	IGLV	q11.1-11.2	Lambda variable region

κ light chain genes on 6, and the heavy chain genes on 12.

The germ line cells of vertebrates carry all the genes necessary to encode the large number of different antibodies necessary for B lymphocytes to contend with many foreign antigens. A given lymphocyte stem cell can produce only one antibody type, however. Once it has this capacity its clonal derivatives will produce this and only this antibody to a specific antigen (Fig. 5.40). This process is called **clonal selection**, and it is assumed but not proven. It is generally accepted because a better explanation is not available. The presence of the antigen activates the reproduction of the cell with the specificity to the antigen, and a clone of cells with sufficiently large population to deal with the antigen is generated. This is one of the several kinds of **immune response** that vertebrates are capable of.

A stem cell that produces an immunoglobulin with a particular specificity originates by the reorganization of the genes that encode the variable parts of the light and heavy chains (reviewed by Tonegawa, 1983; Honjo, 1985). The process has best been worked out in the mouse, which is used here as the example. Figure 5.41 shows sections of mouse chromosomes 16, 6, and 12, which bear the genes for the λ light, and the κ light and heavy chains, respectively. The left side of Fig. 5.41 shows the structure of the regions before reorganization and the right side after reorganization in a B cell to become dedicated to the production of a particular antibody. The λ-chain genes are of several subtypes, but only one type is illustrated here (Heiter et al., 1981).

Light-chain variant chains are created by the extirpation of sections of DNA so that V regions and J regions are joined as shown in Figure 5.41. In the B-cell lineage, in which a λ1 light chain is expressed, the spacer between $V_{\lambda 1}$ and $J_{\lambda 1}$ is deleted so that V and J segments are fused (Fig. 5.41a). The V segment encodes most of the V portion of the light chain except for the FR4 part, which is encoded by the J segment. The C (constant) segment encodes the remainder of the light chain. The diversity results from what is deleted between the V and J segments.

A similar rearrangement by V–J joining results in many different types of light chains. Lewis et al. (1985) have made a detailed analysis of V_{κ}–J_{κ} joining in transformed (neoplastic) B-cell lineages. By using retrovirus vectors containing unarranged mouse germline V_{κ} and J_{κ} segments, they were able to recover coding joints after rearrangement. While they were not able to make the exact mechanism of V–J joining clear, they were able to show that the joints from different lines were from a variety of different junctions formed by joining of different kinds of ends modified by nucleotide base loss and inversions. It has been suggested that unequal sister chromatid exchanges may also be involved in creating these modifications, but there are no substantive data to support this. However, there is good evidence that certain types of inversions are involved (Feddersen and van Ness, 1985). Double-recombination products of the same κ chain allele have been found. The nucleotide sequences of double-recombination products of two plasmacytoma lines are illustrated in Figure 5.42 along with the relevant portions of the V_{κ} polypeptide chain. In Panel a of Figure 5.42 the V_{κ} segment is joined to J1 at the position indicated by the heavy arrow so that a normal reading frame is maintained. In Panel b the V_{κ} segment is joined to J2 (with the elimination of J1) in such a way that the normal reading frame is not maintained, producing a frame shift with the result that a completely different amino acid sequence is generated starting at the point of joining (indicated by the arrow). The two V_{κ} nucleotide sequences differ because they are from different genes. The

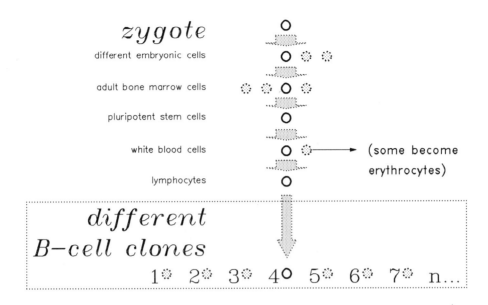

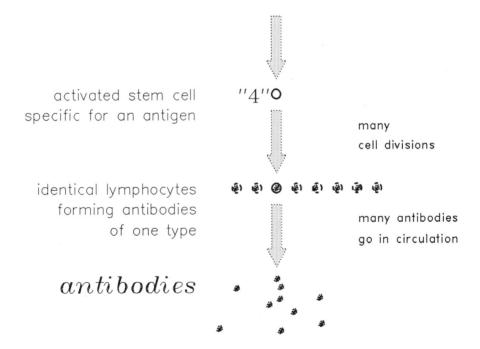

Fig. 5.40. *Origin of a B cell with specificity for an antigen by clonal selection.*

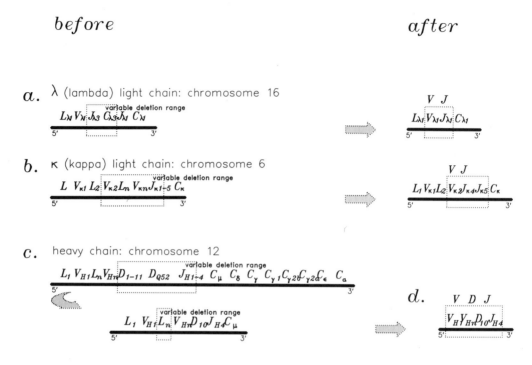

Fig. 5.41. *The sections of mouse chromosomes 16, 6, and 12 that encode λ and κ light chains and the heavy chains. See text. Data from Hieter et al. (1981).*

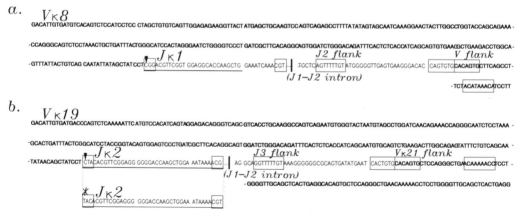

Fig. 5.42. *The nucleotide sequences of double-recombination products from two mouse plasmacytoma lines. See text. Data from Feddersen and Van Ness (1985).*

proper-reading J2 frame is denoted by the asterisk in Panel b. The J2 and J3 segments of Figure 5.42a,b are joined to V segments back to back, suggesting that multiple recombinations have occurred

that in each case have resulted in reciprocal product formation.

As shown in Figure 5.41 the heavy chain gene family on mouse chromosome 12 has a region designated **D** that is composed of a cluster of about 12 D segments in the H chain sequence before reorganization. These types of chain segments are not found in the light-chain families. In addition the H family contains a cluster of 100–200 different L and V segments, as well as a cluster of 4 J_H segments and a cluster of eight copies of C_H segments, one for each of the five basic classes of H chains listed in Table 5.5 plus three subclasses for γ (Liu et al., 1980; Givol et al., 1981). A functional H chain in a B cell is formed by the joining of one each of V_H, D, and J_H. Thus the V_H of the final-product H chain is encoded by three gene segments whose joining in various combinations can generate a diversity even greater than that for light chains. Furthermore, it is also evident that after V–D–J joining in heavy-chain gene clusters another type of rearrangement may occur in which a V–D–J segment is joined to another V segment upstream. This adds considerably to the possible number of variants.

There is a third source of diversity of immunoglobulins, namely somatic mutation. It is apparent that hypermutation occurs in the process of reorganization and this may be related to the V–J and V–D–J events themselves (Selsing and Storb, 1981). A somatic mutation mechanism operates in variable region genes to generate single-base-pair mutations that may result from error-prone DNA synthesis during V–J joining.

4. Four roads to diversification in the soma

The diversification in B-cell immunoglobulin production derives from four sources (Tonegawa, 1983). (1) The V_L and V_H regions are encoded by two and three different gene segments: V and J, and V, D, and I, respectively. Each of these must exist in multiple copies of different sequences, and if, for example, mouse chromosomes 16 and 6 carry $2v\lambda$ and $3J\lambda$, and 300 $V\kappa$ and $4J\kappa$ segments, respectively, the maximum number of V_L will be 1,206 (2×3 plus 300×4). For H_L chains the maximum number for 200 V_H 12D and HJ_H segments would be 9.6×10^3 ($220 \times 12 \times 4$) different combinations of V_H regions. (2) The second type of diversity originates at the V_L–J_L, V_H–D, and D–J_H junctions because the joining ends are not always the same. (3) The third type occurs at the V_H–D and D_H junctions, where one to several nucleotides are inserted in a template-independent fashion. (4) Added to these combinatorial events are the somatic mutation events. Taken together these mechanisms easily account for the millions of immunoglobulins in the B-cell repertoire.

It is quite possible that phenomena of these types are not restricted to the immunoglobulin immune system response. Other systems such as the nervous system may have related events occurring in the genes of their cells. Thus immunoglobulin formation may represent a general paradigm for response of an organism to its environment. Genetic systems incorporated in chromosomes are flexible.

B. Antigenic Variation in Trypanosomes

Trypanosomes are parasitic protozoa in the Class Flagellata. They infect various vertebrate species including *Homo sapiens*. In man they cause serious diseases such as African sleeping sickness and Chagas's disease. The African trypanosome *Trypanosoma brucei* is able to express a large variety of different surface antigens while in its human host. This enables it to overcome the immune response of its host, because when the host produces an antibody specific to one surface antigen of the parasite, the antibody becomes ineffective because the parasite changes the antigen. Thus the parasite is able to protect itself against not only the host's natural immune system, but also vaccines designed to protect against infection. The ability of *brucei* to change its surface antigen composition derives from its capacity to rearrange genetic elements producing these antigens (Williams et al., 1979). The antigenic properties are possessed by glycoproteins generally referred to in the literature as the variant-specific surface glycoproteins (VSG) (Gardiner et al., 1987). Each VSG is encoded by a separate gene and it is estimated that 10^3 of these genes exist in a gene family. Activation of one of these results in the production of a specific antigenic glycoprotein. Several modes of VSG gene activation are known. These center around a silent VSG gene being duplicated and the duplicate being moved by transposition to a point near a telomere of the chromosome. This **active telomeric expression site** contains an active transcription unit with a promoter. The inactive genes sit in a region devoid of enhancer plus promoter, and must be transposed to a region containing these sequences of DNA to be transcribed. A silent gene may also be activated by being moved by reciprocal translocation, which exchanges it with an active gene in an active telomeric expression site. Hence a VSG that may have become subject to the immune response of the host can be changed to an alternative form that the host cannot cope with immediately.

In addition to the process of activation by transposition, a process of discontinuous transcription may also be involved (reviewed by Borst, 1986). This occurs after activation of a VSG encoding gene by the introduction of alternative modes of transcription of the encoding region of the gene.

C. Mating-Type Switching in Yeast

Saccharomyces cerevisiae can reproduce vegetatively by mitosis either in the haploid or diploid state. Figure 5.43 illustrates the life cycle of this yeast, showing that the haploid cells are either of mating type a or α. These mating types are determined by alleles of a gene, *MAT*, on chromosome 3. Although the haplotypes a and α can form stable colonies, the predominant phase is the diploid one. Diploid cells are formed by the fusion of haploid *MATa* with *MATα* cells. Only haploid cells of opposite mating type will fuse. Consequently, a diploid cell is always of the genotype *MATα/MATa*. (*MATα/MATα* and *MATa/MATa* cells may form by chromosome doubling, but they cannot undergo meiosis and sporulate). Under certain conditions, starvation being the main one, a diploid culture will have some of its cells undergo meiosis with the consequent formation of a four-spored ascus. These four haploid cells, or ascospores, will be in the ratio $1a:1\alpha$ with respect to mating type.

The *MAT* alleles are codominant in the diploid and have different DNA sequences called Ya and $Y\alpha$ (Nasmyth and Tatchell, 1980). Transcription of these results is encoded polypeptide products designated as $a1$ and $a2$ and $\alpha1$ and $\alpha2$, respectively. They act as signals. In diploids the α and a products interact to repress haploid-specific genes. As shown in Figure 5.44, another gene, *HO*, on chromosome 4 determines the stability of *MATa* and *MATα*. When its recessive allele, *ho*, is present homozygous, *MATa* and *MATα* are stable, but when *HO* is present the *MAT* alleles are interconvertible at a high rate. This is called switching and can occur in a haploid colony consisting of only α or a cells, thus resulting in both mating types and the ability to form α/a zygotes.

The interconversion of the *MAT* alleles is difficult to explain unless there exists in the yeast genome inactive copies of Ya and $Y\alpha$ DNA sequences that can be used as donors during the interconversion. This proves to be the case (Hicks et al., 1977, 1979; Nasmyth and Tatchell, 1980). Two other genes widely separated from *MAT* are involved: *HMRa*, near the telomere on 3R and *HMLα*, near the telomere on 3L.

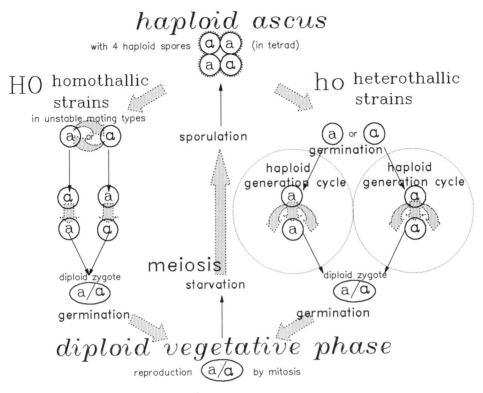

Fig. 5.43. *Life cycle of Saccharomyces cervisiae.*

Figure 5.44 diagrams the relationship of the chromosome 3 genes and their activities in the interconversion under the influence of yet another set of genes, which are designated as *SIR 1, 2, 3,* and *4.* These regulate *HML* and *HMR* and are not linked to them.

In addition to the Y*a* (642 bp) and Yα (747 bp) sequences in *HML*, *HMR*, and *MAT* other DNA sequences are present and are identified in Figure 5.44 as W, X, Z1, and Z2. All the latter are included as *HML*α and *MAT*, but *HMRa* has only Z1, and X. The WXYαZ1ZZ and XY*a*Z1 sequences are referred to as cassettes. The *HML*α and *HMRa* cassettes are said to be silent, whereas the *MAT* cassette will be active and of one type, either α or *a*. In Figure 5.44 it is shown as α. Mating-type switching occurs by substituting information present at *MAT* with information

copied from *HML*α or *HMRa* by unidirectional transposition (Klar and Strathern, 1984). The DNA sequence Y*a* or Yα replaced at *MAT* is apparently lost. The transpositions are induced by the *HO* gene on chromosome 4. When the recessive allele *ho* is homozygous, transpositions do not occur and mating type is stable. The *SIR* genes (also known as *MAR* or *CMT*) regulate *HML*α and *HMLa* (Nasmyth, 1982). The dominant alleles of these genes cause a repression of transcription and transposition of the Y*a* and Yα segments of *HML* and *HMR*. It appears that the chromatin structure of the *HML* and *HMR* regions is different in *SIR* and *sir* strains. The possibility exists then that *SIR* acts by inducing chromatin changes that cause inactivity of *HML* and *HMR* regions of DNA (Nasmyth and Shore, 1987). In *sir/sir* lines the configuration of the chro-

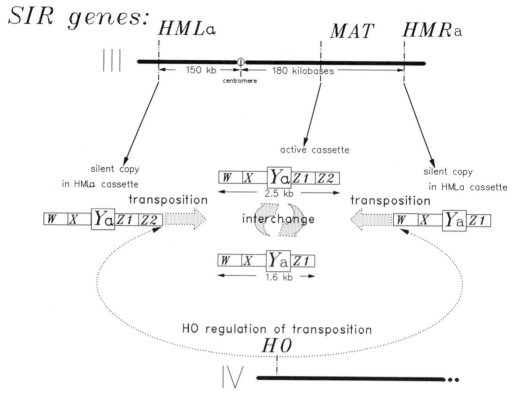

Fig. 5.44. *Mating-type switching yeast. Data from Klar and Strathern (1984).*

matin allows for switching. The switching appears to be the result of gene conversion (Klein and Petes, 1981; Jinks-Robertson and Petes, 1985).

V. RANGES IN SIZE OF EUKARYOTIC GENOMES

The genomes of eukaryotes exhibit an astounding and puzzling array of differences in the amount of DNA and the number of chromosomes per genome. Perhaps one can logically explain why fungi have about 10^7 bp, whereas higher eukaryotes such as humans and tomato have about 10^9 bp, by assuming that higher degrees of complexity of form and function in the latter require more genes. But how can one explain that within a single taxon,

such as a genus, individual species may vary by a factor of 10 or more in the amount of DNA in their nuclei? Or explain why within a single group of insects such as the butterflies (Rhopalaocera) haploid numbers may range from 7 to 221 without a great effect on basic phenotype (White, 1973)? A further puzzle is that animals are essentially obligate diploids, and the absence of a single chromosome, or even a small part of it from a diploid set can lead to serious, deleterious phentoypic effects despite the presence of its normal homologue. It is these questions and related ones that we consider in the second part of this chapter, in which the primary objective is to present problems remaining to be solved that are probably among the most important and pressing in the area of eukaryotic biology.

A. Number of Base Pairs per Genome

Table 5.7 presents several examples of the amounts of DNA per genome expressed as the number of base pairs at G1. These numbers also can be expressed in the form of picograms (1 p = 0.965×10^9 bp), micrometers (1 μ = 3.0×10^3 bp), or daltons (1 D = 1.6×10^{-3} bp) of DNA per nuclear genome. The few generalizations that can be made about these data would seem to be as follows: Among the vertebrates, the mammals, birds, and reptiles have a relatively narrow range, of $(2-37 \times 10^9$ bp). But the Amphibia have a very broad range, in the region of 10^9 to 10^{11}. The lungfish (Dipnoi) has about 10–100 times more base pairs than other fishes and most other vertebrates except Amphibia. Among the invertebrate phyla the range is relatively constant within phyla except for the Arthropoda, which show a wide range especially among the Insecta. Fungi have a relatively narrow range, around 10^7. Not shown in Table 5.7 are data for the nucleated green algae, which have a wide range, of 10^7 to 10^{11} bp. The mosses and other primitive multicellular plants have a wide range, of 10^9 to 10^{11} bp, but the ferns have a fairly narrow range, of 10^{10} to 10^{11} bp. With a few exceptions the gymnosperms are also fairly conservative with respect to base pair numbers. It is among the angiosperms (flowering plants) that the greatest variation of base pair numbers are to be found among the vascular plant groups. A good part of the differences in DNA amounts per genome can be accounted for in the number of repeated sequences particularly in the heterochromatic regions of chromosomes.

TABLE 5.7. Examples of Ranges in Amounts of DNA per Genome Expressed as Number of Base Pairs

Taxon	No. of species	No. of base pairs
Fungi	4	$(1.5-8.6) \times 10^7$
Coniferales	5	$(2.6-3.7) \times 10^{10}$
Angiospermae	71	$(0.12-5.1) \times 10^{10}$
Invertebrates		
Porifera	2	0.6×10^8
Coelenterata	6	$(0.7-1.0) \times 10^9$
Nematoda	2	$(0.8-4.3) \times 10^8$
Annelida	1	0.7×10^9
Mollusca	4	$(0.5-2.3) \times 10^9$
Arthropoda	82	1.7×10^8 to 10×10^9
Echinodermata	5	$(6.5-9.6) \times 10^8$
Chordates		
Tunicates	2	1.5×10^8
Agnatha	1	2.5×10^9
Elasmobranchii	2	$(2.3-3.3) \times 10^9$
Teleosti	157	$(0.6-6) \times 10^9$
Dipnoi	1	5.0×10^{10}
Amphibia	38	$(0.3-13.5) \times 10^{10}$
Reptilia	7	$(1.4-2.6) \times 10^9$
Aves	29	$(0.7-2.2) \times 10^3$
Mammalia	15	$(1-3.5) \times 10^9$

Figures from data base collected by H.S. Shapiro.

B. Number of Chromosomes per Genome

Similarly to numbers of base pairs per genome, no logical pattern of chromosome numbers per genome is apparent. Table 5.8 lists some examples of chromosome numbers in various taxa. Although some groups such as the conifers among plants, and the amphibia, reptiles, and birds among the vertebrates show a narrow range of numbers, the mammals have an extremely wide range even though they have a narrow range of base pair numbers. Clearly the insects have the greatest range in chromosome numbers among the animals.

A good part of the variation in chromosome numbers within certain taxa is the result of chromosome fusions or dissociations. The centric fusions of acrocentrics reduces the chromosome numbers, as already pointed out (p. 187). In general there is a close association between the number of acrocentrics and the total number of chromosomes in the genomes of members of a family or genus. Species with high numbers of acrocentrics tend to have more chromosomes than those with high numbers of metacentrics. Frequently the number of arms, counting two for each metacentric and one for each acrocentric, is equal or close to the number of chromosomes in species in the same taxon that otherwise have widely different chromosome numbers per genome. Table 5.9 presents some examples of this relationship for cats and dogs. All species of cats that have been karyotyped have either 18 or 19 autosomes per genome, all or most of which are metacentrics. Therefore they have about 36–38 arms. This number corresponds closely to the number of arms found in the dog family with a few exceptions. Note that the dogs fall into two groups, those with $n = 32-38$, and those with $n = 18-25$. The numbers of arms for dogs are quite similar with the exception of those of the crab-eating fox (71), kit fox (49), and arctic fox (49). Most of these differences can be explained by assuming centric fusions or dissociations.

However, other types of rearrangements are also involved. The three foxes with the large number of arms listed in Table 5.9 have a preponderance of metacentrics. Rearrangements other than centric fusions are apparently also involved. All of the species listed in Table 5.9 have a high degree of similarity of banding patterns (Wayne et al., 1987a; Wurster-Hill and Gray, 1973, 1975; Wurster-Hill and Centerwall, 1982; reviewed by O'Brien and Seuanez, 1988). Comparisons among the various species make it evident that nonreciprocal end-to-end translocations have led to a reduction of chromosome numbers in some of these dogs. An extreme example of this is found in the deer family genus *Muntiacus*. A number of species of this small deer exist in Southeast Asia that are clearly related morphologically, but have widely different karyotypes. The Chinese muntjac (*Muntiacus reevesi*) has a haploid number of 23 acrocentrics and

TABLE 5.8. Ranges in Chromosome Numbers per Genome for Various Taxa

Fungi	2–18
Algae	6–45
Pteridophytes	7–720
Gymnosperms	11–12
Angiosperms	2–~250
Decapod Crustacea	41–188
Insecta	1–221
Mollusca	5–36
Fish	9–56
Amphibia	10–22
Reptila	15–33
Aves	26–41
Mammalia	
Prototheria	27–32
Metatheria	7–10
Eutheria	3–52
Muntiacus muntjac	$n = 3$
Anatomys leander	$n = 46$
Tympanoctomys barrerae	$n = 52$

TABLE 5.9. Chromosome and Arm Numbers for the Cats and Dogs in the Order Carnivora

Species	Autosomal haploid number[a]	Total arms
Cats, genus *Felis*[b]	18 or 19	~36
Dogs, Canidae[c]		
Maned wolf	38 (0)	38
Bush dog	37 (0)	37
Crab-eating fox	37 (34)	71
Grey wolf	38 (0)	38
Bat-eared fox	36 (2)	38
Fennec	32 (2)	34
Grey l.c.	33 (1)	34
Red fox	18 (16)	34
Kit fox	25 (24)	49
Arctic fox	25 (24)	49
Racoon dog	21 (13)	34

[a]Numbers in parentheses represent the number of metacentrics in the dog genomes.
[b]Number of species = 37. Data from Wurster-Hill and Centerwall (1982) and Wayne et al. (1987a).
[c]Data from Wayne et al. (1987a,b).

is considered to be the ancestral species (Liming et al., 1980). The other known members of the genus have a lower number of chromosomes per genome. The most extreme sample is the Indian muntjac (M. muntjak vaginalis), which has $n = 3$ in the female. The evidence is clear that it and the Chinese muntjac have a high degree of homology in their banding patterns, and that tandem and centric translocations have resulted in fusions that have led to the reduced chromosome numbers in the Indian muntjac. This has been verified by showing that regions of Indian muntjac chromosomes outside of the functional centromeres react readily with antibodies to kinetochores produced by scleroderma CREST patients, as described on page 66 (Brinkley et al., 1984, 1985).

Another noteworthy example is found in the cotton rats of the genus *Sigmodon* (Elder, 1980). As shown in Table 5.10, three known species in the United States have different chromosome numbers. These three are almost morphologically identical and can be separated with confidence only by the differences in their karyotypes. The use of banding patterns to determine homologies leads to the conclusion that *S. hispidus* and *mascotensis* are related primarily by tandem fusions; *hispidus* acrocentrics have fused to form *mascotensis* acrocentrics. If one assumes a hypothetical *mascotensis*–arizonae intermediate, then it can be postulated that *arizonae* arose by centric fusion to form eight metacentrics in addition to three small acrocentrics already present. A centric fusion in the $n = 12$ form apparently produced the $n = 11$ form to create a polymorphism in the population.

From these few examples taken at random from many representing not just mammals but other vertebrates, insects, and plants it can be concluded that changes in chromosome numbers are effected in part by tandem and centric fusions.

C. Some Factors Involved in Variations in Amounts of DNA per Genome

The data in Table 5.7 show that the variation in the total amount of DNA per ge-

TABLE 5.10. Chromosome Numbers per Genome of Three Species of Rodents (*Sigmodon*)

S. hispidus	$n = 26$ (26 arms; 1 metacentric)
S. mascotensis	$n = 14$ (14 arms; 0 metacentric)
S. arizonae	$n = 11$ or 12 (19 arms; 1 centric fusion to give $n = 11$)

nome may be large even among closely related species. These differences can logically be the result of several different causes, among them the degree of ploidy, or the amounts of repetitive DNA, or both.

It is reasonable to assume that all eukaryotes possess a basic set of genes necessary for encoding the essential proteins needs for enzymatic activity, structure, and regulation of basic metabolic activities and development. Beyond this one can also assume that the highly differentiated, multicellular forms will have more active genetic material acting in control of development than the lower, less-differentiated organisms such as fungi, and this does appear to be the case. But it is also true that related higher eukaryotes with essentially identical degrees of complexity of structure and function can have significantly different amounts of DNA per genome. A germane example is to be found among the salamanders of the genus *Plethodon*, which have been extensively studied by MacGregor (1982) and his associates.

Plethodon is a North American genus that comprises 26 known species distinguishable by their adult body size, differences in proteins (Highton and Larson, 1979), and geographic distribution. Otherwise they are very similar morphologically—a condition found in most amphibian genera, and even families and orders. The members of the genus *Plethodon* can be separated into species groups based on C values. For example, *P. cinercus* has 19.5×10^9 bp, *P. veniculum* 35.7×10^9 bp, and *P. vandykei* 67.6×10^9 bp per genome, respectively. This difference in C values is not, however, accompanied by a change in karyotype other than chromosome size.

These and other species of *Plethodon* have consistent differences in C value variation, but no significant karyotypic differences. Otherwise one might guess that the differences are the result of polyploidy (which is not uncommon among the Amphibia) because the ratio of the C values is close to 1:2:3:4. All species are $n = 14$. The logical conclusion is that the differences are ascribable to different amounts of repetitive DNA and this appears to be the case. These species differ primarily in the amounts and, to a certain extent, the kinds of moderately repetitive DNA (Mizuno et al., 1976).

The presence of repetitive sequences in eukaryotic DNA, and the methods to detect and analyze it, have been discussed in Chapters 1 and 2. The point has been made that eukaryotes vary significantly in the amounts of repetitive DNA in their genomes. The lack of correlation between chromosome numbers and C values can be presumed to be the result in part of differences in repetitive DNA rather than in the number of so-called functional genes. That only repetitive DNA is involved, however, is not to be assumed. Other factors are considered in the following sections of this chapter.

VI. GENE AMPLIFICATION

Gene amplification is a term originally coined to describe the production of multiple copies of a specific gene (Brown and Dawid, 1968). The process is different from the duplication process considered in the previous section, although the two have overlaps. Amplification applies to a magnification in numbers of a specific gene and some of its adjacent DNA on either side. Duplications are most often thought of in terms of accidental, spontaneous occurrence. Admittedly, however, there is no sharp line to be drawn between the two. We consider these phenomena separately, but it should be understood that overlaps in construct and meaning are possible, and furthermore that each is not homogeneous. Each may consist of subphenomena that may not even be closely related in form and mode of formation, although all may function in producing more gene product(s), and hence have related phenotypic effects.

Some types of amplification can be shown to be in response to a factor in the environment not previously present before amplification. These types were first noted in cultured cells *in vitro*. We consider some examples of these first, since they may help explain what happens *in vivo*.

A. Amplification in Cultured Cells

When mammalian cells in culture are subjected to a toxic substance, they will frequently become resistant to it, especially if the initial concentration is low, and the surviving cells are subjected to multiple rounds of selection with increasing concentration of doses (Schimke, 1984; Stark and Wahl, 1984; Schimke, 1988). The development of resistance can be caused by four different kinds of events: (1) mutation resulting in impermeability of the cell membrane to the cytotoxin; (2) mutation(s) resulting in the modification of a protein (enzyme or receptor) that is no longer sensitive to the inhibitor; (3) overcoming at least in part the lethal effects of the cytotoxin by enhancement of the rate of transcription of the gene that encodes the sensitive protein; (4) amplification of that gene. It has been unequivocally demonstrated that any one of these can bring about resistance to toxic drugs

and that gene amplification is a cause of resistance in a number of instances. Some examples of resistance by amplification are listed in Table 5.11. It should be noted that one or more of the other events may be imposed upon a primary amplification event. Amplification itself results in the production of more of the target protein, and this results in overcoming the effects of the poison, at least in part.

The frequency at which amplification occurs in the presence of the cytotoxic agent varies depending on the conditions. Generally it occurs at a frequency of 10^{-3} to 10^{-6} cells per generation. Various treatments in addition to the presence of the selective poison can enhance these rates. For example, the tumor promotor TPA (a phorbol ester) increases the frequency of amplification of the gene for dihydrofolate reductase (DHFR), an enzyme involved in thymidine synthesis, about tenfold (Varshavsky, 1981). In the presence of methotrexate, an inhibitor of DHFR, the gene *DHFR* is amplified. In addition to TPA a tenfold increase is also effected in the presence of hydroxyurea, ultraviolet light, and various other carcinogens (Tlsty et al., 1984). Agents such as these could be inducers of amplification events.

As shown in Figure 5.45, gene amplification can be manifested in cultured cells as two distinct phenomena. One leads to the formation of "homogeneously staining regions," or **HSRs**, and the other to the formation of "double minutes" or **DMs** (reviewed by Cowell, 1982). The HSRs are not always homogeneously stained; the name is a misnomer, but is retained nonetheless to avoid confusion because of the initial reports in the older literature. HSRs are more appropriately described as extended regions in which the staining patterns are different from the same regions in normal chromosomes (Fig. 5.45a). They incorporate chromatin stretches where amplification has occurred. The amplification may or may not occur at or near

TABLE 5.11. Examples of Gene Amplification in Drug-Resistant Mammalian Cells in Culture

Drug	Activity inhibited	Cell line	Degree of gene amplification	DMs	HSRs	Reference[a]
Methrotrexate	Dehydrofolate reductase (DHFR)	Mouse Hamster Human	10^2 to 10^3; (3×10^4 to 10^5 times more resistant)	Yes	Yes	1
Phosphoacetyl L-aspartate	Aspartate transcarbamylase of CAD	Hamster	10-fold increase in CAD gene number	?	Yes	2
Cadmium salts	Metalliothionien	Mouse, hamster, human	3.3×10^2 times more resistant	Yes	Yes	3
6-Azauridine pyrazofurin	UMP synthetase	Rat hepatoma	—	—	—	4
"HAT" (hypoxanthine aminopterin plus thymidine)	Hypoxanthine guanine phosphoribosyltransferase	Mouse	50-fold increase in gene number; 20-fold increase in HPRT mRNA	No	No	5
Mevastin (compactin)	3-Hydroxy-3-methylglutaryl coenzyme A reductase	Hamster	10- to 50-fold increase in gene number; 500-fold increase in enzyme activity	?	?	6
Methionine sulfoximine	Glutamine synthetase	Mouse				7

[a]Figures indicate the following sources:
1. Schimke (1984).
2. Wahl et al. (1984).
3. Crawford et al. (1985).
4. Kanalis and Suttle (1984).
5. Brennard et al. (1982).
6. Chin (1982).
7. Young and Ringold (1983).

the site of the original nonamplified gene. Some of the amplified genes may, in fact, be situated on different chromosomes. HSRs may bear the amplified genes either in tandem head-to-tail linkage or in head-to-head linkage (Looney and Hamlin, 1987).

The DMs are extrachromosomal elements consisting of chromatin bearing the amplified genes (Figure 5.45b). They are acentric and may be lost during cell division. DMs and HSRs with the same kind of amplified genes do not seem to coexist in the same cell, but they can if they bear

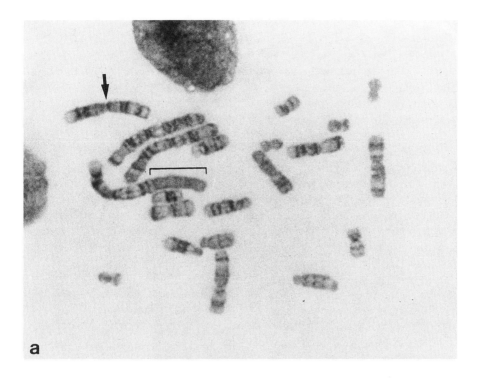

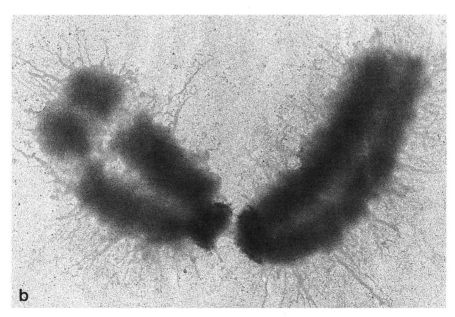

Fig. 5.45. a. *An HSR in chromosome 2 of the Chinese hamster cell line resistant to methasquin is indicated by a bracket. The normal 2 homologue is indicated by the arrow. Courtesy of J.L. Biedler of the Memorial Sloan-Kettering Cancer Center, New York, NY.* **b.** *DM in cultured cells. Courtesy of Barbara Hamkalo, University of California, Irvine, CA.*

different genes. The DMs in methotrex-ate-resistant cells of the protozoan *Leish-mania* are supercoiled circles (Garvey and Santi, 1986). This also appears to be the case for the DMs in mammalian cells (Hamkalo et al., 1985).

When amplified sequences are treated with restriction endonucleases, restriction fragments unique to them are found that not only do not match those found in normal cells, but differ from one another in different amplified populations. These are called **novel joints** (Stark and Wahl, 1984), and are expected if the amplified sequences are accompanied by variable lengths of flanking nongenic sequences. The amount of DNA in a single amplified gene, the **amplicon**, is commonly about 1,000 kb, but it may range upward to about 3,000 kb. Thus there is ample reason to expect HSRs to vary in size and show different restriction fragment patterns.

B. Amplification *In Vivo* in Whole Organisms

The ribosomal and tRNA genes are reiterated in probably all eukaryotes. Amplification of rDNA occurs especially above the normal reiteration level in the nucleoli of the eggs of many fish, molluscs, annelids, amphibians, and insects. In some, multiple nucleoli are formed, and the extrachromosomal rDNA molecules may be compared to the DMs of resistant cultured cells, as described in Ch. 6.

1. Amplification in tumor cells

One of the more interesting examples of amplification in whole organisms is to be found in tumor cells isolated from mammals. Tumors may have cells with DMs or HSRs that are identical to those found in cultured cells under stress from cytotoxins. In fact, these structures were first discovered in tumor cells. They are not found in normal cells of body tissues. It is now apparent that a definite relation exists between amplification and certain types of cancer. Oncogenes may be amplified in them. In a mouse cell line established from an adrenocortical tumor the oncogene *c-Ki-ras* is amplified 30-fold to 60-fold and resides on DMs and HSRs in these cells (Schwab et al., 1983). Also in human retinoblastomas, an intraocular tumor condition that is clearly inherited, the oncogene *N-myc* is amplified 10-fold to 200-fold (Lee, 1984). The *c-myc* oncogene has been found to be amplified in malignant neuroendocrine cells from a human carcinoma (Alitalo et al., 1983). These are but a few examples from among many now known in which amplification of an oncogene has been associated with a neoplasm.

2. Development of resistance to toxic substances

It has long been known that insects will become resistant to insecticides and hence require increasing doses to be effectively controlled. It is now established that whole organisms may, like cells in culture, become resistant to toxic substances in their environment by the simple expedient of amplifying the genes encoding the enzymes inhibited by those substances, or by amplifying genes encoding enzymes that can detoxify them. Table 5.12 gives some examples of cases in which animals and plants have developed resistance to the indicated substances. It should be emphasized that resistance in these cases is not the result of mutation *per se*, because resistance occurs within a single individual, or within a few generations of a population. Resistance resulting from a mutation should take many generations to become spread through the population by inbreeding.

3. Programmed gene amplification during development

The amplification of the rDNA genes of *Xenopus* and *Tetrahymena* occurs at spe-

TABLE 5.12. Examples of Gene Amplification in Whole Organisms

Organism	Product or gene	Reference[a]
Drosophila melanogaster	rDNA	1
Culex sp.	Esterase	2
Musca domestica	Esterase	3
Myzus persicae	Esterase	4
Petunia hybrida	5-Enol-pyruvyl shikimate 3-phosphate synthase	5
Mouse	MT-1	6
Xenopus laevis	r-DNA	7
Tetrahymena	r-DNA	8
Drosophila melanogaster	Chorion-encoding genes	9

[a]Figures represent the following sources:
1. Endow and Atwood (1988).
2. Hyrien and Buttin (1986).
3. Jenness et al. (1983).
4. Field et al. (1983).
5. Shale (1986).
6. Koropatnick et al. (1985).
7. Kafatos et al. (1985).
8. Pan and Blackburn (1981).
9. Spradling and Mahowald (1980).

cific periods of their life cycles. Amplification of *Xenopus* oocyte rDNA occurs during early meiotic prophase. The increase is 2,500-fold so that more than 70% of the nuclear DNA encodes rDNA. In both *Xenopus* and *Tetrahymena* amplification is extrachromosomal, with production of elements analogous to DMs.

Programmed amplification also occurs in *Drosophila melanogaster*. The chorion, or shell, of the egg consists of about 20 proteins. These are encoded by genes that are present in single copy in the haploid genome clustered on the X and third chromosome (reviewed by Kafatos et al., 1985). They become amplified 20-fold to 80-fold in the follicular cells of the ovary just before and during chorion formation (Spradling and Mahowald, 1980).

4. Permanently amplified gene sequences in response to environmental stress

The amplification phenomena discussed in the previous sections are generally maintained while the obtrusive toxic substances or environmental conditions promoting amplification are present. Otherwise the amplified regions will usually revert to the preamplified condition. Some amplifications are permanent, however.

Prominent examples of what appear to be reiterated genes in clusters arising from gene amplification are to be found in fish living in northern latitudes where the seawater becomes ice-laden during winter. The winter flounder, *Pseudopleuronectes americanus*, found in the North American Atlantic region, produces a set of antifreeze proteins (AFPs) (Scott et al., 1985; Davies and Hew, 1990). These are present at a concentration of about 10 mg/ml blood serum during the winter months, but in summer their concentration is decreased by at least three orders of magnitude. The AFP genes encoding these proteins exist as a family of about 40 members that are present in 7- to 8-kb segments of DNA tandemly linked. Each transcribed element is about 1 kb long and shows a certain degree of difference from its neighbors, but despite these polymorphisms they are highly homologous. It is suggested that this AFP copy number of 40 was established by gene amplification perhaps a million years ago in response to the conditions during the Cenozoic ice age. During the summer months these genes seem to be turned off to a considerable extent by a hormone produced by the pituitary (Fourney et al., 1984). Extirpation of this gland causes the AFPs to be produced at a high level during the summer as well as winter months.

A similar condition exists in the northern Newfoundland ocean pout (*Macrozarces americanus*), a fish distantly related to the flounder (Hew et al., 1988). On the other hand, a fish closely related to the winter flounder, the yellowtail flounder (*Limanda ferruginae*) lacks the tandemly highly amplified set of 20–40 AFP genes. It has, instead, a set of 10–12 linked but irregularly spaced AFP genes. During summer it has a peak serum level of antifreeze protein less than half that found in the winter flounder (Scott et al., 1988). Although both flounders occupy approximately the same geographic region of the northeast coast of North America, the winter flounder more often occupies shallow waters exposed to ice, whereas the yellowtail lives at depths up to 80 m, where the water is warmer than at the surface. It would thus appear that the degree of amplification maintained correlates with the degree of danger of serum freezing.

C. Mechanisms of Gene Amplification

Various models have been proposed to explain the mechanism(s) of gene amplification (reviewed by Stark and Wahl, 1984; Schimke, 1988). They may be grouped under the following categories: (1) unequal sister chromatid exchanges, (2) recombination events involving strands of overreplicated DNA (Figures 5.46 and

5.47); (3) repeated transpositions such as those giving rise to *Alu* and *LINE1* repeats; and (4) rolling circle synthesis of circular extrachromosomal DNA of DMs.

1. Unequal sister chromatid exchange

This may occur when two chromatids become misaligned during prophase. If reciprocal recombination then occurs, one chromosome will be generated with a duplication, and its sister will have a deficiency. In order for amplification to continue, this type of event must be reiterated many times unless other factors enter into play (reviewed by Hamlin et al., 1984). It has been proposed that amplification of the 18s–28s rRNA genes of *Drosophila melanogaster* results from unequal exchanges of this type (Endow and Atwood, 1988). The occurrence of thousands of *Alu*- and *LINE1*-type elements throughout the euchromatin of mammalian chromosomes might be expected to lead to misalignment of chromatids by pairing of homologous segments of these elements at different regions of the chromatids. Although this hypothesis for the generation of repeated units of function has merit, there is no definite proof of its occurrence at present.

2. Overreplication and recombination

Models for amplification by unscheduled DNA replication leading to tandem repeats have been described by a number of workers (Hamlin et al., 1984; Schimke et al., 1986; Stark and Wahl, 1984). These models are collectively referred to as the "onion skin" mechanism for amplification. One of these is illustrated in Figure 5.46. The bubbles formed as shown in Panels a and b theoretically continue and result in the formation of tandem repeats (Panels c and d), extrachromosomal circles (Panel e) or linear segments of extrachromosomal DNA (Panel f).

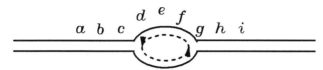

a b c d e f g h i

a b c d e f g h i

a. replication

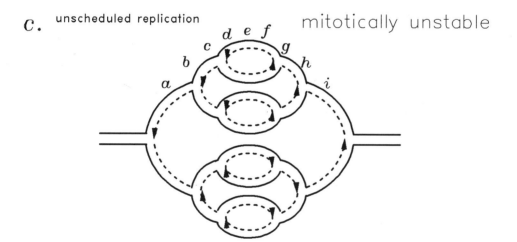

b. unscheduled replication

c. unscheduled replication   mitotically unstable

recombination:

d.   e   *f.*

abcdcdedefgdefgh

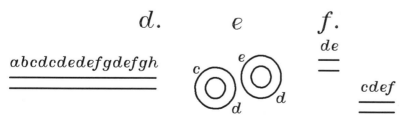

Fig. 5.46. *Model for unscheduled DNA replication leading to gene amplification.* **a–f.** *See text. After Stark and Wahl (1984).*

Amplification in some cultured cells is associated with the occurrence of chromosomal aberrations. For example, large inverted repeats occur in cells in which the *CAD* gene (a complex locus encoding carbamoyl phosphate synthetase-2 plus aspartate transcarbamylase and dehydroorotase) and the *myc* oncogene are amplified (Ford and Fried, 1986). In addition, some cases of amplification appear to be associated with overreplication of DNA and recombination events (Schimke et al., 1986). The treatment of mouse lymphoma cells with hydroxyurea, an inhibitor of DNA synthesis, results in an increase in incidence of polyploidy, extrachromosomal DNA, and endoreduplication after the removal of this poison. Also, in the presence of both methotrexate and hydroxyurea, amplification of the *dhfr* gene is increased ten-fold or more over what would be expected in the presence of methotrexate alone (Hill and Schimke, 1985). Since treatment with hydroxyurea leads to overreplication of DNA, it has been proposed that DNA overreplication followed by recombination leads to chromosomal abnormalities and, if conditions are right, to gene amplification (Schimke et al., 1986). This is supported by the observation that only those cells treated with hydroxyurea that have additional DNA over the normal 4C in the G2 phase have the aberrations and amplification. A model, shown in Figure 5.47, to explain these phenomena has been proposed that is based on the occurrence of amplification and overreplication (Schimke et al., 1986). It assumes that overreplication leads to the generation of free-ended DNA strands, and that these strands undergo nonreciprocal homologous recombination. In Figure 5.47 each thin line represents the double-stranded DNA of a sister chromatid and the heavy lines represent the overreplicated double-stranded DNA. If the overreplicated DNA strands circularize, they should form DMs (Figure 5.47b). If recombination occurs between the overreplicated strand and the chromatid DNA, various types of results could eventuate, one of which is shown in Figure 5.47c. In this example only one end of the overreplicated DNA recombines into the chromatid, resulting in a break in the chromatid. On the other hand, if both ends recombine, an inversion could result (Figure 5.47d). Furthermore, a double-ended recombination involving both parental chromatids (Figure 5.47e) could result in a variety of aberrations such as breaks and duplications. Amplification can be expected to accompany these aberrations simply by the peeling-off of the overreplicated DNA. Other workers have reported results indicating that aberrations and amplifications may occur in Chinese hamster cells without overreplication (Hahn et al., 1986). However, there is general agreement that chromosome aberrations of various types always seem to accompany amplification, at least in cultured cells.

3. Size of amplified units

The question of the size of the unit of amplification, the **amplicon**, has received some attention by a number of workers. For example a Chinese hamster cell line (CHO) with the *dhfr* gene amplified 1,000- to 1,200-fold yielded two types of amplicons. Type I amplicons were 260 kb long and were tandemly arranged head to tail. Type II were 220 kb long, but arranged both head to head and tail to tail. These latter types constituted 85%–95% of the amplified regions (Looney and Hamlin, 1987). Since the *dhfr* gene itself is only about 25 kb long, the amplicons obviously contain a considerable amount of nongenic DNA.

An amplicon may contain more than one gene. Two examples of this have already been discussed: the chorion genes of *Drosophila* and the antifreeze proteins of the

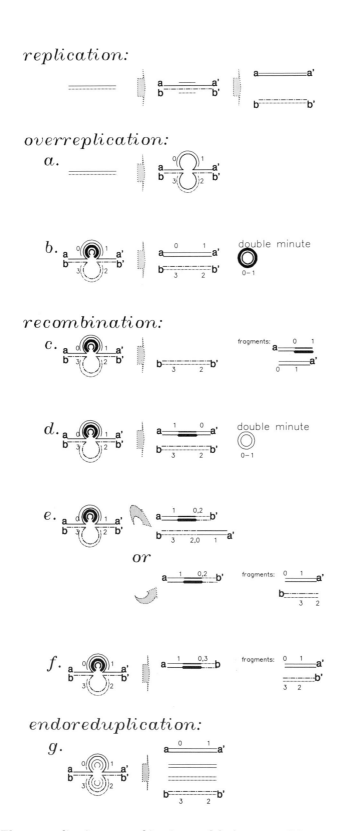

Fig. 5.47. *The overreplication–recombination model of gene amplification.* **a–g.** *See text.*

flounders. The genes encoding the *Drosophila* chorion proteins are amplified during oogenesis, and shown to be the result of multiple rounds of replication in the specific regions where the genes are clustered (Kafayan et al., 1985; Martinez-Cruzado et al., 1988). This may be an endoduplication of the type that causes puffs in chromosomes. A coordinate amplification has also been reported for the two metallothionein genes of the Chinese hamster (Crawford et al., 1985). These coordinate amplifications of groups of clustered genes may be more common than presently realized.

VII. CHROMATIN DIMINUTION AND CHROMOSOME ELIMINATION

In some animals chromatin is eliminated from their germ cells during meiosis, or from the early cleavage cells after fertilization of the egg (reviewed by Tobler, 1986; Goday and Pimpinelli, 1984; Pimpinelli and Goday, 1989). The phenomenon was first noted by Boveri (1887) in the nematode *Ascaris megalocephala*, which is parasitic in horses. Subsequent to Boveri's discovery a similar phenomenon has been discovered in many other species ranging from protozoa to vertebrates and even some plants. Some examples of these are listed in Table 5.13. In some cases, as in *Ascaris*, only parts of chromosomes are eliminated, a process called **chromatin diminution**, whereas in others whole single chromosomes or haploid sets of them are eliminated.

A. Diminution in Ascarid Nematodes

Diminution as originally described by Boveri is illustrated in Figure 2.12. The species of *Ascaris* used by Boveri is now named *Parascaris univalens*, since it has only one chromosome when haploid. A closely related species, *P. equorum*, is distinguishable from *univalens* only because it has $n = 2$ chromosomes. The diminution process in the two forms is similar. Figure 5.48, taken from Boveri's original drawing, illustrates the process of diminution in *Parascaris* from the egg to the 32-cell stage, five cell divisions after fertilization. Diminution begins with the second cleavage, when the chromosomes of one of the two daughter cells from the first cleavage begin to fragment at metaphase. During anaphase the fragments are left behind, while the remaining parts of the chromosomes proceed to the poles. These cells with diminished chromatin become somatic cells. The other daughter cell from the first cleavage contains the full complement of chromatin found in the zygote. In each of the four successive cleavages of this cell with the full chromatin complement, daughter cells with diminution are produced, but one cell line remains intact and forms a germ line, while the others form more somatic cells. The results is that at the 32-cell stage two cells are present that will form the germ line, while the remaining 30 go on to form the soma (Boveri, 1910).

Chromatin diminution in both species of *Parascaris* consists of the elimination of all heterochromatin without regard to its chromosomal location (Goday and Pimpinelli, 1986). Both species have holocentric chromosomes but differ in the distribution of the constitutive heterochromatin that is eliminated (Goday et al., 1985; Goday and Pimpinelli, 1989). *P. univalens* chromosomes have only terminal heterochromatin, whereas *P. equorum* has both intercalary and terminal hetrochromatin. The two species are sexually isolated because hybrids produce lethal offspring. They may differ primarily in the differences in the repetitive DNA of their heterochromatin (Goday and Pimpinelli, 1986). After diminution the somatic cells of both species contain only euchromatin

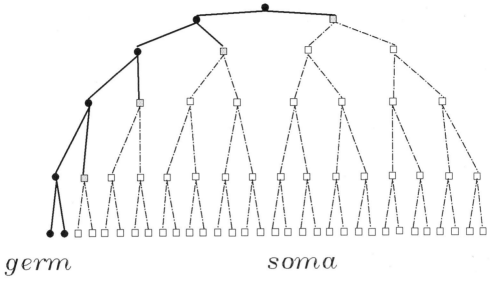

germ *soma*

Fig. 5.48. *Early cell divisions of zygote of Parascaris to show segregation of germ line and soma. The filled circles represent the germ line cells. Cells undergoing* diminution *of chromatin are represented by stippled squares. The open squares represent chromatin-reduced somatic cells.*

that is fragmented into about 60 small segments. The *n* number then becomes 30.

B. Diminution in Ciliate Protozoa

The macronuclei of the ciliates differentiate from division products of their micronuclei after conjugation, as described in Chapter 2. This differentiation consists of a great deal of reorganization of the DNA of the chromosomes of the micronuclei accompanied by the elimination of 10%–98% of the DNA, depending on the species (Blackburn and Karrer, 1986). The micronuclear chromosomes fragment and part of their DNA is eliminated. That which is not is reformed into many linear DNA molecules with new telomeric ends. The eliminated DNA is of two general types: repetitive sequences and **internally eliminated sequences** (IESs), which are not members of repetitive sequence families and could be functional genes (Blackburn, 1987).

C. Diminution in Arthropods

DNA diminution in the copepod crustacean *Cyclops* consists of the excision of intercalary heterochromatic segments, but unlike *Parascaris* there is no fragmentation of the chromosomes (Beerman, 1984). Like *Parascaris*, however, the three species of *Cyclops* investigated by Beerman have different distributions of heterochromatin in their respective chromosomes.

Diminution of chromosome numbers has been extensively investigated in three families of Diptera, the Sciaridae, Cecidomyiidae, and the Chironomidae. These flies have most unusual and bizarre life cycles in which whole chromosomes are eliminated during early cleavage division (reviewed by White, 1973). In the cecidomyid *Miastor*, for example, the somatic cells contain 12 chromosomes and the germ cells 48 (White, 1946).

The sciarid fly *Sciara coprophila* has a life cycle that is fairly representative of those Diptera that have chromosome elimination as a way of life (reviewed by Gerbi, 1986). Figure 5.49 is a diagram of its life cycle as described by Metz (1938). The genome consists of three autosomes (two acrocentrics and a metacentric), sex chromosomes, and "L" chromosomes, which are germ-line-limited and have the properties of B chromosomes. The karyotypes of the germ cells of males and females are similar; they have three pairs of autosomes, two X chromosomes, and generally two L chromosomes (Figure 5.50). Two kinds of X chromosomes exist, one designated simply as X and the other as X'. The X and X' carry demonstrably different genes. There is no Y chromosome. Oogenesis is orthodox insofar as a haploid egg results with one copy of each autosome, one X, and one L (Figure 5.49a).

Spermatogenesis is not orthodox. It results in sperm with two sister chromatids of a single X, a haploid set of autosomes, and usually two L chromosomes (5.49a). As a result the zygote after fertilization has three X chromosomes—two identical paternal and one maternal in source (5.49a).

Two kinds of females result from these zygotes: those that produce only male progeny and those that produce only females. Genetic analysis reveals that female-producing mothers are always XX' and male producers are always XX. Since males are always XX like female-producing males, what determines sex? The answer is shown in Figure 5.49b. In this somatic line the L chromosomes and two of the three Xs are eliminated, resulting in

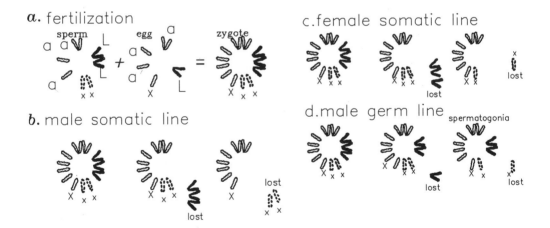

Fig. 5.49. *Diagram of chromosome elimination in the fly Sciara coprophilia.* **a.** *The fertilization of the haploid egg by a sperm results in a zygote with three pairs of autosomes, three L chromosomes, and three sex chromosomes. The paternal sex chromosomes are indicated by lower case x's, and the maternal by upper case X's.* **b.** *A male somatic line of cells derives from a zygote by successive elimination of L chro-* *mosomes and the paternal x's.* **c.** *The female somatic line of cells derives from a zygote by the elimination of the three L chromosomes and one paternal x.* **d.** *Spermatogonia are derived from a zygote by the elimination of one or two L's and the paternal x to give sperm the chromosome complement shown in a. These x's are derived from a maternal X, but are now indicated as paternal x's because that is what they become.*

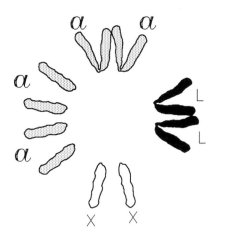

Fig. 5.50. *Karyotype of the male and female germ cells of Sciara.*

XO male somatic cells. These two eliminated Xs are always paternally derived. In the female somatic line the L chromosomes are also eliminated, but only one X is giving a somatic genotype of XX as shown in Figure 5.49c. Both male and female primary gonial cells are XX and generally have two L chromosomes (Fig. 5.50). This results from the elimination of one L and one of the paternally derived Xs, as shown in Figure 5.49d for the male. As noted, the haploid egg ends up with one X as expected from a normal meiosis. But haploid sperm result from an abnormal meiosis in which the paternally derived autosomes and X are eliminated in the first meiotic division, and in the second division the sister chromatids of the autosomes separate while only one of the two products of the division results in the two sister chromatids of the X. This unequal division results in the sperm karyotype shown in Figure 5.49a. Note that the sperm has female derived Xs now indicated as paternal Xs. However, once fertilization occurs these two Xs are the ones eliminated in the male germ and somatic lines, and one is eliminated in the female somatic line, as shown in Panels b and d.

D. Elimination in Vertebrates

Chromosome elimination occurs in a number of species of vertebrates, as noted in Table 5.13. Two species of hagfish have been reported to eliminate chromosomes during early cleavage stages. The Japanese hagfish, *Eptatretus burgeri*, has a model chromosome number of 36 in its somatic cells, 53 in spermatogonia, and 25–26 in its primary spermatocytes. Its haploid number is therefore about 26, and assuming its eggs have the same number, the diploid number is 52. The conclusion is that 16 chromosomes are eliminated during early cleavage (Kohno et al., 1986). Another hagfish, *Myxine garmani*, has been reported to have 14 chromosomes in its somatic cells, 16 in its spermatogonia, and 8 in its primary spermatocytes (Nakai and Kohno, 1987). The largest pair of chromosomes present in the spermatogonia is absent in somatic cells.

While hagfish show chromosome elimination, a holocephalan fish, *Hydrolagus collieri* (Chondrichthyes), shows chromatin diminution. This is apparently a rare phenomenon among insects and vertebrates, as compared to chromosome elimination (Stanley et al., 1984). Diminution consists in eliminating a mass of heterochromatin from the chromosome complement during metaphase I of spermatogenesis. This mass constitutes about 10% of the total DNA of the spermatogonia.

A frog, *Rana esculenta*, is a hybrid between *R. ridibunda* and *R. lessonae*. In both males and females of the hybrid, premeiotic elimination of the *R. lessonae* chromosomes occurs followed by a duplication of the *ridibunda* genome. Meiosis then results in gametes with only *ridibunda* chromosomes (Heppich et al., 1982). The *ridibunda* chromosomes can be distinguished from the *lessonae* by differences in their centric heterochromatin. It is this difference that might account for the elimination of one set and the reten-

TABLE 5.13. Occurrence of Chromatin Diminution or Chromosome Elimination

Taxon	Stage at which loss occurs
Ciliate Protozoa	During macronuclear formation
Nematodes	During early cleavage divisions
Copepod Crustacea	During early cleavage divisions
Insects	
Coccids	Elimination of paternal chromosomes in males during blastula stage
Cecidomyidae	During early cleavage
Sciaridae	During spermatogenesis and early cleavage
Chironomidae	During early cleavage
Siphonaptera	Elimination of one of the two Y chromosomes from male somatic cells
Heteroptera	Elimination of parts of X chromosome during anaphase I of Meioses
Acarina	Half of chromosomes eliminated from germ line and from most somatic cells in male embyro
Vertebrates	
Agnatha	During early cleavage
Chondrichthyes	During spermatogenesis
Marsupialia	Elimination of the Y in males and one X in females in somatic cells
Angiosperms	
Sorghum	Elimination of B chromosomes in root cells
Hordeum hybrids	Elimination of chromosomes
Scilla	Diminution in cells in culture

tion of the other. A type of genetic imprinting may be involved, as discussed on p. 264.

The only mammals that appear to have a significant incidence of chromosome elimination are the marsupials (reviewed by Hayman and Martin, 1969; Close, 1984). The bandicoots, like most other marsupials, have XX female and XY male germ cells. However, some somatic tissues have an XO condition in both sexes, indicating a loss of an X in females and a Y in males. The X chromosomes eliminated are always of paternal origin in the bandicoots; in the kangaroos the paternal Xs are simply inactivated.

E. Elimination in Plants

Chromosome elimination in plants has been noted in monocot angiosperms. The grasses *Poa alpina*, *Secale*, and *Sorghum*

purpurea have B chromosomes eliminated in root cells but not in stem cells (Muntzing, 1949; Darlington, 1956). Crosses between different species of barley produce hybrids in which chromosome numbers per cell decrease during successive mitotic divisions in the embryos or in parts of the mature plants (Thomas and Pickering, 1983; Finch, 1983). In general the chromosomes of one of the contributing parents are eliminated as a set. For example, hybrids of the cross *Hordeum vulgare* × *H. bulbosum* generally lose *bulbosum* chromosomes by the next generation. Elimination does not occur in one mitotic division but in successive divisions. In the cross *H. marinum* × *H. vulgare*, on the other hand, *vulgare* chromosomes are eliminated in the endosperm, but the complete *marinum* genome in the embryo.

Changes in the DNA content by selective diminution have been reported for the triploid Siberian squill, *Scilla siberica* (Deumling and Clermont, 1989). During callus culture *in vitro* chromosome elimination occurred in some triploid cells to produce diploids, but no plants could be regenerated from them. In other cases reduction of heterochromatin resulting in the formation of small chromosomes was observed. Plants regenerated from these, when grown in the garden, showed selective gain of chromatin depending on the environmental conditions.

F. Factors Involved in Diminution and Elimination

The action of genes can be suppressed by a variety of known means, some of which have been discussed in previous chapters. It has already been noted that the entrance of transposing elements into their regulatory or transcribed regions of genes can silence them (p. 211). Likewise the failure of a binding protein to interact properly with regulatory sequences can have a similar effect and methylation undoubtedly plays a role in some species in silencing some genes, as discussed in Chapter 6. In this chapter the silencing of genetic elements by chromatin diminution and chromosome elimination has been described. The latter means of restricting gene action are certainly more drastic than the former, but they are certainly effective even though in many ways they are less well understood.

The first question that comes to mind when considering the phenomenon of DNA elimination by whatever means is Does the eliminated DNA play a role in cell function before it is discarded? It does appear to in some cases. While it is true that in all known cases of diminution the DNA discarded is highly repetitive, this does not mean that it has not played a role before being eliminated. It may be impor-

tant in the packaging of the DNA in the prior stages of germ line cell development. Also functional genes may be incorporated as unique sequences among the repetitive. The DNA eliminated from the genome of *Ascaris lumbricoides* produces a satellite upon gradient density centrifugation. It has an 11-bp repeated oligonucleotide with the consensus sequence $5'GCA(^T)TT(^T)TGAT$ (Muller et al., 1982). If this is junk DNA, it has kept a remarkable degree of conservation of structure. Evidence exists for a functional role of germ-line-limited chromosomes from experiments with certain species of Cecidomyidae. Embryos of these species have 40 chromosomes in the germ cells, but lose 32 of them in the presumptive somatic cells. This has led to the conclusion that the chromosomes normally retained in the germ line are probably indispensible for gametogenesis (Geyer-Dusznska, 1966).

One of the more remarkable features of chromosome elimination is that the paternally derived ones are generally discarded in animals. How are the paternal ones distinguished from the maternal? A possible explanation exists in the phenomenon of genetic imprinting. This was described by Crouse (1960) and used by her to explain the anomalous behaviour of the X chromosomes of *Sciara*. She showed that a segment of the centric heterochromatin of the X is alone involved in the elaborate series of events in the flies' life cycle such as the equational nondisjunction of the X sister chromatids in secondary spermatocytes and the elimination of the paternal chromosomes in the embryo (Figure 5.49). Without this heterochromatic segment the X behaves like an autosome, and when attached to an autosome by translocation this segment causes the autosome to behave like an X. This behavior of the intact X is thus controlled not by its genic content but by its heterochromatin (Crouse, 1979). A chromosome that passes through the male germ line "acquires an **imprint** which will result in behavior exactly opposite to the imprint conferred on the same chromosome by the female germ line" (Crouse, 1960; see also Crouse et al., 1971). The element responding to the male or female line influence is the heterochromatic element, which need not be close to the centromere.

Granted that this heterochromatic segment is involved in **imprinting** for reasons unknown, there also remain the reasons for the elimination and disintegration of the paternal X once it is recognized. Sager and Kitchin (1975) have hypothesized that restriction endonucleases may be involved. The restriction enzymes used in laboratories for cutting DNA are isolated from bacteria. Their existence in bacteria is generally assumed to serve as a protection against foreign DNA that invade in the form of viruses, plasmids, and so on. In the context of chromosome elimination for *Sciara* we can imagine that the imprinting of paternal chromosomes as described by Crouse makes them susceptible to the action of the flies' endonucleases, followed by the action of less specific endonucleases that complete the destruction of the X-chromosomal DNA. Restriction endonucleases have been reported to be present in eukaryotic cells (Lao and Chen, 1986).

Enzymes such as these in eukaryotic cells could be active at certain times during the cell cycle and stage of differentiation, and could recognize foreign DNA in the form of chromosomes in hybrids, or release and destroy hetrochromatin no longer needed. In a mouse–human hybrid as described in Chapter 8 the mouse genome may produce a restriction enzyme that cuts human DNA at the centromeric regions, for example. The result would then be the preferential loss of human chromosomes as acentric fragments.

VIII. NUMBER OF GENOMES PER NUCLEUS

Ploidy is the term used to refer to the number of chromosome sets or genomes per nucleus. The haploid set is the genome with the qualification that in those species with heterogametic sexes, both sex chromosomes are included. Thus the human genome has 243 chromosomes: 22 autosomes plus the X and Y. If each set in a genome is complete, and contains each heterologous chromosome, unaltered with respect to duplication and deletion, that properly belongs in the set, the condition is called **euploidy**; If not, the condition is called **aneuploidy**. Two sets of genomes per nucleus constitute **diploidy**, and more than two, **polyploidy**.

In Chapter 3 polyteny and endopolyploidy are discussed. Polyteny results from the replication of chromatids that remain bundled together to form a polytene chromosome. It is different from polyploidy in that the chromatids do not become chromosomes, and generally the replications are not equal in different parts of the polytene chromosome. Endopolyploidy is the result of **endomitosis** or **endoreduplication**, a form of somatic polyploidization. It takes place within an intact nucleus and results in duplication of chromosomes without nuclear division. Once formed a polyploid cell can undergo a normal mitosis, resulting in polyploid daughter cells and tissues.

A. Euploidy

A euploid nucleus may be haploid, diploid, or polyploid and undergo normal karyokinesis whatever its degree of ploidy. A true haploid cannot undergo meioosis, however, and a polyploid may have problems as discussed below. Some plants that appear to be haploid and do carry out meiosis are probably cryptic polyploids.

1. Haploidy

Cells with a single genome are haploid. These may be gametes, cells of plant gametophytes, or vegetative cells of some fungi and some algae. Animals that reproduce sexually have a **diplontic** life cycle, in which the **haplophase** consists only of gemetes, and the **diplophase** dominates (Figure 5.51). Higher plants and some algae have a **diplohaplontic** life cycle, which includes a multicellular haplophase, the gametophyte, and a diplophase, the sporophyte (Figure 5.52). In the Bryophytes (mosses, liverworts, and hornworts) the haplophase dominates. In the vascular plants (ferns and conifers and their relatives, and flowering plants) the sporophytic diplophase dominates. Many fungi and primitive algae have a **haplontic** cycle, in which the zygote, once formed by fertilization, does not undergo mitosis but enters directly into meiosis (Figure 5.53). Variations on these general themes exist, some of which are discussed in what follows.

Haploid flowering plants occasionally exist in what should be the diplophase. One example is a haploid strain of *Datura stramonium* that was first reported by Blakeslee et al. (1922). This plant, like others since described, was somewhat weaker than normal diploids and produced a high proportion of aborted pollen grains. Sporophytic haploids derived from polyploids and referred to as **polyhaploids** are widely distributed among the families of the angiosperms (reviewed by Kimber and Riley, 1963). These should probably not be considered true haploids. The tolerance of the haploid state by plants is not shared by animals. Haploid frog's eggs can be induced to start cleavage, but the resultant embryo dies early, unless diploidy arises spontaneously by the simple doubling of chromosomes in an early phase of embryogeny. The completely homozygous embryo then continues development

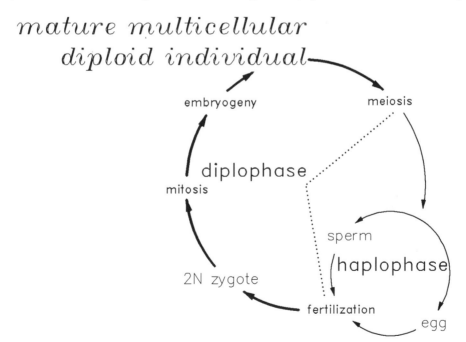

Fig. 5.51. *The diplontic life cycle as found in animals and some algae.*

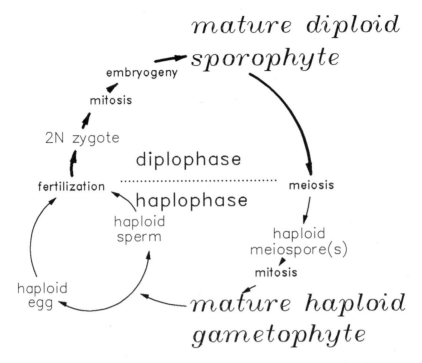

Fig. 5.52. *The diplohaplontic cycle as found in plants with an alternation of generations. These include some multicellular algae and bryophytes, pterido-phytes, gymnosperms, and angiosperms.*

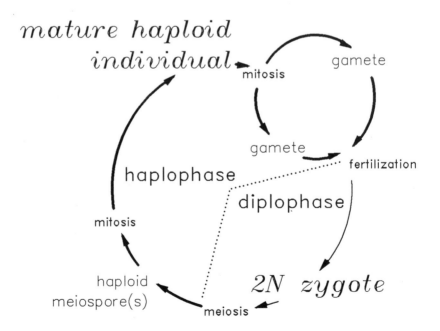

Fig. 5.53. *The haplontic life cycle characteristic of fungi.*

to a normal adult, unless a lethal gene is present in the egg genome. Homozygous plants have also been produced by doubling the chromosomes in haploids. In the case of plants a lethal gene might be expected to be selected against in the haploid gametophyte stage.

In some animals the males normally develop from unfertilized eggs. This is generally referred to as male parthenogenesis or **arrhenotoky** (from the Greek, "bearing male offspring"). It occurs in four insect orders. Almost all Hymenoptera, some coccids of the Order Homoptera, some Coleoptera, and possibly all Thysanoptera (thrips) are arrhenotokous. (See White, 1973, for a comprehensive coverage of this phenomenon.) It is clear that the nuclei of arrhenotokous male early embryos are haploid, but most of the adult somatic tissues are endopolyploid. The tissues of the male honey bee have about the same DNA content as the female tissues (Merriam and Ris, 1954). In the thysanopteran *Haplothrips* some of the male tissues are haploid; others are polyploid (Risler and Kempter, 1962). In general, then, an animal embryo that starts out as a haploid eventually attains diploidy or a higher degree of ploidy in some or all of its cells. Except for certain of the protozoa, animals eschew haploidy. The coccid insects as a group have among the most unconventional life cycles known for animals. In some of them the haploid state may arise from the diploid by the elimination of the paternal set of chromosomes during early embryogeny (Hughes-Schrader and Monahan, 1966). However, they never seem to abandon diploidy completely in all parts of their life cycles (Brown and Chandra, 1977).

2. Diploidy

Diploidy in animals is not only dominant, but to a considerable degree obligate for proper development. At some point in the development of the line leading to animals as we know them now, the diploid state superseded the haploid. This haploid-diploid transition was an important point in their early evolution and raises an important question regarding the functioning of chromosomes, not only in animals but in higher plants as well. Why do their chromosomes function best in pairs?

3. Polyploidy in plants

Polyploidy is a common state in plants (reviewed in Lewis, 1980). It occurs in many species of algae, mosses, Pteridophytes, and flowering plants. It is uncommon among the gymnosperms and liverworts, and of doubtful occurrence among fungi. The homosporous ferns are commonly polyploid (Wagner and Wagner, 1980; Walker, 1984). Some genera are all diploid, but some are 100% polyploid and many range from 45% to 78% polyploid. The n values range from $2n$ to $16n$, and are probably the most extreme found in any eukaryotic group. Various estimates of the occurrence of polyploidy among current species of flowering plants have been made, ranging from 30% to 55% (Averett, 1980). Whatever the exact number, there is general agreement among plant cytologists that most angiosperms either are polyploid now or have polyploidy in their ancestry, as explained in Chapter 10. The number of additional sets of genomes above $2n$ may range from $3n$ to $10n$ or more. For example, the genus *Chrysanthemum* appears to have a basic chromosome number of 9, and the different known varieties have $n = 9, 18, 27, 36$, and 45, forming a series of multiple genomes ranging from diploid (18), tetraploid (36), hexaploid (54), octoploid (72), and decaploid (90). The assumed basic number is generally designated x in most genera, and is generally thought to be less than 9 by many observers (Goldblatt, 1980). The x number may not be the same as the observed n number. The observed n genome may contain homeologous chromosomes. Plants are rather carefree about their genomic constitutions. The high chromosome numbers for plants shown in Table 5.8 are for the most part the result of polyploidy. Thus, among the ferns of the genus *Ophioglossum* species with assumed *n* numbers ranging from 30 to 720 are known (Khandellwal, 1990). It has been suggested that $x = 15$ for these species, in which case *O. reticulatum*, with 1,260 chromosomes, is 84 ploid (Love and Kapoor, 1966, 1967). Grant (1971) classifies species with *n* numbers of 14 or more as polyploids. He calculates that 47% of the angiosperms and 95% of the pteridophytes are polyploids. The estimates given by experts vary, but they are all in agreement that among certain plant groups polyploidy is extensive.

Some species of flowering plants show a high degree of polymorphism for polyploidy. An extreme example has been found for *Aster ageratoides*, native to Japan. Of 791 plants collected from various localities about 35% were diploid, 0.8% were triploid, 51% were tetraploid, 0.8% were pentaploid, and 13% were hexaploid (Irifune, 1990).

Two basic types of euploid polyploidy occur in plants: **autopolyploidy** and **allopolyploidy**. In an autopolyploid plant such as an autotetraploid (4*n*) pairing at the time of first meiotic prophase can occur at any point between two of four homologous chromosomes to form **quadrivalents**. An allotetraploid plant, on the other hand, contains a 4*n* or higher complement of chromosomes derived from two or more plants of dissimilar origin. Its chromosomes pair as in a diploid to produce bivalents in the meiotic prophase. They are polyploids that function as diploids; hence they are sometimes referred to as amphidiploids. Some *octoploid* (8*n*) cultivated plants such as **Dahlia** form quadrivalents at meiosis and therefore function as amphitetraploids (Lawrence and Scott-Moncrieff, 1935).

Plants of the genus *Datura*, particularly *D. stramonium*, can cope with extreme differences in ploidy. The normal condition is diploid, $2n = 24$, but haploid, 3*n*, 4*n*, 6*n*, and 8*n* plants also occur. Blakeslee and his coworkers (reivewed by Avery et al., 1959) did extensive work with polyploid *Datura*. These plants were for the most part derived by crossing normal diploids. Tetraploids were all found to be autopolyploids that formed quadrivalents at meiotic prophase. A tetraploid crossed to a diploid produced triploids, which as expected formed trivalents. Polyploid plants have larger cells than diploids. This may result in larger plants and flowers, but not always.

Treatment of plant cells with colchicine or its derivative, colcemid, results in an arrest of mitosis at metaphase; sister chromatids are not distributed to daughter nuclei, and the doubling of the chromosome number is a frequent occurrence. Chromosome doubling of triploid *Datura* produced hexaploid plants (6*n*) and the doubling of the chromosome complement of a tetraploid produced an octoploid (8*n*). Both hexaploid and octoploid autopolyploids were found to be weak and sterile. Hence in *Datura* there is a practical limit to the degree of autopolyploidy.

This limit may be the result of the production of unbalanced meiospores during meiosis. Autopolyploids are known to be characterized by meiotic irregularities (de Wett, 1980). Those polyploids that produce gametes function as diploids, as explained below. Many polyploid plant species reproduce asexually by **apomixis**, and thereby avoid the meiotic hurdle. Two main types of apomixis exist. First is **agamospermy**, in which new sporophytes are produced from asexually formed embryos and seeds. Fertilization does not occur, or if it does the pollen nuclei to not participate, and the embryo has the same chromosome number and genotype as the mother. It is comparable to parthenogenesis in animals. (Actually the terms parthenogenesis and apomixis are used interchangeably for both plants and animals.) The second type is the well-known ability of plants to form clones from cuttings, from root rhizomes, and in the laboratory even from single cells.

It is not uncommon for two different species of flowering plants, even in different genera, to hybridize and produce viable offspring. Such hybridizations occur naturally in the wild as well as in the laboratory. It is almost certain that interspecific hybridizations of this type have been an important factor in the evolution of plant species (Stebbins, 1950; Lewis 1980). One consequence of hybridizing two different species is that at the time of meiosis in the hybrid the parental chromosomes may fail to pair to form bivalents. The hybrid is technically a diploid, but functionally is a haploid and the consequence of meiosis will be sterile products except in the unlikely event that the first division results in two genomes with full complements of chromosomes. A fertile allopolyploid can result, however, if the chromosomes double. The resulting tetraploid is a functional diploid with the formula AABB, for example, in which A and B represent the different genomes.

Some allopolyploids may have enough pairing so that they are intermediate between the autoploid state, with complete multivalent pairing, and the complete alloploid state in which no pairing occurs between homeologues. Such intermediates may have enough pairing for viable meiocytes to be formed by meiosis. A functional allopolyploid may contain more than two different genomes and can be assigned the formula of AABBCC, etc. Specific examples of this are the cultivated wheats. The principle cultivated bread wheat is *Triticum aestivum*. It is a hexaploid with 42 chromosomes, and has three different genomes, A, B, and D, each with seven chromosomes. Functionally it is a diploid with the constitution AABBDD. Its A genome probably derives from a wild species,

T. monococcum, the B genome from *T. searsii*, and the third (D) genome from another wild species, *T. tauschii* (Kimber and Sears, 1987). Each of these presumed parental species is a diploid with $2n = 14$.

Two questions arise from the studies and experiences with allopolyploids such as wheat. First is the question of how strongly differentiated the chromosomes of the participating parent genomes are. Are they really so different in all of their genes as to prevent pairing? The second question is how the triplication of genes in plants such as *aestivum* is coped with. It has been demonstrated that many of the same genes are indeed present in triplicate. We deal with the second question in the following sections and in Chapters 9 and 10. The first question has been given a partial answer by Sears (1976) and coworkers. The corresponding (homeologous) chromosomes of *T. aestivum* are closely related genetically (Sears, 1952). The reason that they do not pair is that homeologous pairing is suppressed mainly by a gene or genes on the long arm of chromosome 5 in the B genome called the Ph gene system (reviewed by Sears, 1976).

The three genomes of *T. aestivum*: A, B, and D, have DNA contents in the ratio of 1.14:1.2:1 (May and Appels, 1987). The total DNA content of the three is about 16 $\times 10^9$ bp, which is approximately three times the $\sim 5 \times 10$ bp per genome found for the diploid wheat. Chromosome banding (Gill, 1987) and genetic linkage analyses (McIntosh and Cusick, 1987; Milne and McIntosh, 1990) confirm that the three parental genomes have similarities in structure, but also differences in linkage and banding are noted as expected, since the three are from different species. Both translocations and inversions are evident.

4. Polyploidy in animals

Present-day animals that have been karyotyped have nearly always been found to be diploid, but the exceptions are notable enough to warrant some description and discussion. They bear upon probable significant differences in the functioning of animal genomes and their evolution as compared to plant genomes.

Polyploidy occurs in nematodes (Triantaphyllou, 1984), earthworms (Muldal, 1952), and certain Crustacea such as shrimp, in which tetraploid and hexaploid species have been described (Benazzi, 1957). Probably only about 100 species of insects out of about one million identified species have polyploid members. These are found in the Orders Orthoptera, Homoptera, Lepidoptera, Diptera, Coleoptera, and Hymenoptera. Among the Coleoptera, species with $3n$, $4n$, $5n$, and $6n$ karyotypes are known, and all are autopolyploid (Virkki, 1984). Some of the polyploid Hymenoptera have tetraploid females and the males develop from diploid eggs instead of the usual haploid. All of the polyploid insects now known reproduce parthenogenetically (Astaurov, 1969). Among the vertebrates polyploid species have been identified in fishes, amphibians, and reptiles (Schultz, 1980; Bogart, 1980).

A sharp difference exists between the occurrence of polyploidy among obligate, bisexually reproducing animals and those known to engage in parthenogenesis. It is rare or absent among the former and common among the latter. If there are any advantages to polyploidy, it may be achieved indirectly by somatic polyploidy, endomitotic polyploidization, and polyteny, all of which can be found among bisexual animals to one degree or another.

Muller (1925) made two cogent observations about the scarcity of polyploidy among animals. First he pointed out that bisexual animals have a diplontic life cycle, and produce germ cells after meiosis and fertilization that result in zygotes that immediately develop into adults. This obligate zygotenic reproductive mode can only lead to tetraploidy, for example, if both sperm and egg are diploid—an unlikely

occurrence except in a tetraploid population. A zygote could become tetraploid and then produce diploid gametes. But these are expected to result in tetraploids only if selfing occurs. Unlike in plants selfing is essentially nonexistant in bisexual animals. Muller's second hypothesis posits that the chromosomal mechanism for sex determination is such that it would be upset by additional genomes above the diploid level. Both of these Mullerian strictures have some merit, but it is also true that they have been overcome by animals in some groups. One way is to resort to the parthenogenetic life style.

Three general types of parthenogenesis occur among animals. In **arrhenotoky** the feritilized egg develops into a female and the unfertilized one into a male. In **thelytoky**, on the other hand, the unfertilized eggs develop into females and there are no males. Finally there is **deuterotoky**, in which unfertilized eggs develop into either males or females. The matter is really quite complex because each of these three modes has subdivisions (see White, 1973). Details aside, the point to be made is that parthenogenesis is amenable to polyploidy in animals. Parthenogenesis is rarely obligate and therefore an animal may develop polyploidy in an asexual phase and carry it over into a sexual phase.

Thelytoky occurs among these vertebrates that have polyploid members: the teleost fishes, salamanders, and lizards. A poeciliid fish, *Mollienisia formosa*, has a population in south Texas that consists entirely of females, all of which are apparently diploid. The eggs of the female will not develop to embryos, however, unless the mother is mated to one of several related species which are truly bisexual. The sperm enter the eggs, but the pronuclei do not fuse and the offspring inherit only the maternal genes (Kallman, 1962). Occasionally a true fertilization occurs and triploid offspring result (Rasch et al., 1970). Triploid sunfish in the genus *Lepomis* have been described. These arise by diploid eggs being fertilized by haploid sperm (Dawley et al., 1985). Evidence exists that some species of Cyprinid and Salmonid fishes are or have been true tetraploids and are in the process of diploidizing (Leipoldt and Schmidtke, 1982). Some species in each family have about twice the chromosome number and DNA content per nucleus as others.

Thelytoky occurs among salamanders in the genus *Ambystoma*. *A. tremblaya* and *platineum* are triploid thelytokous forms that carry chromosomes of two bisexual species, *A. laterale* and *A. jeffersonianum* (Uzzell and Goldblatt, 1967; Sessions, 1982). Hence *tremblaya* and *platineum* are probably hybrid in origin. MacGregor and Uzzell (1964) have shown that quadrivalents as well as bivalents occur during meiosis, indicating a probable polyploid condition in these thelytokous forms. (It should be noted that translocation heterozygotes can also produce quadrivalents at meiosis).

Frogs that appear to be obligate bisexuals may have diploid and polyploid members within the same species, and closely related species groups (Bogart and Tandy, 1976). Diploidy and tetraploidy are common in frogs and may appear in any genus; 19 polyploid species representing five families have been identified, including some that are octoploid (Schmid et al., 1985). All appear to be autopolyploids. Thus polyploidy appears to have arisen spontaneously and been maintained in different bisexual forms of Anura.

Triploidy exists among several families of lizards (Order Squamata); as in the case of the Ambystoma salamanders and the poecilliid fishes, it occurs among those lizards that are thelytokous (Bickham, 1984). The lizard genus *Anemidophorus* has both bisexual and unisexual forms. The unisexual ones are thelytokous and may be either diploid or triploid (Lowe et al., 1970). The lizards are the only reptilian group

that have polyploid species, although a diploid–triploid mosaicism has been reported in a turtle, *Plotemys platycephala* (Bickham et al., 1985). The triploid condition appears to arise somatically only, since only somatic cells are triploid, and all meiocytes are diploid prior to meiosis.

If the effects of triploidy in humans is typically mammalian, the mammals in general can be assumed to be intolerant of autopolyploidy. It has been reported that 17% of spontaneously aborted human fetuses with chromosomal aberrations are triploid (Carr and Gideon, 1977). Only one in ten thousand triploid zygotes develop to term and are born alive. Nearly all of these die as infants within a few days, although a very few have survived for a few months (Schroecksnadel et al., 1982). Triploid infants are usually found to be somatic mosaics or mixoploids with both diploid and triploid cells. Tetrasomic human zygotic products generally abort early with the probability of live births essentially zero.

As in the case of mammals no polyploid species of birds have been described. When mammals and birds arose from their respective reptilian ancestors, they may have done so as obligate diploids.

5. Significance of polyploidy

Clearly polyploidy occurs throughout the plant and animal kingdoms with a number of exceptions. It is relatively common among plants, especially the ferns and flowering plants, but it is less common in animals except for those that have parthenogenetic life styles. In all forms where it occurs it may have great evolutionary significance, as discussed in Chapter 10. The absence of polyploidy in the germ lines of some groups such as the mammals should not be taken to mean that their present diploid karyotypes have not descended from a polyploid state. Good evidence exists for some vertebrates that an originally polyploid state has become cryptic by a conversion to diploidy (Ohno, 1970; Schmid et al., 1985).

Polyploidy that arises in somatic cells and results in diploid–triploid–tetraploid mosaicism is fairly well distributed even in mammals, where it is confined to certain tissues. Mammals seem to have eschewed inherited polyploidy, and the phenomenon may have played an unimportant role in their more recent evolution. Since all present-day mammals have roughly the same amount of DNA per genome, the differences in chromosome numbers within the Class Mammalia are probably the result of translocations. The apparent ubiquitous presence of somatic polyploidy in itself indicates possible advantages to polyploidy even when not inherited as such through the germ line, although its achievement in the soma may be programmed in the germ cells.

B. Aneuploidy

Any disturbance of the condition of euploidy in which parts of chromosomes or whole chromosomes are absent from a genome, or present in excess, is defined as aneuploidy. Figure 5.54 diagrams some of the types that are categorized as **chromosomal** or **segmental** following the nomenclature of Dyer et al. (1970).

Chromosomal aneuploidy can result from **nondisjunction** of homologous chromosomes or their chromatids during meiosis or of sister chromatids during mitosis (Figure 5.55). If a chromosome is in excess of the normal euploid number, the condition is called **hyperploidy**; if it is deficient it is called **hypoploidy**. The same terms apply for segmental aneuploidy, which, however, arises by a somewhat different manner than chromosomal. Rather than arising from simple nondisjunction of whole chromosomes, segmental aneuploidy often arises as the result of heterozygosity for chromosomal rearrangements, for example, the adjacent types of disjunction of translocated chromosomes described in Figures 5.17 and 5.18 leading to unbalanced genomes. The distinction between the aneuploid state and a duplication or deficiency is a matter of degree. In general, if either of these is gross enough

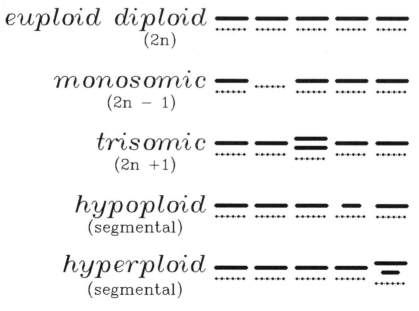

Fig. 5.54. *Some types of aneuploids.*

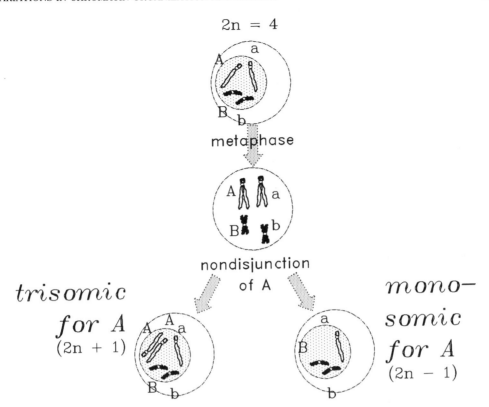

Fig. 5.55. *Aneuploidy resulting from nondisjunction. This can occur either during mitosis, as shown in this figure, or during either the first or second meiotic divisions.*

to cause a phenotypic effect, the condition is called aneuploidy.

1. Aneuploidy in plants

Besides being useful in the early studies of euploidy, the jimson weed, *Datura*, has also been a provider of material for studies in aneuploidy. The normal $2n$ number of *D. stramonium* is 24, but it was found that mutant strains produced gametes with 13 chromosomes. Therefore plants with the cytological designation of $2n + 1$ occurred as a result of fertilization by such gametes of 12 chromosome eggs (reviewed by Avery et al., 1959). Such **trisomics** were found to exist for each of the 12 different chromosomes in the *Datura* genome. But it was also found that there were more than 12

$2n + 1$ types. These others had an extra chromosome composed of a doubled half-chromosome. All these many types of *Datura* aneuploids are slower-growing than normal euploids, and each one has morphological characteristics that differentiate it from the others.

Aneuploid lines of vegetative cells may also arise spontaneously in euploid *Datura* plants. These mosaics have euploid cells in some tissues and $2n + 1$ or $2n - 1$ cells in others. It was found that deficiencies were limited to only certain chromosomes; deficiencies of others are probably lethal. Since in *Datura* a haploid cell cannot survive unless all 12 chromosomes are present, a monosomic cannot be expected to produce monosomic progeny. The intervening male or female gametophyte acts as a filter in this and other angiosperms to eliminated the $n - 1$ genome.

Some plants are highly polymorphic for chromosome number in their natural state. An extreme example for an angiosperm has been described for a South African species, *Themeda triandra*. The diploid number of this species is 20, but one population has 33% aneuploid members (Fossey and Liebenberg, 1987). Trisomics and a possible hexasomic have been identified with $2n = 21$, 22, and 24 chromosomes, respectively. No apparent phenotypic differences were noted. The extra chromosomes are functional autosomes and not B chromosomes.

The various species of the hawkweed, *Crepis*, have genomes varying from 3 to 7 in chromosome number. Another composite genus, *Chaenactis*, has species with 6 and 5 haploid numbers (Kyhos, 1965). A cytological analysis of several of these species has led to the conclusion that the five-chromosome species are derived from the six-chromosome ones by aneuploid reduction. There seems to be a general trend among some angiosperms for perennials with high numbers of chromosomes to convert to low-chromosome-number annuals (Grant, 1971). In the case of *Crepis* species

the decrease in numbers involved unequal reciprocal translocations (Babcock, 1947).

2. Aneuploidy in animals

More is known about the occurrence and effects of aneuploidy in the human population than any other group of animals. This is not necessarily because it occurs more frequently in humans, but because of its readily recognized clinical effects, which are always severe except when it occurs in some mosaics. Trisomy has been identified for all the human autosomes except chromosome 1 (Morton et al., 1982; Hassold, 1986). Generally trisomies result in spontaneous abortions. Table 5.14 summarizes some of the available data. Appreciable numbers of live births occur only for the trisomies 13, 18, and 21, and these children are always abnormal (Hook et al., 1989). Mosaic trisomies of these chromosomes are more likely to survive to term than those that are not. Postzygotic loss of a trisomic 13 and 18 that resulted in placental mosaics increased the survival rate of these trisomic fetuses to term.

About one in every 200 liveborn children is chromosomally abnormal, most with significant physical and mental impairment. By the time of birth, however,

TABLE 5.14. Chromosomal Abnormalities Occurring in Recognized Human Abortuses

Type of abnormality	Percentage of abortuses detected with abnormality	Percentage survival to term
All abnormalities	~50	5
Trisomies of autosomes		
16	7.5	0
13, 18, and 21	4.5	15
All others	13.8	~0
Trisomies of sex chromosomes XXX, XXY, XYY	0.3	75
Monoploid X (XO)	8.7	1
Structural abnormalities (e.g., translocations)	2.0	45

Data from Hassold (1986).

over 95% of chromosomally abnormal conceptuses have been aborted spontaneously. Fully 50% of aborted fetuses are chromosomally abnormal. Most of these are a result of aneuploidy. Chromosomal structural rearrangements such as translocations comprise only about 2%–3% of the abortions. As shown in Table 5.14, most of the aberrencies are trisomics. Chromosome 16 is the most likely to be trisomic, and none of the conceptuses with this trisomy reach term. Trisomies are to be expected to result from nondisjunction during meiosis, and this is borne out by the observations that the incidence of disomies and other chromosomal abnormalities in sperm from healthy males is 8%–9% (Martin et al., 1987) and even higher in human oocytes (Wramsby et al., 1987). Nondisjunction should result in an equal number of monosomic and trisomic zygotes after fertilization, but monosomics are rare, which indicates either the monosomic lethality may be expressed earlier than the trisomic before conception is recognized or that gametes lacking a chromosome may be nonfunctional to begin with. Also trisomy of chromosome 1 has not been noted, indicating that abortions may be occurring before the conception is noted, since disomy of this chromosome has been noted in gametes.

Among the three human trisomies that reach term, trisomy 21 is the commonest. About 70% of these achieve live birth and are described phenotypically as Down's syndrome (Hook et al., 1989). Down's syndrome children are mentally retarded and have relatively minor physical abnormalities, but they have a predisposition to cancer and a shortened life span. Like trisomy 21 children and abortuses, those with trisomy 18 and 13 have a high incidence of brain abnormalities (Gullotta et al., 1981).

Epstein (1985, 1988) has made extensive analyses of human aborted aneuploid clinical material and arrived at a number of conclusions based on his observations. Principally it is evident that different human aneuploid types show different patterns of phenotypic effect just as has been noted for *Datura*. The effects are distinct from one another, although there may be some overlap such as the fact that human trisomics that survive to term are all mentally retarded.

Trisomy and other aneuploid conditions occur in all animal groups. Mammalian, like human, trisomics die *in utero* or shortly after birth in most cases. The incidence is especially high in those that are polymorphic for chromosome rearrangements, and hence have heterozygotes for these rearrangements in their populations.

The effects of ploidy changes on the phenotype, whether they be of an autoploid, alloploid, or aneuploid type, clearly tell us that genomes are functional units. This consideration leads us logically to the discussion in the following sections dealing with the effects of imbalance in gene dosage.

IX. CHROMOSOMAL AND GENIC BALANCE AND IMBALANCE

The effects of euploid polyploidy and aneuploidy raise some interesting questions about the functioning of the genetic material. These questions bear directly on chromosomes as genetic elements above the level of genes. In the case of trisomy, for example, is the deleterious effect the result of an imbalance in one or several genes, or all the genes in the trisomic set? What is "imbalance" in the first place? Why does the addition of one complete genome or more to a normal diploid set in a mammal such as a human result in lethality, whereas an amphibian can tolerate $4n$ and many plants up to $12n$? Plants are by no means as sensitive to these imbalances as mammals, but it should be noted that a haploid *Datura* will grow and flower only if it is euploid. If one chromosome is added to the haploid complement, the resultant aneuploid disomic ($1n + 1$) with 13 chromosomes is much weaker than the euploid haplont, and of much smaller size.

A. Gene Dosage

When the haploid set of chromosomes is multiplied to $2n, 3n, 4n, \ldots, xn$ in plants and animals that tolerate autopolyploidy, the phenotypic effect is relatively minor despite the fact that the dosage of all the genes is multiplied accordingly. In a few plants a slight increase in size may result, but for the most part the cell size, but not the plant size, increases. Polyploid amphibians are the same size as diploid ones, but they have larger and fewer cells.

Plants that are amphidiploid alloploids in consitution, such as $4n$ cultivated cotton and $6n$ cultivated wheat, have different genomes represented. All the component genomes in each case presumably have similar sets of genes, because the relevant diploids from which they were derived have similar phenotypes. Hence one should expect that the alloploids carry large numbers of duplicated genes and this does turn out to be the case. This is also true of plants such as maize, which shows no signs cytologically of being a polyploid. However, maize has many duplicate genes within its genome. This has been demonstrated both by genetic (Rhoades, 1951) and biochemical means (Wendell et al., 1986) and by the use of cloned fragments as probes (Helentjaris et al., 1988). It is possible, therefore, that present-day maize is a cryptic or veiled polyploid with alloploid ancestors. It may also have arisen from an autoploid state, which we discuss for animals in Chapter 10, or even by transpositions.

The increase in dose of a single gene does not always change the phenotype noticeably. But there are many exceptions to this. In maize the endosperm content of

carotenoids is increased by the substitution of the gene Y for its inactive allele y. An approximate linear increase in carotenids occurs with the increase in dose of Y in the triploid endosperm (Mangelsdorf and Fraps, 1931). A practically linear correlation between gene dosage and the amount or activity of an encoded protein has been noted for a variety of organisms ranging from yeast (Ciferri et al., 1969; Johnstone and Hopper, 1982) to higher plants (Carson, 1972) and animals (Lucchesi and Rawls, 1973), including humans and mice (Epstein, 1988). A particularly notable example has been described for allohexaploid wheat, *T. aestivum* (Aragoncillo et al., 1978). The increase in the amount of certain proteins in the endosperm was measured in plants that had special aneuploid additions of homeologous chromosomes in the three different genomes A, B, and D. Increasing the doses of chromosomes 4D, 7B, and 3B in the endosperm from 0 to 6 caused an increase in concentration of the proteins encoded, in some cases linearly. In others the increase was not linear, and showed a reduction at the $6n$ ploidy levels. Gene dosage responses such as these have been interpreted as evidence that transcription is a rate-limiting factor in gene expression, and therefore the more genes the more transcription and the more product. However, there may also be a rate-limiting effect associated with the activity of the product, if, for example, it is an enzyme.

A dominant gene may not change the phenotype with increasing dose, but its hypomorphic alleles may (Muller, 1932). Hypomorphic alleles by definition produce less phenotypic effect than the dominant or wild-type alleles of the gene. Figure 5.56 diagrams the effects of increasing the dose of a hypomorph until the phenotypic effect resembling or equal to the dominant effect is achieved. A reasonable explanation of this can be made if it assumed that the gene in question encodes an enzyme E

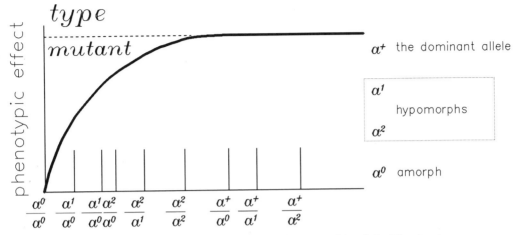

Fig. 5.56. *Effect of increasing the dose of a hypomorphic allele. The dominant allele (a^+) produces the maximum type effect when present once. The hypomorphs, a^1 and a^2 produce toward type, but less of it than a^+. The allele, a^0, designated an amorph, produces no effect and could be a deletion at this locus.*

which converts a substrate S to a product P. The amount of P determines the phenotype; it may be, for example, a pigment. The dominant allele produces sufficient E so that the substrate S concentration is limiting, and therefore increasing [E] will not produce more P. On the other hand, the hypomorphic alleles encode less enzyme or a less effective enzyme than the dominant, and as a consequence [E] is rate-limiting rather than [S]. Thus increasing the dose of a hypomorph will cause more P to be produced until the substrate again becomes limiting. A detailed discussion of this hypothesis and its consequences is provided in the numerous papers on the subject by Wright (1977). It follows from this that the euploid polyploids that are $3n$ or higher in degree of ploidy and show no phenotypic differences from $2n$ other than increase in cell size may be assumed to have gene complements constituted of dominant alleles. This may be true of plants and some animals, but it is not a ready explanation for the results of polyploidy in animals such as mammals.

The mystery remains that changes in the dosage of a small section of chromatin, a single gene for example, may be tolerated, but not a large section or a whole chromosome. A gene deletion heterozygous with its active counterpart may have no obvious phenotypic effect, but when the deletion encompasses a larger section of chromosome, lethality results. Lindsley et al. (1972) have shown that in *D. melanogaster* most small deficiencies encompassing one-eightieth of an autosome survive in heterozygotes but may cause sterility in one or both sexes. They synthesized segmental aneuploids with autosomal segments translocated to Y chromosomes for 85% of chromosomes 2 and 3. In general, hyperploids with small segments were viable with the exception of the region 83D to 83E on chromosome 3, which was triplolethal. Extensive hyperploidy is lethal, however. This is interesting in view of the fact that triploid *Drosophila* are viable. Hypoploidy is definitely less well tolerated in *Drosophila* than hyperploidy even for small deficiencies. Effects of this type are described under the general terms of **chromosomal and genic imbalance**, and they are certainly related to the dominance of the diploid state in higher animals. But it is to be noted that the terms are descriptive and explain nothing about the phenomena they describe.

Mammalian cells in culture frequently become polyploid after a certain number of passages in culture. This may be followed by a condition called **heteroploidy**, defined as an elevated variability of chromosome number (Kraemer et al., 1983). The condition may be either euploid or aneuploid, but it is generally the latter and is associated with a transition to the neoplastic-like state. The hypothesis that cancer is the result of some kind of genic imbalance derives from observations such as these. Of course it is also possible that the heteroploidy is an effect and not the cause. In any case aneuploidy for cells in culture can enhance viability rather than be deleterious. A further point to be made is that in animals that are mosaic for otherwise lethal aneuploidy, or even haploidy, the aberrant cells may be maintained in a viable state.

B. Dosage Compensation

1. Of sex-linked genes

Dosage compensation is a regulatory mechanism for sex-linked genes that differ in dose between males and females in organisms that have a **heterogametic** sex, that is, one that has two kinds of sex chromosomes. The most familiar examples are the XY male and XX female in most insects and all mammals in which the males produce X- and Y-bearing sperm and the females only X eggs. In some fishes and reptiles and all birds the males are **homogametic** (ZZ) and the females heterogametic (WZ). In these species the Z chromosome is analogous to the X in the XX:XY types. In some animals the Y-type chromosome is not present and the males are

XO. On the other hand many insects have sex-determining mechanisms in which types involving more than one X and Y are found, so that the combinations X_1X_2Y, $X_1X_2X_3Y$, X_1X_2O, XYY_2, and so on occur. Most of these have arisen by translocation between sex chromosomes and an autosome (reviewed by White, 1973; Bull, 1983). In those plants in which sex chromosomes are demonstrable the heterogametic sex is the male.

Since the Y chromosome in *Drosophila* and mammals is essentially devoid of most transcribable DNA, and the XX females are generally phenotypically equivalent to the males with respect to quantitative expression of the sex-linked genes, some method must exist to adjust the expression of these genes. It appears that there are two types of mechanisms: One has been studied intensively in *Drosophila* and the other in mammals.

2. Drosophila-type compensation

The males of *Drosophila melanogaster* with one dose of X-linked genes usually produce the same amount of X-encoded gene product as females with two doses (Muller, 1950). This regulation is effectively accomplished by hyperactivity of the X chromosome of the male rather than by repression of X-chromosome activity in the female (reviewed by Lucchesi, 1983; Baker and Belote, 1983; Jaffe and Laird, 1986). The capacity of X-linked genes to compensate by increasing the transcription rate appears to be autonomous, because genes in small segments of the X chromosome inserted into an autosome continue to be dosage-compensated (Bowman and Simmons, 1973). On the other hand, an autosomal gene moved into the X was not compensated (Roehrdanz et al., 1977).

The transcription of X-chromosomal DNA of cytologically normal male and female *melanogaster* has been compared in aneuploid larvae containing duplications of the X-chromosome regions 9A–11A and 15D–18D. In the aneuploid male with the 9A–11A region hyperploid transcription for that segment and its homologous region on the intact X was found to be 70% more active than in the normal diplo-X female (Prasad et al., 1981). In the aneuploid female the 15D–18D segment present in three doses showed 100% more transcriptional activity than the normal diplo-X transcriptional activity. Since the segments seem to be autonomous in their hyperactivity, it has been suggested that autosomal factor(s) mediate the X-transcription rate. But regulatory sequences for transcription are also definitely located on the X chromosome itself as well (Chatterjee, 1985). Point mutations on the X chromosome, that reduce the rate of transcription of the X are male-specific lethals (Belote and Lucchesi, 1980).

Even though transcription rates are now known to be involved, the mechanism of dosage compensation is far from being understood. A further complication to its understanding is that not all X-linked genes of *Drosophila* are compensated in the male. For example, the sex-linked α-chain gene of LSP-1, a larval serum protein, is present in male larval hemolymph at one-half the concentration found in female hemolymph (Roberts and Evans-Roberts, 1979). Two other LSP-1 polypeptides, β and γ, are immunologically similar to the α, but are encoded by genes on autosomes. Since the three chains are apparently encoded by members of a gene family, it is possible that after arising on an autosome and not being compensated, the α-chain gene was translocated to the X, where it retained its autonomy from control of whatever factors control transcription of the X.

The nematode *Caenorhabditis elegans* has a sex-determining system in which males are XO. The XX individuals are hermaphrodites. A dosage compensation mechanism exists that equalizes the X-specific transcript level in the XO males and XX hermaphrodites (Meyer and Casson, 1986; Meyer, 1988). By using duplications and deficiencies of regions of the X with four known linked genes it has been shown that, not unlike the situation in *Drosophila*, the transcript levels of these genes are directly proportional to the gene copy level in the XX worms. Dosage compensation was noted for three of the four genes, since the levels of RNA transcripts are equal in the XO and XX. A further similarity to the *Drosophila*-type compensation was found in the effects of mutations at three autosomal loci. These mutants had an overexpression of the X-linked genes subject to dosage compensation so that compensation was effectively eliminated. The similarities between compensation in *Drosophila* and *Caenorhabditis* are striking enough to lead to the inference that the *Drosophila*-type dosage compensation may be of widespread occurrence among invertebrates.

3. Mammalian-type compensation

The XX–XY-type of sex determination is common to both *Drosophila* and mammals, but there are several important differences between them, the type of dosage compensation being one. In *Drosophila* the single male X has its rate of transcription enhanced so that its output activity is equal to the female's two Xs. In the eutherian mammals one of the two female Xs is silenced so that the male and female transcription outputs are essentially equal. The inactivation is random for the most part, so that either one of the female's Xs is inactivated, making the female a somatic mosaic for those genes that are heterozygous. This inactivation process was proposed first by Lyon (1961), and all subsequent studies have supported the original conjecture (reviewed by Gartler and Andina, 1981; Lyon, 1986, 1988).

The inactivation is chromosomal in the sense that nearly all genes are inactivated uniformly. The inactivated chromosome is not eliminated but is retained and can be detected cytologically as a **Barr body** in somatic cells. The Barr body is heteropycnotic and is found only in cells with more than one X, as the cytological manifestation of the inactivated paternally or maternally derived X. In males that are XXY one Barr body will be present; in aberrant females with more than two Xs correspondingly more Barr bodies will be found in the somatic cells.

Not necessarily all genes on the inactivated X are silenced. In particular those on the short arm of the X may not all be silenced. It is known that the blood type gene, *XG*, located at or near the tip (Xpter–p22.3) is not activated, nor apparently are those near it on the short arm. The mechanism of inactivation of the long arm of the X is not understood. However, the following facts have been established for mouse and human X chromosomes (reviewed by Lyon, 1988). Inactivation of one of the Xs occurs during early embryogenesis of the female. Once established, it is apparently irreversible in the somatic line, but not the germ line. Reactivation occurs at a specific stage of germ cell development. In the mouse both the paternally derived X^P and the maternally derived X^M chromosomes are active up to the 9- to 16-cell stage, but one becomes inactive as differentiation sets in. This begins first with the formation of the trophectoderm layer of cells, which are involved in the process of implantation and the formation of other extraembryonic membranes. The cells destined to become the fetus (the inner cell mass) do not show inactivation until they begin to differentiate. This begins about 4.5 days after fertilization in the late blastocyst stage. By 6.5 days most of the embryonic cells show X inactivation in the female embryo. In general the inactivation process is random except in the trophectoderm and primary endoderm, in which only the X^P is inactivated. This corresponds to the situation in the marsupials, in which the X^P only is inactivated in all cells.

It has been proposed that the inactivation of the X is a consequence of the methylation of its genes, rendering them untranscribable (Gartler and Riggs, 1983). Some credence has been given this hypothesis by the demonstration that some of the genes of the inactivated X can be reactivated by treatment with 5-azacytidine (Mohandas et al., 1982). However, only partial reactivation of the activity is achieved, and not all genes are similarly affected. That some alteration of the DNA is involved in inactivation is certain, but that it is methylation is not presently proven.

The regions of chromosomes that have the most constitutive heterochromatin are the centric and telomeric regions. There are also the regions that appear to have the fewest active genes. Is there a relationship between heterochromatization and gene activity? If there is, then the falcultative heterochromatin of the Barr bodies may be the important factor in the inactivation, although methylation may also play a role.

Inactivation of the mammalian X is presently not understood at the molecular level, but it is known that the X possesses an **inactivation center** (reviewed by Gartler and Riggs, 1983). Deletions of parts of the short arm of the X have no effects on its inactivation, but an inactivation center may exist on the long arm proximal to the centromere in the region Xq13 (Therman and Sarto, 1983; Mattei et al., 1981). Just how this region prevails in the inactivation process is not known.

4. Autosomal compensation

Many autosomal genes duplicated in whole-arm trisomies of *D. melanogaster* have their activities compensated so that only diploid levels of product result (Devlin et al., 1982, 1984). But, significantly, autosomal trisomy can also affect the expression of genes elsewhere in the genome outside the hyperploid region (Devlin et al., 1988). Some genes unlinked to those in the hyperploid chromosome arm are unaffected, but about as many have their activities repressed by one third. This type of observation may lead to a beginning of an understanding of the deleterious effects of aneuploidy as an upset in genic balance.

In the previous section on gene dosage it was pointed out that the increase in the number of certain genes in *Triticum aestivum* resulted in an increase in the amounts of proteins they encoded (Aragoncillo et al., 1978). A similar study was done on the same cultivated wheat species by altering the doses of chromosome 1B so that the endosperm contained one, two, or six of this chromosome, which has the gene clusters encoding the glutenins and the gliadins of the endosperm. In addition certain other combinations of the homeologous chromosomes, 1A and 1D, were made and their effects noted (Galili et al., 1986). A gene-dosage response of a number of seed proteins of the glutenin and gliadin type was noted. However, the response was nonlinear, especially for the gliadins, and the conclusion was reached that dosage compensation is involved. But it is almost certainly not the same type of compensation as that noted in *Drosophila*. Rather, rate-limiting factors related to the supply of precursors and the activities of elements involved in transcriptional and posttranscriptional processes different from those found in *Drosophila* should be considered as possibilities. For example, what controls the dosage of binding proteins that are so important in the regulation of transcription?

5. *Absence of compensation*

Animals with the XX–XY type of sex-determining mechanism have a means of coping with the dosage problem; however, this does not appear to be the case in animals with the ZZ–ZW type. A number of birds have been investigated to establish whether compensation exists for the Z-linked gene *Acon*, which encodes the enzyme aconitase. No compensation was found for this gene in the heterogametic females of the house sparrow and spotted turtle dove (Baverstock, 1982). The activity level of aconitase in male liver homogenates was approximately twice that found in the female. Absence of compensation of Z-linked genes also appears to be the case for butterflies (Johnson and Turner, 1979).

It is clear from linkage data that the Z and X chromosomes are not homologous. Mammalian X chromosomes invariably have a special set of genes on the X. These types of genes are not found on Z chromosomes.

C. Aneuploidy and Sex Determination

The determination of sex in plants and animals is one of the more puzzling phenomena in the eukaryote world. The familiar XX–XY and ZZ–ZW genotypic patterns make the sex ratio of 1:1 easy to understand for most eukaryotes, but the determination of the final phenotypes—the functional male and female—is not. It is not the intention here to delve into the innermost mysteries of the action of genes that determine sex, or even why there are different sexes. Reviews by Bull (1983), Ghiselin (1988), and Bell (1988) deal with the more recent and esoteric ideas involved with these questions. Here a more simplified approach is taken by commenting on the obvious fact that sex determination is involved with aneuploidy. In addition a euploid–haploid–diploid condition is prevalent among those animals that are arrhenotokous. Ploidy differences, it would seem, are fundamental to sex determination, although male haploidy is modified by endopolyploidy.

For the obligate diploid animals in which it seems to be important that the sex ratio be 1:1, it is obvious that heteromorphism of a pair of homologues (or homeologues) such as XY or ZW solves the problem. Apparently just having a set of autosomal genes with male- or female-determining alleles would not do because of the occurrence of three different genotypes. With the heteromorphic sex chromosome system only two are possible, with certain exceptions that are due to nondisjunction—other factors being equal.

It is the latter exceptions that reveal some interesting facts. Two types of XX–XY systems have been identified. Those in which sex depends upon the presence of one or two Xs without respect to the presence of a Y. This type, called by Bull (1983) the **recessive X system**, results in XO males and XXY females. The Y has no apparent effect. The second type, which Bull calls the **dominant Y system**, depends primarily on the presence of the Y. The system of two or more Xs with a Y present still gives a male, and XO individuals are female. The first type is characteristics of *Drosophila* sp. and the second of mammals including humans and some insects. An XO *Drosophila* is always a male, but a female in the human dominant Y-type system. In neither case are they completely normal, however, since among other things they are sterile.

In *Drosophila* the ratio of X chromosomes to autosomes, rather than the presence of a Y, is a primary determinant of sex. Flies with equal numbers of X chromosomes and autosomes (e.g., 1X:1A, 2X:2A, 3X:3A, etc.) are female. On the other hand 2X:3A flies are intersexes. This finding prompted Bridges (1921) to propose the **genic balance theory** of sex determination. He posited that the autosomes have a preponderance of male-determining genes, while the X has a majority of female-determining ones. Subsequent findings have shown this hypothesis to be too simplistic (Bull, 1983; Baker and Belote, 1983).

In *Drosophila melanogaster* at least 11 regulatory genes have been identified that have functions relating to sex determination and dosage compensation (reviewed by Baker and Belote, 1983). Mutations at these loci can drastically affect sex determination and dosage compensation. Only one of these, *Sxl*, is X-linked; the rest are scattered among the autosomes.

Mutations of Sxl^+ result in a variety of alleles with phenotypic effects ranging from embryonic lethality to sexually normal females with malformations in various body parts. The gene is not normally expressed in males, but in females it functions to prevent hypertranscription of the X. In this role it is involved with a class of genes identified as male-specific lethal mutations. The wild-type products of these genes are essential for male viability. They occur at four loci, $msl-1^+$, $msl-2^+$, mle^+, $mle-3^+$, and they apparently function to regulate the transcription of the genes of the male single X chromosome so that their products are maintained at a level comparable with the levels maintained by the two Xs present in females.

An interaction between the *Sxl* and *msl* genes is noted in females heterozygous for a nonfunctional *Sxl* allele and homozygous for a male-specific lethal. These flies frequently develop as interesexes with patches of male tissue. Two other autosomal genes, tra^+ and $tra-2^+$, have mutant alleles that when homozygous in 2X:2A females cause them to develop as phenotypic males. But they cannot function as males because their gonads are poorly developed and spermatogenesis does not occur. Also XY individuals homozygous for *tra* do not produce functional sperm.

These are just two examples of gene function in sex determination and dosage compensation that illustrate that the processes controlling sex determination and transcription regulation are related and complex, and probably depend upon the interaction of numerous genes only some of which have so far have been identified.

The Y chromosome is male-determinant in humans, other mammals, and some plants such as *Melandrium*. No matter how many X chromosomes are present a human with a Y is a male, although if he has more than one X he will be abnormal and sterile. The Y male-determining element is identified as *TDF* (testis-determining factor) and is located on the short arm of the Y (Ypter-p11.2). Individuals with the short arm missing are phenotypically XO (Rosenfeld et al., 1979). A 230-kb segment of the Y chromosome that contains the TDF activity has been cloned. Probes derived from this segment of DNA have a close sequence homology with the X chromosomes of other mammals (Page et al., 1987). Also similar sequences are found in the DNAs of other vertebrates.

The TDF product(s) along with other genes on the short arm of the X (Xp) regulate the activity of an autosomal gene that produces the so-called H-Y antigen. This is the actual testis-inducing principle that starts an embryo on the path to becoming a male. In females there are two doses of the Xp gene(s) and the *H-Y* gene is suppressed. In males the TDF product(s) presumably suppress all Xp gene activity. Thus a regulatory balance exists in the Y-dominant mammalian type of sex determination just as in the X-dominant *Drosophila* type. In both cases an aneuploidy is involved, since the Y in both types is almost barren of protein-encoding structural genes. But the problem is even more complex because there are male mice that lack the H-Y antigen (Burgoyne et al., 1986).

As in *Drosophila* a single gene can cause a sex reversal, transforming XX mammals into phenotypic males. In mice the dominant *Svx* factor located at the distal end of the X chromosome reverses the sex of XX individuals (Cattanach et al., 1982;

McClaren and Monk, 1982). In humans sex reversal can occur in XY and XX individuals. A sex-linked gene, *Tfm*, causes XY individuals to develop female characteristics (Migeon et al., 1986). XX males have been described that have normal levels of H-Y antigen but are sterile. One of the more curious examples of anomalous mammalian sex determination has been found in the wood lemming, *Myopus shisticolor*. This animal has two types of females: XX and XY. Both are fertile and phenotypically indistinguishable from each other. YY males also exist (Fredga et al., 1976).

X. SUMMARY

1. Genomic alterations are necessary for the emergence of new life forms. They provide the variants that are the material basis for evolution. Their rate of production must, however, be maintained at a level consistent with the ability of life to survive. An equilibrium must operate to ensure that sufficient organismal variants survive and reproduce to maintain the continuity of life.

2. A major source of genomic variation in eukaryotes is the formation of new arrangements of chromatin that result from breakage and reunion of chromosome segments. The major variations produced by breakage and new reunions are inversions and translocations.

3. Deficiencies or deletions and duplications of chromatin can result from unequal crossovers between homologues, by crossing-over within an inversion loop in inversion heterozygotes, and by formation of unbalanced translocation heterozygotes.

4. Chromosome breakage frequency is difficult to estimate because much of it may occur in the interphase nuclei and restitution by repair will not be evident in the mitotic phase chromosomes unless it

eventuates in recognizable rearrangements.

5. Instability of chromosome structure that leads to elevated rates of breaks can be inherited. Fragile sites have also been identified *in vitro* by treating chromsomes with various chemical agents. These agents are also inherited.

6. Inversions and translocations are commonly found in natural populations of plants and animals. They are probably of universal occurrence among eukaryotes. Rearrangement polymorphisms are frequently observed within populations of the same species. Some closely related species may be distinguished cytologically by the possession of different kinds of gross rearrangements.

7. Germ cells heterozygous for translocations or inversions will upon meiosis generally produce some haploid inviable gametes that contain contain incomplete, deficient genomes. Hence, in general there is selection against rearrangements becoming fixed in populations, but nonetheless some do become fixed. The one type of rearrangement that occurs commonly and has a high probability of becoming fixed is the Robertsonian type, which results from the fusion of two nonhomologous acrocentrics at their centromeric regions.

8. In some plants and animals a type of conserved translocation heterozygosis exists. This may result in the conservation of specific genotypic patterns that have a selective advantage.

9. Mobile elements, or transposons, are sequences of DNA capable of transposing to different positions in the genome. Two categories of them have been identified: (1) those that transpose as DNA (Class II), and (2) those that transpose via an RNA intermediate involving a reverse transcriptase enzyme (Class I).

10. Elements that transcribe via an RNA intermediate are of two types: (1) those that have a structural relationship to the RNA retroviruses (the retrotransposons) and (2) those not related to retroviruses—retroposons that are related to RNAs transcribed from endogenous DNA.

11. Retrotransposons may encode reverse transcriptase from the OFRs in their DNA phase. These have a high degree of sequence homology with the sequences in the genomes of retroviruses that encode this enzyme.

12. The nonviral retroposons are transcripts of RNA POLII or POLIII. The POLII retroposons include the *L1* family of repeats and processed pseudogenes in the mammals. The *Alu* family of repeats found in the primates are transcripts of POLIII.

13. A considerable portion of the DNA of *Drosophila* and humans is constituted of transposable element sequences repeated many times. At least in humans these sequences are confined primarily to the euchromatic regions and are scarce or absent in the centric heterochromatin.

14. *L1* elements have been shown to transpose in the human genome and cause malfunctioning of the genes into which they are transposed. *Alu* sequences have not been shown to transpose, although it is highly probable that they did at one time.

15. Mobile elements can transpose into the regulatory regions of genes, or their exons and introns. In either case they can cause gene mutations, or chromosome breaks. It is possible that many chromosome aberrations observed in the eukaryotes of all taxa are the result of transpositions.

16. The maize elements of the *Ac* and *Spm* type are most likely to invade a gene that is hypomethylated and hence active. Hypermethylation may also maintain these elements in the cryptic, inactive state. It is possible that this is also true of transposing elements in general.

17. Somatic rearrangements of DNA by transposition may occur either in meiocytes or somatic cells. Specialized pro-

grammable types of rearrangements may also occur with the participation of transposons. A prominant type of specialized somatic rearrangement involves the formation of a great variety of immunoglobulins by V−J joining. The transposition of inactive genes to regions of chromosomes that activate them also occurs, as in the examples of antigenic variation in trypanosomes and mating-type switching in yeast.

18. Separation of the transcribed part of a gene from its enhancer−promoter regulatory region may occur either as a result of gross rearrangements by inversion or translocation or by transposition. A gene may then become inactive, or it may be moved next to a different regulatory region and assume new properties of expression.

19. Tranposing elements may play important roles in evolution by contributing to genetic variation, leading to new forms of gene expression with adaptive value in response to environmental stress.

20. The DNA content of eukaryotic genomes has a wide range not only over the whole of the Kingdom, but even within taxa with closely related species. In terms of numbers of base pairs per genome the range is about 10^7 to 10^{11}. Essentially no relationship exists between the numbers of base pairs and the degrees of complexity in form and function of the different species, whether they be "low" or "high" on the phylogenetic tree. In addition the same is true for the numbers of chromosomes per genome, and there appears to be no correlation between DNA content and the numbers of chromosomes per genome. Some taxa have species with approximately the same haploid number of chromosomes, others have a wide range.

21. Variations in the DNA content of genomes appear for the most part to be the result of amounts of repetitive sequences with no apparent function, whereas the differences in chromosome numbers appear to be the result of tandem and centric fusions or dissociation.

22. Chromosomes are subject to random deletions and duplications of segments of their DNA. Deletions are generally lethal homozygous, but may be viable heterozygous if only a relatively small segment is deleted. Duplication and deletion of segments may arise jointly by unequal crossing-over. But they also arise from crosses involving parents heterozygous for inversions and translocations.

23. The generation of multiple copies of a gene by amplification is widespread among eukaryotes. It can be induced in somatic cells in culture by treating them with increasing doses of toxic substances to which they become resistant. Resistant cell lines that arise are found to have a gene amplified that encodes a protein sensitive to the toxic substance. Excess amounts of the protein thus overcomes the inhibitor. Amplification is generally visible cytologically as HSRs and DMs. Gene amplification also occurs in tumor cells of mammals with cancer with the production of HSRs and DMs. Oncogenes are amplified in certain neoplasms. In addition amplification of specific genes may occur in plants and animals that become resistant to toxic substances in the environment. Programmed gene amplification may occur during embryogeny as a response to the needs of the embryo for specific functions, such as the need for a high rate of protein synthesis at particular stages of development. Certain types of environmental stress may result in permanent gene amplifications in the whole organism. Otherwise amplification generally reverts to a preamplified condition of gene dosage when the stress factor is removed.

24. The exact mechanism(s) of gene amplification are not known, but unequal sister chromatid exchanges, and various types of overreplication associated with recombination have been implicated. In addition repeated transpositions and roll-

ing cycle synthesis of DMs have been suggested as possible mechanisms. Amplification in cultured cells is generally accompanied by chromosomal aberrations. Amplification of more than one gene is found in some cases in which the genes with related functions are clustered on the chromosome.

25. In some animals chromatin is eliminated from germ or early-cleavage cells after fertilization of the egg. This is called chromatin diminution. The DNA eliminated is generally in heterochromatin, but functional genes may also be cast out. In some animals and plants whole chromosomes may be eliminated. The elimination of either segments or whole chromosomes is programmed and specific. Recognition of specific chromatin as segments or whole chromosomes is possibly programmed in some cases by genetic imprinting. Hybrids between two species in some plants and animals may result in the casting out of the chromosomes of one of the parents.

26. The numbers of genomes per nucleus, described as the ploidy, shows a wide range among the different taxa of the eukaryotes. With respect to the dominant euploid phase in any group, three basic types of life cycles are recognized: (1) the haplontic, with a dominant multicellular haploid and a single-cell diploid phase that undergoes meiosis without prior cleavage; (2) the diplontic, with a dominant diploid phase and a single-cell haploid phase consisting of gametes that fuse to form zygotes that cleave; and (3) the diplohaplontic, with both multicellular haploid and diploid phases in which meiosis results in the formation of a haploid gametophyte that produces gametes by mitosis. The vascular plants (pteridophytes and angiosperms) have diplohaplontic cycles with the alternation of dominant diploid sporophyte generation with a haploid gametophyte. The opposite is true for the mosses

and liverworts; they have a dominant gametophyte.

27. The animals have the diplontic type cycle and hence have complete diploid dominance. In general the dominant diploid state with chromosomes in pairs is found in the most highly differentiated organisms, whether plant or animal. Disturbance of diploid-type euploidy by aneuploidy is tolerated better by plants than animals, but in either case it is generally phenotypically deleterious.

28. Euploid polyploidy is tolerated by most plants but only a few animals. Mammals and birds in particular are intolerant of ploidy levels above the diploid. Those rare invertebrates that tolerate polyploidy are primarily found among a few insects and crustacea that reproduce parthenogenetically. In general this is also true of polyploid fishes, salamanders, and lizards. Polyploidy among bisexually reproducing species has been found only in certain groups of frogs that are autotetraploids.

29. Polyploid plants may be either auto- or allopolyploid. In the latter case the ploidy is the result of a cross between two different species. The resultant hybrids have somewhat different genomes and the homeologous chromosomes may not pair to form bivalents in meiosis. The meiotic behavior of chromosomes in a polyploid is prescribed by the homology of chromosomes it contains. If chromosome doubling occurs to form a diploid, the identical homologues pair and the plant is described as an amphidiploid, since it has the meiotic processes characteristic of a normal diploid. If the chromosomes do not double, the plant may reproduce parthenogenetically by apomixis or it may reproduce sexually while producing a high frequency of inviable zygotes. These result from partial pairing among homeologous chromosomes. Some bivalents may form but also multivalents and univalents so that aneu-

ploid gametophytes are produced. As in polyploid animals many polyploid plants reproduce by apomixis and avoid the meiotic obstacle course.

30. A euploid genome consists of all chromosomes necessary to produce a functional organism present once. Equal numbers of whole sets result in the usually chromosomally balanced diploids and polyploids. When whole chromosomes or parts of chromosomes are absent from the complete genomic set, or some are present in excess, an unbalanced condition of aneuploidy results. It is of common occurrence in both plants and animals, generally resulting from nondisjunction or from meiosis in inversion or translocation heterozygotes. Plants are more tolerant of aneuploidy than animals. Generally it is lethal in animals and results in spontaneous abortions in mammals.

31. The deleterious effects of aneuploidy are apparently the result of imbalance in gene dosage. However, certain types of aneuploidy are normal for the determination of sex, and dosage imbalance is overcome by dosage compensation. In animals two principal types of compensation exist in those species with XX:XY homogametic and heterogametic sexes. In mammals one X becomes inactive in the female; in *Drosophila* the single X in the male transcribes at a rate equal to the total transcription rate of the two Xs of the female.

6 Transcription and Its Regulation

The building of a definite cell product, such as a muscle fiber, a nerve process, a cilium, a pigment granule, a zymogen granule, is in the last analysis the result of a specific form of metabolic activity, as we may conclude from the fact that such products have not only a definite physical and morphological character, but also a definite chemical character. In its physiological aspect, therefore, inheritance is the recurrence, in successive generations, of like forms of metabolism; and this is effected through the transmission from generation to generation of a specific substance or idioplasm which we have seen reason to identify with chromatin.

E.B. Wilson, 1896

This quotation from E.B. Wilson is probably one of the first statements, if not the first, that the chemical substance inherited through the chromosomes has the function of directing the building of cell products. To put his thoughts into modern biological English we can say that there is a substance called DNA that replicates itself and directs the course of metabolism within cells. In doing so, it makes cells and organisms. Wilson labored with the disadvantage that he was unaware of Mendel's units of inheritance at the time he wrote this passage, but it is quite clear that he had grasped the important point that an inherited physical substance directed cell processes. After the rediscovery of mendelism, that something was called the **genotype** by Johannsen (1909), and what it directed to come about eventuated in what he called the **phenotype**. The formation of the phenotype by the action of the chromosomal DNA is a complex of processes sometimes referred to as **gene action** (Fig. 6.1). The fundamental first step in this series of processes is **transcription**, the transcribing of nucleotide sequences in the DNA of the chromosomes into RNA.

Figure 6.1 is an attempt to convey the complexity of the processes involved in the formation of an organism. (1) The first step is the transcription of a structural gene encoding a polypeptide. The production of the primary RNA transcript involves the activity of RNA polymerase II and a host of other enzymes. (2) For example, the addition of a poly (A) tail at the 3′-OH end requires polymerase, and the addition of a methylated **cap** at the 5′ end involves the enzymes guanyl transferase guanine-7-methyl transferase and 2′-0-methyltransferase. Each of these enzymes, of course, requires the expression of one or more genes. Also DNA-binding proteins play an important role in the regulation of transcription by acting in the processes of enhancement and promotion, as we discuss in the following sections. In addition

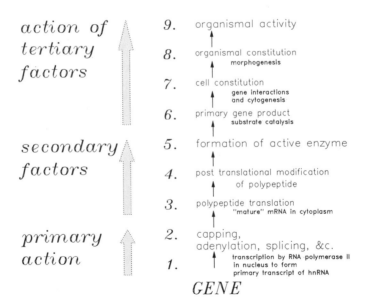

action of
tertiary
factors

secondary
factors

primary
action

9. organismal activity

8. organismal constitution
   morphogenesis

7. cell constitution
   gene interactions
   and cytogenesis

6. primary gene product
   substrate catalysis

5. formation of active enzyme

4. post translational modification
   of polypeptide

3. polypeptide translation
   "mature" mRNA in cytoplasm

2. capping,
   adenylation, splicing, &c.
   transcription by RNA polymerase II
   in nucleus to form
   primary transcript of hnRNA

1.

GENE

Fig. 6.1. *Overview of factors involved in the formation of a phenotype in a multicellular eukaryote, using a "structural" protein-encoding gene as an example. The primary action of a gene is considered to be its transcription to produce the primary transcript, which is then modified by action of secondary factors involving* *many other genes. The primary gene product is the result of the catalytic activity of the encoded protein upon a substrate that may be a product of another gene. The following tertiary factors result in a functional organism. All the other genes in the genome may participate.*

to capping and adenylation most primary transcripts contain introns, which must be spliced out to produce the "mature" mRNA, which is ready to undergo translation in the cytoplasm. (3) The process of translation involves probably hundreds of genes such as the ones encoding the various ribosomal and transfer RNAs, the aminoacyl transferases for charging the tRNAs, the proteins of the ribosomes, and the various enzymes involved with the formation of peptide bonds, and so on. (4) Genes encode enzymes that modify the polypeptides posttranslationally. (5) The assumption of the various conformations of polypeptide chains, such as α-helices, β-sheets, and the folding to form the tertiary structures characteristic of globular proteins, may occur spontaneously in the proper environment, but the products may be subunits that function as parts of multimers that may be homo- or heteromultimers. The latter would involve at least another primary gene encoding another polypeptide. This multimer may not operate alone but as part of an enzyme complex or holoenzyme such as the DNA polymerases discussed in Chapter 3. (6) The functioning of the enzyme in catalysis may thus involve other gene products to finally produce what might be called a primary gene product, which may be only a precursor for the final product (e.g., an amino acid). The process of building a cell requires the participation of many thousands of gene products at this level. Some of these are structural material, others play a regulatory role, and some may play both (e.g., receptors). Morphogenesis results in an organism with certain morphological and behavioral characteristics.

In conclusion, no phenotype is the product of the action of a single gene. The mutation of a single gene may alter a phenotype but it does not eliminate it, for the other factors (genes) involved in its existence continue to act. A lethal mutation, of course, eliminates all; it need only be in a gene involved in replication of DNA, for example.

I. GENERAL FEATURES OF TRANSCRIPTION

The only known way for RNA to be synthesized in eukaryotic cells is by the action of RNA polymerases transcribing or copying the nucleotide sequence in a single strand of DNA into an RNA polynucleotide (Fig. 6.2). The process involves at least three basic steps: **initiation, elongation**, and **termination**, in which the polymerase attaches to the double-stranded DNA of the chromosomes, proceeds to transcribe along a single strand, and then drops off with the release of a single strand of RNA. RNA is also synthesized outside of the nucleus in mitochondria and chloroplasts, and it becomes involved in the synthesis of polypeptides within these organelles.

Transcription occurs only in the $5' \rightarrow 3'$ direction. In this it is similar to the replication of DNA by DNA polymerase. The $3' \rightarrow 5'$ strand "read" by the RNA polymerase is called the "sense" strand. It performs a function analogous to the $3' \rightarrow 5'$ DNA template strand read by the DNA polymerase in DNA replication. The RNA transcript produced by action of RNA polymerase will be $5' \rightarrow 3'$, as shown in Figure 6.2. By common agreement the direction of transcription defines what is called "upstream" and "downstream" on the transcriptional unit and its associated sequences. Upstream is toward the $3'$ end of the sense strand, and toward the $5'$ end of the transcript. As in DNA replication, both strands of DNA may be involved in polymerase activity. Also, as in the case of replication, the DNA helix is known to be unwound in order for transcription to proceed as shown in Figure 6.2. Therefore a strain should result within the helix as described on p. 28. The supercoiling can be negative to the left of strand separation,

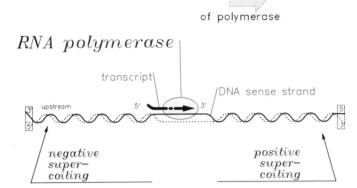

Fig. 6.2. *A diagram visualizing RNA transcription, and the supercoiling produced by the unwinding. The ends of the DNA helix are assumed to be fixed. See text.*

and positive to its right in the direction of transcription. Evidence from observations on two yeast mutants that have deficiencies in topoisomerase I (the nicking and closing enzyme as shown in Fig. 3.11) and topisomerase II (double-stranded cutting and rejoining) indicates that these enzymes are involved in the relaxation of the helix as it rotates to allow for strand separation (Brill and Sternglanz, 1988; Gilmour et al., 1986; Futcher, 1988). Thus, it appears that topoisomerases are functionally important in both transcription and replication. The natural state of DNA, at least that associated with the nucleosomes, is to be negatively supercoiled, but positive supercoiling should result from the unwinding during transcription and replication (Graever and Wang, 1988). How transcription proceeds on DNA packaged on nucleosomes is presently not understood, but a number of interesting schemes have been proposed (Thoma, 1991).

A. RNA Polymerases

Three nuclear RNA polymerases are known in the eukaryotes in addition to the polymerases that function in mitochondria and chloroplasts. The nuclear ones are RNA polymerases I, II, and III, which are active in the transcription of nucleolar ribosomal RNA, messenger RNA, and small RNAs (including tRNAs and 5SRNA), respectively. Polymerase I (POLI) is confined to nucleoli, while the other two are presumably closely associated with chromosomal segments other than the NORs.

The single RNA polymerase of *E.coli* has been well characterized and found to be a holoenzyme with a molecular weight of about 480 kD with four types of subunits. The eukaryotic polymerases are also holoenzymes with molecular weights in the range of 500 kD. Like the prokaryotic polymerase, they are made up of a number of subunits.

RNA polymerase II (POLII), which transcribes mRNA, has been the most extensively studied of the eukaryotic polymerases. It consists of at least 10 subunits and is strongly inhibited by α-amanitin, a toxin produced by the mushrooms of the genus *Amanita*. In contrast, polymerase I is not inhibited. The largest subunit of polymerase II has a molecular weight of about 215–220 kD. It is this subunit that is sensitive to amanitin, as shown by the fact that mutants resistant to amanitin have mutations in the gene locus encoding it (Voelker et al., 1985; Sanford et al., 1983). The nucleotide sequences for the genes *RP021* and *Pr11215*, which code for this large subunit in *Saccharomyces cerevisiae* and *Drosophila*, respectively, have been determined (Allison et al., 1985; Biggs et al., 1985). A comparison of these sequences shows a close homology between the yeast and *Drosophila* forms, and even more remarkably with the large subunit of the RNA polymerase of *E. coli*. The sequence of the gene *RP031* of yeast that encodes the large subunit of its RNA polymerase III (POLIII) has also been determined and found to have extensive homology to the RNA polymerase II gene *RP021* (Allison et al., 1985). It is now apparent in fact that yeast polymerases I, II, and III all have subunits in common, and this is probably true of eukaryotes in general (Sentenac and Hall, 1982).

POLIII is found in all eukaryotes and is involved in the transcription of 5SRNA, tRNAs, and small RNAs derived from moderately repetitive sequences, and even some viral RNAs. Like POLII and III it is made up of at least 10 subunits, and has a total molecular weight of ~700 kD. Animal POLIII is inhibited by high levels of α-aminitin, but like POLI no inhibition by this toxin is found for the POLIII of yeast or insects.

All RNA polymerases—prokaryotic and eukaryotic—are undoubtedly ancient ones, and over the billions of years that they have existed their genes have had their nucleotide sequences conserved to a remarkable degree.

B. General Features of Transcriptional Regulation

To begin to understand transcription it is necessary to have some appreciation of current knowledge of gene structure. First we may ask a simple question: what is a gene? It can be defined by two different approaches: operational and physical. It can be defined operationally by the ***cis-trans* test**. This is a complementation test to determine whether two mutations are within the same genetic unit of function, sometimes called a **cistron** (Benzer, 1961). If two mutations, m_1 and m_2, give a wild type phenotype when in the *cis* arrangement in heterozygotes (coupled): $m_1 m_2 / + +$; as well as in *trans*, (in repulsion) $m_1 + / + m_2$ they are considered to be alleles of two different genes. However, if a mutant phenotype is produced in the *trans* arrangement, but not in the *cis*, they are considered to be alleles of the same gene.

Physically a gene can be defined as a specific sequence of DNA base pairs that can be shown to function in the formation of a specific phenotype. Ordinarily this is done by comparing the phenotypic effects of a mutant allele to the standard or wild type. Presently a specific DNA sequence can be isolated, reproduced (cloned) in a plasmid, and used in transformation or transfection. Thus, genes can be isolated and analyzed physically so that their properties can be studied as a function of the sequences of their base pairs.

Functionally two different categories of genes can be defined: 1) those that are transcribed, but the transcripts are not translated, and 2) those whose transcripts are translated into polypeptides. The former are the genes that transcribe the various types of RNA such as rRNA, tRNA, etc., that function in translation, and the latter are the protein encoding genes. The possibility remains that there exist genes that are neither transcribed nor translated. These we speculate about in Chapter 10.

In this chapter the primary concern is with the protein encoding or structural genes. Figure 6.3 diagrams a model for a structural gene. It describes what is known about the structure of some but not all structural genes. One of the characteristics of most eukaryotic structural genes is

Gene Organization

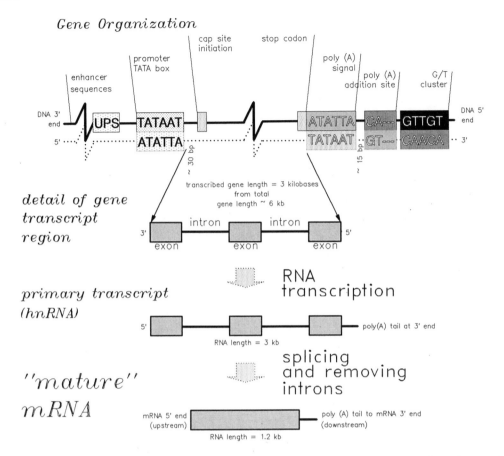

detail of gene transcript region

primary transcript (hnRNA)

RNA transcription

"mature" mRNA

splicing and removing introns

Fig. 6.3. *A diagrammatic representation of a structural gene. The strand transcribed is the 3′ → 5′ one, indicated by the heavy line. This is the sense strand, which acts as template for the 5′ → 3′ transcript. The enhancer and other regulatory sequences upstream from the cap site are not transcribed, but have sequences that are bound by the proteins that regulate the transcription process as described in the text.*

the presence of sequences known as **exons** and **introns** within the transcribed region of the genes. Transcription results in a primary transcript. This is processed by splicing out the introns to produce the mature mRNA that contains the exon sequences and functions as the encoder of the polypeptide.

The transcribed region of a gene is only a small part of the total gene. Figure 6.3 represents a small gene of 6kb of which only 3kb is transcribed. This 3kb transcript after splicing is then only 1.2kb in

length. Part of the sequences upstream and frequently downstream from the transcribed region are regulatory in function.

1. cis-Acting elements

Eukaryotic genes have *cis*-acting DNA sequences that are required for their regulation. These are oligonucleotide sequences located upstream or downstream on the transcribed strand that function as recognition sites for *trans*-acting transcription regulatory factors. These factors are proteins encoded elsewhere in the genome and they bind to *cis*-acting DNA sites (Dynan and Tjian, 1985; Mitchell and Tjian, 1985). Genes that are transcribed by POLIII may have binding sites within the transcribed area. These are referred to as ICR-containing genes because they have internal control regions (Murphy et al., 1987).

Figure 6.4 gives highly schematic diagrams for the major classes of sequences for rRNA genes, structural genes, and small-RNA-coding genes for the eukaryotes, and, for comparison, the analogous sequences for a prokaryote such as *E. coli*, which has only one RNA polymerase.

Two general types of *cis*-acting regulatory sequences have been identified for the various kinds of known genes: **promoter** and **enhancer sequences**, or **elements**. Promoters are required for the efficient initiation of transcription. Essentially they "turn genes on." The polymerases probably interact indirectly with the promoter regions and start transcription because of the alterations brought about in the DNA helix by interactions of binding protein and DNA. Enhancers increase the rate of initiation by promoters. The distinction between the two types is not always sharp, however.

Three categories of promoters have been identified: TATA boxes, upstream regulatory sequences or elements (generally referred to by the acronyms URS and URE, respectively), and the ICRs previously mentioned. Some structural genes have TATA boxes located about 25–30 bp upstream from the site of transcription initiation called the CAP site; others have no identifiable similar sequence. Those with TATA boxes have URS sequences 50–100 bp upstream that have conserved elements such as GG(C or T)CAATCT, the so-called CAAT box, and GGGCGG, the GC box. The TATA box sequences (such as ATATAA etc.) are AT-rich. Genes without TATA boxes have a series of URS sequences that act alone as promoters.

Enhancer sequences are also recognized by binding proteins. They are generally located further upstream than promoters, up to 1 kb or more, and they may also be located downstream from the transcribed region. Two functional types of enhancers have been identified: inducible ones, which respond to environmental changes, and temporal or tissue-specific ones. Examples of inducible enhancers are those that respond to exposure to heavy metals, viral infections, heat shock, or specific chemical substances such as steroids. Some may be close to the promoters of their genes rather than far upstream, some may be within the introns of the transcribed region, and some may be downstream from that region. One of the interesting characteristics of some enhancers is that they function equally well whatever their orientation, that is, whether inverted or not.

2. *trans-Acting elements*

The cellular binding proteins that play a role in transcriptional regulation are generally referred to as ***trans*-acting factors**. They function by interacting with the *cis*-acting promoter DNA sequences described in the previous paragraphs. They may function either to inhibit or induce gene expression. A number of them have been described that have rather unusual domains such as the **helix−loop−helix**, the **zinc finger motif**, and the **leucine zipper** (reviewed by Busch and Sassone-Corsi, 1990).

Both positive and negative regulation of transcription may be the result of the action of factors that interact with promoters and enhancers. Although inducible enhancers by definition act positively, this action can be achieved by interaction with a binding protein to enhance transcription directly, or by the interaction of the binding protein with a repressor to inactivate it and allow transcription to proceed (Fig. 6.5). Both types have been

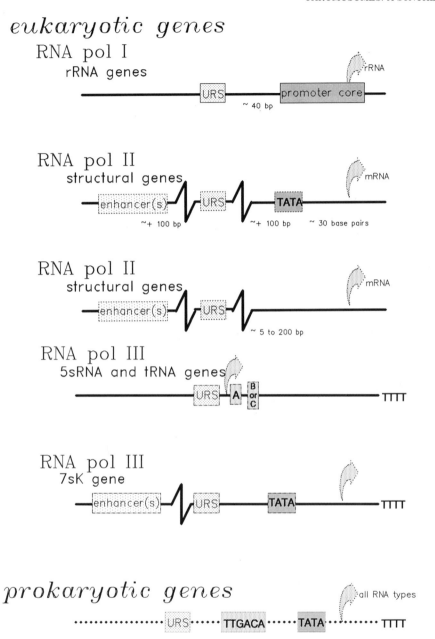

Fig. 6.4 *Schematic representations of some of the various kinds of RNA and structural genes of eukaryotes, showing the elements involved in the regulation of the activities of their polymerases. The scheme for prokaryotic genes is also shown to illustrate the close resemblance between the eukaryotic and prokaryotic plans. Some genes have TATA boxes upstream, others do not, but nonetheless they do have promoter sequences. The cap initiation sites are indicated by the bent arrows. Note that the 5sRNA and tRNA genes have promoters within the transcribed region; 5s and tRNA genes have sequence type A in common, but tRNA genes have ICR sequence B and 5sRNA genes have sequence C. These interact with binding proteins just as do the upstream elements. The TTTT sequences at the ends of genes transcribed by POLIII and the prokaryotic RNA polymerase are transcription terminators.*

a. positive action

b. inactivation of a negative action

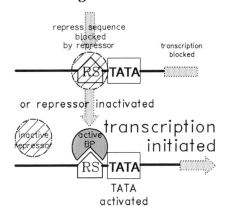

Fig. 6.5. *Schemes for positive action* **(a)** *and negative action* **(b)** *for initiation of transcription on the part of an inducible DNA-binding protein (BP). The BP is inactive until modified by the inducer. Once activated, it can bind to the regulatory sequence (RS) for which it is specific, either directly* **(a)**, *or after first removing a repressor of an RS* **(b)**.

verified. The induction of the heat shock protein gene, *hsp70*, of *D. melanogaster* by elevated temperature treatment is a simple positive action. On the other hand, the human β-interferon gene, β-*IFN* is apparently regulated negatively.

C. Initiation and Termination of Transcription

Precisely how the nucleotide sequences such as promoters and enhancers function to initiate and modulate transcription is not known. It is clear, however, that the interaction between these sequences and the binding proteins is essential for activating genes. It is probable that initiation of transcription is not a simple, direct enzyme–substrate interaction between

DNA and polymerase. For example, PO-LII does not initiate transcription in *Drosophila* unless two proteins, one of which binds specifically to promoter regions, are present (Parker and Topol, 1984). As stated previously, polymerases probably interact indirectly with the promoter regions by binding protein–DNA interaction such as these. These alterations may prepare the CAP sites to receive the polymerase.

Transcription starts at the CAP site with a ribonucleotide triphosphate, usually a guanine or adenine, which retains its 5′ triphosphate group and makes a 3′ phosphodiester bond with the 5′ position of the next nucleotide, thus: 5′pppApNpNp-–-. Almost immediately this 5′ end reacts with a Gppp in the presence of the enzyme guanyl transferase to give the structure: GpppApNpNp-–-. The G residue is present at the 5′ in reverse orientation from the other nucleotides (Figure 6.6). This is the CAP. The guanine is methylated at position 7 in the presence of guanine-7-methyl transferase. All eukaryotes have this CAP, but the multicellular eukaryotes have further methylations of the 2′O positions of the nucleotide(s) first added as triphosphates (Fig. 6.6). These methylations of the deoxyribose are catalyzed by 2′O-methyltransferase. The degree of methylation of the 2′Os and the bases varies with the species.

After initiation and capping, the actual transcription with the production of an RNA chain proceeds at a rate that varies with the species and the gene type, as discussed in the following sections. All DNA nucleotide sequences of a single DNA sense strand transcription unit are transcribed. Most structural genes contain introns, and if so, the intron sequences are transcribed along with the exon sequences.

Transcription of ribosomal RNA by POLI is generally terminated at the 3′ end of the transcript close to the enhancer elements of the next rRNA repeat (Sollner-Webb and Tower, 1986). This may be several hundred base pairs beyond the 28S endpoint, bringing the POLI molecules that have just completed one transcript to a position where the poly (A) tails are added. The processing of the pre-mRNA or hnRNA is discussed in the next section.

Transcription of *Xenopus* tRNA genes by POLIII is terminated in downstream regions that have four or more T residues generally imbedded in GC-rich regions

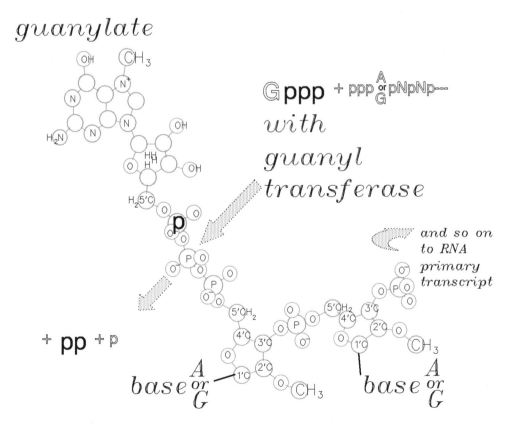

Fig. 6.6. *Initiation of transcription at the cap site, which consists of Gppp-ApNpNp. See text.*

(Bogenhagen and Brown, 1981). The termination process does not appear to involve *trans*-acting factors at least for 5S gene transcription in *Xenopus* (Cozzarelli et al., 1983).

II. REGULATION OF TRANSCRIPTION

The regulation of transcription in the eukaryotes is a critical step in the formation of the phenotype, for it is at this point that most primary gene action is presumably controlled (as pointed out in Fig. 6.1). In the higher eukaryotes the events controlling development from a zygote emanate from regulation of transcription. Other events follow, of course, but not unless something is transcribed to begin with. When we understand regulation at the transcription level, we will have taken a big step toward the understanding of differentiation. A complete understanding of transcriptional regulation will also lead to a fuller understanding of certain disease states, such as neoplasia. At this writing, however, we can only say that we have made a beginning to this understanding.

A. Regulation in *Saccharomyces*

The *S. cerevisiae* genome has about 1.5×10^7 bp of DNA and is capable of encoding about 5,000 polypeptides. The discussion of regulation in yeast makes a fitting start to understanding transcription in higher eukaryotes. In the mouse, POLI terminates transcription 565 bp downstream from the 3' end of the 28S transcript. The termination region contains DNA sequence repeats that interact with factors that may be DNA-binding proteins (Grummt et al., 1985).

The termination of transcription of a protein-encoding gene by POLII is more complicated than termination of the RNAs transcribed by POLI and III. Processing at the 3' end of a potential functional mRNA transcript is only recently beginning to be understood, but what is now known indicates a considerable degree of complexity, and the process is of major importance in the regulation of gene activity (Platt, 1986).

Termination apparently can occur at multiple sites in the DNA extending over hundreds to thousands of nucleotides downstream from the last exon nucleotide. The transcription of the last exon nucleotide is at the 5' side of a stop DNA codon, which may be ATT, ACT, or ATC. The sequences that follow downstream continue to be transcribed up to a sequence known as the polyadenylation {pol(A)} signal in some but not all genes (Fig. 6.3). In the higher eukaryotes and in some yeast genes this signal is a hexanucleotide with the sequence AATAAA. Downstream from this sequence at an average of about 15 nucleotides there occurs in many genes a type of sequence called the G/T cluster which invariably starts with the sequence CA or TA. It is at CA or TA that polyadenylation is started and introduction to a similar discussion of regulation in higher eukaryotes with genomes in the range of 10^8 to 10^9 bp that control the development of very complex phenotypes compared to yeast.

Yeast genes generally have three basic types of *cis*-acting nucleotide sequences involved in the regulation of transcription: upstream activation sites (UAS), TATA boxes, and initiation sites. In order for transcription be be initiated both the UAS and TATA elements must be present (reviewed by Guarante, 1987; Struhl, 1989). Some genes also have operator sites that bind to proteins that repress transcription.

In this discussion of yeast transcription, we focus primarily on three genes on chromosome 15, and a series on chromosome 2, as chronicled by Struhl et al., (1985) (Fig. 6.7a). The genes, *pet56*, *his3*, and *ded1* are adjacent but have no obvious relation-

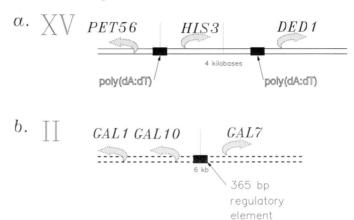

Saccharomyces cerevisiae

Fig. 6.7. a. *The pet56–his3–ded1 region on yeast chromosome XV. The pet gene is transcribed in a direction opposite to the his and ded genes.* **b.** *The galactose genes, gal1, gal10, and gal7 on yeast chromosome II. Data from Citron and Donelson (1984).*

ship in the functioning of their products. The *gal* genes *gal1, gal10,* and *gal7* are closely linked and coordinately regulated (St. John and Davis, 1981) (Fig. 6.7b). They do have related functions insofar as they encode enzymes that function in galactose metabolism.

The *his3* and *ded1* genes are transcribed in the same direction, in contrast to *pet56,* which, although adjacent to *his3,* is transcribed on the complementary DNA strand (Fig. 6.7a). These genes are expressed constitutively, that is, they are always turned on. When the cells are subjected to amino acid starvation, however, the *his3* gene is transcribed at a fourfold higher rate than its constitutive rate. The level of transcription of *gal1* and *gal10* is essentially zero in the absence of galactose, when the sole carbon source is glucose. In the presence of galactose, the cells have ~40 molecules of GALmRNA per cell at the steady-state level.

A diagram of the yeast gene *gcn4,* which codes for a regulatory protein, is given in Figure 6.8. It has certain features in common with a structural-type gene in a higher eukaryote (Fig. 6.3). It is somewhat sim-

pler in structure but its structure and function possibly are related to what may be happening in the more complex higher multicellular forms.

1. Enhancement of initiation

In yeast as well as in other eukaryotes, a TATA box sequence exists about 35 bp upstream from the initiation sites of many of their genes. This and a regulatory sequence some 94–99 bp from the initiation sites for the *his3* are essential for the efficient transcription of the gene. The AUG site for the start of translation is about 25 bp downstream from the initiation sites. There are actually two transcription initiation sites about 12 bp apart, and initiation may start at either one for this gene. It should be mentioned, to avoid possible confusion when consulting the literature, that the regulatory sites of the type described here are also referred to as enhancers, modulators, or upstream activating sites (usually indicated by the acronym UAS in the yeast literature). To add to the confusion, the terms are sometimes used as synonyms and sometimes not. The site-specific sequences are anal-

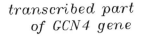

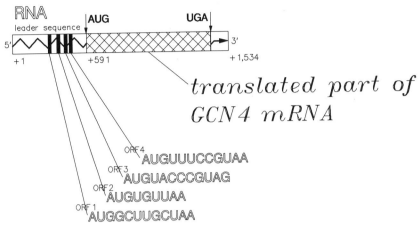

Fig. 6.8. *Representation of the yeast gene, gcn4, that encodes a binding protein. The transcribed part of the gene has a leader sequence upstream from the part that ultimately encodes the GCN4 binding protein.*

ogous to the URS and URE sites of the higher eukaryotes.

By using deletions and insertions located between TATA and the initiation sites, it has been shown, at least for the *his3* gene, that the spacing between TATA and initiation is relatively unimportant. TATA can be moved to the initiation sites or farther upstream and still be active, provided it is not moved farther away than 90–100 bp upstream. These and other mutants described below are synthesized by transforming yeast host cells using modified forms of lambda-related vectors. In addition, "swap" mutants have been synthesized in which the promoter–enhancer regions of the *his3* and *ded1* genes are exchanged. In general, the promoter–enhancer regions permit normal levels of transcription of *his3*.

2. Positive and negative regulatory elements

The galactose genes *gal1, gal10,* and *gal7* are tightly linked in that order on chromosome 2 (Fig 6.7b) and are coordinately expressed in response to galactose (Citron

and Donelson, 1984). They encode the enzymes in the Leloir pathway: galactokinase, uridine diphosphoglucose 4-epimerase, and galactose-1-phosphate uridyltransferase, respectively. The direction of transcription is not the same for all three; *gal7* and *10* are transcribed on one strand and *gal10* from another. Each gene has its own promoter. A number of other *gal* genes, among them *gal14,* and *gal80,* are unlinked and act as regulators of the enzyme-encoding genes.

The divergently transcribed *gal1* and *gal10* have between them an upstream activating site, that is functional only in the presence of the TATA boxes of these genes. A 365-bp fragment of this region has been isolated in a vector and then manipulated so as to fuse it to various derivatives of the *his3* gene as listed in Table 6.1. These have various segments of the *his3* upstream enhancer sequences deleted. Their fusion products are then reintroduced into the yeast cells and the phenotypes of the transformed cells are determined. Fusions of the UAS 365-bp element to *his3* segments that lacked various segments up-

TABLE 6.1. Results of Fusion of 3656 bp *gal* UAS Segments to Upstream *his-3* Regions[1]

Allele	End Point	Glucose	Galactose
his3-G1	−8	−	−
his3-G2	−23	−	−
his3-G3	−35	−	−
his3-G16	−55	−	+ + +
his3-G16	−55	−	+ + +
his3-G5	−80	−	+ + +
his3-G6	−136	±	+ + +
his3-G15	−157	±	+ + +
his3-G14	−173	±	+ + +
his3-G13	−192	±	+ + +
his3-G12	−204	±	+ + +
his3-G7	−253	+	+ +
his3-G8	−330	+	+ +
his3-G9	−357	+	+ +
his3-G11	−389	+	+ +
pet56-G1	−86	±	±
pet56-G2	−66	±	±
pet56-G3	−22	−	−

[1]Reproduced from Struhl et al., 1985, with permission of the publisher.

stream of the initiation sites yielded some most interesting and informative results. Fusion products without the TATA box failed to express *his3* under any growth conditions, but those that contained the TATA box placed *his3* under *gal* control (Struhl et al., 1985). The HIS3mRNA was transcribed at a level 20–27 times above the normal level of two molecules per cell in the presence of galactose. The *his3* gene under the influence of the *gal* UAS sequences is an inducible gene but *only* in the presence of galactose. In the presence of glucose it is not. Thus it would appear that UASs may do more than increase the transcription roles of genes; they may also control the gene's susceptibility to environmental factors. This may be a most important element in differentiation.

The *gal* 365-bp UAS activates the transcription of the *his3* gene when present in either orientation. If, however, this UAS element is placed downstream from the *his3* initiation sites, no *his3* transcription is activated. Thus, the UAS is active only

when 5′ to the initiation site. It is important to note that it must not be too far upstream for maximum activation. The UAS located more than 250-bp upstream from the initiation site causes only a three- to fourfold increase in the number of *his3* mRNAs per cell, and again this is only in the presence of galactose.

When the *gal* UAS segment is 131 bp upstream from the *his3* initiation site, it gives maximum enhancement of *his3* mRNA production in the presence of galactose. Even though the *gal* UAS is also upstream from the *pet56* gene (which is transcribed in the opposite direction from *his3*) and less than 100 bp from its promoter box, no enhancement of *pet56* mRNA occurs. Thus it is not necessarily a universal enhancer.

When yeast cells are grown in the presence of glucose as the carbon source, the *his3* gene is transcribed at a steady-state rate of 2 molecules of mRNA per cell. However, certain of the *gal–his3* fusion cells have a depressed mRNA production in the presence of glucose even though the *his3* mRNA level is elevated in the presence of galactose. Thus a catabolite repression by glucose is expressed which is quite independent of enhancement of *his3* mRNA production in the presence of galactose.

3. Amino acid starvation effects

The transcription of genes involved in amino acid biosynthesis in yeast is enhanced greatly when the cells are grown in the absence of amino acids. The starvation for any *single* amino acid in the group: histidine, tryptophan, arginine, isoleucine, and lysine causes the derepression (reactivation) of the biosynthesis pathways of all. This general enhancement is known as *general amino acid control*; it involves transcription of at least 24 different genes in the biosynthesis pathways.

It has been found that mutants with deletions in the region 83–103 nucleotides upstream of the initiation site do not respond to amino acid starvation with enhancement of transcription. The sequence ATGACTC, 94–99 bp upstream from the initiation site is the most important element for induction of the histidine biosynthetic genes. Certain point mutations in this sequence or its deletion result in the cessation of transcription of these genes (Donahue et al., 1983). The protein encoded by the yeast gene *gcn4* binds specifically to the ATGACTC oligonucleotide of *his3* and other *his* biosynthetic genes as well as those involved in the other amino acids cited above in connection with general amino acid control (Hope and Struhl, 1985). Thus a specific protein, GCN4, stimulates transcription of a large number of genes by binding to the ATGACTC sequence(s) *cis* and upstream from their initiation sites (Fink, 1986).

The *gcn4* gene transcribes an mRNA with a ~600-bp sequence 5′ to the polypeptide coding sequence (Fig. 6.8). This leader sequence contains four short **open reading frames** each complete with an initiation codon (AUG) and a termination codon (UAA or UAG). (An open reading frame, usually written **ORF**, is a sequence upstream from the actual initiation site that is transcribed and all or part of the transcript is potentially translatable into protein.) Mueller and Hinnebusch (1986) have shown that these upstream mRNA sequences mediate translational control of *gcn4*. They are necessary for the repression of *gcn4*. Hence, when GCN4 protein is not synthesized, the genes encoding the amino acid biosynthesis enzymes are not derepressed. Missense mutations that produce base substitutions in the AUG triplets of the open reading frames result in *gcn4* expression. The degree to which the gene is expressed depends on the number of AUG triplets that are altered and hence the degree to which the amino acid biosynthesizing genes are derepressed. When all four open reading frames have mutations in the AUG codon, the effect is the same as a deletion of the section of the gene including them.

Since the four upstream AUG codons in the open reading frames can repress *gcn4* expression under conditions of nonstarvation of amino acid, it has been proposed that these sequences are recognized as initiation codons in translation. The inhibitory effect of these upstream AUG codons is then the result of depressing the reinitiation by yeast ribosomes at internal AUG codons (Mueller and Hinnebusch, 1986). The reasons for this are obscure but may be related to the stalling of the ribosomes at the upstream sequences that also have termination codons. In any case, what we have here is a translational phenomenon regulating transcription of a group of genes, albeit indirectly through direct control of the translation of a positive regulating protein, GCN4. This may be an isolated case of regulation of transcription by translation, but it is more probable that it is a general mode of control. Thus, although in the eukaryotes transcription and translation are separated by a nuclear membrane, a feedback or cybernetic control does exist via products of metabolic activity in the cytoplasm. Of these products the DNA-binding proteins may be by far the most important.

4. Inducible versus constitutive states

Some cellular proteins are inducible, some are constitutive, that is, produced at a constant rate, and some are both. The enzyme encoded by the *his3* gene (and others like it) is both. The constitutive expression of *his3* depends on the poly (dA:dT) sequence 115–130 bp upstream from the initiation site. Deletion of this region reduces transcription of *his3* below the normal constitutive level, and also the adjacent gene, *pet56*, which is transcribed from the complementary strand in the opposite direction. Thus the poly (dA:dT) sequence appears to operate bidirectionally. Similarly a poly (dA:dT) region 88–121 bp upstream from the initiation site of the *ded1* gene when deleted causes a drastic reduction of the constitutive level of transcription of this gene (Struhl, 1989). Micrococcal nuclease preferentially digests DNA in the spacer regions between nucleosomes (see p. 56). Hypersensitivity to the nuclease is seen in all mutant strains lacking in sequences upstream of bp 155 to the initiation site but not in those that lack the ATGACTC regulator region to which GCN4 protein binds. This suggests that DNA binding by GCN4 protein induces a change in DNA conformation. Also, it is to be noted that genes for amino acid synthesis, such as *his4, arg4*, and *trp5*, which are coregulated with *his3*, also have GCN4 binding sites in their upstream regions. Genes not subject to general control, such as *ura3, trp1, gal1*, and *ded1*, do not (Hope and Struhl, 1985). Thus it is apparent that a single binding protein can regulate all genes that possess its specific binding site upstream from their TATA boxes.

B. Regulation in Animal Viruses

Just as a considerable amount of our knowledge of eukaryotic DNA replication has been gained by studying the replication process in viruses, so too our knowledge of transcription has been enhanced by studying it in these same viruses. One of the first found and most studied enhancer sequences is in the SV40 virus. This papovirus has a genophore containing nucleosomes. Thus it approximates the status of a chromosome. The DNA molecule is 5.2 kb long and contains two units of transcription, one of which is expressed early and the other late. Interestingly, the two units are transcribed in opposite directions (Fig. 6.9). Each has a promoter located in a nucleosome-free area in which the replication origin site is also located. The enhancer region for the early genes expressed during the initial period of viral expression is located upstream from the TATA box. It overlaps with the enhancer for the late genes (Serfling et al., 1985; Cereghini et al., 1983). When excised and introduced into other cell types, it has been shown to be relatively nonspecific. It is active in all mammalian cells, in amphibian cells, and even in algal cells (Wasylyk and Chambon, 1983). However, not all viral enhancers are so nonspecific. The mouse polyoma viral enhancer is about four times more active in mouse cells than primate cells. The enhancer of the papovirus JCV is 20 times more active in glial cells of the brain than in HeLa cells.

Transcription of the early gene of SV40 is dependent upon two 72-bp repeats that, as shown in Figure 6.9, lie 107–250 bp upstream from the initiation site for the early RNAs. In addition, the 21-bp repeat region between the latter sequence and

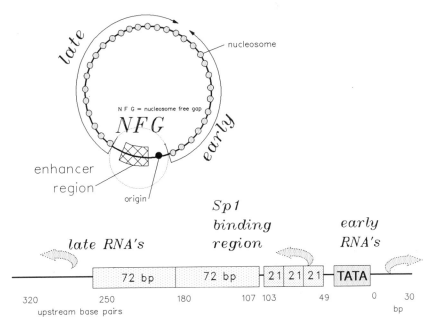

Fig. 6.9. *SV40 genophore organization. NFG indicates the nucleosome-free gap. See text.*

the TATA box is also necessary for efficient transcription. The **T antigen**, which is encoded by the early transcription unit, is a protein that binds to three sites in the DNA surrounding the TATA box and indicated by the black bars in Figure 6.9. The T anitgen is also necessary for the initiation of replication at the origin indicated in the figure. However, once the concentration of T antigen reaches a certain level it represses its own transcription in the early gene sequences. Thus, it is a negative regulator like the *lac* repressor of *E. coli*, and the binding sites are functionally analogous to the *lac* operator site.

The 21-bp repeat regions have a high affinity for the Sp1-binding protein (Fig. 6.9) principally because of the CCGCCC hexanucleotide stretches that occur within them. As shown in Figure 6.10, these same hexanucleotide stretches occur in enhancers in the mouse metallothionein-I gene

and in the herpes thymidine kinase gene, but not in precisely the same arrangement as shown in Figure 6.9. Also to be noted is that not all genes have the same type of enhancer sequences, e.g. compare human interferon-α_1 and rabbit β-globin with the other three and with one another in Figure 6.10.

McKnight (1983) has made a thorough study of the thymidine kinase gene, *TK*, of the herpes virus and accurately delineated those parts of the sequences of the gene upstream from the CAP site by generating and introducing deletions. He found that the only regulatory sequences were between 95 and 100 nucleotides upstream from the cap site of the gene (Fig. 6.10). No sequences 3′ or downstream from the coding region affected the transcription of the gene when deleted. By an ingenious method of introducing base change mutations into the enhancing sequence region at different positions upstream from

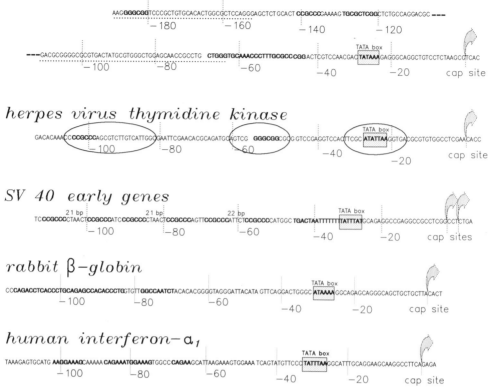

Fig. 6.10. *A comparison of enhancer and promoter regions for some animal and animal virus genes. The metal-responsive enhancer regions of the mouse metallothionein-I gene are boxed. Note the inverted CCGCCC elements in the herpes thymidine kinase gene. Reproduced from Serfing et al. (1985), with permission of the publisher.*

them, he was further able to accurately pinpoint the three regions within the 100-nucleotide region that are important. Mutations within the regions between have no effect on transcription.

The herpes simplex *TK* gene is not fastidious. It has been introduced into the genomes of a wide variety of vertebrate cells and shown to be capable of encoding functional thymidine kinase of the herpes variety in all that have been tested. This finding in itself is indicative of the fact that by studying transcription in the easily manipulated virus we can learn some-

thing substantive about transcription in general in the higher eukaryotes.

C. Regulation in the Higher Eukaryotes

Like the lower eukaryotes, higher eukaryotes also have *cis-* and *trans-*acting elements that control the rate of transcription.

1. *trans-Acting regulation*

Some of the earliest evidence for the existence of *trans*-acting factors came from somatic cell genetic studies in which somatic cell lines expressing genetic loci responsible for highly specialized or differentiated gene products, such as dopa oxidase, albumin, or tyrosine amino transferase, were fused to nondifferentiated cell lines (Davidson et al., 1966; Brenda and Davidson, 1971; Darlington et al., 1982). The expression of highly differentiated genes could be altered in the nuclei of hybrids. In general, the expression of more specialized gene products, such as pigment, is extinguished in hybrid cells.

Other *trans*-acting factors have been identified by *in vivo* competition assays. For example, a positive-acting factor was found to control expression of the metallothionein gene family (Seguin et al., 1984). In these experiments, *trans*-acting factors were titrated by the introduction of *cis*-acting sequences into cultured cells by DNA-mediated gene transfer. The *cis*-acting sequences are regions that bind specific *trans*-acting elements. Usually the *cis*-acting sequences are fused to a "reporter" gene, such as chloramphenicol transferase (CAT), whose expression is easily monitored in eukaryotic cells, since CAT is not an endogenous eukaryotic gene. Different sequences can be fused to the CAT gene in order to determine the effects of different sequences on gene regulation. In the experiments performed by Seguin et al., it was determined that when large amounts of sequences 5′ to the mouse MT gene were introduced into cells, the level of CAT enzyme was lowered, presumably because a positive-acting factor present in limited amounts was being titrated. A negative *trans*-acting factor (a repressor) would presumably lead to an increase in CAT enzyme when the factor is titrated by an excess of transcription factor binding sequence. Numerous types of sequence-specific DNA-building proteins, such as zinc-finger domains, leucine zippers, and helix–turn–helix, have been identified (Johnson and McKnight, 1989).

2. *cis-Acting regulation and methylation*

An important *cis*-acting effect on gene regulation involves DNA methylation (Cedar, 1988). Hypermethylation and hypomethylation of cytidine bases have been identified as important in the regulation of DNA replication in the slime mold *Physarum* (Cooney and Bradbury, 1990). This role of methylation in replication and repair may be in effect in most organisms, although at present it is established in only a few (Grafstrom et al., 1984). It is now certain that methylation also plays a role in transcription in both prokaryotes and eukaryotes. The methylation state of DNA can influence the binding affinity of proteins (Saluz, et al., 1991).

Although DNA replication involves only the nucleotides: dATP, dCTP, dGTP, and dTTP, it has been well documented, with S-adenosylmethionine as a methyl donor, that postreplication methylation of cytidine and adenine bases occurs in bacteria in the presence of methyl transferases. In animals and plants only 5-methylcytidine has been identified. Methylated cytosines are identified with CpG sequences in species that have methylated DNA. Some organisms such as *Drosophila melanogaster* appear to have no methylated DNA (Rae and Steele, 1979; Urieli-Shoval et al., 1982). It is of some interest that CpG sequences are also sites for Z-DNA formation (Rich, 1984).

The distribution of methylated cytidines in chromosomal DNA is not random (Razin, 1984). This has been established by using restriction enzymes that recognize CpG sequences (Cedar, 1984). The pattern of distribution of methylated bases in the different cell types of a eukaryote is characteristic for each cell type. Methylated cytidine appears to be several times higher in concentration in repetitive DNA than in unique sequences. The DNA of nucleosomes, especially those with histone H1, is more highly methylated than the DNA of the spacer regions (Conklin and Groudine, 1984; Cooney and Bradbury, 1990).

A strong correlation exists between the level of activity of a gene and its degree of methylation. Genes that are actively transcribed in all tissues such as the rRNA genes are hypomethylated. That hypermethylation blocks and hypomethylation releases blocks to transcription is a reasonable conclusion based on a considerable amount of data from vertebrate animals in particular. Specifically hypermethylation at the 5' ends of genes appears to be the main blocking mechanism (Cedar, 1984).

It is not to be concluded, however, that the changes in methylation patterns are the sole controlling factor in gene regulation. Hypermethylation may block transcription, but other regulatory elements must dictate the patterns of methylation leading to blockage, and then to relieve the degree of methylation to permit the onset of transcription. Metaphorically, conversion to hypomethylation is like unlocking a door, but the door must be opened for transcription to start, and this involves the *trans*-acting elements discussed in the previous sections among other possibilities. The methylation patterns of DNA found in the germ and somatic cells of all eukaryotes thus far examined except for some insects is definitely a factor in gene regulation, but it is not the only factor (Otto and Walbot, 1990).

3. DNA methylation and X-chromosome inactivation

Methylation may play an important role in the inactivation of the mammalian X chromosome. In the XX female mammal both X chromosomes are functional in zygotes and during the very early stages of embryogenesis. But once cytodifferentiation starts in the embryonic and extraembryonic tissues the derivative somatic cells have only a single X chromosome active (reviewed by Lyon, 1986, 1988; Grant and Chapman, 1988). In the mouse inactivation begins at about the 40-cell stage with a preferential inactivation of the paternally contributed X chromosome (X^P) in the trophectoderm and the primitive endoderm. The other cell lineages of the embryo show a random inactivation of the maternally derived X (X^M) and the x^P (Frels and Chapman, 1980). This random pattern of inactivation is apparently true for the eutherian mammals, making the females functional mosaics for heterozygous X-linked genes. Marsupials, however, show an absolute inactivation of the X^P in certain tissues of the female (Sharman, 1971).

Inactive X chromosomes of somatic cells are highly heterochromatic and appear as visible Barr bodies during interphase. Like heterochromatin, in general, they replicate late in the S phase. The role of X-chromosome inactivation in mammals is probably to compensate for X-linked gene dosage. The XX female should produce twice the amount of X-linked gene products as the XY male, unless one X is inactivated. This has been shown to be the case in development (Chapman, 1986).

It is known that specific genes on the human X chromosome are hypermethylated with respect to their methylation status on the active X chromosome (Mohandas et al., 1981). Treatment of cells with 5-azacytidine, an analogue of cytidine that cannot be methylated, can lead to hypomethylation and reactivation of genes on both the X chromosome and on autosomes (Mohandas et al., 1981).

Once methylation is initiated, inactivation appears to be stabilized and maintained through successive cell generations by hypermethylation of the cytidine bases. A methylase exists that is involved with the methylation of hemizygous sites following DNA replication.

4. DNase I hypersensitivity and transcriptionally active chromatin

Hypomethylation of DNA is also strongly correlated with other general features associated with transcriptional activation, such as DNase I sensitivity. Genes that are transcriptionally active are generally hypersensitive to DNase digestion.

It has previously been mentioned that chromatin in the neighborhood of transcriptionally active genes has an unusually high sensitivity to cleavage by DNA nuclease I. This is not generally true for genes that are unexpressed. Tuan et al. (1985) have mapped the distribution of major and minor DNase I sites in the β-globin-like gene cluster of humans (Fig. 6.11). They determined the DNase I-hypersensitive (HS) sites in the flanking DNA of the β-cluster in a leukemic cell line (K562) in which the embryonic ε-globin gene is mainly expressed, in a human erythroleukemia cell line (HEL) that produces primarily the fetal globin genes, in normal bone marrow adult cells in which

β-globin is expressed, and finally in a human promyelocytic leukemia cell line (HL60) that expresses none of the β-globin genes. Major DNase I-hypersensitive sites were found in the 5' and 3' boundary areas of the cluster far upstream and downstream of the expressed genes for the erythroid cell types, but not in the HL60 cells (Fig. 6.11). The major site V is presumably outside the cluster. Several interesting conclusions derive from these data. There are four β-cluster major nuclease-sensitive sites as represented by the bold arrows in the figure. These are numbered I, II, IV, and VI. The I, II, and IV sites in the area 5' to the ε-gene are located in DNA regions containing enhancer-like sequences. The VI site is downstream over 30 kb from the β-gene but it is apparently important, since it does not occur in the HL60 cells. Upstream from the ε-gene four minor sites, designated by light arrows, occur. Again these are not found in HL60 cells. Also minor sites occur 5' to the Gγ-, Aγ-, δ-, and β-genes. These have been identified as being about 200 bp 5' to the respective cap sites (Grondine

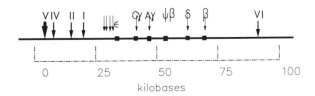

HS	V	IV	II	I	ε	γ	δ+β	VI	predominate globin gene expressed
HL 60	+	−	−	−	−	−	−	−	none
K 562	+	+	+	+	(++)	+	−	+	ε
HEL	+	+	+	+	+	(++)	−	+	γ
marrow			+	+			(++)	+	β

Fig. 6.11. *The distribution of the major and minor DNase-hypersensitive sites within the human β-gene cluster region. + and − indicate presence and absence, respectively, of hypersensitive sites. The HS V is indicated by a larger arrow. Data from Tuan et al. (1985).*

et al., 1983). The minor δ and β sites do not occur in the K562 and HEL cells, but do in the marrow cells, which produce δ- and β-globins predominantly. The fact that all three of the erythroid cells have major sites 5′ to the ε-gene, whether they express ε or not, indicates the presence of a general regulatory region. The same is true for the VI site. The existence of common general regulatory regions, as well as specific ones, may be one reason why these genes remain linked as they have over millions of years of evolution.

The study of the location and significance of DNase I-hypersensitive sites has led to a better understanding of eukaryotic gene action (Gross and Garrard, 1988). In the previous chapters we have referred to and discussed activities of chromosomes such as replication, compaction, and decompaction during the cell cycle, and pairing of homologues, recombination, and segregation. All of these activities involve the participation of *trans*-acting binding proteins interacting with DNA. For example, the *cis*-acting DNA regulatory sequences of genes interact with these proteins. Presumably this can occur only if the DNA sites are accessible. This accessibility apparently occurs primarily in the nucleosome-free regions. These are the nuclease-hypersensitive sites, which are at least two orders of magnitude more sensitive than other regions of chromatin and probably represent only about 1% of the total genome.

HS sites have been mapped at specific loci as described for the globin gene cluster region discussed in the previous section. They have been shown to occupy specific positions of known function, such as promoters, URSs, and enhancers of transcription, origins of replication and recombination, sites around telomeres and centromeres, topoisomerase cleavage sites, and terminators of transcription.

Several types of HS sites have been identified. In all cases it appears that the structures of the DNA are atypical. The available evidence is that histones are not present within the sites and that nucleosomes are excluded. Furthermore, the sites do not ever seem to occur within exons or even introns, but always outside the transcribed part of the genes. DNA within the sites appears to be the non-β-form with unpaired bases.

One of the many interesting characteristics of HS sites is that their distribution in different tissues may be different for the same gene. A good example of this is found in the pattern of sites around the chick lysozyme gene (Fig. 6.14). Note that the patterns are different for inactive and active states of the gene, and also for activity in different tissues (e.g., activity for oviduct shows a different pattern than for macrophages).

Findings such as these make it apparent that a proper state of chromatin structure outside the transcribed region is important, if not essential, to the proper expression of genes by transcription. The regulatory parts of genes function because their DNA can assume configurations that enable them to receive messages from *trans* elements that can then be transmitted *cis* to enable chromosomes to carry out their essential functions.

5. Examples of gene regulation in higher eukaryotes

The vertebrate globin genes. Two types of vertebrate globin genes, α-like and β-like, are encoded by two related families of genes, with members clustered on separate chromosomes. The members of each family function in transcription at different periods in the animals' development from zygote to adult. It was by observations on the products of these genes during the course of human fetal development that it was first realized that genes are turned on and off in eukaryotes as well as in *E. coli* (Ingram, 1959). Indeed it soon became evident that as an animal embryo devel-

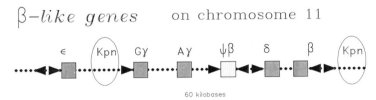

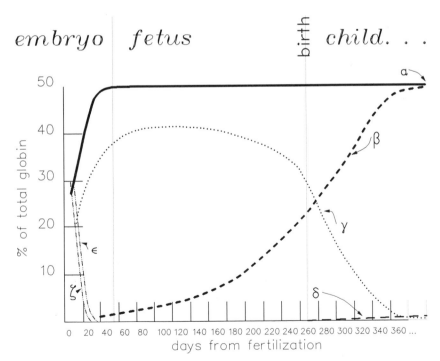

Fig. 6.12. *Organization of the human globin gene clusters (α on 16; β on 11). Shaded boxes represent genes that become active; lightly stippled boxes represent pseudogenes. The triangles indicate locations of Alu sequences. Ovals indicate approximate locations of the Kpn sequences in the β-cluster. Data from Karlsson and Nienhuis (1985).*

Fig. 6.13. *Activities of the human α- and β-globin genes starting shortly after fertilization up until about 100 days after birth.*

ops, some sets of genes are actively transcribed while others are not, and this fact alone makes it certain that these phenomena of differential activation and inactivation are important if not crucial factors in the process of cell differentiation (Maniatis et al., 1987).

Switching among the globin genes in humans during ontogeny occurs sequentially following the order of the genes on the chromosomes (Karlsson and Nienhuis, 1985). Two basic switches occur in the globin gene clusters: the embryonic to the fetal globin and the fetal to the adult globin. Involved are the α-like gene cluster on chromosome 16 and the β-like cluster on chromosome 11. The organization of these gene clusters is diagrammed in Fig-

ure 6.12, and the activities of various members of the clusters represented in Figure 6.13. A fourth member of the α-globin gene cluster, θ, has been identified and shown to be active in cells of erythroid origin (Hsu et al., 1988; Clegg, 1987). It is located downstream from the α_1-globin gene, but it does not appear to be an essential member of the cluster, since humans without the gene are seemingly normal.

Not only are genes turned on and off as shown in Figure 6.13, but different tissues are involved. The megablast nucleated cells in the primitive yolk sac seem to be the first centers for the synthesis of the embryonic globins. When the liver begins to form, switches occur; the embryonic genes

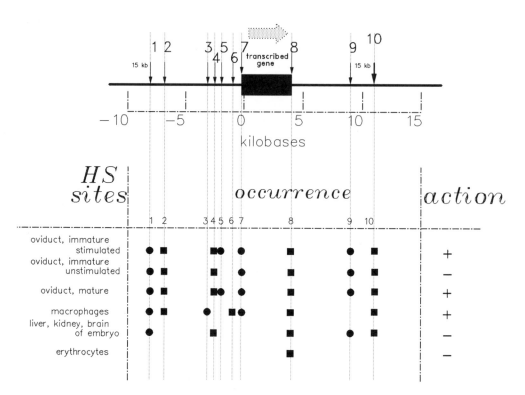

Fig. 6.14. *Patterns of HS sites around the chick lysozyme gene in various tissues and functional states of the gene. The filled circles and boxes indicate HS sites. Note that these may be downstream as well as upstream from the transcribed regions.*

are turned off and the α- and γ-globins begin to be encoded by their respective genes. The spleen becomes involved at the end of the embryonic period and the dominant globins are α and γ, with a small percentage of β. This condition persists until about 2 months before birth, when the red bone marrow activity significantly increases its production of β-globin, and the γ-globin rate of synthesis begins to fall. Shortly after birth γ-globin synthesis is reduced to an insignificant amount, and the dominant hemoglobin becomes HbA ($\alpha_2\beta_2$), with a small percentage (<2%) of HbA_2 ($\alpha_2\delta_2$). Table 6.2 lists the kinds of functional hemoglobins present at the various stages in a phenotypically normal embryo, fetus, and adult.

The β-globin gene cluster spans a 60-kb region on chromosome 11 and all of it has been sequenced. Within this stretch of DNA are five known functional genes (Fig. 6.12), a promoter, a pseudogene, eight *Alu* sequences, and nine segments homologous to the *Kpn* family of interspersed repetitive DNA sequences. The α-globin cluster on chromosome 16 spans about 30 kb and contains three functional genes, two pseudogenes, and three *Alu* repeats.

Members of the *Alu* family of repeats are generally found dispersed in the euchromatin of the chromosomes of primates rather than present as tandem clusters, as are many heterochromatin repeat sequences. The repeats are about 300 bp long, and in humans they are dimers consisting of two repeated 130 bp monomers with a

31 bp insertion in one of them.

The *Kpn* family also called the *LINES* or *LI* family is a second family of repeats. In humans and other primates 50×10^3 or more of 7- to 8-kb-long *Kpn* sequences are interspersed in the genome. Shorter sequences ranging from 70 bp to a few kilobases with definite homologies to the longer ones have also been found, and some of these are in and around the β-globin cluster (Fig. 6.12). A 6.4 kb *Kpn* sequence is found downstream from the β-globin gene. A second long *Kpn* sequence with two members 3–5 kb long in tandem array but inverted as compared to the 6-kb member downstream is found between the ε- and ^{G}γ-genes. Several 70 bp sequences homologous to the 3' end of the *Kpn* consensus sequence are found in the cluster in the region of the γ-genes. Like the *Alu* family, the *Kpn* sequences are polymorphic, but nonetheless obviously related. They may or may not be transcribed, and like *Alu* their role remains a mystery.

Within its transcribing region each globin gene has three exons separated by two introns. Their exon sequences are similar enough to indicate that both the α-like genes and the β-like genes have arisen from an ancestral gene by duplication. The α-cluster probably became separated from the β-cluster possibly 200 million to 300 million years ago, since in *Xenopus* the two clusters are closely linked (Hosbach, 1983). The clusters are unlinked in both birds and mammals, but their linkage in reptiles at present is not known, and it would be of great interest to know the status of this relationship in them.

The DNA outside the exon coding regions of the β-globin-like genes shows homologies just as do the exons. Conserved splice sequences are present between intron 2 and the 3' end of exon 2 and the 5' beginning of exon 3. Intron lengths are fairly uniform among the β- like genes, but not between the α- and

TABLE 6.2. Human Functional Hemoglobins at Different Stages

Embryonic	Fetal	Adult
Hb Grower$_1$ ($\xi_2\epsilon^2$)	HbF ($\alpha_2\gamma_2$)	HbA ($\alpha_2\beta_2$)
Hb Grower$_2$ ($\alpha_2\epsilon_2$)		HbA (α_2S_2)
Hb Portland ($\xi_2\gamma_2$)		

ζ-genes of the α-cluster. Also, regions upstream from the transcribed part of the gene show some degree of homology in sequences, as described in the next section.

As in yeast and the animal viruses, each globin gene has a cap site, a promoter sequence, and enhancer sequences upstream from the 5′ initiation of the transcription site. The transcription-regulating sites have been investigated extensively in humans and mice. β-Globin genes are no exception to the general rule that structural genes have upstream promoter sequences. Figure 6.15 shows sequences upstream from the cap site for the mouse β- and the human β- and $^A\gamma$-genes. Note the identical or similar sequences in the TATA box, and the so-called CACA and CAT boxes. These regions are more highly conserved than the intermediate sequences. Base changes by mutation in the sequences between the boxed areas have little or no effect on the transcription rate, and there is little conservation of sequences between the boxes. This is in marked contrast to the sequence homology in the transcribed portion of the gene.

The amino acid sequences of β- and γ-globins are 73% in correspondence. Thus it would appear that genes that have arisen by duplication and have assumed similar, but not identical functions, may differ significantly in the regions from which their transcription-regulating signals emanate. These external regions have some conserved regions in common but also certain sequences not in common. Those not in common may be the ones that make the $^A\gamma$- and β-globins functionally different, because they interact with different binding proteins and not because of the differences in the globins they encode. γ-Globin does in fact function quite well in humans in place of the β-globin, a condition found in certain inherited thalassemias and in sickle cell anemia. The HbF levels in the latter may be as high as 30%–40%, but it is the HbS that causes the profound pathological effects in homozygotes, not the HbF. It may be hypothesized that the turning-on of the β-globin may be a signal received and transmitted from the upstream CTTGA sequences not present in the β-globin upstream region (Fig. 6.15).

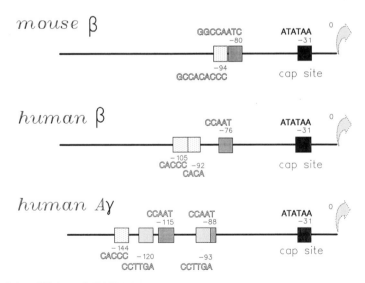

Fig. 6.15. *URS and TATA box sites for the mouse β-globin and the human β- and $^A\gamma$-globins. Data from Myers et al. (1986) and Karlson and Nienhuis (1985).*

The procedures used to demonstrate that the upstream sequences illustrated in Figure 6.15 are of importance in regulation of the β-globin-like genes in general are basically the same as those used for yeast. Mutations and deletions have been introduced into HeLa cells by transformation and also in transgenic mice, as described in the next section. In addition, point mutations and deletions have been identified within the upstream regions of the β-cluster in individuals with thalassemias, a hereditary anemia found in relatively high frequency among certain ethnic groups whose gene pools originate in the tropical and Mediterranean regions. When the globins of thalassemic persons were first sequenced, it was found that they did not have point mutations or deletions within the exon regions. It is now obvious that they have mutations within the regulatory sequences and may have normal β-globin otherwise (Karlsson and Nienhuis, 1985).

Most of the experimental work has been done with cells of patients with thalassemias (Orkin and Nathan, 1981). In these persons the amount of HbF in the erythrocytes ranges from 0% to 15% in the heterozygotes, but in most the maximum is 4%–5%. This is in contrast with normal individuals who have a range of 0.3% to 1.2% HbF. Other persons with HbF elevated beyond the normal range up to 25%–30% may show no clinically abnormal symptoms (unless homozygous). This kind of hereditary persistence of fetal hemoglobin (HPFH) is generally the result of point mutations in the β-globin-regulating regions, whereas the pathological thalassemic state is a result of deletions at various parts of the β-cluster regions.

The removal of the distal section of $^A\gamma$ down to upstream nucleotide 131 reduces transcription 75% in HeLa cells, and deletion down to about nucleotide 110 to remove the distal CAT box reduces activity 90%. On the other hand, deletion of the proximal CAT box while leaving the upstream sequences intact *increases* transcription of γ-globin in HeLa cells twofold to fourfold (Karlsson and Nienhuis, 1985). Thalassemic persons who produce no HbF are presumably heterozygous for a deletion of a good part or all of the β-gene cluster.

A great deal has been learned by studying the sequences in the regulatory regions of $^A\gamma$- and $^G\gamma$-globin of persons who have HPFH. At least four cases have been found in which single base pair substitutions occur. These include a G-to-A substitution at nucleotide 117 in the distal CCAAT box of the $^A\gamma$-globin gene (Gelinas et al., 1985; Collins et al., 1984). Since this point mutation results in the persistence of γ-globin production into the adult stage, it has been suggested that the CCAAT box region of the γ-genes is the binding site of a repressor protein that turns off the γ-genes in the normal adult. Altering the base sequence of the CCAAT box may inhibit the binding of the repressor protein (Gelinas et al., 1985; Collins et al., 1984).

A point mutation in the $^G\gamma$-gene at position 202 upstream from the cap site has been found in a family in which HbS was also segregating. The HPFH condition was found to be a result of a C-to-G substitution that occurs in a sequence GGGGCGCC, which occurs in the region 208–200.

This is closely related to upstream sequences in the herpes thymidine kinase gene and a 21 bp repeat in SV40 that have been shown to be important for efficient transcription (Collins et al., 1984). Two other cases of point mutation in the upstream regions of the genes both involving C-to-T transitions—one at position 196 (Giglioni et al., 1984) and the other at position 158—have been shown to cause HPFH. It is thus apparent that regions 100–200 bp upstream of the γ-gene coding regions must be important in the γ-gene regulation. But even mutations downstream to the coding region of a globin gene may cause disruption of transcription. A globin gene isolated from a patient with β-thalassemia had a T-to-C transition within that conserved AATAAA polyadenylation signal region. Persons with this thalassemic syndrome and heterozygous for this transition and β^S had a HbA content of 20%–25% and HbS content of 70%–74% with 1% HbF (Orkin et al., 1985).

Mice are a handy tool for the study of mammalian genetics for a variety of reasons. One of their important advantages is the ease with which their embryos can be used in the creation of **transgenic** mice. The production of a transgenic mouse involves the introduction of a foreign bit of DNA into a fertilized egg nucleus and then implanting the genetically transformed egg into a uterus. The DNA introduced may be in the form of a whole chromosome or purified pieces of DNA introduced by microinjection. Whole chromosomes can be taken up by cells in suspension, and in rare cases fragments of the chromosomal DNA are integrated into the host genome. The best procedure for consistent results makes use of the donor cDNA of specific genes introduced into plasmids. The plasmids can be injected directly into an egg zygote nucleus, or excised from the plasmid by the proper restriction enzymes and the donor-purified DNA can be introduced into a nucleus. The donor DNA will be integrated into the host genome with a fairly high frequency, and a transgenic animal produced as a consequence. This process of genetic transformation is also called transfection, but this is in reality a misuse of the term transfection. Genetic transformation is not the same as cellular transformation, which in mammalian systems refers to the uncontrolled growth exhibited by "transformed" neoplastic cells. (A confusing usage is the designation *transformation* in *Drosophia* versus *transfection* in mammals. Both mean the same thing.)

The regulation of human globin genes during development has been studied by injecting purified genic DNA's into mouse eggs just prior to the fusion of the male and female pronuclei (Constantini et al., 1985, 1986; Magram et al., 1985; Chada et al., 1986). Human cloned β- and ^{G}γ-genes have been excised from plasmids, and their integration into transgenic host animal DNA has been verified by collecting blood samples from the mouse and determining the presence of human β- or ^{G}γ-globins by using fluorescent antibodies against human β-globin chains. The mRNAs for both human genes were also isolated and identified. Five transgenic mice were examined for the presence of the human β-globin genes. The results are given in Table 6.3. These mice were all

TABLE 6.3. Presence and Expression of Human β-Globin Genes in
Adult Transgenic Mice[1]

Mouse	Estimated no. of gene copies per genome	mRNA levels (pg/μg)	Anti-human β-globin fluorescent staining
Hβ 56	50–100	2000	all cells strongly positive
Hβ 58	10–20	2000	strong, but heterogeneous
Hβ 64	5–15	<1	medium, homogeneous
Hβ 66	50–100	10	5% of cells strongly positive
Hβ 81	1	<1	negative

[1]Data from Constantini *et al.* (1985), with permission of the publisher.

derived from eggs transformed with β-globin DNA. They differed in the number of β-globin gene copies integrated, as determined by estimating β-globin DNA on Southern blots. They also differed in the amount of mRNA produced; for example, Hb56 and Hb66 both appeared to have the same number of β-globin gene copies but showed a 200-fold difference in mRNA production. The use of fluorescent staining of blood cells indicated that Hb58 and Hb66 might have been mosaics, but this is not certain. In any event it is clear that the human β-gene is expressed in transgenic mice, and expressed only in erythroid cells, since other cell types did not contain detectable β-globin. In addition it is activated in the mouse at the same stage of development as when *in situ* in the human.

One approach to analyzing the regulatory sequences in the β-globin gene by using transgenic mice has been made by Chada et al. (1985) with a hybrid mouse/human gene. This hybrid gene contains 1.2 kb of the 5' flanking region of the mouse β major-globin gene and 3.8 kb of the flanking 3' portion of the human β-globin gene (Fig. 6.16). The hybrid was originally constructed and integrated in the plasmid pBR322, and is expressed in transgenic mice in blood, bone marrow, and spleen

cells, as determined by measuring β-globin mRNA production (Chada et al., 1985). When the hybrid was treated with DNase I, three hypersensitive sites were detected (arrows, Fig. 6.16). One site is in the upstream region of the mouse gene, the second is in the third exon of the human gene, and the third is in the 3' flanking region of this gene. These sites were not detected in chromatin from brain cells. The site 3' to the human gene is apparently the same as the VI hypersensitive site identified by Tuan et al. (1985), as shown in Figure 6.11.

The fetal γ-globin genes are found in both the anthropoid apes and *Homo*. The $^{G}\gamma$- and $^{A}\gamma$-globins are not found in other mammals, including the nonsimian primates. In these the embryonic β-like globins are replaced in the fetal stage by the "adult" β- globins. It is believed on the basis of sequence homology analysis that the simian γ-genes as well as the nonsimian β-like embryonic genes descended from a hypothetical gene present in an ancestral mammal (Fig. 6.17). To investigate this possibility Chada et al. (1986) introduced a cloned human $^{G}\gamma$-globin gene into the mouse germ line. They found that the human $^{G}\gamma$-gene reverts to an embryonic pattern of expression in these transgenic mice during their development (Fig. 6.18) as measured by the appearance and

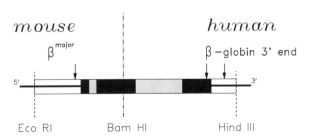

Fig. 6.16. *Approximate structure of the human/mouse hybrid, showing the location of the DNase I-hypersensitive sites (arrows). This gene contains upstream 5' regulatory regions of the β^major-globin and 3' downstream regions of human β-globin (open bars). Exons are indicated as solid bars and introns by stippled bars. Data from Chada et al. (1985) and Constantini et al. (1985).*

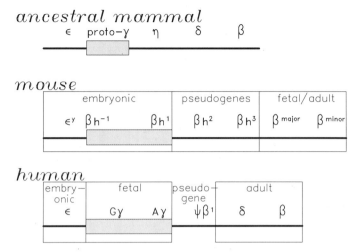

Fig. 6.17. *Possible evolutionary origins of mouse and human β-gene clusters. Data from Chada et al. (1986).*

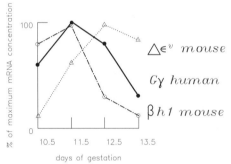

Fig. 6.18. *Rates of accumulation of human $^G\gamma$ and mouse $\Delta\varepsilon^V$ and βh1- globin mRNAs in transgenic mouse blood cells during embryonic development.*

disappearance of the mRNAs of the human γ-gene and the mouse ε-gene and β-genes. These data have been interpreted as indicating that a shift has occurred in the evolution that led to the simians in the timing of the turning-on of the proto-γ-gene. In the ancestors of the simians this gene became inactive in the earliest stages of embryogeny, and activated later in the fetal period to become the present-day γ-genes (Chada et al., 1986). We can posit that an element produced by another gene,

possibly a binding protein acting as a *trans-acting* signal, is active early in mouse embryogenesis and the γ-gene responds to this signal, which otherwise is latent until the fetal period in the simians. If so, it follows from this that the mouse embryonic β-like genes, and the human and ape β-genes, have enhancer sequences that react to the same types of signals. This leads to the important albeit tentative conclusion, which we shall return to discuss further in Chapters 9 and 10, that the control of the timing of gene activation may be a most important factor in the evolutionary process. Mutations and changes in topography of the genome involving the regulatory elements of genes may in fact be more important in evolution than mutations within their exons.

Transcription of histone genes. One of the most interesting groups of genes in relation to transcription is the histone group (reviewed in Stein et al., 1984). As discussed previously (p. 49), five different basic types of histones exist along with a number of their variants, and these are in various ways involved in chromosome replication as well being important constituents of chromatin (p. 134) (Zweidler, 1984). In most animals the histone genes are organized in quintets that encode each of the five histone genes. In birds and mammals there is no quintet organization. The genes are still clustered but in a somewhat unorganized fashion. For example, in the human genome histone genes are found on chromosomes 1 and 7, and in the mouse on chromosomes 3 and 13. In both they are present as multigene families as discussed on p. 381.

Histone synthesis occurs mainly during the S phase but some cultured cells, such as mouse lymphoma cells, have high rates of synthesis in the G1 phase. In the slime mold, *Physarum polycephalum*, the initiation of H4 transcription does not occur in S, but in G2. Thus transcription of histone genes and replication are not necessarily coupled in the S phase (Jalouzot et al., 1985).

In the sea urchin *Strongylocentrotus purpuratus* three major classes of histone genes have been described (Old and Woodland, 1984). Early- acting histone genes, which are tandemly arranged as quintets, are active up to the late blastula stage. A "switch" occurs at this stage and the early forms are replaced by a second, different set of histones that are encoded by a dif-

ferent set of genes. At the onset of gastrulation a third, late set of genes is turned on (Maxson et al., 1983). This switching of histone types is reminiscent of globin switching. The difference is that whole new sets of proteins are transcribed rather than a successive set of proteins as in the globins. In the case of the sea urchin a master promoter system must exist for each set of histone types, unlike the globin case, in which each gene has its own transcriptional controls.

***The Adh gene of* Drosophila melanogaster.** The alcohol dehydrogenase-encoding gene of *melanogaster* is transcribed at different levels during development, and the single copy of the gene in the genome has two different promoters. One of them, the adult promoter, is engaged at high levels in the adults, but not in embryos and larvae (Posakony et al., 1985). The same POLII recognizes both promoters, but not simultaneously. Obviously different *trans*-acting factors must be active in the embryo–larva cells than in the adult cells. The adult promoter, TATTTA, allows for initiation of transcription some 700 bp upstream from the embryo–larva promoter, TATAAATA (Fig. 6.19). The adult transcribed region contains three introns shown by the blank spacers in the transcript. The

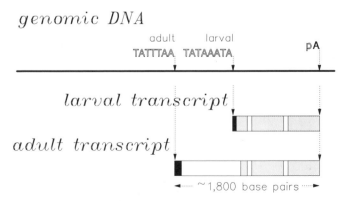

Fig. 6.19. *The Adh gene of Drosophila melanogaster, showing the two different transcribed regions for the embryo–larva and adult forms of the enzymes. Data from Posakony et al. (1985).*

larval transcript has two introns and three exons, in contrast to the four exons of the adult version. Thus, although the same gene encodes both the larval and the adult alcohol dehydrogenase, the enzyme has partially different structures in each.

This example forms an interesting contrast with the human globin situation. A different globin family member exists for each of the different globins. In *melanogaster* one gene, not a family, exists to produce two different forms for the different stages of development.

6. *Transcription of both strands of DNA*

Transcription in opposite directions of the two complementary strands of DNA has been previously described for yeast (p. 300). This is a common phenomenon, in which two different genes are transcribed in opposite directions. However, it has also been shown that complementary strands of the same segment of DNA can transcribe and encode unrelated proteins. The *Gart* gene of *D. melanogaster* encodes three of the enzymes involved in purine synthesis (Henikoff et al., 1986). In addition, a functional pupal cuticle protein is encoded by a gene located within the opposite strand of the first intron of the *Gart* locus (Henikoff et al., 1986). A similar situation has been found in the mouse (Adelman et al., 1987). The gonadotropin-releasing hormone (*gnrh*) gene segment also encodes a second protein on its opposite strand. The function of this second protein is not known, but its RNA transcripts are found in the heart. Just how common this parsimonious phenomenon is, is not known, but it obviously may add significantly to the transcriptional capacity of eukaryotic genomes.

7. *Antisense RNA*

Gene expression is controlled not only by regulatory genes encoding *trans*-acting protein factors that bind to specific DNA sequences, but possibly also by genes that direct synthesis of regulatory RNAs that act as repressors of gene action. These RNAs, called **antisense RNAs**, have sequences that are complementary to the RNAs transcribed by the protein-encoding structural genes. The binding of the antisense RNA to the sense RNA inhibits gene expression (Moffat, 1991). The exact mechanism of this inhibition or repression of gene activity is not understood, but it is believed to also be important in prokaryotic regulation.

The phenomenon of naturally occurring antisense RNA genes has yet to be demonstrated in eukaryotes, but some experimental results do nevertheless indicate that such genes exist in *Xenopus*. For example, *Xenopus* β-globin mRNA and antisense RNA were synthesized *in vitro* with globin cDNA serving as template. Antisense RNAs complementary to the globin RNA were then injected into oocytes along with globin RNA. Specific inhibition of globin production resulted when the antisense RNA included sequences complementary to the 5' initiation site of the globin mRNA (Melton, 1985). Just how important antisense RNA is in regulation *in vivo* in eukaryotes is yet to be established.

8. *Replication, transcription, and recombination*

Two of these three most basic activities of chromosomes, replication and transcription, occur primarily during the S phase of the cell cycle. The question immediately arises, Are they completely independent? The answer appears to be an unequivocal no. It has been pointed out in Chapter 3 (p. 130) that a close relationship has been found between replication and transcription in certain cell lines in culture. It has also been shown that transcriptionally *cis*-acting elements can function as components of eukaryotic origins of DNA replication (DePamphilis, 1988).

The *ori* regions that function as origins of replication frequently contain two components: one involved in DNA replication and a second possessing promoter and enhancer elements that may be involved in both transcription and replication. Most of the experimental work in this area has been with animal viruses and mitochondrial DNA. Here we describe for illustrative purposes some of the factors found to be involved in the polyoma virus.

The polyoma virus, PyV, requires an enhancer element for *ori*- dependent replication in mouse cells. The enhancer activity resides about 80–200 bp upstream from the *ori*-core region where bidirectional replication begins. The enhancer element stimulates both replication and transcription of mRNAs. Replication of this virus and its relatives requires the presence of the T-ag antigen. T-ag binds both to the *ori* region and to part of the enhancer region, and apparently initiates both transcription and replication.

Early hypotheses concerning the mechanism of crossing-over, such as proposed by Belling (1933), might be described as copy-choice. It was assumed that crossing-over occurred at the time of chromosome replication. Since it is now apparent that crossing-over occurs in prophase I of meiosis, it is expected that this kind of recombination is effectively separated from transcription occurring in the S phase. However, it will be recalled that premeiotic S phase meiocytes do not completely replicate their DNA (p. 167). The residual unreplicated DNA is finally synthesized in prophase I at the time recombination is occurring. Accompanying this replication, a certain amount of transcription occurs.

III. SUMMARY

1. Transcription of DNA is the fundamental essence of gene action, although the path leading from a gene's active transcription and its expression in the final phenotype is a long and complex one. This is because no phenotype is the product of action of a single gene.

2. Transcription is catalyzed by the action of three RNA polymerases: POLI, POLII, and POLIII. These transcribe RNA only in the $5' \rightarrow 3'$ direction by reading the sense strand of DNA in the $3' \rightarrow 5'$ direction. Either complementary strand of DNA may be read to produce a complementary RNA strand.

3. POLI, which is active in the nucleoli, transcribes ribosomal RNA. POLII transcribes messenger RNAs, and POLIII transcribes 5SRNA, transfer RNAs, and a number of other small RNAs.

4. The concept of the gene as a unit of structure and function contributing to the phenotype was an abstraction when first stated. Now it is an established fact that a physical entity, a stretch of DNA, functions with elements of control within it that modulate its activity.

5. Only a relatively small region of a gene is transcribed. The untranscribed regions contain *cis*-acting regulatory sequences that interact with *trans*-acting binding proteins encoded by regulatory genes. These interactions are the regulators of transcription.

6. The *cis*-acting regulatory sequences are described as promoters and enhancers. Promoters are short sequences generally less than ten nucleotides long located upstream from the transcription initiation sites. In some cases they may be within the transcribed region itself. Promoters are required for initiation of transcription; enhancers modulate the rate of transcription. This is a generalization and is not always true.

7. Transcription starts at a cap site generally some 30–40 bp downstream from a promoter, which in some cases may be a TATA box. Both the exons and the introns of genes are transcribed, if the genes have introns.

8. The process of regulation of transcription in yeast may also serve as a model for transcription in eukaryotes in general. The transcription, like the replication process, is an ancient one, and its mechanisms have probably been highly conserved. A very important observation made in connection with transcription of a yeast gene, *his3*, is that it can be caused to respond to the presence of galactose by transcribing at a high rate when a segment of DNA that regulates galactose genes is placed upstream from its initiation site.

9. The T-antigen, T-ag, of SV40 is a regulatory binding protein, necessary for initiation of replication, that represses its own transcription once it reaches a certain concentration. It is an example of a negative regulator.

10. The human globin genes occur as two related families in the human genome: the α-globin and the β-globin. Each family has its own regulatory apparatus to switch the various members on and off during the development of the human embryo and fetus. The switching during ontogeny occurs sequentially, following the order of the genes on the chromosomes. Two basic switches occur: the embryonic to the fetal, and the fetal to the adult. The β-globin family appears to have a master sequence controlling the transcription of the entire complex.

11. Chromatin has DNase-hypersensitive sites that appear to be those sequences that are being actively transcribed.

12. Human and ape β-globin genes cloned in transgenic mice react to the promoter and enhancer sequences of the mouse to encode human and ape globin production in erythroid cells of the mouse. Two different orders of mammals thus appear to have similar transcriptional controls.

13. Histone genes and replication are not necessarily coupled in the S phase of animals.

14. In *Drosophila melanogaster* a single *Adh* gene produces two different forms of alcohol dehydrogenase.

15. Methylation of cytidine bases in most animals and plants inhibits transcription, and in some cases replication of DNA. The inactivation of one of the X chromosomes in female mammals is accompanied by hypermethylation.

16. Antisense RNAs have been shown to inhibit transcription *in vitro*, but not *in vivo* in the eukaryotes. Along with methylation they may be important in regulation.

7

Chromosome and Gene Activity in the Interphase

The nucleus consists of a mass of substances which are peculiar to it, and to a certain extent different from protoplasm, and may be distinguished from it. On this account in all definitions of the nucleus, more importance should be attached to the properties of its structural components than is usually the case.

O. Hertwig, 1892

Hertwig was right. The contents of the nucleus are indeed different from the cytoplasm (which in his time was called protoplasm). Not only are they different in their organization, but they are packed into an extremely small volume, compared to the relatively dilute cytoplasm.

A diploid mammalian cell in G1 or G0 may contain, depending on its source, a spheroidal nucleus with a radius of ~5μ and a total internal volume of ~522μ³. In this space is packed ~2m of DNA weighing about 6 pg, ~12 pg, of protein, and ~6 pg of RNA. These three constituents will thus be present at a concentration of about 46 mg/ml. During interphase DNA is both replicated and transcribed, hnRNA is spliced, proteins and other molecules of various types enter and leave the nucleus, and different chromatin regions of the chromosomes assume various conformations at different times. How do these highly ordered events succeed in occurring within such a limited space filled with a material so concentrated that it is essentially in the form of a coagulum? In this chapter we attempt to bring together some of what is known about the structure and function of the nucleus as a prelude to an understanding of the expression of the chromosomal genes of the genome.

I. THE GROSS STRUCTURE OF NUCLEI

Figure 7.1 presents a schematic rendition of a typical nucleus for an animal cell. The **nuclear envelope** segregates the nuclear internal components from the cytoplasm. It is composed of a double membrane. The outer membrane is continuous with the **rough endoplasmic reticulum**, to which ribosomes are attached. The inner membrane is continuous with the outer membrane at the site of the **nuclear pores**. It is a smooth membrane that is associated with the **nuclear lamina**, a fibrous net-

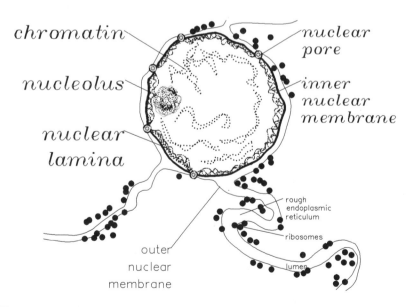

Fig. 7.1. *A cell nucleus showing constituent parts as seen at interphase.*

work composed of three lamin proteins, A, B, and C. The nuclear pores are not simply holes in the envelope. They have a complex structure constituted of a large number of different kinds of proteins that appear to be involved in the regulation of nucleocytoplasmic transport exchange.

II. THE NUCLEAR MATRIX AND ITS CONTENTS

When nuclei are treated sequentially with nonionic detergents, nucleases, and high-salt buffers under certain conditions, a preparation described as the **nuclear matrix** is obtained (reviewed by Nelson et al., 1986; Verheijen et al., 1988). Nuclear matrix preparations contain residual elements of the envelope such as the pore and lamina constituents, residues of nucleoli, and an **internal matrix** component, the structure of which is largely dependent on the type of isolation procedures used. If only DNase is used in its preparation, a chromatin-free preparation is obtained that consists of a network of proteinaceous filaments associated with 70% of the nuclear RNA, from which hnRNA and precursor ribosomal RNA can be isolated. This is called the internal matrix. The **scaffold-bound proteins** are the tightly bound proteins in the M-phase chromosomes after the dissociation of histones and other less tightly bound proteins. There are two major scaffold proteins: Sc1 and Sc2. Sc1 is topiosomerase II. It appears that hnRNA that has been formed by transcription associates with this matrix of ~10-nm core filaments to produce ribonucleoprotein (RNP) particles or granules about 20–25 nm in diameter. Many different proteins are associated with these RNPs, some of which may be nucleic acid-binding proteins. Also present are small RNAs that may play a role in the processing of hnRNA. Some of these are restricted to the nucleolus, while others are mainly located in the reticular network of the internal matrix between the nucleolar surface and the nuclear envelope (Spector, 1990). These RNAs are associated with proteins to form **snRNPs**. It has been suggested that transcription of hnRNA and its processing proceeds in some way under the guidance of these snRNPs, since the euchromatic DNA of the nucleus is distributed in areas rich in snRNPs (Spector, 1990).

The reader should bear in mind that, as in the case of studies on the structure of chromosomes in the mitotic phase (discussed in Chapter 1), the procedures used to study the interphase nucleus and its chromosomes are highly conducive to the formation of artifacts. The nucleus is an extremely highly organized entity with its ionic contents in a delicate balance. Subjecting it to the crude and rude manipulations of the human observer is necessarily going to give uncertain results that will lead to uncertain conclusions about its actual state *in vivo* (Cook, 1988).

Numerous enzymes involved in DNA and RNA metabolism are present in nuclear matrix fractions. These include, in part, DNA polymerases and primase, topoisomerases, RNA polymerases, and DNA methylases. The precise arrangement of these in the nucleus is not established. However, it is becoming clear that the RNA components of the nucleus are important elements in the organization of the factors that are involved in replication and transcription (Spector, 1990).

III. ORGANIZATION OF CHROMATIN IN THE INTERPHASE NUCLEUS

A. The Nuclear Matrix and Its Association With Chromatin

The different structures that chromatin assumes in the interphase chromosomes are largely unknown. The rigid scaffold of nonhistone protein to which DNA is attached to form loops (as described in Chapter 1) for mitotic chromosomes does not appear to be present in interphase chromosomes (Earnshaw et al., 1985). When nuclei are subjected to inhibitions of RNA synthesis, and the enzymatic digestion of RNA results are obtained that suggest that the maintenance of the normal structure of interphase chromosomes is dependent on active RNA synthesis (Nickerson et al., 1989).

The further observation that the nuclear matrix collapses with RNase A treatment simultaneously with the collapse of chromatin organization suggests that RNA may be a structural component of the nuclear matrix and that it is important for the maintenance of chromosome structure (Bouvier et al., 1982). Numerous investigators have in the past presented data that indicate that chromosomes are attached in the normal state to the nuclear membrane, nucleolus, and matrix (reviewed by Comings, 1980; Hancock and Hughes, 1982). More recent data indicate that RNA may be an important component of these complexes (Nickerson et al., 1989).

Although the metaphase scaffold of mitotic chromosomes, containing topoisomerase II as a component, does not appear to be present in interphase chromosomes, DNA loops are present just as in metaphase. These loops are approximately 30,000–100,000 bp long and are anchored to a nuclear matrix (Pienta and Coffey, 1984; Cockerill and Garrard, 1986). The points at which DNA is anchored to the matrix have been localized in mouse lymphocytes by Cockerill and Garrard (1986) and designated as **MAR** sites (nuclear matrix association regions). Topoisomerase II is associated with these sites in the κ immunoglobulin gene of the mouse. Whether these MARs are homologous to the SARs described in Chapter 1 is open to question. It has been reported that topoisomerase II is a major polypeptide component of the *Drosophila* salivary gland nuclear matrix fraction (Berrios et al.,

1985). It also appears to be present in all other cell types.

Thus although DNA loops are present in both mitotic and interphase chromosomes, the nature of the attachment to a scaffold may be different. The rigid scaffold of protein seen in mitotic chromosomes is replaced by a nuclear or matrix scaffold, which may also involve topoisomerase II but in a different and as yet unknown state. The matrix is diffuse and apparently more complex in structure than the scaffold of the metaphase chromosomes. A major part of it may be the peripheral lamina located against the nuclear membrane as shown in Figure 7.1 (Mirkovitch et al., 1987).

B. Arrangement of Interphase Chromosomes

The three-dimensional organization of chromosome parts in the nucleus has been a subject in interest for over a hundred years. The pioneer cell biologists Rabl (1885) and Boveri (1887) both presented evidence soon after the discovery of chromosomes that interphase chromosomes occupy specific domains within nuclei. The nonrandom localization of chromosome arms and centromeric regions with their associated heterochromatin has since been verified by numerous observers (Hilliker and Appels, 1989; Hadlaczky et al., 1986). Rabl's and Boveri's observations indicated that the positions of chromosomes at telophase are preserved at the subsequent interphase. These conclusions have been supported by more recent observers such as Cremer et al. (1982), who used a variety of techniques such as the following:

1. Interphase chromosomes can be induced to condense prematurely in G0, G1, S, and G2 nuclear phases by a variety of means such as incubation with inactivated Sendai virus, and anoxia (Foe and Alberts, 1985). This premature chromosomal condensation (PCC) makes the

chromosomes visible in the nucleus (Sperling and Ludtke, 1981).

2. The use of banding to identify interphase chromosomes has been useful under certain conditions (Burkholder, 1988).

3. *In situ* hybridization with specific DNA probes can locate specific centromeres, and euchromatic regions in arms (Manuelidis and Borden, 1988).

4. Immunofluorescence staining with CREST antikinetochore autoantibodies has been used to locate centromere locations (Hadlaczky et al., 1986).

5. Irradiation of cell nuclei with laser microbeams results in unscheduled DNA synthesis (UDS) in the irradiated parts of chromosomes. The UDS DNA is identified by pulse labeling with [³H]thymidine (Cremer et al., 1982).

6. Confocal scanning laser microscopy and imaging microscopy have been used to study the arrangement of chromosomes in nuclei along with computer processing methods (Oud et al., 1989; Rawlins and Shaw, 1988; Shotten, 1988; Puck et al., 1991).

By using these various methods, observers of interphase nuclei have been able to identify a number of organizational characteristics of chromatin during this period in the cell cycle.

1. Positioning of NORs

The NORs of chromosomes in most eukaryotes are frequently positioned so that the nucleoli derived from them are located at or near the nuclear envelope (Bourgeois et al., 1987). It has been suggested that its proximity to the envelope expedites the transport of rRNA into the cytoplasm as ribosomal particles.

2. Heterochromatin and centromeres

Constitutive heterochromatin is for the most part associated with the centromeric regions that, as noted previously, form clusters called **chromocenters**. These may involve all the chromosomes, as in the *Drosophila* polytene chromosomes, or just a few to form a number of centers as shown in Figure 7.2. Facultative heterochromatin, as found in the Barr body of one of the Xs in a female mammal, does not generally aggregate with constitutive heterochromatin.

The polarization of centromeres as the aggregation of centric heterochromatin in Rabl orientation has been described in both plants (Church and Moens, 1976) and animals (Hadlaczky et al., 1986). The nonrandom distribution of centromeric chromatin of interphase chromosomes may be a general phenomenon.

3. Disposition of telomeres

As noted in Chapters 2 and 3, the telomeric regions of eukaryotic chromosomes have specific repeat sequences that are apparently required for replication of their linear DNA without loss of significant genetic material. Like centromeres telomeres are important structural elements of chromosomes, and the disposition of both in the interphase appears to be nonrandom. The so-called Rabl orientation, with centromeres facing one end of the nucleus and the telomeres at the opposite, has been observed (Avivi and Feldman, 1980; Sperling and Ludtke, 1981; Bennett, 1982). However, this may not be prevalent in all cell types, and in those in which it does occur it may only be transitory. In some plants, such as the onion *Allium cepa* and *Crepis*, telomeres may associate end to end to form chains (Wagenaar, 1969). These associations are observed in telophase and early prophase and are inferred to be present in interphase.

4. The three-dimensional arrangement of whole chromosomes

Although the clustering of centromeric regions of chromosomes noted by Rabl has been identified by more modern tech-

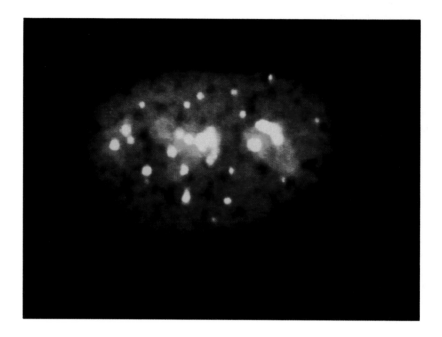

Fig. 7.2. *Aggregations of centromeric regions in human lymphocyte nuclei as identified by CREST immunoantibodies. Courtesy of M. Bartoldi, Los Alamos National Laboratory.*

niques (Fig. 7.2), the evidence for somatic pairing of whole homologues in somatic interphase has not been observed except in a few cases. It seems to be the rule in the Diptera but not in other animals. It apparently does occur in some plants (reviewed by Avivi and Feldman, 1980). Pairing between homologues in the prophase of meiosis is, of course, universal. It must be appreciated, however, that somatic pairing may indeed occur in some differentiated cells, since somatic crossing-over certainly requires some sort of pairing of homologous chromosomes in diploid cells. Homologous chromosomes in nuclei of rat-kangaroo and Indian muntjac cells do appear to occupy adjacent territories (Hadlaczky et al., 1986). Whether this is proof of actual pairing is open to question. Pairing of homologues may be under genetic control, and in some cases it may actually be suppressed (Avivi and Feldman, 1982a,b).

The spatial ordering of whole chromosomes, whether homologous or not, is a subject with an extensive literature that in some cases supports order in disposition of chromosomes in the interphase, and in other cases does not. Clearly a large part of the controversy surrounding this question is the result of the methods of preparing cells at the appropriate stage for observation. On the other hand, it must also be appreciated that there may be an order that is different for different stages of the interphase, and that the disposition of chromatin may be different in different tissues in multicellular eukaryotes.

Chromosome-specific DNA probes have been used extensively to identify the position of mammalian interphase chromosomes. Probe sequences that have been purged of high-frequency repetitive sequences, such as *Kpn* and *Alu*, and been hybridized to human lymphocyte chromosomes have clearly indicated that a territorial organization of these chromosomes exists (Lichter et al., 1988). Furthermore, it has been shown that different cell types have different patterns of chromosome spatial organization (Manuelidis and Borden, 1988).

Subjection of human central nervous tissue to hybridization with alphoid satellite sequence probes that are specific for certain chromosomes gave results showing that in large neurons of the cortex centromeric regions are clustered close to the nucleolus. This is the same configuration as that found in mouse Purkinje brain cells (Manuelidis, 1984). In other neuron classes in different parts of the brain, however, other types of aggregations of centromeric regions have been found. Use of probes specific for 9q12 and 1q12 showed that their homologues were not paired. One homologue generally had its paracentric region adjacent to the nucleolus; the other was on the nuclear membrane. On the other hand, the astrocytes, or glial cells, which are not nerve cells, had centromeres of both 1q12 and 9q12 homologues preferentially located at the nuclear membrane. The distance between homologues was variable: Sometimes they were close, sometimes at opposite sides of the nucleus. The probes that hybridized to the euchromatin in the arms distal to the centromeres showed that, unlike the centromeres, the arms are located in domains toward the center of the nucleus.

The polytene chromosomes of Diptera are considered to be in interphase state, since they are decondensed and are active in transcription. In *D. melanogaster*, chromosomes of this type are found in the larval salivary glands, in the prothoracic gland, and in the hind- and midgut. A study of the organization of these chromosomes in the four tissue types shows that each type has a characteristic arrangement different from the others (Hochstrasser and Sedat, 1987a,b). Specific parts of certain chromosomes are attached to the nuclear envelope. No attachment between the nucleolus and autosomes was noted.

5. Genome separation in plant polyploids and intergeneric hybrids

Allopolyploid plants generally show an interphase pattern in which chromosomes of each different genome are clustered together. For example, in the bread wheat *T. aestivum* the chromosomes of each of the three genomes: A, B, and D, cluster together in different regions of the interphase nucleus (Feldman and Avivi, 1973). Hybrids between species of *Secale* (rye) and *Hordeum* (barley) show genome separation in the hybrid metaphase plates. The *Hordeum* chromosomes tend to assemble at the center and the *Secale* at the periphery (Bennett, 1982; Schwarzacher-Robinson et al., 1987). A similar type of segregation has been noted in human–hamster cell hybrids (Manuelidis, 1985; Zelesco and Graves, 1988). However, the wheat–rye hybrid *Triticale* does not show a significant separation of interphase chromosomes (Cermeno and Lacadeno, 1983).

Pairing of homologues is absent in some plants. An extreme example of nonpairing is found in diploid barley, in which the two haploid sets appear to be separated as shown by the distribution of centromeres at metaphase (Bennett, 1982). Whether this arrangement is also present in interphase is not proved, but it is possible, since separation is apparent in early prophase.

IV. CHROMATIN ORGANIZATION AND ACTIVITY

There must exist a certain degree of spatially organized structure within the nucleus just as there is in the cytoplasm with its endoplasmic reticulum, mitochondria, plastids, lysosomes, and so forth. Complex processes such as transcription and DNA replication do not occur in a bag of enzymes in which Brownian movement is the only kinetic factor, but nuclear organization cannot be in a consistently invariant state; certainly it is in a dynamic state, changing as the nucleus proceeds through G0, G1, S, and G2 to the mitotic phase.

A. The Structures of Interphase Chromosomes

Figure 1.29 from Manuelidis and Chen (1990) illustrates one model for an interphase chromosome. It assumes that the 30-nm solenoid formed by the helical organization of the 10-nm nucleosome-containing fibers (shown in Fig. 1.21) is in turn folded to form a **radial array** constituted of radial subunits (Fig. 7.3). This is only one possible model based on current knowledge. Many of its assumptions are reasonable, but not all are in agreement with the findings of others.

B. Replication in the S Phase

Some of the details of replication of DNA *in vitro* have been presented in Chapter 3. Here we consider the matter of replication of whole chromosomes *in vivo*. Experiments with regenerating rat liver and exponentially growing fibroblasts in culture indiate that DNA polymerase is tightly bound to the nuclear matrix, and newly synthesized DNA remains tightly attached to the nuclear matrix (Smith and Berezney, 1980; Pardoll et al., 1980). That the growing points of replicating DNA are attached to the nuclear matrix is supported by kinetic studies, and the experimental evidence supports the anchoring of the DNA replication complexes, variously called **replisomes** or replicases, to the nuclear matrix (Huberman, 1987).

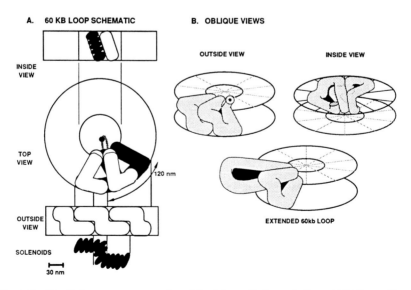

Fig. 7.3. *Radial arrays as postulated by Manuelidis and Chen (1990).* **a.** *Two radial subunits shown in three different views.* **b.** *Appearance of a 60-kb loop of solenoids from the outside and inside of the array shown in* **a.** *Reproduced from Manuelidis and Chen (1990), with permission of the publisher.*

DNA is replicated as it is reeled through these replisomes (Vogelstein et al., 1980). It may be that the nuclear matrix organizes DNA so that it has the topological properties of supercoiled covalently closed DNA. These supercoils are the loops, and they may function as replicons of 10–180 kb of DNA, with an average size of 63 kb (reviewed by Nelson et al., 1986). Thus an average mammalian cell should have about 3×16^6 kb/63 kb = 50,000 loops or replicons per genome.

C. Chromatin Structure and Transcription

1. SARs and loops

As described in Chapter 1 specific scaffold-associated regions (SARs) have been identified in metaphase chromosomes. These appear to be attached to a nonhistone protein core, one of whose constituents is topoisomerase II. The peripheral lamina of the nuclear matrix appears to be closely associated with the chroma-tin of the interphase nucleus and it has been suggested that these SARs attach to topoisomerase II within this part of the matrix (Lebkowski and Laemmli, 1982a,b). By use of lithium-3',5'-diiodosalicylate (LIS) at low concentrations in isotonic buffers, it has been found possible to remove histones and other proteins, leaving behind a residual matrix containing what appears to be DNA with specific scaffold associated regions (reviewed by Mirkovitch et al., 1987). The extracted nuclei are then treated with various restriction enzymes and the fragments containing SARs are separated from unattached loop material by centrifugation, since SARs cosediment with the residual insoluble matrix containing the scaffold to which they are presumably attached. Material from 18 *Drosophila* SARs situated near a variety of genes that are transcribed by RNA POLII has been mapped on chromosome 3. Part of the region containing about 320 bp is diagrammed in Figure 7.4. It contains a number of genes, among them rosy

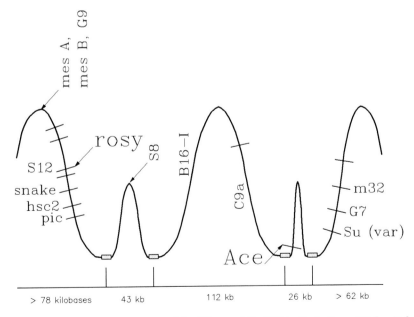

Fig. 7.4. *Loops and genes in 320 kb of Drosophila DNA. Data from Mirkovitch et al. (1987).*

(*ry*) and piccolo (*pic*), and it has four SARs. Note that the loop sizes are not all identical. *ry* has been located on the genetic map at 52.0 and *pic* at 52.1. (This puts them 0.1 cM or 60 kb apart, if lcM = ~6 × 10^5 bp for *Drosophila melanogaster* as shown in Table 8.2). Mirkovitch and associates conclude that SARs in *Drosophila* can be mapped to DNA fragments 0.6–1 kb long, and that these may contain multiple sites of scaffold interaction. These sites are in nontranscribed regions, the free loops between them being 4.5–112 kb long. Between two SARs one to several differentially regulated genes may exist along with their regulatory sequence.

All five histone genes in *D. melanogaster* are included in a 5-kb loop repeated tandemly about 100 times. The regions between the transcribed genes include the SARs and sequences with 70% homology to the *Drosophila* topoisomerase II conserved sequence (Sander and Hsieh, 1985; Gasser and Laemmli, 1986a). All SARs tested contain a large number of sequences related to the topoisomerase cleavage sites: (GTN(A/T)A(T/C) ATTNATNN(G/A). However, the clustering of topoisomerase II box sequences does not appear to follow a simple pattern.

Although SARs do not appear to be transcribed, some do have within them 5′ regulatory sequences upstream from transcription starts. Examples are the *Adh, Sqs-4*, and *ftz* genes of *melanogaster* (Gasser and Laemmli, 1986b). This may not be generally true for all eukaryotes and all genes. The immunoglobulin κ gene of the mouse has a SAR site adjacent to, but separate from, the enchancer sequence (Cockerill and Garrard, 1986).

SARs do appear to occur as families and may in fact play an important role in regulation beyond merely having enchancer regions within them. Although all SARs may be associated with TOPOII, they also have parts with different sequences that hybridize with different specific probes.

Thus a family of SARs is associated with nuclear scaffold regions of the heat shock protein genes (*hsp70*) of *melanogaster* (Mirkovitch et al., 1984). Other families are associated with the actin *5C* gene, *Adh, ftz*, and the histone genes. This apparent specificity of SAR regions leads to the conjecture that they are recognized by regulatory binding proteins. Some may also be bound by histones, and this would not be recognized after histones are removed by the extraction procedures.

Regions of chromatin defined as loops whose boundaries are fixed by SARs probably exist to sustain the operation of the genetic material in three dimensions. A matrix of loops maintained in the nucleus by SARs attached to the nuclear matrix would be conducive to interactions between various parts of the genome both extra- and intrachromosomally. The loops may be held fairly rigidly in place at both top and bottom rather than just at the bottom. A pattern such as this would make possible the rapid interactions between binding proteins and their binding sites, for example. We tend to think of gene activity in two dimensions along a string of DNA, but under these conditions reaction rates would be slow, particularly if (as is often the case) the transacting factors arise from regions of the genome far from their sites of binding. A three-dimensional organization would make rapid kinetic activities possible. If this is the case, then the SARs could be extremely important, for they basically determine the organization of the **loop matrix** and hence gene activity. One can also imagine that repetitive sequences with no obvious genetic activity may play an extremely important role in the formation of these three-dimensional matrices.

Formation of loop matrices as pointed out here depends on the capacity of DNA to bend (Travers and Klug, 1987b). The positioning of at least a subset of nucleosomes in the 10-nm fiber is apparently not random, but sequence-specific. In addition other types of binding not involving nucleosomes *per se* are possible. These may involve binding proteins, which recognize certain DNA sequences and are involved in the bending. Among these may be topoisomerases I and II, which have the capacity to adjust for the strains that double-helix DNA is subject to when it is bent (Travers and Klug, 1987a).

2. RNA polymerases

If indeed transcription does occur within the regions between the SARs, how does it occur? Two basic mechanisms have been advanced for the process. The generally accepted one is that the RNA polymerases are mobile and proceed along the genes during transcription. An alternative to this assumes that the polymerases are fixed and immobilized in the nuclear matrix, and transcription proceeds by the DNA reeling through the polymerase complex (Jackson and Cook, 1985). The experimental evidence supports the former model. RNA POLI and II are not nuclear matrix proteins and move along DNA loops during transcription (Roberge et al., 1988). Thus the nuclear matrix scaffold may have less to do with organizing RNA polymerases in space than with organizing the DNA so that it can be transcribed by them.

3. Topoisomerase I

Although topoisomerase II appears to be located primarily within SARs and MARs, topoisomerase I is associated with the transcriptionally active region of the genome (Fleischmann et al., 1984). Along with TOPOII, TOPOI is one of the major nonhistone chromosomal proteins. Antiserum to TOPOI has detected TOPOI at regions of *Drosophila* polytene chromosomes that contain transcriptionally active genes. Transcription of the heat shock gene, *hsp70*, of *melanogaster* is accomplished by cross-linking to TOPOI. Only the transcribed regions of the gene are in-

volved. Since both positively and negatively supercoiled DNA is relaxed by TOPOI, it can be hypothesized that it plays an essential role as a swivel factor in the alleviation of topological constraints occurring during transcription, as described previously in Chapter 6.

4. Nucleosomes

The role of nucleosomes in the transcriptional process is a matter of considerable interest. Transcription of "free" DNA can be carried out *in vitro* and apparently occurs *in vivo*, but what about DNA "bound" in nucleosomes? Like replication, transcription of DNA requires separation of the DNA strands. If nucleosomal DNA is transcribable, then strand separation must occur, and the polymerase must be able to access the nucleosomal DNA and follow the template around the nucleosome. Certain physical constraints enter, however. RNA polymerases are huge, being about 500 kD, compared with the ~262 kD of the nucleosome, and they bind tightly to a region of the template about 50 bp long. It seems likely that this makes it impossible for transcription to occur on the surface of nucleosomes without some alteration of the relationship between the DNA and the histone core. Nucleosome structure may be transiently altered so as to accommodate transcription by the polymerase. Thus, nucleosomes may be present during transcription of an active gene.

The Balbiani ring genes in the salivary glands of the Diptera *Chironomus tentans* are about 37 kD long in their actively transcribed regions. According to Bjorkroth et al. (1988) the actively transcribed genes appear to be in three configurations: as 5-nm, 10-nm, or 30-nm fibers. The most abundant form is approximately 5 nm in diameter, which probably indicates that nucleosomes are partially unfolded in the transcribed regions. Nucleosomes are present in genes being actively transcribed, but the chromatid of the active

Balbiani ring genes is apparently in a dynamic state and can be extended and be packed and then unpacked in a nucleosome and a solenoid.

D. Gene Activity and Chromosome Disposition in Interphase

Is transcriptional activity related to the arrangement of chromatin in the nucleus? If the nonrandom arrangement of chromosomes in the nucleus is indeed a fact, does this mean that it has a functional significance with respect to gene activity? Is the position of a gene within the genome important to its functioning, and does the way in which a gene is packaged in its chromosome determine its activity? It would appear so, since if the enhancer promoter regions of the gene are so blocked in its packing to prevent binding of *trans*-acting regulatory proteins, it would be inactive. Therefore differentiation of a cell line would then be a consequence of the way in which its chromatin is packaged or arranged in the nuclear matrix. Furthermore, transcription by the polymerases would be blocked if the transcribable regions were inaccessible. The relevance of these kinds of questions becomes more and more apparent as we learn more about the structure and function of the nucleus and its matrix.

Aside from a gene's position on its chromosome, there is the further question of whether there is a relationship in the activities of the chromosomes within a single nucleus. In a diploid nucleus many possibilities of what has been called "chromosome cross talk" (Lewin, 1981) exist. Pairing of homologues such as occurs in the Diptera is one possible example of cross talk. Another is the possibility of a three-dimensional relationship between homologous chromosomes if the domains of each chromosome are coordinated with other chromosomal domains.

V. THE NUCLEAR PORES AND THEIR FUNCTION IN GENE EXPRESSION

The **nuclear pores** are complex structures that, along with the nuclear lamina, play an important role in the control of molecular traffic into and out of the nucleus (Feldherr et al., 1984). Blobel (1985) has proposed that active domains of the chromatin-containing transcribing gene are attached to the pore complexes. This he calls **gene gating**. The pore complex (Fig. 7.5) contains three major constituents, the central plug (C), the spokes (S) and the rings (R). The spokes are attached to the rings. The complex is imbedded in the nuclear envelope (NE), and its maximum radius is ~60 nm. The nuclear lamina is interposed between the NE inner membrane and the chromatin except for those regions occupied by the nuclear pores. Parts of the chromatin can be seen to be attached to the lamina. Blobel has hypothesized that the inactive chromatin surrounding the transcribing parts of the chromosome is anchored to the lamina holding the active parts in place at the pore where it can communicate and exchange with the cytoplasm. The outer membrane of the NE is continuous with the endoplasmic reticulum, and the association of the active chromatin with the pore opening would place its products, that is, mRNAs, in close proximity to the reticulum where translation occurs. Blobel has also posited that the arrangement of pores is attuned to the disposition of the chromatin so that the transcribing parts have access to the pore exit.

VI. CHROMOSOME REARRANGEMENTS AND GENE ACTIVITY

As noted in Chapter 5, the transposition of segments of DNA categorized as transposons can alter phenotypes, sometimes

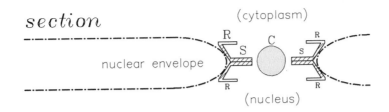

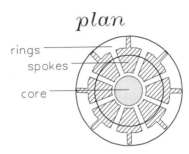

Fig. 7.5. *Diagram of nuclear pore complex. C, central plug; S, spokes; R, rings.*

drastically. In this section we deal with the effect of moving genes about in the genome by inversions and translocations. The question being addressed is: Does a gene's activity depend upon it position in the genome? This question is germane to our inquiry into the events occurring within the nucleus, because it is in the interphase that genes are active.

A. Variegating Position Effects

1. In Drosophila

Position effect variegation (V-type position effect) is a phenomenon first discovered in *D. melanogaster* by Muller (1930). Later it was shown by Schultz (1936) that genes normally present in the euchromatic regions of the chromosomes, when moved by inversion or translocation into the regions of centric heterochromatin, lose their activity in at least some of the cells in which they are active prior to being moved (reviewed by Lewis, 1950;

Baker, 1968; Spofford, 1976). A classic example involves the white (*w*) gene on the X chromosome of *melanogaster* (reviewed by Tartof et al., 1989). When the normal allele, w^+ is placed next to the centromeric heterochromatin of the X by inversion [$In(1)w^{m4}$], the eyes of the fly have pigment in some of the ommatidia but not in others; hence they have a variegated pattern of red on white. The variegated phenotype is not the result of a mutation *per se*, because reversion to the wild-type phenotype occurs when the gene is moved back away from the heterochromatin (Judd, 1955). *Drosophila* heterochromatin is of three types: α, β, and intercalary. α-Heterochromatin consists of highly repetitive sequences such as satellites located adjacent to the centromeres. These regions contain few genes, are relatively under-replicated in polytene chromosomes, and are highly compacted. β-Heterochromatin is less compact and contains moderately repetive sequences at a higher concentration than found in the euchromatin (Miklos et al., 1988). Intercalary heterochromatin is dispersed in patches in the euchromatic arms. Of these three types, only α-heterochromatin induces variegated position effects. However, it should be noted that genes such as light (*lt*) and cubitus interruptus (*ci*), both of which are normally near or in centric heterochromatin (*lt* on 2 and *ci* on 4), appear to variegate when moved into euchromatin (Spofford, 1976). Intercalary heterochromatin also is found interspersed in the euchromatic long arms of *Drosophila* chromosomes; when euchromatin is moved to positions adjacent to it, the time of replication is delayed, and gene inactivation occurs in the region (Zhimulev et al., 1989).

This *cis*-inactivation effect may include more than one gene. The activities of as many as 60 genes in the euchromatic segment placed in juxtaposition with the heterochromatin may be affected (Hartmann-Goldstein, 1967). The closer a gene in the segment lies to the heterochromatic point of attachment, the more pronounced is the variegated effect on its activity. In addition, a cytological effect is noted in the polytene chromosomes; the areas in which the variegating genes are located lose their normal banding pattern (Cole and Sutton, 1941; Henikoff, 1979). Since the loss of the banding pattern does not appear to be the result of gene deletion or underreplication of DNA in the polytene chromosome, one is led to the conclusion that the physical association of euchromatin with heterochromatin may alter the structure of the euchromatin, and perhaps vice versa. Condensation of the chromatin of active genes in the euchromatin can lead to a reduction or total loss of their transcriptional activity. Rearrangements involving the rosy (*ry*) gene of *Drosophila* support this (Rushlow and Chovnick, 1984; Rushlow et al., 1984). Rosy encodes xan-

thine dehydrogenase (XDH), an enzyme involved in the synthesis of eye color pigments. It has been shown that the position effect does not alter the quality of the encoded product but does cause a reduction in its transcription. This reduction is not the result of underreplication, but possibly of condensation that results in inactivation.

Further support for this conclusion comes from the observation that variegation in *Drosophila* is temperature-sensitive; low temperatures ($\sim 17°$) enhance it and high temperatures ($\sim 29°$) suppress it. Cytologically detectable condensation at the *Sgs-4* locus is enhanced at $17°$ and reduced at $29°$. *Sgs-4* loci appear to be one third less polytenized in the translocation $T(1;4)^{wm258-21}$ (Kornher and Kauffman, 1986).

The loci of the breakpoints of three variegating white gene mutants: w^{m4}, w^{mMc}, and w^{m5lb} were identified by Tartof et al. (1984) and shown to be between w^+ and the telomere. The euchromatic–heterochromatic junctions formed were identified and cloned in λ vectors. The euchromatic breakpoints for the three are clustered and lie about 25 kb distal to the locus of w^+. The α-heterochromatic DNA to which they are joined in the inversion has the properties of mobile elements that lie well within the heterochromatic domain of the X. It is logical to assume that a potentially transcribable gene may be subjected to inactivation by a DNA condensation process extending from adjacent heterochromatin. The decondensation of euchromatin itself may be related to its having $(CA/GT)_n$ sequence repeats. These are not present in heterochromatin, and their absence may have a role in the compacting of the euchromatic chromatin injected into the heterochromatin.

It must also be recognized that moving a gene to a new position may induce a variegated position effect in the absence of heterochromatin. Transposition of the white gene in *Drosophila* by *P*-element-mediated DNA transformation enables one to inject the gene into many different chromosomal locations. Of the 20 transformations induced by *P*, 17 had no apparent effect on the gene expression, but 3 did and showed variegated position effect. Mutant *A4-3* had the white gene located near the heterochromatin of 2L, mutant *A4-4* had the gene at the end of the right arm of 3 (Levis et al., 1985), and *P(w^{vas})* had the gene at the left end of chromosome 2 (Gehring et al., 1984). The *A4-3* transformant had the gene inserted at the 24CD site of the polytene chromosome. There is no evidence for intercalary heterochromatin at this site. The effect in this case could be the result of moving the gene next to a transposable element of the type found in maize.

The expression of variegating genes can be modified by dominant suppressor and enhancer mutant genes elsewhere in the *Drosophila* genome. These include genes that suppress position effect variegation (Sinclair et al., 1983) and those that enhance it (Locke et al., 1988). It has been suggested that these genes may have a role in heterochromatin formation. And, in fact, a gene at or near a suppressor of variegation encodes a nonhistone protein that is associated with the heterochromatic regions of the genome (James and Elgin, 1986). Histones may also be involved. The dominant suppressor, *Suvar(2)I^{01}* has been found to enhance acetylation of histone H4 (Dorn et al., 1986), and the amount of variegating gene expression is a function of histone gene multiplicity (Moore et al., 1979). Suppressors may be either cytologically visible duplications or deficiencies and the same appears to be the case for dominant enhancers (Locke et al., 1988). The use of *P*-element mutagenesis made it possible to induce 12 dominant enhancers of variegation, and

these have been mapped to four loci on the second and third chromosomes.

Modifiers have been categorized into two classes by Locke et al. (1988); Class I modifiers enhance variegation when duplicated and suppress it when deficient, and Class II enhance when deficient and suppress when duplicated. Gene dosage thus appears to be involved. The suggestion is made that Class I modifiers code for structural proteins that influence the assembly of heterochromatin, while Class II genes by their action inhibit heterochromatin formation or enhance euchromatin formation. This hypothesis is supportive of the idea that gene activity is controlled in part by compacting and decompaction as discussed in the previous sections of this chapter.

In addition to these suppressor elements on the autosomes, the Y chromosome is also involved in suppression. The addition of a Y suppresses variegation toward wild type and its deletion enhances (Lindsley et al., 1960). Apparently these are discrete genetic units suppressing variegation on the *Drosophila* Y (Brosseau, 1964).

While a great deal of the research into variegated position effects in *Drosophila* has involved the white gene, many other studies of variegation have involved a wide variety of genes, some on chromosomes other than the X, and some involving translocation between the X and autosomes. For example, at least 72 genes normally present in the euchromatin have a variegated expression when placed next to the centric heterochromatin.

The gene *pgd* encoding 6-phosphogluconate dehydrogenase (PGD) of *Drosophila* is X-linked and its expression in flies with various types of chromosomal aberations has been monitored by Slobodyanyuk and Serov (1983, 1987). The *pgd* gene has two alleles, *pgd^a* and *pgd^b*, the protein products of which can be distinguished electrophoretically in pgd^a/pgd^b individuals: AA, AB, and BB. This makes it possible to estimate the relative amounts of PGD encoded by pgd^a and pgd^b in heterozygotes, since BB, AB, and AA products are separated on electrophoresis. Enzyme activity was determined in pgd^a/pgd^b adults and third-instar larvae with 21 different rearrangements. Ten rearrangements showed a suppression of activity of the pgd^a allele in comparison with the activity of pgd^b, while three showed an enhancement, and six showed essentially no difference from the wild-type ratio. Two translocations, one involving the X and chromosome 4 and the other the X and chromosome 2 such that pgd was juxtaposed with the heterochromatin of these autosomes, showed definite depression of the activity of the pgd^a allele in both imagos and larvae as a percentage of the pgd^b activity in heterozygotes. Furthermore, the effects were greater in the adult flies than in the larvae. From these findings one can conclude that variegation may be induced in a sex-linked gene by heterochromatin in other chromosomes, and that its degree may be determined by the stages of cell differentiation and the allelic state of the gene.

2. In other organisms

The α-heterochromatin of *Drosophila* is constitutive. This is in contrast to the facultative heterochromatin of the Barr bodies of XX female mammals. X-autosomal translocations in the mouse may cause variegation in the adjoining autosomal loci (Gartler and Riggs, 1983). A Barr body must be present for variegation to be expressed. XY males and XO females carrying the translocations are wild-type, but XXY males show variegations. Translocations between autosomes have been observed to cause variegation in the mouse (reviewed by Baker, 1968). It would appear from this that heterochromatin whether constitutive or facultative is effective in causing variegation.

Balanced heterozygous translocations in mice have been reported to have definite phenotypic effects. In a review of the literature dealing with radiation-induced skeletal mutants in mice Selby (1979) found that 3 of 37 mouse lines with dominant skeletal mutations were heterozygous for dominant reciprocal translocations. It has also been reported that a semisterile male translocation heterozygote exhibited neurological defects (Rutledge et al., 1986).

A position effect similar to the variegated effect in *Drosophila*, accompanied by a rearrangement involving chromosomes 3.4 and 11.12, has been noted in the evening primrose *Oenothera blandina* by Catcheside (1947). Two genes, *P* and *S*, showed variegation when placed near the site of the rearrangement. When moved back out of the rearrangement, they became stable again in their activity.

B. Other Types of Mutant Effects

1. Lethal effects

One of the more interesting examples of the effects of inversion is to be found in a prokaryote, *Salmonella typhimurium*. It and *E. coli* are closely related in the enterobacteria group. They can be crossed and the order of genes on their genophores is similar despite the fact that they may have diverged from a common ancestor some 150 mya (Sankoff et al., 1990). Inversions have been induced in the *Salmonella* genophore and found to be of two types, permissive and nonpermissive (Segall et al., 1988). Apparently there are limits to viable inversion formation at certain regions of the genome. One can infer from this that a selective advantage exists in maintaining the order of genes found in the present-day strains of these bacteria and it is the reason for the conserved linkage between them.

Nonpermissive inversions and translocations also exist for *D. melanogaster*. For example, many inversions on the X are male lethal even though no deficiencies or duplications are noted in the polytene chromosomes. This does not mean they do not exist, of course, and have a lethal effect in the X-hemizygous males in at least some cases. Also the use of ionizing radiations and other mutagens can be expected to sometimes cause lethal damage to vital genes at the site of breaks or even within the section of chromosome to be inverted or translocated. Therefore conclusions about what is permissible with respect to rearrangements and what is not must be tempered by consideration of this fact.

Inversions induced in the X chromosome of *melanogaster* under the influence of the high-mutation-rate gene, *hi*, nearly always had one breakpoint in the heterochromatin (Hinton et al., 1952). Of 12 inversions, 7 were male lethal, and the remaining 5 produced sterile males with an extremely low survival rate. Among the latter was an inversion that had both breakpoints well distal to the centric heterochromatin.

Lindsley et al. (1960) made an extensive genetic and cytological study of 36 mutants bearing X-chromosome inversions and X-autosome translocations, and found that all resulted in very low relative viability or no viability of XO males. Interestingly the relative viability of XY males was better. Only four XY males had 0–0.02 viability, while 12 had essentially normal viability. These findings are in agreement with the previously noted effects of the Y on expression of the V-type position effect (p. 337). Since all of the proximal breakpoints of the inversions and the autosomal breakpoints of the translocations were in the heterochromatin, it may be considered that V-type position effect is involved in these effects. However, the data also definitely indicate that some

inversions and translocations involving the X are nonpermissive in the absence of the heterochromatin of the Y.

These nonpermissive rearrangements involve autosomes as well as the X. Inversions within chromosome 2 and chromosome 3 are frequently lethal when homozygous (Lindsley and Grell, 1968). The same is true for T (2:3) translocations. Some are viable and phenotypically normal, but other viable ones may show male or female sterility and V-type position effects. A study of 160 T(2:3) translocations in *melanogaster* has been made by Hilliker and Trusis-Coulter (1987). They determined the effects on viability of these translocations in homozygotes and found about 60% to be lethal and about 9% to be semilethal. Only 31% of the translocations were viable when homozygous. The breakpoints for both lethal and viable translocations were found to be distributed throughout the two autosomes (Fig. 7.6). A number of inversions were also tested for viability and some were found to be semilethal or completely lethal. Other independent investigators have found homozygous lethal translocations at approximately the same frequency. Breakpoints in the heterochromatin of chromosomes 2 and 3 are about 64% lethal and semilethal. Since heterochromatin is mostly repetitive DNA and has few genes, one is led to the conclusion that most or all of this lethal effect is probably not the result of damage to vital genes and that the organization of DNA in heterochromatin is important in some way presently not understood.

2. Viable nonlethal effects

The amplification of the *CAD* gene in Chinese hamster ovary (CHO) cells at various sites within the genome has been studied by introducing cloned copies of this gene from Syrian hamster into the chromosomes of *CAD⁻* CHO cells. Transformants with the donated *CAD* gene were

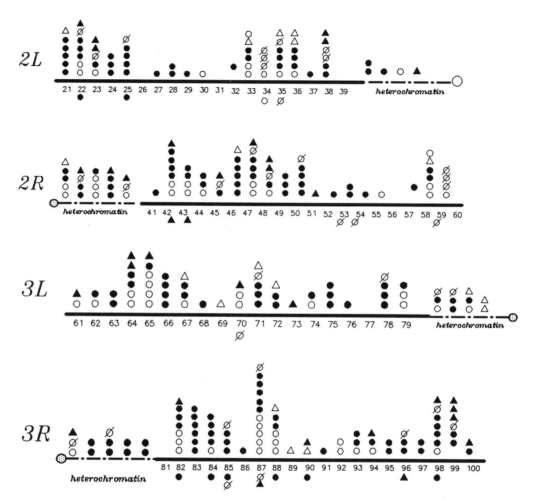

Fig. 7.6. *Breakpoints for translocations and some inversions involving chromosome 2 and 3 of Drosophila melanogaster. Filled circles indicate homozygous lethals, and open circles homozygous viables. Open circles with diagonal slash denote semilethals. Triangles represent stocks that were lost but were either homozygous lethal (filled) or viable (open). After Hillicker and Trusis-Coulter (1987).*

identified by their ability to grow in the absence of uridine (Wahl et al., 1984). The transformants were sensitive to N-phosphonacetyl-L-aspartate (PALA), an inhibitor of the aspartate transcarbamylase component of the CAD multienzyme complex. In the presence of PALA some of the transformed cell lines became resistant and grew in the absence of uridine while others did not. It was found that the position of the gene in the CHO genome was the important factor in the induction of amplification by PALA. In one cell line amplification was 100–fold more than the others; in another line the amplification was minimal and accompanied by chromosomal rearrangements. Of most interest was the finding that the line with most amplification activity had the transduced *CAD* gene in the middle of a chromosome arm in the euchromatin, while the gene in the line with low activity and chromosome instability was situated near the centromere.

The replicons in a eukaryotic nucleus are not activated all at the same time. As pointed out in Chapter 3, some genes replicate late and other early in the S phase. The timing appears to be important in development. A controlling factor for this timing appears to be the position of the gene in the genome. This has been shown to be true for an α-globin gene and the immunoglobulin genes of the mouse (Calza et al., 1984).

When human β-globin genes are introduced into a mouse egg, the resultant transgenate may express the human genes specifically in its erythroid cells without regard to the position of the transduced gene in its genome (Chada et al., 1985, 1986; Grosveld et al., 1987). One problem with drawing conclusions about position effects from these results is that we have no way of knowing how many loci occupied by the human gene in the transgenate are not active. Many may be, but this has not been proven. Rabbit β-globin genes have been transduced into mice, and the loci checked by *in situ* hybridization. In some transgenates no rabbit β-globin was formed; in others rabbit β-globin transcripts were found, but in inappropriate tissues such as skeletal muscle in one case and testis in another (Lacy et al., 1983). Chicken transferrin genes in transgenic mice were active. However, in some cases expression was two to four times greater than for the mouse gene (McKnight et al., 1983).

The methylation of cytosines may be involved in regulation, as discussed in Chapter 6. It may also play a role in position effects. For example, the chromosomal position of a gene affects its methylation status in sperm (Jaenisch and Jahner, 1984). A plasmid, pMK, has the promoter-regulatory regions of the mouse metallotheionein-1 (*MT-1*) gene fused to the thymidine kinase (*TK*) gene of herpes simplex virus. This artificial gene, designated *MK*, is regulated by cadmium (Palmiter et al., 1982). The gene is differentially regulated depending on its locus. Methylation is apparently involved.

C. Oncogenesis and Rearrangements

A close correlation exists between chromosomal aberrations and the development of neoplasms (reviewed by Rowley, 1980; German, 1983; Pathak, 1986; Croce, 1987; Rabbitts et al., 1988). Extensive karyotypic studies of cells from persons with leukemias, lymphomas, and solid tumors show the involvement of either translocations or cytologically demonstrable deletions. Most, but not all, neoplasms have cells with consistently similar types of abnormalities; the association of a particular aberration with a particular type of cancer is nonrandom (Yunis, 1983). Furthermore, in many neoplasms there is a correlation between a specific type of rearrangement and an oncogene. These findings all indicate that the development of many, if not most, neoplasms in humans and other mammals are the result of somatic position effects. Table 7.1 lists examples.

Burkitt's lymphoma is a relatively common cancer in certain areas of Africa. One hundred percent of the cases that have been studied show that the *c-myc* proto-oncogene is involved in reciprocal translocations between chromosome 8, on which the *myc* gene is located at q24, and one of chromosomes 14, 22, or 2 (Croce et al., 1985). In all three types of translocations one break is in the region of 8q24, while in the case of chromosomes 22 and 2 the second breaks appear to be in the centric heterochromatic regions. Eighty percent of persons with the lymphoma have the

TABLE 7.1. **Human Neoplasms Associated With Chromosomal Aberrations**

Chromosomal aberrations	Neoplastic condition	Genes involved in breaks	Reference[a]
t[14;8][q11,q24]	T-cell lymphoid tumor	TCRδ/α at 14q11 c-MYC at 8q24	1, 6
inv14[q11;q32]	T-cell lymphoid tumor	TCRδ/α at 14q11.1 IGH at q32.3	1, 5
t[14;18][q32;q21]	B-cell lymphoma	IGH at 14q32 BCL2 at 18q21.3	1
t[8;14][q24.1,q32.3]	Burkitt's lymphoma	c-MYC at 8q24.1 IGH at 14q11	2
t[8;2][q24.1;2p12]	Burkitt's lymphoma	c-MYC at 8q24.1 IGK at 2p12	2
t[8;22][q24.1;q11]	Burkitt's lymphoma	c-MYC at 8q24.1	2
t[9;22][q34;q11] (Philadelphia)	Chronic myelogenous leukemia	c-ABL at 9q34 BCR at 22q11	3
del18q21	Colorectol cancer	DCC	4
inv16[p13.3;q22]	Myelocytic leukemia	HBA at 16p13.3 FRA at 16q21.1	

[a]Figures indicate the following sources:
1. Rabbitts et al., (1988)
2. Croce (1987)
3. Dobrovic et al. (1991)
4. Fearon et al., (1990)
5. Denny et al., (1986)
6. Haluska et al., (1986)

second break at 14q32, which is near the distal tip of 14 and not involved with centric heterochromatin. However, in all three types of translocations the second break is in the region occupied by immunoglobulin genes. Thus, putting *myc* next to one or another of the many immunoglobulin genes appears to result in lymphoma. This type of position effect, involving an association of an oncogene with a specific region of a homeologous chromosome, is quite common. Another example is found in 95% of patients with chronic myelogenous leukemia. They have translocations of the type t(9;22)(q34;q11), in which the gene *c-abl*, normally at 9q34, is relocated proximate to the heterochromatic region of 22. Again an oncogene is placed in the region of immunoglobulin genes.

Inversions may also be involved in neoplastic conditions. A pericentric inversion in human chromosome 16; inv(16p13.2; 16q22), is associated with the development of a leukemia, and another involves chromosome 14: inv(14q11.2; q32.3), a paracentric inversion that places *IGH* genes proximal to the centric region in the region of TCR genes (Denny et al., 1986).

Persons born with translocations such as Robertsonian fusion types may have a predisposition to the development of cancer. This is especially true of persons with t(13q;14) fusions, in which the breakpoint in 14 is in the region of the T cell, a receptor locus. Also a congenital tandem translocation between 13 and 14 with a breakpoint in the region of the immunoglobulin locus may predispose to B-cell malignancies (Pathak, 1989). In general persons with congenital translocations or inversions should be monitored for early tumor detection.

It has been demonstrated that Chinese hamster cells in culture show specific alterations in phenotype and karyotype as they progress from the normal to the neoplastic condition (Kraemer et al., 1983; Cram et al., 1983). This mimics what occurs in human lymphocytes with the onset of leukemias. Aberrations with breakpoints at particular sites may predispose to neoplasms whether they are introduced from the germ line or occur spontaneously in the soma.

That chromosome mutations involving oncogenes may be associated with and apparently cause the development of neoplasms is supported by many examples in addition to those listed in Table 7.1. It is also now established that there are tumor suppressor genes whose loss or inactivation results in the development of tumors (reviewed by Sager, 1989). The wild-type alleles of these genes play regulatory roles in cell growth as apparently do the proto-oncogenes. Their respective roles are apparently antipodal. Proto-oncogenes act positively to stimulate, whereas tumor suppressor genes act negatively to prevent excessive activity of oncogenes. In the normal, nonneoplastic state, cells are apparently maintained in growth rate homeostasis by the interaction of the positive and negative activities of these types of genes.

D. Transvection

Transvection is a term coined by Lewis (1954) to describe a type of position effect that occurs in certain combinations of mutant alleles in *D. melanogaster* (reviewed by Judd, 1988). It is a phenomenon that is exhibited when two different alleles in heterozygous combination are also involved with a heterozygous rearrangement so that allelic somatic pairing is disrupted. It was originally discovered when it was found that the *cis–trans* phenotypic effects of some of the heterozygous combinations of *BX-C* genes (or alleles), as noted in Table 9.12, could be greatly enhanced by incorporating one of them in a rearrangement. For example, the heterozygote $bx^{34c}+/+Ubx'$ has the third thoracic body segment modified toward second thoracic segment morphology (AT3-AT2). This mutant transformation is enhanced considerably when one of the alleles is involved in a third chromosome rearrangement with a break between the centromere and a region just proximal to *BX-C*. Thus $R(bx^{34e}+/+Ubx)$ has a more mutant phenotype than $bx^{34e}+/+Ubx'$ (where R = either an inversion or translocation involving bx^{34e}. The fact that flies homozygous for the same rearrangement in both chromosomes $(R(bx^{34e}+)/R+Ubx)$ has the phenotypic of $bx^{34e}+/+Ubx$ leads logically to the conclusion that the disruption of somatic pairing in the R heterozygote enhances the mutant effect. As Judd (1988) has put it, "allelic cross talk" is disturbed when certain alleles of gene complexes are separated spatially by heterozygous rearrangements that are expected to interrupt their pairing in somatic cells.

Transvection has been described for other combinations of *BX-C* genes (Mathog, 1990) as well as for interactions between the sex-linked zeste (z) and white (w) genes. (reviewed by Wu and Goldberg, 1989). Zeste encodes a DNA-binding protein (Biggin et al., 1988). One of its mutant alleles, z^1, when homozygous, produces eyes with yellow pigment in otherwise w^+/w^+ female flies. In $z^1w^+/z^1Df(1)w$ flies in which the $w/^+$ gene is deleted on the X but inserted on an autosome the eyes are phenotypically normal red. (Fig. 7.7). Thus a transvection-type position effect results, since the two alleles should not be able to pair, situated as they are on homeologous chromosomes. When paired as shown in Figure 7.7c, the phenotype is as in Figure 7.7a. The repression of w^+ function by z^1/z^1 is effective only when $w+$ alleles

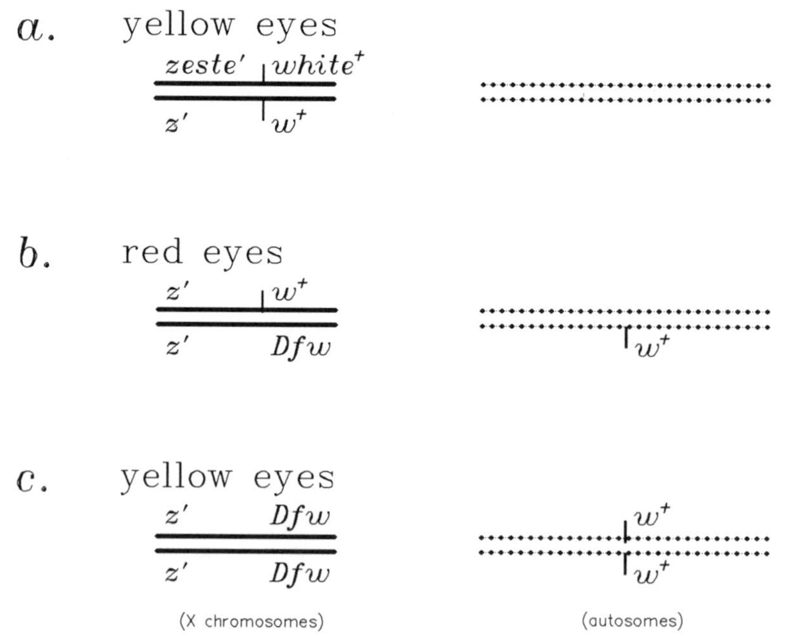

$a.$ yellow eyes

$$\frac{zeste' \mid white^+}{z' \quad \top w^+}$$

$b.$ red eyes

$$\frac{z' \quad \mid w^+}{z' \quad Dfw}$$

$c.$ yellow eyes

$$\frac{z' \quad Dfw}{z' \quad Dfw}$$

(X chromosomes)   (autosomes)

Fig. 7.7. *Interactions of zeste and white genes of Drosophila melanogaster.*

are paired, (as in Fig 7.7a and c), not when they are separated (Jack and Judd, 1979).

Zeste also interacts with alleles of the decapentaplegic (*dpp*) gene and the *BX-C*. The w^+, dpp^+, and $BX-C^+$ genes have no obvious functional relationship in common. When, then, is their common relation with z^+? The zeste gene has been cloned, and its protein product identified and characterized. Its mRNA transcript translates a polypeptide about 570 amino acids in length (Pirrotta et al., 1987; Mansukhani et al., 1988). The DNA-binding region is about 200 amino acids long at the amino end. The RNA transcript is found at all developmental stages of the fly from egg to imago. Since antibodies specific to zeste bind to about 60 bands on the polytene chromosomes (Pirrotta et al., 1988), zeste may, by action of its binding protein

product, be an important and perhaps essential regulator. However, flies with the zeste gene deleted are viable.

Transvection at present is known only in *Drosophila*. Whether it exists in forms other than those with somatic pairing of homologues as found in the Diptera is open to question. The fact that it does occur in species with somatic pairing does support the concept that the three-dimensional organization of chromatin in interphase is important. The allelic cross talk exhibited in *Drosophila* may be but one aspect of chromosomal cross talk that is a possibility in eukaryotes in general, particularly in those with dominant diploidy. We have yet to explain why the diploid condition is practically obligate in animals and to a considerable extent necessary in the vascular plants with their diplohaplontic life cycles.

VII. GENERAL CONCLUSIONS ABOUT THE ACTIVITY OF GENES AND THEIR POSITIONS IN THEIR GENOMES

The importance of a gene's position on its chromosome or on chromosomes other than its original locus cannot be ignored. Does proper activity depend on position?

A. Heterochromatin, Euchromatin, and Position Effects

The partitioning of the chromatin of higher eukaryotes into euchromatin and heterochromatin appears to be fundamental property of their chromosomes, but it is far from being understood. It is the centric chromatin that seems to be of most importance, and the fact that it can vary greatly in amount even among closely related species without gross phenotypic effects makes the mystery even deeper. Furthermore, one cannot ignore the presence and possible importance of intercalary heterochromatin. It seems that every chromosome of a genome must have a sector that remains highly compacted throughout the cell cycle; if not, at least one of the chromosomes must have a detectable amount of heterochromatin, but why? One possibility is that heterochromatin represents a type of regulatory mechanism not found in bacteria, the cells of which do not undergo differentiation of the type found in the higher eukaryotes. (The fungi have the least centric heterochromatin among the eukaryotes and are among the least differentiated. Protozoa such as the ciliates have a great deal of heterochromatin in the micronuclear chromosomes, but much is cast out by diminution when the macronuclei are formed.)

A unique property of higher eukaryotic systems is the ability to store stable somatic gene states of activity in the chromatin. Each different cell type, while it may have the same housekeeping genes turned on as other cell types, does have its own specific set of genes turned on. Although feedback loops such as found in bacteria are also present in eukaryotes, bacteria do not maintain different stable states of gene expression as do eukaryotes. There may be a link between euchromatin and heterochromatin as regards this stable type of regulation (Tartof et al., 1989). But what is it?

B. Permissive and Nonpermissive Rearrangements

A considerable body of evidence now exists showing that at least in some cases a gene's position in the genome is important for its proper functioning in space and time. The fact that a gene from a eukaryote can be isolated and introduced via a vector into a bacterial cell where it functions to produce its expected proper protein product is no proof that in its native environment its genomic position is inconsequential. The eukaryotic gene from a higher form such as a mammal is steadily transcribed in the bacterial cell without regard to time and space. But genes are constantly being activated and deactivated in eukaryotic embryos and the timing and space (cell type and position) factors are important. The results obtained with transgenic mice are not conclusive even though some of them indicate that position is unimportant. Other results show that some transformed genes act inappropriately in the host mouse.

Some geneticists seem to be strongly inclined to the view that a gene's position is unimportant. They opine that moving it and changing its linkage relationships by rearrangements is of no consequence with regard to its functioning. The available data do not support this conclusion for all rearrangements. There appear to be permissible and nonpermissible rearrangements. At least in *Drosophila* a majority of rearrangements are nonpermissive in homozygotes. These must be reckoned with as being manifestations of position effects about which we currently have little understanding.

One should also address the problem of what happens to zygotes that carry a translocation or inversion in the heterozygous state that alters the position of proto-oncogenes. We know that when rearrangements occur in somatic cells to change the linkage relationships of these genes, the result is generally the development of a neoplasm. Proto-oncogenes are involved in the regulation of important and basic functions. What is the effect of such rearrangements on growth regulation in a fertilized egg containing one? It is quite possible that they act as dominant zygotic or very early embryonic lethals, and would not be recognized. Dominant zygotic lethals are commonly observed in *Drosophila*, but detecting them in mammals, for example, is difficult. In plants they might be expected to be eliminated in the gametophyte and never get to the zygote stage.

VIII. SUMMARY

1. The interphase nucleus is packed with DNA, RNA, and proteins, all three of which are involved in DNA replication and transcription and its regulation, leading to the synthesis of rRNAs, tRNAs, URNAs, and hnRNAs. The latter are processed in the nucleus, leading to the formation of mature mRNAs that leave the nucleus via the nuclear pores along with the rRNAs and tRNAs to function in translation in the cytoplasm.

2. The parts of the nucleus about which we have some knowledge are the double-membrane nuclear envelope, the nuclear pores, a nuclear lamina associated with the inner membrane, and a nuclear ma-

trix, which constitutes the bulk of the interior of the nucleus along with the chromatin and a large amount of RNA.

3. Freed of chromatin the nuclear internal matrix acts in part as a scaffold. In some preparations it is a network of proteinaceous filaments associated with hnRNAs and ribonucleoprotein (RNP) particles as well as a wide variety of proteins, some of which may be binding proteins involved in regulation. The RNPs may be involved in the organization of the internal matrix scaffold, and they appear to be involved in replication and transcription. The removal of RNA by RNAase causes a collapse of the matrix.

4. Interphase chromosomes do not have the rigid scaffold associated with topoisomerase II found in the metaphase chromosomes. DNA loops are present as well as topoisomerase II in the interphase chromatin, but the association between the two appears to be different from that in the metaphase. The attachment of the DNA to the matrix appears to be diffuse and more complex than that in the metaphase.

5. Although a considerable amount of disagreement exists in the literature about the exact nature of chromosome ordering in the interphase nuclei, agreement does exist that ordering is nonrandom. The difficulties involved in evaluating much of the data result from the effects of preparing the nuclei for observation, and the fact that different cell types have different types of organization of chromatin. Also the organization may change from one period of the interphase to another. In general nucleoli appear to be located near the inner membrane of the nuclear envelope; centric heterochromatin and associated centromeres appear to be polarized in specific parts of the nuclei; in some nuclei telomeres appear to be located away from the centromeres so that the chromosome arms are free; and in some plants the two genomes of the diploid nucleus are separated. In some plants and in the Diptera

homologues pair somatically. Somatic pairing does not, however, appear to be a general phenomenon.

6. The interphase chromosomes of the higher eukaryotes are about 240 nm thick, and may be at their maximum length.

7. The DNA replication machinery, or replisomes, in interphase nuclei are anchored in the nuclear matrix, and the DNA is replicated as it is reeled through the replisomes.

8. Transcribed DNA is apparently in the free loops and not in the SARs or MARs, to which the loops are attached. The RNA polymerases involved in transcription are free to move along the DNA to be transcribed, and are not part of the rigid matrix scaffold to which replisomes are attached. Topoisomerase II is associated with the MARs, and topoisomerase I is apparently associated with the transcribed part of the DNA in the loops.

9. Variegating position effects as manifested in *Drosophila* are the result of positioning a gene that is ordinarily located in the euchromatin in or close by the centric heterochromatin. Other types of position effects are also manifested by moving a gene to a different place in the genome by rearrangement. If the rearrangement involves an oncogene, a neoplasm may be induced.

10. Some translocations and inversions are viable homozygous and therefore permissible; others are not, and are therefore nonpermissible.

11. Heterochromatin is probably not excess repetitive baggage but in some way not understood as an important element in the functioning of eukaryotic chromosomes.

8 Gene Linkage and Chromosome Maps

Genes in populations do not exist in random combinations with other genes. The alleles at a locus are segregating in a context that includes a great deal of correlation with the segregation of other genes at nearby loci.

R. Lewontin, 1974

The number of identified gene sequences assigned to the chromosomes of man and many other organisms has greatly increased since 1970 as a result of somatic cell hybridization and recent innovations in recombinant DNA technology and DNA hybridization. The development of detailed genetic maps has many practical benefits, which include an increased understanding of the molecular basis of human genetic disorders such as cystic fibrosis and muscular dystrophy. In fact, the segment of DNA responsible for cystic fibrosis was isolated as a direct consequence of knowing it's chromosomal location (Rommens et al., 1989). Revolutionary ways of performing clinical medicine may come from present-day mapping efforts, since a molecular understanding of disease etiology in many instances will lead to better ways of dealing with diseases. In more general terms, mapping genomes is essential to understanding how genes are regulated in different tissues and at different times during development. Genes

349

do not function as discrete entities apart from their environment. Their regulation is controlled by both *cis*-acting regulatory sequences and *trans*-acting regulatory factors. Genes can be regulated over long distances (classical position effects are examples of alterations of gene expression that result over long distances). The largest number of position effects have been observed in *Drosophila*. For example, the white locus is one of several loci that undergo alteration in expression when it is translocated next to heterochromatin (Chapter 7). The chromosome breakpoints that place the white locus next to heterochromatin are clustered approximately 25 kb downstream from the white promoter (Tartof et al., 1989). A complete map of a species genome would greatly facilitate our understanding of long-range regulatory effects of chromosome rearrangement.

Genes can be mapped by either sexual or parasexual approaches. Sexual approaches take advantage of the variation generated during meiosis by crossing-over events between homologues, while parasexual methods use somatic cell hybrids and other techniques to map genes. In this chapter, we will briefly review some of the classical sexual and parasexual approaches to mapping and then move on to more recent approaches to physically mapping chromosomes.

I. ASSIGNING GENES TO CHROMOSOMES

One of the largest undertakings in modern biology is the present effort to completely map and sequence the entire human genome and the genomes of several reference organisms such as *Mus musculus, Caenorhabditis elegans, Drosophila melanogaster,* and *Saccharomyces cerevisiae* (Lewin, 1986). The massive effort to map and sequence each of these genomes has become known as the *human genome project*. The effort to understand the genetic maps of these organisms, however, had an earlier origin. *Drosophila* was the first organism to which genes were assigned to specific chromosomes (Morgan, 1910; Morgan and Lynch, 1912). The first gene assigned to a human chromosome was assigned to the X chromosome. In 1911 E.B. Wilson coupled the observation of Guyer (1910) that human males are a heterogametic sex (producing an X and Y sperm) with the known fact that color blindness was inherited in a sex-linked fashion. Since color blindness was a recessive trait, Wilson correctly deduced that the higher frequency of this trait in males over females resulted from the trait being located on the X chromosome. Following Wilson's initial observation, many genes for other traits or disorders, such as hemophilia, were assigned to the X chromosome (Bell and Haldane, 1937).

Assignment of a gene to an autosome of man, however, remained elusive, the first gene being assigned to an autosome in 1967. A long time passed during which no autosomal genes were assigned to a specific autosome because individual autosomes in humans and other mammalian species could not be unambiguously distinguished and because there was no means of observing segregation of an autosome with a particular phenotype. Certainly a number of linkage groups could be identified in man and other mammals that were known to be not X-linked and thus autosomal, but no means were available to place these linkage groups onto their respective chromosomes. The history of the development of various chromosome banding techniques that allowed the unambigous identification of human chromosomes has been described by Hsu (1979).

The first gene assignment to an autosome of man was accomplished by Donahue et al. (1968) and this involved the assignment of the Duffy blood group (*FY*) to human chromosome 1 by a combination of linkage and cytogenetic analysis. Donahue and colleagues took advantage of a genetically inherited cytogenetic variation called uncoiling (*UN*). Persons with this trait possess a chromosome 1 that is different from most chromosomes 1 in the general population. The *UN* phenotype causes a region of chromosome 1 near the centromere to decondense so that the chromosome becomes noticeably longer and thinner than a normal chromosome 1. No clinical abnormality is associated with the *UN* phenotype. Since *UN* is inherited in a mendelian fashion, linkage analysis can be performed between this trait and any other inherited trait. It was determined that the Duffy blood antigen group (*FY*) was linked to *UN* by genetic linkage analysis (Renwick and Lawler, 1963). Since prior linkage of *FY* and the gene for congenital cataracts (*CAE*) had been demonstrated, *CAE* could also be assigned to chromosome 1.

II. ASSIGNING SYNTENY AND MAPPING BY PARASEXUAL METHODS

Soon after the *FY* and *CAE* loci were assigned to human chromosome 1, several developments led to a much more direct method of assigning genes to human and nonhuman chromosomes. First, as reviewed in Chapter 2, several methods of banding chromosomes allowed for each chromosome in any mammalian species to be unambiguously identified. Second, the discovery that cells could fuse in culture and that chromosomes could be segregated opened the door to assigning genes to chromosomes.

A. Development of Somatic Cell Hybrids

By **segregation,** we mean the loss of one species's chromosomes from the hybrid cell. It should be noted that linkage

implies more than the fact that two genes are on the same chromosome. The term linkage implies that two genes are close together on a chromosome and the distance between the loci can be measured. Two genes known to be located on the same chromosomes are syntenic.

In 1960, Barski et al. discovered that on rare occasions cells grown in culture would fuse, forming a **somatic cell hybrid.** A short time after this discovery it was found that cells treated with the inactive Sendai virus could be made to fuse at a higher frequency (Harris and Watkins, 1965) (at a later date it was discovered that polyethylene glycol could also do this). A very important development was added to the isolation of hybrids by Littlefield, who in 1964 developed a method of selecting for somatic cell hybrids by genetic complementation. Littlefield took cells of a mouse cell line that was deficient in thymidine kinase (TK1) and fused them to cells from a second mouse cell line deficient in hypoxanthine phosphoribosyl transferase (HPRT). Using growth media containing hypoxanthine, aminopterin, and thymidine (usually designated as **HAT** medium), hybrid cells could be selected for. Aminopterin is an inhibitor of the enzyme dihydrofolate reductase, which is a key enzyme in the *de novo* synthesis of purines. If this enzyme is inhibited by aminopterin, the cell can still make purines by a salvage pathway using the enzymes HPRT and TK and the substrates hypoxanthine and thymidine. However, in Littlefield's experiment, one set of parental mouse cells were TK-deficient and the other set were HPRT-deficient. When exposed to the HAT media, both parental cell types would die. Hybrids between the two cell lines, however, could survive in HAT media because of complementation. The hybrid cell used HPRT from one line and TK from the other.

Since Littlefield's first demonstration that somatic cell hybrids could be selected, many different types of hybrids between different species have been developed. Additional biochemical selection schemes were also developed that involved other genetic loci such as adenine phosphoribosyl transferase (Kusano et al., 1971) and diphtheria toxin sensitivity (Creagan et al., 1975).

B. Use of Somatic Cell Hybrids in Gene Mapping

The use of somatic cell hybrids in gene mapping followed the discovery that the chromosomes of one species are always lost in interspecific somatic cell hybrids (Ephrussi et al., 1963). The mechanism by which this is accomplished is not well understood. However, what can generally be predicted is the species from which the chromosomes will be lost, or in somatic cell genetics jargon, will segregate. In every known instance when primary cells (cells taken directly from the organism, such as spleen lymphocytes or fibroblasts) are fused to established, transformed cell lines, the chromosomes from the primary cells are segregated from the hybrid. Thus, chromosome segregation appears to be independent of the species. It may be significant that no hybrid line has ever been identified in which chromosomes of both species are lost. This may be true for the obvious reason that the loss of essential genes could be lethal.

Mapping of genes by means of somatic cell hybrids takes place by concordant segregation analysis. In order to perform such an analysis, it is necessary to create a panel of hybrids that have segregated different sets of chromosomes. For example, in Chinese hamster × mouse somatic cell hybrids, the hamster chromosomes segregate and it is therefore possible to map genes to hamster chromosomes (Stallings and Siciliano, 1981). Mapping genes by means of interspecific somatic cell hybrids takes advantage of interspecific genetic

variation. For example, one can frequently detect variations between isozymes of different species by **starch gel electrophoresis** and histochemical staining. By separating enzymes based on differences between net electrical charge, one can distinguish species-specific forms of the enzyme. Figure 8.1 shows the resolution of hamster and mouse forms of

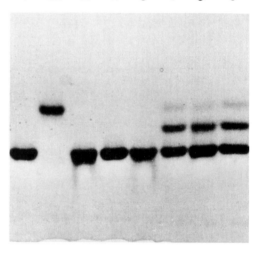

Fig. 8.1. *Example of the analysis of somatic cell hybrids by starch gel electrophoresis and histochemical staining. The enzyme stained for is 6-phosphogluconate dehydrogenase (PGD). Lanes designated CH and M contain supernatant from Chinese hamster and mouse cells, respectively. Chinese hamster and mouse forms of PGD possess a difference in electrical charge and can be separated by electrophoresis. Lanes 1–6 are hamster × mouse somatic cell hybrids that retain all mouse chromosomes and segregate hamster chromosomes. The hybrids in lanes 4, 5, and 6 (hybrids C1/4, C8/1, and C9/2—see Table 8.1) have retained hamster chromosomes 2 along with the hamster gene for PGD. The enzyme PGD is dimeric, so a three-banded pattern is observed (a hamster homopolymer, a hamster–mouse heteropolymer, and a mouse homopolymer). Hybrids in lanes 1, 2, and 3 (C2/ E1, C12/9, and C17/A1) have lost hamster chromosome 2 along with the gene for PGD. Adapted from Stallings and Siciliano (1981).*

6-phosphogluconate dehydrogenase (PGD) and the loss of the hamster enzyme in some hybrids that have lost the hamster chromosome containing the gene for the enzyme.

Table 8.1 illustrates segregation of Chinese hamster chromosomes from a hybrid clone panel of mouse × hamster somatic cell hybrids. One can correlate the presence or absence of a specific gene, or gene product such as an enzyme, with the presence or absence of each chromosome. As seen in Table 8.1, the hamster *ENO1*, *GALT*, *PGD*, and *PGM1* genes are always present when hamster chromosome 2 is present and always missing when the chromosome has segregated. This allows these genes to be assigned to this chromosome.

It is possible to map other phenotypic traits, such as drug resistance, by using somatic cell hybrids, in addition to enzymes that are detectable by starch gel electrophoresis and histochemical staining. For example, the gene for cytosolic thymidine kinase, *TK1*, was originally mapped on the basis of its phenotypic expression in somatic cell hybrids. *TK1* was the first genetic locus to be mapped in humans by this approach. Weiss and Green (1967) constructed somatic cell hybrids by fusing *TK1*-deficient mouse CL1D cells to human fibroblast cells and selecting in HAT media for hybrids. In this hybridization scheme, human chromosomes segregate. In a panel of hybrids constructed in this manner, one would expect the chromosome containing the human *TK1* gene to always be present. All other human chromosomes can be missing from hybrid clones and subclones, but in order for the hybrid to be viable, the human *TK1*-bearing chromosome must remain. By removing the hybrids from HAT selection, Weiss and Green permitted a small subpopulation of cells to arise that had segregated human *TK1*. Using BudR (an analogue of thymidine that is used to select for *TK1*-negative mutants), they selected for hybrid cells that had lost the human chromosome containing the *TK1* gene. When they analyzed these cells cytogenetically, they found that a chromosome in the 6-12-X group was missing. They

**TABLE 8.1. Segregation of Hamster Isozymes in Chinese Hamster ×
Mouse Somatic Cell Hybrids**

Hybrid clone	Hamster chromosomes segregated	Hamster isozymes present (+) or absent (−)			
		PGD	PGM1	ENO1	GALT
C1/4	X	+	+	+	+
C2/10	None	+	+	+	+
C2/A1	X	+	+	+	+
C2/E1	1,2,X	−	−	−	−
C3/E1	None	+	+	+	+
C4/3/E1	None	+	+	+	+
C5/4	X	+	+	+	+
C6/1	2	−	−	−	−
C11/3	1,X	+	+	+	+
C11/3B5	1,2,5,8,X	−	−	−	−
C11/3E6	1,8,X	+	+	+	+
C12/9	1,2,3,X	−	−	−	−
C14/2/A1	1,2,X	−	−	−	−
C17/A1	2,X	−	−	−	−

could not pin down the chromosome further because banding had not yet been invented. It later turned out that chromosome 17 bears *TK1*, but the authors can hardly be blamed for their mistake, since human chromosomes at that period could not be unequivocally identified.

The concordant segregation analysis of hybrids is a statistical method based on many hybrid clones and subclones because one usually observes some level of discordance between a marker and a chromosome in hybrids. If a chromosome that bears a particular gene breaks and the rearrangement goes undetected, the hybrid clone might be scored as positive for presence of the gene product and negative for the chromosome. This can lead to discordant clones (which may have contributed to the failure of Weiss and Green to recognize that *TK* was located on chromosome 17). It is possible to take advantage of recognizable chromosome break points to regionally localize genetic loci by somatic cell hybrids. The number of hybrid clones containing different rearranged human chromosomes is quite extensive.

C. Radiation Fragmentation Hybrids

In addition to making hybrids with cell lines containing preexisting chromosome translocations in order to make regional mapping assignments, it is also possible to induce chromosome breaks in somatic cell hybrids with radiation. Goss and Harris (1977) made the first set of radiation fragmentation hybrids. Essentially, hybrid cells are irradiated and then fused to cells from a cultured cell line (the irradiated cells die and must be rescued by fusion). Small pieces of DNA from the irradiated cells' nuclei will be retained and incorporated into the chromosomes of the hybrid cells' nuclei. Because interstitial deletions and inversions occur at a high frequency following radiation treatment, this method of mapping was abandoned

for a long period of time. Cox et al. (1989) has recently introduced concepts that have greatly revitalized this approach to mapping. He has shown that although many types of scrambling occur as the result of radiation, two genes that are normally very close together on a chromosome will generally remain close together following radiation. But genes far apart may be brought close together by deletions, inversions, and translocations. A statistical analysis has been developed that shows that if a large number of hybrids are examined, genes that are very close together originally will be distinguishable from rearrangements that bring distant genes close together. By Cox's approach, the distance between loci can be measured in centirads.

III. DIRECT MAPPING OF GENE SEQUENCES

Improvements in technology often lead to new understandings in biology and this is particularly true of mapping chromosomes. The advent of recombinant DNA technology has dramatically increased the number of genes that have been assigned to chromosomes in different species. **Southern blotting** (Southern, 1975) and DNA hybridization (see box 8.1) has allowed many genes and arbitrary DNA sequences to be mapped. This has facilitated the mapping of genes for which no assay exists for protein activity. Southern blot analysis has been extensively used in conjuction with somatic cell hybrids to map DNA sequences.

The use of recombinant DNA technology and DNA hybridization has also led to a more direct method of assigning DNA sequences to chromosomes. This method is called *in situ* **hybridization** and was first demonstrated by Pardue and Gall (1970). These authors made a probe from mouse satellite DNA by labeling it with a radioactively tagged nucleotide. The denatured mouse satellite DNA probe was then hybridized directly to mouse chro-

BOX 8.1 SOUTHERN BLOTTING

The basic method of detecting specific sequences in DNA is Southern filter hybridization analysis (Fig. 1). High-molecular-weight DNA is isolated from cells, such as somatic cell hybrids, digested by an endonuclease restriction enzyme, and electrophoresed through an agarose gel. The electrophoresis will separate DNA molecules on the basis of size. The DNA in the gel is denatured and then transferred out of the gel onto a nylon membrane by capillary action. It is permanently attached to the nylon membrane. A radioactively labeled probe, such as the sequence for a gene, can then be hybridized to the DNA attached to the membrane. After hybridization, unhybridized probe can be washed off the membrane and the membrane is placed over an X-ray film to visualize sequences that hybridized to the labeled probe. Only identical or nearly identical sequences on the membrane will hybridize to the probe. Because of mutations occurring during evolution, gene sequences from different species will contain different restriction enzyme recognition sites that can be visualized by Southern blotting.

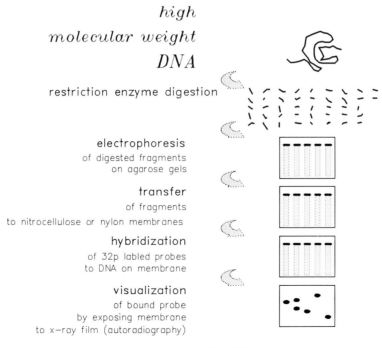

high
molecular weight
DNA

restriction enzyme digestion

electrophoresis
of digested fragments
on agarose gels

transfer
of fragments
to nitrocellulose or nylon membranes

hybridization
of 32p labled probes
to DNA on membrane

visualization
of bound probe
by exposing membrane
to x-ray film (autoradiography)

Fig. 1. *Detection of specific DNA sequences.*

mosomes fixed on a microscope slide. The chromosomal DNA on the microscope slide was denatured, and the places where the labeled probe annealed could be visualized by autoradiography (Fig. 2.10). It was discovered by these methods that satellite DNA mapped to the centromeric regions of mouse chromosomes, and that other repetitive DNA sequences and single-copy sequences that had become amplified in cell lines following drug selection could also be mapped by this procedure. By a combination of G-banding and *in situ* hybridization, it was later determined that single-copy gene sequences could be assigned directly to specific regions of chromosomes (Harper and Saunders, 1981). This protocol requires a statistical approach, since not every metaphase spread contains a set of homologous chromosomes that display hybridization signals. Typically, only 20%–30% of the chromosomes on a slide that contain a particular sequence will show hybridization to the sequence. However, if enough metaphases are scored, silver grains will begin to accumulate at one location on a chromosome, while other chromosomes will have silver grains from the background randomly distributed over them.

Leary et al. (1983) introduced the use of biotin-labeled probes as an alternative to radioactively labeled probes. This was a major step forward, since it eliminates the long autoradiographic exposure periods. Since this initial innovation, biotin-labeled probes have been used extensively for *in situ* hybridization (Lichter et al., 1990). The hybridization signal is observed by fluorescence or colormetric indicators. For fluorescence-based detection, an anti-avidin-fluoroscene antibody is allowed to react with the biotin-labeled probe that is hybridized to the chromosomal DNA (antiavidin is an antibody that binds biotin). The signal is then observed by fluorescence microscopy (Figure 8.2). Signals can be detected from probes as small as 2 kb.

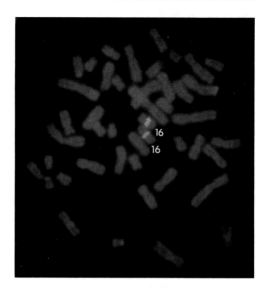

Fig. 8.2. *Example of in situ hybridization with biotin-labeled probe. Normal human chromosomes were hybridized with a biotin-labeled satellite repeat sequence specific for the centromere region of chromosome 16 (Moyzis et al., 1987). Only the two chromosome 16 homologues show hybridization signal (yellow). Other human chromosomes appear red because of propidium iodide counterstaining. Courtesy of Dr. Julie Meyne, Los Alamos National Laboratory.*

IV. LINKAGE ANALYSIS IN MAN

The advent of molecular biology techniques also dramatically increased the utility of classical linkage methods for mapping genes in man. Since many excellent reviews of linkage analysis have been written (White and Lalouel, 1988; Ott, 1985), we will only briefly review the fundamental concepts of linkage analysis and some of the latest methods that have not yet been published in texts.

A. General Principles of Linkage Analysis

Synteny for human genes can be demonstrated in a number of ways, the most common being through concordant segregation analysis of somatic cell hybrids and by *in situ* hybridization analysis. Genes that are linked are also obviously on the same chromosome, but linkage implies much more information than simple synteny. Linkage between two genetic markers can be demonstrated by genetic analysis only if their loci are within a certain distance of each other. As seen in Figure 8.3, the genetic analysis of linkage is totally dependent on the recombination that takes place during meiosis. In Chapter 4 we indicated that linkage was first observed in crosses between peas with different traits when the traits failed to show independent assortment. Figure 8.3a shows that independent assortment results when loci are on different chromosomes, and 50% of gametes are recombinant.

The centimorgan (cM) is the map unit of measurement of linkage distance and represents 1% recombination resulting from crossing-over between two loci on homologous chromosomes. Loci that are very close together may not undergo measurable recombination. Therefore, only two types of gametes are expected for tightly linked loci possessing two alleles (Fig. 8.3b). Alleles of two tightly linked loci will be inherited as a unit or **haplotype** within a pedigree unless a very rare crossover event occurs.

As shown in Figure 8.1c, if two gene loci are far enough apart for crossovers to occur at a measurable level between them, recombinant gametes will be produced. In this example the recombinant types, aB and Ab, constitute 20% of the total (0.1 + 0.1/1.0 × 100) while 80% are parental types. Therefore, these two loci are by definition 20 map units or 20 cM apart. These distances of 20 cM or less give a reasonable estimate of physical distance between two loci on a chromosome. If the distance, X, between two linked loci is equal to the recombination frequency, θ, then an increase in θ is directly proportional to X unless the two loci are too far apart even though linked. Recombination frequen-

cies of θ from 1% to about 15% are proportional to map distance X measured in centimorgans, but as θ between two markers increases beyond 15%, depending on the organism, X decreases until it reaches a limit of 50 cM. This is 50% recombination and is undistinguishable from independent assortment, as shown in Figure 8.3d. Thus two genes may be syntenic but not show linkage. To understand this, one must appreciate the fact that more than one cross-over may occur between two linked genes, and further, more than two chromatids may be involved. The 50% recombination frequency is a composite of effects from singles, doubles, and so forth. Map units actually equal half the percentage crossover frequency between two genes. Observed recombination frequencies in test cross progenies do not decrease beyond 15%; they just don't increase at a linear rate (physical distances do not necessarily correspond well to recombination frequencies). Most organisms are remarkably similar in this respect. The major reason for the departure from linearity is the occurrence of invisible double recombinants (distinguished from crossovers) so correction for these invisible doubles (where effects cancel) yields reasonable genetic maps. This is what mapping functions usually do (see Box 8.2). Figure 8.3d illustrates the consequences of the four types of double crossovers between two genes that are spaced far enough apart on a chromosome to result in 50% recombination. Two-strand double crossing-over will result in products that show no recombination—indicated as PT (parental type). The three-strand doubles will produce 1/2 PT and 1/2 NPT (nonparental type), and the four-strand double crossovers will produce only NPT. Therefore the number of NPT equals the number of PT expected, and θ is therefore 50%. The genes are syntenic but not linked by the genetic test.

Over large distances genetic distances between more than two different loci are

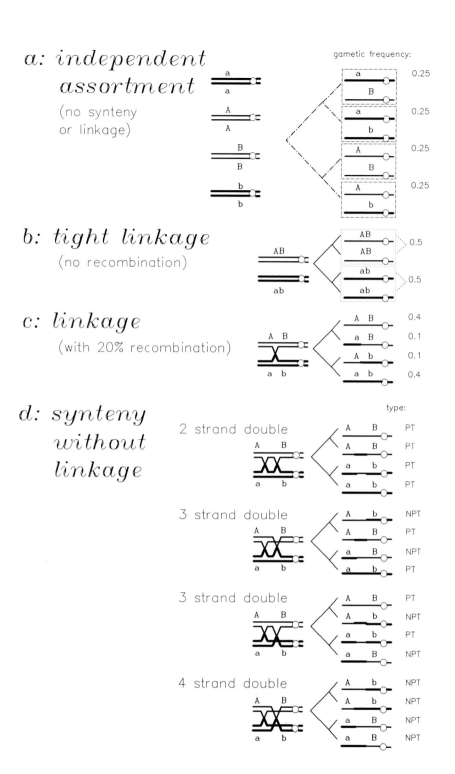

Fig. 8.3. *Results expected from crosses involving unlinked, nonsyntenic genes (**a**), tightly linked genes with no recombination (**b**), linked genes giving less than 50% recombination (**c**) and syntenic genes far apart on the chromosome (**d**). Double crossovers are likely to occur between genes that are more than 12 cM apart, with frequencies that have to be reckoned with in calculating map distances. Panel **d** illustrates the outcome (in equally frequent categories) from meiocytes in which exactly two crossovers occur between two genes. The most common class is usually none; the next is 1, then 2, 3, and so forth, but more than 3 almost never happens in most organisms.*

BOX 8.2 MAPPING FUNCTIONS

One of the first mapping functions (general reference—Crow and Dove, 1990) to be derived was that of Haldane (1919). It is still widely used, and the one generally applied in student textbooks of general genetics.

Let r = fraction of recombinants, m = map distance in centimorgans, and c = coincidence = (actual double crossovers observed)/(number expected with no interference).

For closely linked genes, $c = 0$ and $r = m$. Haldane assumed, however, that $c = 1$ and derived

$$m = -\frac{1}{2} \ln(1 - 2r). \quad (1)$$

For an r of 26%, then, $m = \frac{1}{2}$ ln $0.48 = 0.37 = 37$ cM.

Since Equation 1 does not take positive interference into account, Kosambi (1944) made some specific assumptions about interference and arrived at a more widely used function that has been found to give reasonable

results with data from higher eukaryotes:

$$m = \frac{1}{4} \ln(1 + 2r)/(1 - 2r) \quad (2)$$

For $r = 26\%$, $m = \frac{1}{4}$ ln $3.17 = 0.29 = 29$ cM, which is a more reasonable figure for map distance, since it takes some interference into account. But note that in neither case does $r = m$. If we assume that $r = 6\%$, then we get a value of 0.06 cM with both Equation 1 and Equation 2 because in this range (between 0% and 15%) interference is negligible and $r \sim m$.

Many more-refined functions have been developed. For example, see Felsenstein (1979), Ott (1985), Stahl (1989), and Lyon (1990). All give approximations because of the vagaries of coincidence, but developing them keeps some geneticists happy, and their efforts are important in helping to relate genetic maps to physical maps.

---

additive if corrections for invisible double recombinants (not crossovers) are made. Map distances of 10 cM or less are additive without the necessity of corrections because doubles do not occur within 10 cM (in most organisms).

Positive interference results from the fact that once a crossover occurs in a region, a second crossover is less likely to occur nearby. Because crossovers are not uniformly distributed over a chromosome, calculations require complex mapping functions. For distances >10 cM, more complex mapping function equations are required to correct for double-crossover events and positive interference (Box 8.2).

B. Problems Associated With Studying Genetic Linkage in Man

As mentioned above, linkage analysis in humans is quite difficult because planned crosses cannot be made. Progeny from large, three-generation families, however, can be examined. Quite frequently, genotypes of grandparents cannot be determined, so the phase of the alleles cannot be assessed directly. Whether, for example, A is coupled with B and a with b (*cis* phase) or whether they are in repulsion Ab and aB (*trans* phase) is critical for determining if progeny represent a recombination event.

1. The LOD Score

In order to overcome these shortcomings, some rather clever statistical approaches to linkage analysis have been devised to study linkage in humans. The most commonly used method for determining linkage in man is the method of likelihood or odds (Haldane and Smith, 1947). In this method, the probability of obtaining a particular set of data if two loci are linked is compared with the probability of obltaining the same data set if the loci are unlinked. For convenience, the logarithim of the probability is obtained, so that the ratio of the two probabilities is then called a **lod score**. The higher the lod score, the higher the odds are in favor of linkage. A lod score of 3 is the minimum value that must be obtained to assign linkage. A lod score can be obtained for each estimate of recombination and the recombination fraction with the highest lod score is the best estimate of the strength of linkage.

2. Physical versus genetic distance

In some organisms, such as mammals, recombination frequencies are very different in males than females. In most instances rates of recombination are higher in females and thus genetic maps derived from recombination events in ooocytes are longer than those derived exclusively from male meiosis. There are also recombination hot spots in the genome so that it is

difficult to correlate genetic distance precisely with physical distance. In general, 1 cM is equivalent to 1,000 kb of DNA, but there are regions of the genome where a centimorgan is equal to only 100 kb. The relationship of genetic and physical distances in some organisms is listed in Table 8.2. Table 8.3 presents some data relating physical and cytogenetic data for human chromosomes. The relative lengths of the 22 autosomes and the chiasma frequencies were determined cytologically. The chiasma frequencies were converted to centimorgans, and the number of base pairs for each chromosome were calculated using 3.3×10^9 bp as the total DNA per genome. The fifth column in Table 8.3 shows that a close relationship exists between the numbers of base pairs and centimorgans. The average of 1.18 bp/cM for 22 autosomes comes close to the generally accepted value of 1 cM = 10^6 bp of DNA.

3. Different types of polymorphisms

Prior to the development of recombinant DNA technology, linkage maps of the human were limited because of the paucity of highly polymorphic genetic loci. The polymorphic information content (or PIC value) of a locus can be determined (Box 8.3—Botstein et al., 1980). Loci with PIC values greater than 0.5 are highly informative in linkage studies. The development of recombinant DNA technology,

TABLE 8.2. Relation of the Genetic Map to the Physical Map

Organism	Total cM	Total bp	bp/cM	Reference
S. cerevisiae	5,000	1.4×10^7	2.8×10^3	Mortimer and Schild (1987)
Caenorhabditis	337	8×10^7	2.4×10^5	Edgley and Riddle (1990)
D. melanogaster	287	1.7×10^8	5.9×10^5	Lindsley and Grell (1972)
Mus musculus	1,100	3×10^9	2.7×10^6	Davisson et al. (1990)
Zea mays	1,201	5.4×10^9	4.4×10^6	Coe et al. (1990)
Lycopersicon	1,063	6.8×10^9	6.4×10^6	Bernatzky and Tanksley (1986)
Homo sapiens	2,577	3×10^9	1.2×10^6	Morton et al. (1982)

**TABLE 8.3. Physical and Genetic Lengths of Human Autosomes
(Chiasma Frequencies for Males Only)**

Chromosome	% Relative length	No. bp $\times 10^9$	cM	bp/cM $\times 10^6$
1	8.44	0.279	195	1.4
2	8.02	0.265	173	1.5
3	6.83	0.225	150	1.5
4	6.30	0.208	138	1.5
5	6.08	0.201	137	1.5
6	5.90	0.195	131	1.5
7	5.36	0.179	136	1.3
8	4.93	0.163	131	1.2
9	4.80	0.158	115	1.4
10	4.59	0.152	127	1.2
11	4.61	0.152	110	1.4
12	4.66	0.154	137	1.1
13	3.74	0.123	88	1.4
14	3.56	0.118	88	1.3
15	3.46	0.114	96	1.2
16	3.36	0.111	108	1.0
17	3.25	0.107	109	1.0
18	2.93	0.097	98	1.0
19	2.67	0.088	97	0.9
20	2.56	0.085	99	0.9
21	1.90	0.063	54	1.2
22	2.04	0.067	60	1.1
Totals	100	3.3	2,577	Av. 1.18

Modified from Ott (1985). Chiasma data from Morton et al. (1982).

however, allowed for the cloning of numerous structural gene loci and thousands of anonymous DNA sequence fragments that show polymorphic variation detectable by Southern blot analysis. These variants are the results of point mutation at restriction enzyme recognition sites and are called **restriction fragment length polymorphisms** (RFLP). Another class of polymorphic marker, called hypervariable minisatellites or **variable number tandem repeats** (VNTR sequences) have been described (Jeffreys et al., 1985) and represent a major breakthrough in genetic linkage analysis because almost any two individuals picked at random will have variation for any VNTR probe. An example of the high level of variability displayed by a VNTR sequence probe may be observed in Figure 8.4. The variability in the size of the restriction fragments observed following hybridization with a VNTR sequence probe results from differences in the number of tandem repeat units found in DNA from different individuals. In this respect, the variability observed from VNTR probes is quite different from RFLP-type probes.

4. Microsatellite polymorphisms

The newest type of polymorphism to be used in linkage analysis is those associated with microsatellite repetitive sequences. Microsatellites are simple repeats of mono-, di-, tri-, . . ., and hexanucleotides. One of the most useful microsatellite repeats for linkage analysis is the interspersed dinucleotide $(GT)_n$ se-

BOX 8.3 POLYMORPHISM INFORMATION CONTENT

One cannot detect the existence of a gene by mendelian genetic techniques of analysis if the gene has no alleles. But it can be recognized and mapped by such techniques if it has at least two alleles and their different phenotypic expressions can be assigned to the same presumed single-marker locus. The question, then, is how polymorphic a **marker** locus must be for it to be easily identified with a particular phenotype, for example, a disease. To determine this a statistic called the **polymorphism information content (PIC)** is calculated for the marker locus. Assuming that the alleles at the marker locus are codominant in their expression, PIC is the expected fraction of offspring that show coinheritance of a specific phenotype produced by a dominant allele (D) associated with a disease at an inferred **index** locus, and the marker locus. Such offspring are called the **informative** offspring. Therefore PIC is equal to the fraction of informative offspring in a population in which the specific phenotype or sequence is segregating. It is an estimate of the influence of the degree of polymorphism at the marker locus on the probability of detection of linkage to the inferred index locus.

In order to derive an equation to estimate PIC we explicitly define informativeness as a property of linked loci. The first locus, the index locus, has two types of alleles: "normal" (standard or wild-type) and deviation from normal (e.g., a dominant disease allele). The second locus, the marker locus, consists of a set of codominant alleles (e.g., RFLPs). Given a pedigree in which one parent is heterozygous and one is homozygous normal at the index locus, one has informativeness when knowledge of the offspring's genotype at the marker locus is sufficient to determine coinheritance with the index locus. If the affected parent is heterozygous at the index locus but homozygous at the marker locus, no determination is possible; thus this combination is not informative (Fig. 1A). If both parents have the same heterozygous genotype at the marker locus, the offspring of each possible type are produced with equal probability, one can make a determination for half of the offspring; thus this combination is informative half of the time (Fig. 1B). All other combinations are always informative (Fig. 1C).

Since PIC is the expected fraction of informative offspring, under the assumption of Hardy–Weinberg equilibrium (or independent frequencies of alleles at the two loci), it follows from the foregoing discussion as shown by Botstein et al. (1980) that

$$\text{PIC} = 1 - \sum_{i=1}^{n} p_i^2$$

$$- \left(\sum_{i=1}^{n} p_i^2 \right)^2 + \sum_{i=1}^{n} p_i^4 \quad (1)$$

where p_i = frequency of the index allele, i and n = number of different alleles.

To take an example from Weber et al. (1990), four alleles of a marker gene on human chromosome 16 that were designated by number with their respective frequencies in a population of 120 chromosomes indicated were the following: 170/.01; 168/.12; 166/.21; 154/.67. The alleles in this instance are CA-GT repeats, and the allele number is the number of base pairs in the repeat.

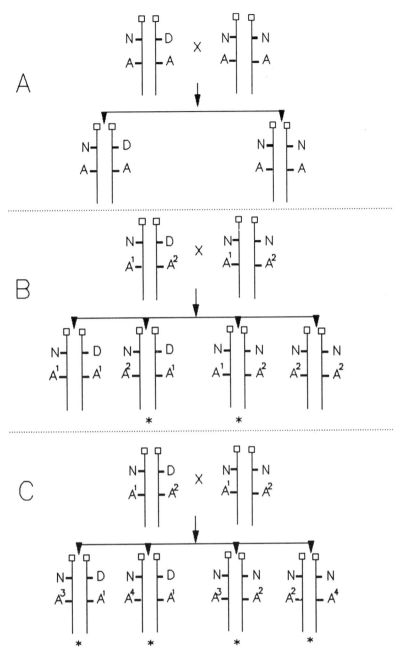

Fig. 1. *Estimation of PIC from various types of crosses. Index locus = rare dominant (e.g., disease) allele (D) and normal allele (N). Marker locus = codominant (e.g., RFLP) alleles A¹, A², A³, A⁴, and so on.* **A.** *Marker locus homozygous in affected parent. Segregation of D with A¹ cannot be used as test for linkage. Informativeness = 0.* **B.** *Marker locus heterozygous in affected parent and heterozygous for same two alleles in unaffected mate. Segregation of D with A¹ can be used as a test for linkage in half of the offspring. Probability that offspring are informative = 0.5. These are indicated by *. * **C.** *Marker locus is heterozygous in affected parent and heterozygous for two different alleles in unaffected mate. Segregation of A¹ with D can be used in test for linkage and segregation of N with A², A³, or A⁴, and also can be used in test for linkage of marker with index locus. Probability that offspring are informative = 1.0.*

(Continued)

BOX 8.3 POLYMORPHISM INFORMATION CONTENT
(Continued)

Using Equation 1,

$$PIC = 1 - (.01^2 + .12^2 + .67^2)$$
$$- (.01^2 + .12^2 + .21^2 + .67^2)^2$$
$$+ (.01^4 + .12^4 + .21^4 + .67^4)$$

$$PIC = 1 - .508 - .258 + .204$$

$$= .44.$$

Thus 44% of offspring should be informative.

PIC can theoretically range from 0 to 1, but generally the highest values found are those in the range 0.7–0.9. These are highly informative values, whereas a value of 0.44 is moderately informa-

tive. A gene with only two alleles will generally have a maximum PIC of 0.375. Clearly the more alleles identified with the marker the better. (The authors are indebted to C.E. Hildebrand, D.C. Torney, and J.N. Spuhler of the Los Alamos National Laboratory for their help in writing this box.)

REFERENCES

Botstein D, White RL, Skolnick M, and Davis RW (1980): Construction of a genetic linkage map in man using restriction length polymophisms. Am J Hum Genet 32:314.

Weber JL, Kwetek AE, and May PE (1990): Dinucleotide repeat polymorphisms at the D16S260, D16S261, D16S266, and D16S267 loci. Nucleic Acid Res 18:4034.

quence (Weber and May, 1989). Because this repeat sequence is interspersed every 50–100 kb in the genomes of all eukaryotes (Hamada et al., 1982; Stallings et al., 1991), it will perhaps become the most widely used polymorphic marker in linkage mapping. There is extensive variation in the number of GT sequences in the $(GT)_n$ repeat at the same locus for different individuals. Any two individuals picked at random will likely have different numbers of dinucleotide repeats at any particular $(GT)_n$ site. These variations can be detected by **polymerase chain reaction (PCR)** products (Box 8.4) that are analyzed on sequencing gels. Two short oligonucleotide primers are required that flank the $(GT)_n$ repeat. These primers must be in opposite orientation on different DNA strands. They are annealed to the homologous DNA sequences that flank the repeat so that the polymerase has a double stranded DNA segment to act on. The polymerase extends the sequence from the 3′ end of the primer, making the entire length of DNA double-stranded. After several cycles of denaturation, primer an-

nealing, and extension with the polymerase, the region between the primers is amplified. Because a $(GT)_n$ repeat sequence of variable length exists in this region, the amplified PCR product will be of different size for different individuals. Figure 8.5 shows the principal of how these polymorphisms are detected. Approximately 90% of all $(GT)_n$ repeat sequence sites display polymorphism in the length of the repeat block, thus making this system extremely useful for the detection of polymorphism.

5. Positional cloning of disease genes

The use of linkage analysis in humans has allowed many of the inherited genetic disorders to be assigned to specific chromosomes and to be even sublocalized on those chromosomes. Genetic traits for which specific enzymes have not yet been identified, such as Huntington's disease, can be mapped to chromosomes only by observing linkage of the phenotype to polymorphic protein or DNA markers (or other phenotypes) that have been previ-

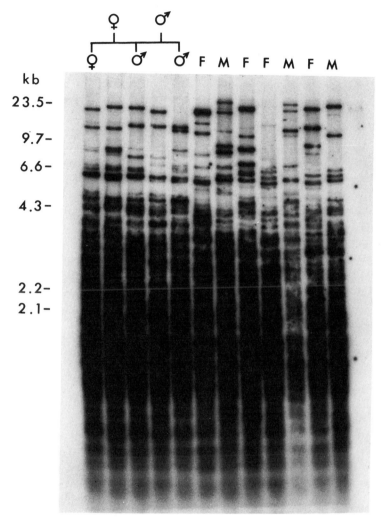

Fig. 8.4. *Southern blot of human DNA from different individuals hybridized to a ^{32}P-labeled region of the M13 bacteriophage DNA that shows homology to some regions of the human genome that are hypervariable. Each in-dividual shows a unique pattern that results from variations in the number of tandem repeat sequences found at specific loci. These types of probes are very useful in linkage mapping. Courtesy of Jonathon Longmire.*

ously mapped. Mapping a phenotypic trait for which no enzyme product is known is considered to be the first step in the cloning of the DNA sequence that is responsible for the phenotype. The general idea is to map DNA markers, such as RFLP, that flank the locus defined by the phenotype. Using the flanking DNA markers as probes, one can screen recombinant DNA libraries for sequences that overlap and extend the sequences that flank the disease locus. Attempting to retrieve overlapping probes from libraries by this method is called chromosome walking. Cloning the DNA for a phenotypic trait by this approach is called positional cloning or reverse genetics (for review see Orkin, 1986). In theory, reverse genetics

BOX 8.4 THE POLYMERASE CHAIN REACTION

The polymerase chain reaction (PCR) is a method of amplifying short segments of DNA. PCR has been applied to linkage mapping, forensic medicine, and other specialized fields. In order to amplify a segment of DNA, one needs a template (the segment of DNA to be amplified) and two short DNA primers (~20 bp) that are inversely oriented and come from opposite strands of the template DNA. As seen in Figure 1, the first step in the reaction is to denature the template DNA. Following denaturation, the primers are allowed to hybridize or anneal to the template. The primers are usually made on a DNA synthesizer and the distance between the primers is usually 300–3,000 bp. After the primers have annealed to the template, a DNA polymerase then adds base pairs that are complementary to the template starting from the 3′ end of each primer. The end result of this procedure is two copies of the template strand of DNA. By repetition of the procedure 20–35 times, millions of copies of the template DNA segment between the primers can be produced. The entire procedure can be automated by loading samples into a thermocycler that changes the temperature so that denaturation (~95 °C), annealing (~65 °C), and extention (~74 °C) can take place in alternate cycles.

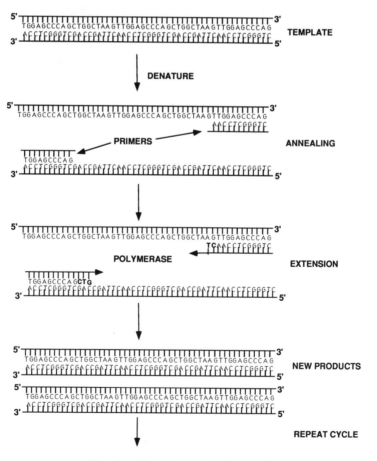

Fig. 1. *The polymerase chain reaction.*

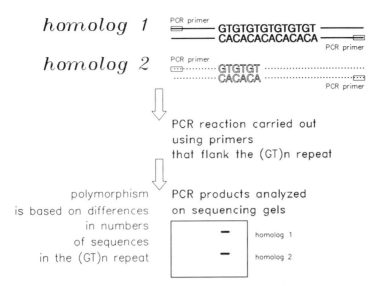

Fig. 8.5. *Polymorphisms can be detected in human DNA sequences near (GT)ₙ repeat sequences by the polymerase chain reaction (PCR) and analysis of the reaction products on polyacrylamide sequencing gels. Two PCR primers, in opposite DNA strands, flank the repeat se-* *quences. The region between the primers is amplified during the PCR process and amplified sequences are analyzed on gels. A high level of variability for the number of GT dinucleotide repeat exists at each $(GT)_n$ repeat site between different individuals.*

seems quite simple, but in practice it is exceedingly difficult. This is because chromosome walking is a very laborious and time-consuming enterprise and because the physical distance between markers flanking the disease locus is usually very large. For example, if one flanking marker is 1 cM proximal and the other 1 cM distal to the disease locus, the distance that must be covered is approximately 2,000 kb. It could take years for a laboratory to walk this distance. In addition, identifying the sequence that may be responsible for the phenotypic trait is also very labor-intensive and involves searching for deletions and point mutations. For this reason, it was decided at numerous conferences (Lewin, 1986) that physical maps, as represented by ordered sets of overlapping cloned DNA sequences, should be constructed for each human chromosome. Such sets of ordered DNA segments, referred to as **contigs,** would provide rapid access to

any region of the genome and would provide the raw input material for DNA sequencing.

V. PHYSICAL MAPS

The term physical mapping has been used to mean different things by different authors. A physical map has been used by some to mean the linear order of loci on a chromosome determined by chromosome breakpoints or by *in situ* hybridization. At a finer level of resolution, a physical map could be a restriction map (see Chapter 2) of a specific segment of DNA. The ultimate physical map of a chromosome is a complete knowledge of its DNA sequence. Prior to sequencing an entire chromosome, however, it will be necessary to develop sets of ordered, cloned DNA fragments that span the entire length of a chromosome. These sets of overlapping, cloned DNA will provide the raw input material for sequencing. Ordered, overlapping cloned DNA fragments are another type of physical map, called a contig map.

DNA fragments can be cloned in a variety of cloning vectors such as phage or cosmids. **Cosmids** are cloning vectors that can typically accept up to 45 kb of insert DNA, while large insert phage vectors can take up to 17 kb of insert DNA. Cosmids contain some elements of phage, such as cloning arms, but are propagated more like plasmids within bacteria as opposed to phage growing as plaques on lawns of bacteria. More recently developed cloning vectors, such as **yeast artificial chromosomes (YACs),** can take much larger inserts, on the order of 200–1,000 kb (Burke et al., 1987). Since certain types of sequences will be consistently deleted from any type of vector (this concept is called **cloning bias**), ordered clone maps that extend the entire length of a chromosome will most likely have to utilize many different types of vectors.

The process of contructing ordered, overlapping clone maps (or contig maps) is analagous to chopping a chromosome into small pieces and then determining the original linear order of the pieces. A group of ordered, overlapping clones representing a single region on a chromosome is called a contig. Eventually, it would be very desirable to have a single contig for each chromosome. Presently, for those chromosomes that have large-scale physical maps being developed, there are hundreds of contigs with many gaps or missing pieces between contigs. The order of DNA fragments within a contig can be determined only by identifying overlaps between different segments. If a region of two clones overlap, then by definition both clones belong to the same contig. If two clones do not directly overlap each other but are connected because they overlap a third clone, then all three clones form part of the same contig. Figure 8.6 is an illustration of a contig.

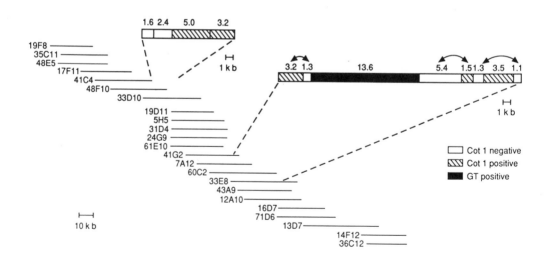

Fig. 8.6. *An example of a contig, or set of ordered, overlapping cosmid clones from human chromosome 16. Reproduced from Stallings et al. (1990), with permission of the publisher.*

A. Fingerprinting Methods

Overlaps between clones within a contig can be detected by a variety of different **fingerprinting schemes** that are based upon either restriction enzyme digestions or DNA hybridization. Numerous strategies have been developed for constructing ordered clone maps and such maps are currently being constructed for the nematode genome (Coulson et al., 1986), the yeast genome (Olson et al., 1986), the bacterial genome (Kohara et al., 1987), and some human chromosomes (Evans and Lewis, 1989; Stallings et al., 1990). Most ordered clone maps use cosmids as vectors, since the insert size of 40 kb is very convenient. All strategies for creating ordered clone maps involve schemes for fingerprinting the cosmid clones so that overlap between clones can be detected. One of the simplest methods for ordering clone maps was first attempted by Olson and colleagues (1986) for mapping the yeast genome. Phage clones containing inserts from the yeast genome were digested with restriction enzymes, and the sizes of fragments from each clone were compared. Clones that overlap contained restriction fragments of similar size. This strategy required a high degree of overlap ($>50\%$) for the overlap to be detected. Olson et al., (1986) were able to generate sets of small, overlapping phage clones that spanned approximately 70 kb on average. Closure or completion of the yeast map could not be accomplished by this mapping strategy so the group is presently using YAC (yeast artificial chromosomes) cloning vectors. As previously mentioned, YACs can take very large inserts. These vectors utilize a yeast centromere, telomere, and origin of replication and an insert from a foreign genome. YACs are propagated in yeast cells as artificial chromosomes.

Coulson et al (1986) have physically mapped a large portion of the nematode genome by a strategy that involves restriction enzyme fingerprinting of cosmid clones (Fig. 8.7). This is an approach that provides more resolution than was obtained by Olson et al. (1986). Their strategy entails digesting cosmid clones with a six-base cutting restriction enzyme, end-labeling the fragments with a radioactive nucleotide, and then digesting the fragments with a four-base cutting restriction enzyme. The small fragments resulting from double digestion with the six- and four- base cutters are electrophoresed on a sequencing gel and restriction fragments containing a radioactively labeled end fragment are visualized by autoradiography. As seen in the diagram, only those fragments that contain an end fragment that was radioactively labeled before the four-base cutter digestion will be visualized by autoradiography. Since a six-base cutting restriction enzyme cuts, on average, every 4 kb and a four-base cutting enzyme cuts every 0.25 kb, many fragments resulting from the four-base cutting enzyme will not be visualized. This makes the restriction patterns less complex and easier to analyze. Cosmid clones that overlap will share some fragments of the same size. Certainly some clones will have fragments of the same size by chance, so cosmids must have considerable overlap to be detected. A variation of this mapping method has been developed that incorporates a fluorescence-based detection system that provides a small amount of DNA sequence information on the lasts four nucleotides of each labeled end fragment (Brenner and Livak, 1989). This requires far less overlap between clones for detection of overlap. At present, approximately 80% of the nematode genome has been mapped in cosmids. The nematode physical mapping group is also using YAC clones to close the map (Coulson et al., 1988).

Strategies that are capable of detecting the lowest amount of overlap have the greatest advantage because fewer clones

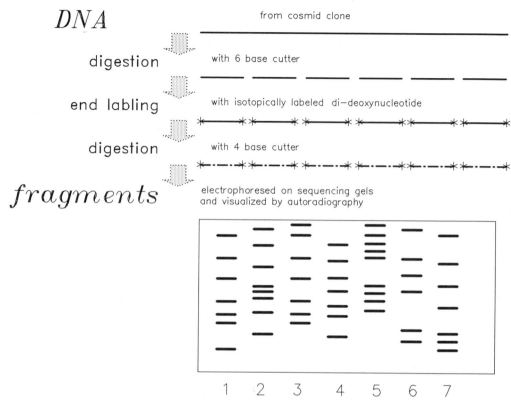

Fig. 8.7. *The physical mapping strategy devised by Coulson et al. (1986) involves cutting DNA clones into cosmids with a six-base cutting enzyme, end-labeling the fragments with a radioactive nucleotide, cutting the fragments with a restriction enzyme with a four-base recognition site, and visualizing the fragments on sequencing gels following autoradiography. Clones that come from overlapping regions of the genome will contain some fragments of similar size, as in lanes 1 and 3.*

need to be fingerprinted. Lander and Waterman (1988) quantitatively showed how the minimum overlap required for the detection of overlap relates to the progress that can be obtained in bridging gaps that exist in the contig map. To date, further progress has been made in the construction of the nematode ordered clone map than any other species. The size of the nematode genome is comparable to a small human chromosome and the progress shown on the nematode physical mapping can be used as a model to evaluate strategies for mapping human chromosomes and the chromosomes of other eukaryotes.

The experimental approach to random ordering of DNA fragments of the nematode genome showed rapid progress in the early phases of mapping. This progress, however, declined dramatically in later stages of the project when the percentage of the target genome reached 70% in contigs. This is because as more clones are fingerprinted, fewer of them extend the map. Instead, new clones fall within contigs that have already been mapped.

Strategies that can detect minimal amounts of overlap between cosmid clones are of the greatest value. The contig mapping strategy (Evans and Lewis 1989) requires only very short overlaps, on the order of only 250 nucleotides for detection. This mapping strategy involves a multiplex hybridization scheme in which pools of probes are obtained from the ends of individual clones and hybridized to large sets of clones individually arrayed on hybridization membranes (Fig. 8.8). Several-fold coverage of the target genome must be arrayed onto nylon hybridization membranes. This strategy requires the use of cosmid vectors that contain phage T3 and T7 promoter sequences that flank the insert so that labeled RNA transcripts can be generated from both ends of the insert DNA. Sets of pooled RNA probes from cosmid clones arrayed in microtiter dishes can then be hybridized (following prehy-

bridization to genomic DNA to mask repetitive sequences) to nylon filters on which the clones have been arrayed in rows and columns. The most logical way of pooling probes is from a row of the matrix followed by a column. Any clones that cohybridize following separate hybridizations by a row and then a column must overlap the clone at the intersection of the row and column. By pooling sets of probes from cosmids in an ordered fashion, fewer hybridizations have to be carried out in order to piece together overlaps. Using this approach to mapping, Evans and Lewis (1989) have identified over 300 contigs on human chromosome 11. This approach to mapping, by itself, does not provide information on the degree of overlap between cosmid clones.

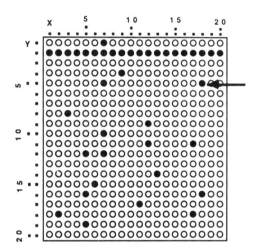 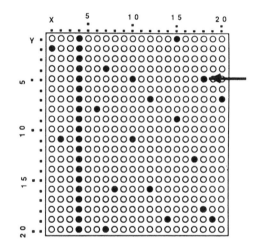

Fig. 8.8. *A hybridization-based strategy for detecting overlap between cosmid clones from a chromosome-specific library (Evans and Lewis, 1989). Probes are made from the end fragments of inserts from cosmid clones and hybridized to sets of cosmid clones ordered on hybridization membranes in rows and columns. Probes from cosmid clones can be pooled to reduce the number of hybridization steps necessary. For example, in hybridization 1, probes from a row of the matrix (row 2) are pooled and hybridized to the clones arrayed on the membrane. All of the clones from row 2 should show positive signal on autoradiographs. Additional clones that*

overlap clones from row 2 will also be positive if they overlap any of the regions covered by the cosmids from row 2. In hybridization 2, clones from an entire column are pooled. Any clones that are positive in both hybridizations (except for clones pooled from the row and column) must overlap with the clone at the intersection of the row and column. In this example, the arrows point to a clone that must overlap the clone at the intersection of the row and column because it was positive in both hybridizations. Reproduced from Evans and Lewis (1989), with permission of the publisher.

At the Los Alamos National Laboratory, an approach to physical mapping has been developed that allows for very small regions of clone overlap to be detected (Fig. 8.9). This strategy takes advantage of the organization (i.e., spacing) of repetitive sequences in the eukaryotic genome to provide "nucleation sites" to accelerate the contig generation. In all physical mapping strategies thus far described, repetitive sequences have been an impediment to physical mapping because they provide the possibility of identifying false overlaps. Rather than being an impediment to physical mapping, the distribution of *Alu*, *L1*,

and $(GT)_n$ repetitive sequences provide abundant "tags" to uniquely fingerprint cloned genomic regions (Stallings et al., 1990). DNA from cosmid clones are digested with three different restriction enzymes, electrophoresed on agarose gels, and transferred by Southern blotting to nylon membranes. The membranes are hybridized to different classes of repetitive sequence probes. The combined information on restriction enzyme fragment size, coupled with whether a fragment contains a particular class of repeat sequence, provides a highly unique fingerprint signature for each clone. This approach is the

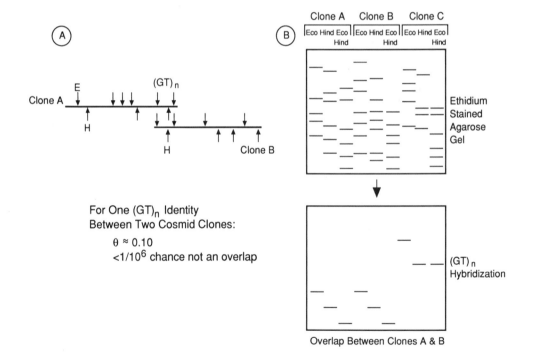

Fig. 8.9. *Schematic showing the rationale behind the repetitive sequence fingerprinting approach developed at Los Alamos. **A**. Two clones, A and B, overlap by approximately 10%. A $(GT)_n$ repeat sequence is shared by both clones in the region of overlap, and information on restriction fragment size, coupled with $(GT)_n$ hybridization, yields an extremely high probability of*

*overlap because of the rarity of $(GT)_n$ repeats in the genome. **B**. Diagram of an ethidium bromide-stained gel and a Southern blot of the gel hybridized to the $(GT)_n$ repeat probe. Clones A and B share a few restriction fragments of the same size and have identical $(GT)_n$-positive restriction fragments. Clone C does not overlap clones A and B.*

only physical mapping approach that allows for the generation of useful biological information, that is, the distribution of repetitive sequences, as the map progresses. By this mapping approach, approximately 400 contigs have been identified on human chromosome 16. $(GT)_n$ repeats are very useful polymorphic markers in linkage mapping. Because the distribution of these repeat sequences is determined by the physical mapping approach of Stallings et al. (1990), the production of a physical map with useful polymorphic landmarks is generated so that the genetic and physical maps can be rapidly integrated.

B. Macrorestriction Maps

In addition to the ordered clone map, another type of physical map based on pulsed-field gel electrophoresis also exists. This is called the **macrorestriction map** and involves the ordering of large DNA fragments and rare cutting restriction enzyme recognition sites. Pulsed-field gel electrophoresis is a special method of electrophoresing very large DNA molecules (50–5,000 kb). By this method of electrophoresis, each individual yeast chromosome can be separated and visualized on an agarose gel (Schwartz and Cantor, 1984). This method differs from conventional electrophoresis because the direction of the electric field alternates. DNA samples are cut with rare cutting enzymes, such as NotI, so that DNA molecules remain quite large. Following electrophoresis, DNA samples can be analyzed by Southern blot analysis. Macrorestriction maps can be produced over large regions of the genome (Poustka et al., 1988; Burmeister et al., 1988). It is expected that macrorestriction maps will provide much useful information in orienting contig maps. Information obtained from pulsed-field gel electrophoresis has been used to correlate genetic and physical distances between markers (Westphal et al., 1987).

It should be emphasized that ordered clone maps allow rapid access to any region of the genome. Ordered clone maps are a very powerful adjunct to linkage mapping, since RFLP, VNTR, and polymorphisms, associated with the GT:AC repeat, can be mapped to disease loci in addition to being positioned on contig maps. The development and interactive use of linkage and physical maps holds great promise for the cloning of DNA sequences responsible for genetic disease.

VI. SUMMARY

1. Some of the earliest genes assigned to human chromosomes were X-linked traits such as hemophilia and color blindness. These traits could be assigned to the X chromosomes because of differences in the frequency with which they were inherited by males and females.

2. Large numbers of genes could not be assigned to individual human autosomes until the development of somatic cell hybrids. Many types of somatic cell hybrids now exist that allow the assignment of genes to many different species. Using somatic cell hybrids that contain rearranged chromosomes has allowed genes to be regionally assigned.

3. The development of recombinant DNA techniques has allowed the detection of many types of polymorphisms at the level of DNA, such as restriction fragment length polymorphisms (RFLP) and variable number tandem repeats (VNTR), that have greatly facilitated linkage mapping in the human and other species. The advent of Southern blotting and polymerase chain reaction are two techniques that have greatly facilitated gene mapping. The use of DNA hybridizatin has also allowed the direct assignment of labeled segments of DNA to chromosomes by *in situ* hybridization. *In situ* hybridization allows DNA segments to be regionally assigned to chromosomes.

4. The present emphasis in gene mapping is on the construction of physical maps. A physical map is a set of overlapping, cloned segments of DNA that span a certain region on a chromosome. The DNA segments can be cloned into many types of cloning vectors, such as lambda phage, cosmids, or yeast artificial chromosomes. Many different schemes have been used to order DNA fragments. These different strategies involve "fingerprinting" clones by various means, such as restriction enzyme analysis or DNA hybridization.

9 Organization of Genomes and Their Chromosomes

We scarcely need add that the contemplation in natural science of a wider domain than the actual leads to a far better understanding of the actual

A.G. Eddington, 1929

The chromosomes of genomes have been viewed in diverse ways by many observers, each of whom generally interprets what he or she sees as consonant with a particular bias supported by preconceived notions. This statement is not designated to be critical of any particular group of observers. It does apply, in fact, to all who contemplate chromosomes by whatever means. Chromosomes have been and still are scrutinized by four basic approaches. Each leads the observer to a particular vision of what chromosomes are, and like the blind men and the elephant, leads to the fallacy of taking a partial view to be a comprehensive picture.

The first approach is the cytological one: the direct microscopical observation of chromosomes during the mitotic and meiotic phases of the cell cycle. This approach begun over 100 years ago and, aided by the use of stains, continues today with improved instrumentation and methods of preparing chromosomes for observation. The staining of prophase and prometaphase chromosomes from vertebrate, insect, and plant sources in particular has made it possible to detect banding pattern relationships and differences within and between species. Differential staining along with the use of chromosome sorting instruments has made it possible to isolate and study single chromosomes from a genome.

The second approach is the genetical one, begun in the first decade of this century with the demonstration that chromosomes are the bearers of Mendel's genetic determinants. The chromosome theory of heredity resulted from the partial fusion of the cytological findings of the previous century and the newly rediscovered mendelian techniques of analysis. The development of new genetic techniques led to linkage analysis and the identification of linkage groups. This, along with banding techniques, allowed for the assignment of the different linkage groups to the chromosomes of genomes such as the mammalian, which are otherwise difficult to identify. The development of somatic cell genetics has made it possible to use cell fusion techniques to greatly increase the rate at which genes can be assigned to chromosomes.

Soon after the identification of DNA as the genetic material the third, or molecular, approach began to be put into use. The ramifications of this approach have been tremendous and have begun to open vistas heretofore unimagined about the organization of genomes and their chromosomes. The development of procedures to isolate genes and other DNA segments by the use of plasmid, cosmid, and YAC vectors; the use of these DNA segments as probes to locate homologous regions on

chromosomes *in vitro*; the determination of nucleotide sequences of the DNA segments and their ordering on the chromosomes; and the elucidation of the structure of histones and other proteins involved in the structure of chromatin by chemical and biophysical methods have extended our understanding of chromosome structure to the submicroscopical level.

The fourth and one of the oldest approaches, but one neglected until recently, is to study that black box, the interphase nucleus. An understanding of the relationships, activities, and orientation of chromosomes in the G1, S, and G2 phases of the cell cycle is critical to our understanding of gene action. The interphase is the period in which the chromosomes do their important work, described in Chapter 7, in the determination of the phenotype. It is the period in which most all scheduled and repair DNA replication occurs along with transcription. We know quite a bit about these processes *in vitro*, but not *in vivo* in the intact interphase nucleus.

To understand genomes and their chromosomes at least all four of these general approaches must be used. In this chapter we concentrate on the second (genetic) approach and the third (molecular) approach in a continuation of the search for order in the structure of genomes.

I. THE GROSS MOLECULAR ORGANIZATION OF GENOMES AND THEIR CHROMOSOMES

The eukaryotic genome is a functional unit consisting of a hierarchy of subunits: chromosomes, genes, and other identified and unidentified sequences of DNA. These elements function within a framework of proteins that, with the DNA, form the chromatin. The genetic activity of the DNA *per se* is to a large degree dictated by the way it is packaged within the chromatin.

This packaging is the *sine qua non* of chromosome structure and function.

A. The Genome as a Whole

One of the more important structural elements of chromatin appears to be the protein scaffold containing topoisomerase II, to which DNA is attached at seemingly specific points to form the many loops of DNA. Presumably the active transcription of DNA occurs in certain regions of these loops, but also within these loops sequences of DNA occur that do not appear to have a direct role in transcription and its regulation. This DNA may constitute well over 90% of the DNA and at least 25% of it consists of various identified repeated specific sequences in human chromosomes. The distribution of these repeated sequences is far from random. Some like the *Alu (SINES)* and *LINE1 (LINES)* families are dispersed completely in the regions between the centromeres and the telomeres that are defined as the euchromatin. It is in these regions that most identified genes are also located. The centric regions and to some extent the telomeric regions have different families of repeats (some called satellites because of their behavior in density gradient centrifugation) that make up the constitutive heterochromatin of the chromosomes. In some organisms, such as maize, the euchromatin has intersitial constitutive heterochromatin. Not all of the repetitive sequences in either the euchromatin or the heterochromatin have been identified, and there still remains to be identified the function(s) of the large number of unique sequences that occupy about 50% of the DNA of the human genome.

Cesium salt density gradient determinations of DNA done under various conditions by Bernardi and his associates (reviewed by Bernardi, 1989) have yielded interesting and important data relating to the genomes of vertebrates including human. The fractionation of DNA fragments by equilibrium centrifugation in Cs_2SO_4 density gradients in the presence of different ligands such as Ag^+ and 3,6–bis(aceto-mercuri-methyl) dioxane enables one to distinguish different families of DNA fragments characterized by their base compositions. Figure 9.1 shows the major components for *Xenopus*, chicken, mouse, and man plotted as histograms with the buoyant densities, and against their GC content. *Xenopus* DNA is significantly different from that of the warm-blooded species with respect to GC content. Chicken, mouse, and man and all other warm-blooded vertebrates analyzed have DNA components labeled H that are high in GC content, whereas the cold-blooded vertebrates have only DNA labeled L for low levels of GC. These different components, distinguished by their buoyant densities, are called **isochores** (equal regions) by Bernardi.

By using specific genic sequences as probes it is possible to indentify the loci of different genes in the isochores. Human genes were found by this method to be mainly concentrated in isochore H3, which is highest in GC content and constitutes only 3.5% of the total genomic DNA. A comparison of the GC contents of genes, introns, and exons revealed that the GC levels of the exons are about 10% higher than the introns. Most genes in warm-blooded vertebrates were in fact found to be GC-rich and therefore in the isochores with the highest GC content, as shown in Figure 9.2, which plots the numbers of rat, mouse, and human genes against the GC levels of their coding regions.

CpG islands, which are regions rich in CpG doublets, are found in sequences over 0.5 kb long, and are associated with the 5′ flanking sequences of many genes. It is in these regions that the regulatory sequences are usually located. These islands are very rare in the genomes of cold-blooded

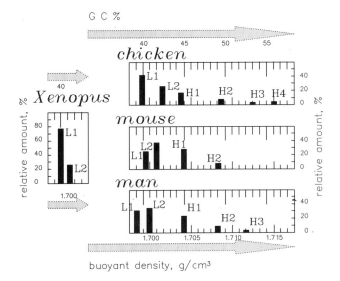

Fig. 9.1. *DNA isochores of Xenopus, chicken, mouse, and man. After data of Bernardi (1989).*

vertebrates such as *Xenopus*. Therefore it cannot be asserted that they are universally essential for gene activity as such. They may, however, be important in dic-

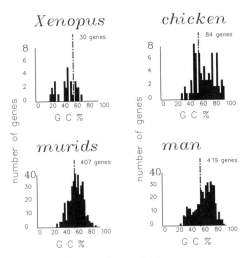

Fig. 9.2. *The numbers of identified genes of rat, mouse, and human that have homologous segments in their GC-rich DNA isochores compared to one another and Xenopus. The broken line at 60% provides a reference. From data of Bernardi et al. (1988).*

tating the different patterns of regulation that determine what is to be a mammal instead of a salamander.

Single-copy sequences of DNA separated from the moderately repetitive sequences by reassociation kinetics followed by hydroxyapatite chromatography has yielded further supporting evidence for the nonrandom distribution of identifiable sequences. Single, copy, unique sequences increase with changes from GC-poor to GC-rich components. This is not inconsistent with the increase of gene abundance in GC-rich isochores, since known genes make up a very minor fraction of the unique class of sequences. This may be telling us that the unidentified single-copy sequences have an identity of their own both structurally and functionally.

Since the *Alu* and *B1 SINES* families of repeats in man and mouse, respectively, are present in GC-rich regions, and the longer *LINES* repeats such as *Kpn* are in the GC-poor regions, it is apparent that the active genes are more closely associated with *SINES* than *LINES* (Soriano et al., 1983). As noted previously (p. 132) the

R bands (Giemsa light) appear to have the heaviest concentration of genes as compared to the G bands (Giemsa dark). This correlates with the fact that the R bands are GC-rich and the G bands GC-poor.

Table 9.1, adapted in part from Bernardi (1989), summarizes some of the data from a variety of sources that show that DNA organization in a chromosome is far from random, if one compares light and dark bands. A more general summary may be made that the more we learn about the distribution of the different sequence forms of DNA, the more it becomes evident that the organization of genomes is on the whole not random.

B. The Genic Part of the Genome

We define a gene as a DNA sequence that is constituted of regulatory sequences upstream and sometimes downstream and the transcribed region between. The transcript may be rRNA, tRNA, and other forms of RNA, or it may be precursor mRNA in which case, in eukaryotes, it will usually contain introns. Various estimates have been made ranging from 30 to 100 thousand for the number of genes in a mammalian genome. The usual estimate for the amount of genic DNA is 2%–3%, but this includes only the exon parts of genes. If the intron parts and the repetitive segments transcribed to produce the rRNAs, tRNAs, and so on are added, this figure becomes about 12%. Twelve percent of the 3 billion bp of the human genome is

$\times$ 10^8 bp. Therefore if the genome has 50,000 genes, each gene on average should be 7.2 kb in length. This includes all the trappings regulatory regions, introns, and so on associated with the whole of the genetic unit. Table 9.2 gives estimates for the sizes of actual human genes. The figures given in the first column for the 15 genes listed have a geometric mean of 18 kb. For 18-kb average gene size a value of total structural gene number of 20,000 is arrived at (3.6×10^5 kb/18 kb). The second column of Table 9.2 gives a geometric mean of 2 kb for the protein-encoding part of the genes. This translates into polypeptides of about 666 amino acids in length. The third column gives the coding fraction, CF, for each gene (mRNA kb/total kb). The mean CF value for these genes is about 0.08. Thus the calculated encoding fraction is 8%. This is higher than the generally assumed value of 2%, but certainly in reasonable range of that figure. The sample of genes in Table 9.2 may be skewed toward the larger-sized ones and not represent the true distribution of sizes. Another method for estimating the number of genes per genome is given in the following section on gene families.

II. ORGANIZATION OF IDENTIFIABLE GENETIC ELEMENTS IN GENOMES

Little attention has been given to the arrangement of recognizable, functioning

TABLE 9.1. Summary of Characteristics of G (Dark) and R (Light) Bands

G Bands	R Bands
GC-poor isochores (At-rich)	GC-rich isochores
Scarcity of CpG islands	Abundance of CpG islands
DNase insensitivity	DNase I sensitivity
Late replication	Early replication
Relative scarcity of *Alu*	Abundance of *Alu* (*SINES*)
Early condensation	Late condensation

TABLE 9.2. Some Estimates of Sizes of Human Genes[a]

Gene	Total kb	mRNA kb	CF[b]
Four in α-globin family	6.3	0.4	0.06
Four in β-globin family	12	0.4	0.03
Insulin	1.7	0.4	0.23
Apolipoprotein	3.6	1.2	0.33
Parathyroid	4.2	1.0	0.24
Protein C	11	1.4	0.12
Collagen			
Pro-alpha-1 (1)	18	5	0.28
Pro-alpha-2 (1)	38	5	0.13
Albumin	25	2.1	0.08
HMG CpA reductase	25	4.2	0.02
Adenosine deaminase	32	1.5	0.05
Factor X	34	2.8	0.08
LDL receptor	45	5.5	0.12
Phenylalanine hydroxylase	90	2.4	0.05
Factor VIII	186	9	0.05
Geometric means	18	2.0	0.09
Medians	25	2.1	0.08

[a]Data in part from McKusick (1986).
[b]CF, coding fraction.

genes in the previous chapters. In this section we concentrate on this aspect of genome organization and to a lesser extent the organization of the so-called nongenic material, which appears to constitute the bulk of the chromatin of chromosomes. The discussion concentrates on a few organisms about which we have the most gene linkage and DNA sequence information, namely, mouse and human among the vertebrates, and some of the cultivated plants such as tomato and maize. The linkage data presently available for vertebrates, particularly the mammals, show certain patterns of organization in common. Some of their genes, if not most, appear to be related in function and structure. And some appear to have been closely linked over long periods of evolutionary time.

In general, if two or more genes show related functions, they are said to constitute a **gene cluster** (Demerec and Hartman, 1959) or **gene complex** (Brink, 1932).

The terms cluster and complex are used more or less synonymously in the current literature. Most recently the term **gene family** has been introduced to describe genes that are related in structure and function and appear to have arisen by duplication of a primal segment of DNA. Obviously all three terms are closely related, since they refer to a common property, namely related function(s). However, they are used in different ways by different authors and there is a considerable degree of ambiguity about their precise definitions. In addition to these three rather nebulous categories, a fourth term, **conservate**, is added in this chapter, as well as fifth term, **singlet**.

In order to make the following discussions about genes and their spatial, functional, and structural relationships as clear as possible, these five terms will be used in the senses defined as follows:

Family: A group of genes linked or unlinked that are related in structure and probably arose by duplication of a primordial, mother gene. A family may consist of two or more genes that show related but not necessarily identical functions. Family members may be linked or unlinked.

Conservate: A group of genes whose linkage appears to have been conserved over a long period of time. Members may be unrelated in structure and primary function, but family members may also be present.

Cluster: A general term applicable to any class of closely linked genes.

Complex: A general term for closely linked genes that have similar or related functions. They may or may not be structurally related.

Singlets: Genes that occur only once per genome and appear to have no identifiable relatives.

It must be admitted that some degree of overlap exists with these definitions. For example, genes in a complex may also constitute a family. A conservate may include members of a family, as well as unrelated genes. But at least the definitions reduce the degrees of ambiguity in the use of the terms. We have only a glimmer of understanding about these relationships, but the possibility is real that they are of importance in the functioning of genomes and their chromosomes.

A. Gene Families

Members of a gene family appear to be homologous in their DNA structure and have related functions. The terms homology and homologous are used here as defined by biologists: **homology** describes a fundamental similarity based on common descent. Thus an arm of a human and the wing of a bird are homologous. This relationship is to be distinguished from that of entities that have similar functions but not a common descent; the wing of an insect and that of bird do not have a common

descent, nor does the eye of an octopus and that of a human. These are said to be **analogous**, since they have arisen independently in evolution. A distinction is usually made between homologous genes present within a species, and those present in different species; the former are designated as *paralogous* and the latter as *orthologous* (Hart, 1987). The general assumption is made that paralogous genes arise either through unequal crossing-over or increase in ploidy of the euploid or aneuploid type. Their evolution is discussed in the next chapter.

Gene families are involved in the synthesis of ribosomal RNAs. These rDNA families are characteristic of all prokaryotes and eukaryotes. Duplication of ribosomal genes is a necessary means of producing enough rRNA to allow for the necessary rate of synthesis of proteins particularly during the embryonic phases of development of an animal and plant. The number of duplicates varies with the demands and the phenomenon is somewhat akin to gene amplification. Here we treat of the gene duplicates constituting families of genes with related but not identical functions that are invariant parts of the genomes that bear them. These genes are in a different category from those duplicate sequences that arise through amplification to form clusters of elements with identical functions.

As of 1989 about 171 groups of two or more genes that have similar functions had been identified in the human genome according to the data reported at the Tenth International Workshop on Human Gene Mapping, published in volume 51 of *Cytogenetics and Cell Genetics*. Three kinds of families can be identified using linkage as a criterion: (1) those in which all members are linked, the **undispersed** families; (2) those in which some members are linked and some unlinked, the **dispersed** families; and (3) the **holodispersed**, in which dispersal is complete and all members are unlinked. Linkage in this context means that so far as can be determined, the genes are in a cluster. Family genes occupying loci on different arms of the same chromosome are considered to be dispersed, but syntenic.

Table 9.3 lists examples of these three types found in the human genome. Members of the same family are considered to be homologous, but unless something is known about their structure, it cannot be said that they are homologous with certainty; they could be analogous. It is important to note that the number of families identified as of 1989 was established by the symbolic acronyms assigned to their genes as shown in Table 9.3. Thus all apolipoprotein-encoding genes are designated *APO*. There also are genes that encode proteins that do not seem to be closely re-

lated in function, but do have obvious amino acid sequence similarities. For example, some of the plasma proteins are clearly related in structure, but their genes have been assigned different symbols. Of the genes listed in Table 9.3 *CGB* and *LHB* are known to be closely related in structure and hence are family members. Also the *CHC* and *GH* genes encode hormones with somewhat different functions, but they have similar structures. (See review by Bowman and Yang, 1987.)

Table 9.4 gives two additional examples of such families: the haptoglobin and albumin families. Undoubtedly many more of these exist, suggesting that these are many genes that belong in families that have not been identified. The 171 families identified have some that are putative members, therefore, but the use of specific

TABLE 9.3. Examples of Families Identified in the Human Genome

Symbol	Family names	Number or genes	Chromosomes	Type[a]
ADH	Alcohol dehydrogenase	5	4	U
AMY	Alpha amylase	5	1	U
APO	Apolipoprotein	9	1,2,3,11,16,19	D
COL	Collagen	14	1,2,6,7,12,13,15,17,21,22	D
CRY	Crystallin	12	2,3,11,17,21,22	D
CYP	Cytochrome P450	15	See Table 9.8	
HLA	Major histocompatibility complex	34	6 (see Fig. 9.5)	U
Ig	Immunoglobulin system	~55	14,15,4,2,22,6	D
HBA	α-Globin	4	16	U
HBB	β-Globin	5	11	U
IL	Interleukin	9	2,4,10,5,7	D
KRT	Keratin-related	10	17,X,9,11	D
CA	Carbonic anhydrose	4	8,16	D
CGB1	Chorionic gonadotropin	1	19	U
LHB1	Luteinizing hormone	1	19	U
CSH2	Chorionic somatomammotropin	2	17	U
GH2	Growth hormone	2	17	U
HOX	Homeobox regions	4	2,7,12,17	D
MH	Myosin heavy chain	8	7,14,17	D
ML	Myosin light chain	4	2,3,8,17	D
LAM	Laminin	2	1,7,18	D
LPC	Lipocortin	5	4,9,10,15	D
MT	Metallothionein	2	16	D
TCR	T-cell receptors	25	7,14	D

[a]D, dispersed; U, undispersed.

TABLE 9.4. The Haptoglobin and Albumin Families

Symbol	Family name	Locus
	Haptoglobin family	
HP	Haptoglobin	16q22.1
HPR	Haptoglobin-like	16q22.1
CTRB	Chymotrypsin B	16q22.3–q23.2
TRY1	Trypsin	7q32–qter
ELA1	Elastase	12
	Albumin family	
GC	Vitamin D-binding protein	4q12–q13
ALB	Albumin	4q11–q13
AFP	α-Fetoprotein	4q11–q13

Data from Bowman and Yang (1987), with permission of the publisher.

DNA probes is reducing doubtful cases to a minimum. Clustered, undispersed members are most likely to be homologous, but these make up only about 22% of the total identified. In many cases probing reveals sequences that are labeled "like." These may be active family members or they may be pseudogenes.

Family clusters of undispersed members are scattered throughout the human genome. Every chromosome has at least one, and chromosomes 1, 2, 6, 7, 8, 11, and X have five or more. Clusters vary in physical size from several thousand to 3 million bp or more for a huge cluster like the major histocompatibility complex (MHC or HLA).

All eukaryotes have gene families. The fungi have fewer than the higher plants and animals. The mouse probably has as many as humans. The last count was 68 (Nadeau and Kosowsky, 1991). However, the number increases almost monthly.

1. Families and singlets

As of 1989 the 171 identified families in the human genome had 682 identified genes and thus an average of four genes per family. In addition every chromosome in the genome has singlets, which occur once and have no apparent relatives. Again emphasis should be put on the word apparent. Some of them may be unrecognized family members. In general these are genes that encode enzymes with activities expected to be found in all heterotropic organisms—prokaryotic as well as eukaryotic. Some examples of singlets in the human genome are listed in Table 9.5. As of 1989 about 154 of these had been identified. The total number of genes in the two categories is 836, of which 82% are members of families. It may not be a coincidence that the number of genes in families identified by 1987 was 370, while the number of singlets was 82. This also indicates that 82% of genes are in families. If it is assumed that this relationship is significant, it provides another possible approach to estimating the number of genes. Singlet genes are housekeeping-type genes for the most part that encode enzymes and other proteins expected to be found in both *E. coli* and humans. Estimates for the total number of genes in *coli* range from 2,000 to 3,000. This a reasonable estimate, since Hahn et al. (1977) detected about 2,300 mRNAs in *E. coli*. Using a possible average of 2,500 for *coli* and assuming that the human genome has a like number of housekeeping genes, it can be calculated that about 14,000 genes are

**TABLE 9.5. Examples of Genes That Apparently Occur
Only Once in the Human Genome**

ACY	Aminoacylase
ADA	Adenosine deaminase
ADK	Adenosine kinase
ADSL	Adenylosuccinate lyase
ADSS	Adenylosuccinate synthetase
ASNS	Asparagine synthetase
ASS	Arginosuccinate synthetase
CAT	Catalase
CBS	Cystathionine-beta-synthase
CCK	Cholecystokinin
CS	Citrate synthase
CTH	Cystathionase
EPO	Erythropoietin
FAH	Fumarylacetoacetate
FDH	Formaldehyde dehydrogenase
FH	Fumarate hydratase
GAD	Glutamate decarboxylase
GALE	UDP-galactose-4 epimerase
GALK	Galactokinase
GALT	Galactose-1-phosphate uridyltransferase
GDH	Glucose dehydrogenase
GLUD	Glutamate dehydrogenase
GPD	Glycerol-3-phosphate dehydrogenase
GPT	Glutamic-pyruvate transaminase
GSR	Glutathione reductase
HPRT	Hypoxanthine phosphoribosyltransfersae
INS	Insulin
IVL	Involcrin
LARS	Leucyl tRNA synthetase
LCT	Lactase
MARS	Methionine tRNA synthetase
MLN	Motilin
MP1	Mannose phosphate isomerase
MB	Myoglobin
OTC	Ornithine carbamoyltransferase
PAH	Phenylalanine hydroxylase
PC	Pyruvate carboxylase
PGD	Phosphogluconate dehydrogenase
REN	Renin
RHO	Rhodopsin
SHMT	Serine hydroxymethyltransferase
SST	Somatostotin
TH	Tyrosine hydroxylase
TOP1	Topoisomerase I
TS	Thymidylate synthetase
UMPK	Uridine-monophosphate kinase
VARS	Valyl-tRNA synthetase

present in the human genome ($685 \times 2{,}500/150$) + 2,500. If the average gene size is estimated at 18 kb, this gives a value of 252×10^6 pb for the part of the total genome occupied by genes. This is 7.6% of the total genome, a value not far from the estimates (given on p. 379) calculated by a different approach, and quite close to an estimate made by Bodmer et al. (1986), using a similar approach.

2. The immune system superfamily of genes

The MHC family may have 40 or more protein-encoding units. It in turn is related to other families to form with them a **superfamily** (Hood et al., 1985; Hunkerpiller and Hood, 1986). These superfamily genes are related in structure and function, since they encode proteins having cell surface functions (reviewed by Williams and Barclay, 1988). Table 9.6 lists the families of genes within the immune system superfamily. The major families: immunoglobulins (Ig), MHC, and T-cell receptor complex (Tcr) are directly involved as the major components of the vertebrate immune system. Members of these families and the others may have arisen at the time of the genesis of the vertebrate line in the Cambrian or before, since an MHC-like system has been identified in a tunicate (Scofield et al., 1982). Genes homologous to these may occur in nonchordate phyla as well. Related molecules have been reported to exist in Drosophila melanogaster.

The members of the MHC, Ig, and Tcr families are dispersed over eight chromosomes, although the MHC genes are all except one concentrated on 6p12.13. Members of the other families such as THY-1 and NCAM are linked with members of the CD3 genes at 11q23. All of the protein products listed in Table 9.6 have some parts with some degree of sequence similarity. The three protein domains designated C1, C2 and V have sufficient conserved sequence runs described as the C1–SET, the C2–SET, and the V-SET to make it conceivable that the molecules that share them are encoded by homologous genes.

The stem cells of the vertebrate bone marrow give rise to a variety of cell types (Fig. 9.3). The macrophages derived from the monocytes that wander about in the lymph and blood systems are involved with the B and T cells in the immune reaction to foreign invaders along with the antigen products of the MHC system.

Klein (1986) has defined the MHC as a "group of genes coding for molecules that provide the context for the recognition of foreign antigens." An MHC has been identified in the Amphibia, Aves, and Mammalia. Because of the large number of investigators that have been involved over the 50 or more years of study of this family of immune responses genes, many synonyms exist for the MHC. In the mouse it is designated H-2, in humans HLA, and in chickens B. At least 19 different synonyms are known (Klein, 1986). In this discussion only the acronym MHC will be used.

The MHC and Ig families have different but related functions. They both provide for an immune response, but they do it in different ways. We have discussed the immunoglobin system in Chapter 5, and pointed out that B lymphocytes circulate in the blood plasma and secrete antibodies that attack carriers of foreign antigens. The immunoglobin system is therefore defined as the humoral immune response to foreign invaders. The MHC system acts with T-type lymphocytes and macrophages, which do not secrete antibodies but have receptors on their cell membranes. Their response is cellular rather than humoral. Several different functional forms of T cells are characterized by the presence of specific surface proteins that determine their phenotype. In addition to these surface markers common to each cell type of T cell, surface receptors

TABLE 9.6. The Immune System Superfamily of Genes

Category	Human chromosome	Functions
I. Immunoglobulins (Ig)		
H chains (IgM)	14q32.33	B-lymphocyte antigen receptors; and in
Kappa L chain	2p12	secreted form, antibodies
Lambda L chain	22q11.12	
II. Major histocompatibility complex (MHC)		
Class I H chain	6p21.13	Produce antigens that deliver foreign
β2-m	15q21–q22	antigens to the Tcr
Class IIα	6p21.3	
Class IIβ	6p21.3	
III. T-cell receptor complex (Tcr)		
Tcr α-chain	14q11.2	Bind MHC antigens to T-lymphocyte
β-chain	7q35	receptors
γ-chain	7p15	
X-chain	?	
CD3 γ-chain	11q23	Associated with Tcr; exact function not
δ-chain	11q23	known
ε-chain	11q23	
IV. β2-m-Associated antigens		
CD1a H chain	1	Not known
V. T-cell adhesion molecules		
CD2	1p13	CD2 of T cells interacts with LFA-3 on
LFA-3	1	other cells
VI. T subset antigens		
CD4	12pte.p12	CD4 and CD8 control bias of T cells
CD8 Chain I	2p12	toward interaction with Class I and
CD8 Chain II	2p12	Class II MHC
CTLA 4	?	
VII. Brain/lymphoid antigens		
Thy-1	11q23	Anti-thy-1 antibody triggers mouse
MRC OX-2	3	lymphocyte division
VIII. Immunoglobulin receptors		
Poly IgR	?	PolyIgR transports multimeric IgA or
Fcγ 2b/γ1R	?	IgM across epithelium
IX. Neural molecules		
Neural adhesion (NCAM)	11q23	NACM mediates adhesion of neural
Myelin-associated (MAG)	?	cells. MAG may function in
Po myelin protein	?	myelination
X. Tumor antigen (CEA)	?	Unknown
XI. Growth factor receptors		
Platelet-derived (PDGF)	5q31–q32	Trigger cell division
Colony-stimulating factor-1 (CSF1) receptor	5q33.2–q33.3	
XII. Non-cell-surface molecules		
α_1-, β-glycoprotein	?	Unknown
Basement membrane link protein	?	

From Williams and Barclay (1988), with permission of the publisher.

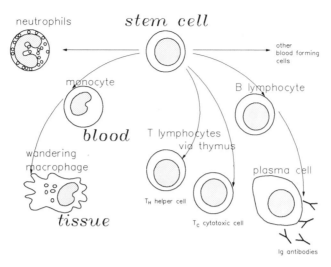

Fig. 9.3. *The origin of macrophages, B lymphocytes and T lymphocytes from bone marrow stem cells.*

that are highly variable are also present. These **T cell receptors**, listed in Table 9.6III, bind specifically only to foreign antigens on the surfaces of other cells. The T-cell system so far as it is understood functions like the immunoglobin system but does not secrete antibodies. It is a highly complex system that is not completely understood, but it clearly enables the bearer to distinguish self from nonself.

Structure of the MHC and its products. Figure 9.4 is a genetic map of the human MHC region on chromosome 6. The map of the mouse MHC on chromosome 16 is closely related in structure to the human, and this is probably true of vertebrates in general (Klein, 1986). The two major groups of genes, designated Class I and Class II, are separated by a Class III region of genes about 400 kb long. (These genes are indirectly involved in the immune response, since some of them produce complement proteins, but we shall not consider their functions here.) The entire region of about 3.3 Mb is within a Geimsa-GC-rich light-band region of 6 and therefore within the GC-rich isochore.

In humans about 16–20 Class I genes have been identified. These encode 44-kd glycoproteins that associate with cell membranes as shown in Figure 9.5. The Class IA, IB, and IC subfamilies of genes encode the α polypeptide chains $\alpha1$, $\alpha2$, and $\alpha3$, which together with the polypeptide $\beta2m$ form the $\vartheta\alpha\beta$-heterodimer illustrated. The $\beta2m$ moiety is a microglobulin with an Mr of about 11,500. It is encoded on chromosome 15 and is not a glycoprotein. The heterodimer is connected by a stalk interdigitated in the lipid bilayer of the plasma membrane and ends in a domain about 40 amino acids long in the cytoplasm of the cell. Note the disulfide bonds characteristic of most of the protein products of the superfamily.

The Class II genes identified in humans encode MHC proteins similar in structure to the Class I (Kappes and Strominger, 1988). They are also $\vartheta\alpha\beta$heterodimers but they do not have the β_2m microglobulin, and have two tails instead of one. They are encoded by the DP, DQ, and DR subfamilies, which together possess about 14 genes (Kaufman et al., 1984; Klein,

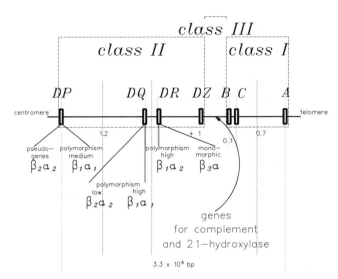

Fig. 9.4. *A genetic map proposed for the human MHC region on chromosome 6. The order for the DQ, DR, and DC loci is not certain. The numbers in parentheses indicate map units in centimorgans. The total region including complement and 21-hydroxylase is about 3.3 Mb.*

The β3 element in the DR region is a pseudogene, as are the β2 and α2 elements. The vertical broken lines indicate possible positions of recombinational hot spots. After Klein (1986) and Bodmer et al. (1986).

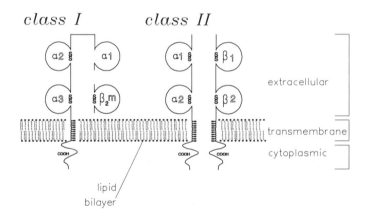

Fig. 9.5. *Models for the Class I and Class II molecules of the MHC. See text.*

1986). The genes of the MHC family are for the most part polymorphic and hence able to produce a large number of different **haplotypes**. The MHC system is apparently designed to produce the many different haplotypes known to exist in the human and mouse populations. The term haplotype refers to polymorphisms in a cluster of closely related and linked genes. Each gene in a cluster may have a number of alleles. The combined polymorphisms are the haplotypes producing the different antigens of the MHC. For example consider two linked genes, a and b, with their respective alleles, a^1 and b^1. For this two-gene cluster four haplotypes are possible: ab, ab^1, a^1b, and a^1b^1. Gene complexes with many polymorphic genes are capable of producing many different haplotypes. For example, four genes each with four alleles segregating in the population will produce 256 different haplotypes, assuming that each allele occurs in equal frequency. The number of possible genotypes in a diploid will be $256(256 - 1)/2 = 3.26 \times 10^4$. The MHC complex in humans encodes about 34 identified antigens; hence it presumably has at least 34 genes. If each gene locus has two alleles, then 2^{34}, or 1.72×10^{10}, haplotypes should be expected, and $1.72 \times 10^{10} (1.72^{10} - 1)/2 = 1.48 \times 10^{20}$ should be circulating in the population provided each allele is present in equal frequency. The fact that each person has a unique histocompatibility identity should come as no surprise, for the chance that any two persons will have identical histocompatibility patterns is essentially zero, unless they are identical twins. As shown in Figure 9.4 some loci are highly polymorphic, some have medium polymorphy, and at least one is monomorphic. Medium polymorphy is shown in the DP and DQ families, while the DR family has a high degree of polymorphy. The Class I genes are the most polymorphic. Altogether a total of 132 different alleles have been identified in this group (Klein et al., 1990). Some of these alleles differ by more than 90 nucleotide substitutions and the proteins they encode by 20–30 amino acid substitutions. It has been shown that recombinatorial hot spots are present in the mouse MHC (Steinmetz et al., 1986). These might be expected to result in somatic hypermutations by gene conversion, as has been suggested for the immunoglobulin genes (Maizels, 1989).

The T-Cell Family T cells neither secrete antibodies, as B cells do, nor produce antigens that attach to the surfaces of cells, as do the MHC. Rather they are lymphocytes that develop antigens of their own on their surfaces that are generally designated CDI–CD8. The antigen carried by the T cell determines its phenotype. It can be a **helper cell** (T_H), in which case it is CD4, or a **cytotoxic cell** (T_C) which carries the antigen CD8. These are also referred to as T4 and T8 cells, respectively, in T-cell jargon, which is probably the most complex known to man. T lymphocytes will react with antigens only under certain conditions. They will not respond to "self" antigens, and they will react to "nonself" antigens but only under certain conditions. From the foreign antigen must be "presented" on the surface of an **antigen-presenting cell** (APC) and, second, the T cell must possess on its surface a **receptor** that recognizes the antigen (Strominger, 1989). The T-cell receptor, TCRn is an $\alpha\beta$-heterodimer with a structure not unlike the Class II MHC molecules. The two chains are not closely related, however, and are encoded in two different families: the α-chain by T$\gamma\sigma\alpha$ on chromosome 14 and the β-chain by the T$\gamma\sigma\beta$ genes on chromosome 7. These receptor molecules, like the immunoglobin antibodies, have variable and constant regions and are formed by processes analogous to those involved in immunoglobulin formation. Hence one expects that there should be an almost unlimited number of different receptors.

When a T cell with a receptor that recognizes a specific foreign antigen meets that antigen, a train of processes is started that eventuates in the MHC–T-cell immune response. MHC Class 2 antigens react with T_H-cell receptors, and Class I antigens with the T_C-cell receptors (also called CTLs) (Kourilsky and Claverie, 1989).

The participants are the macrophages that have engulfed a foreign invader with foreign antigens, the MHC antigens, and the T cells of various types. B cells also become involved. The process begins with the macrophage engulfing a foreign cell (Fig. 9.6). The ingested cell is digested in a vacuole much as an amoeba digests a food particle, and its antigens (i.e., foreign proteins etc.) are released. These migrate to the surface of the macrophage, where they become associated with the host's MHC antigens. The macrophage so endowed is the APC, as shown in Figure 9.7. The APC can now present the antigen to

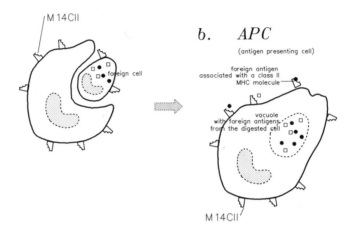

Fig. 9.6. a. *A macrophage with Class I MHC antigens on its surface has engulfed a foreign cell.* **b.** *A foreign antigen from digested cell becomes associated with the MHC antigen and the macrophage becomes an APC (antigen-presenting cell).*

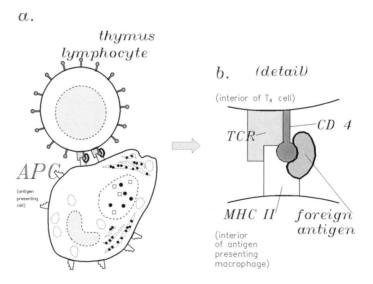

Fig. 9.7. *The APC reacts with a CD4 helper cell (T_H). The TCR is the T_H cell's receptor that is specific for the foreign antigen. See text.*

the T_H cell that has the TCR that recognizes it, as well as its characteristic antigen, in this case CD4. Once this encounter occurs the recognized T cell begins to proliferate and send out chemical messages to other immune system cells.

The chemical instructions induce other self monocytes and macrophages to become active and enhance the ability of the invaded individual to cope at a high level of efficiency. The cytotoxic T_C lymphocytes, also called killer lymphocytes, are induced to proliferate. These carry the antigen CD8, and kill the foreign cells that display foreign antigens and interact with the macrophages carrying the Class II antigens. B lymphocytes capable of secreting immunoglobulin antibodies to the foreign antigens are also stimulated to proliferate and join in on the attack. Some of these activities are summarized in Figure 9.8. To keep things from getting out of hand T cells called suppressor cells (T_S) are activated. They secrete chemical messages to slow the immune response, or stop it, if it is no longer needed.

These considerations concerning the immune system superfamily of vertebrates give us one picture of the functions of a multigene family. The genes of this family collaborate through their products to carry out an extremely complex and important function (Kourilsky and Claverie, 1989). Without this system a vertebrate cannot survive beyond its earliest years.

3. The P450 superfamily

The P450 family of genes encodes the cytochrome 450 enzymes which are involved in oxidative metabolism of a wide variety of compounds such as prostaglandins, steroids, fatty acids, various plant products, drugs, and noxious compounds present in the environment such as carcinogens and pesticides (Nebert and Gonzales, 1987). Table 9.7 lists the 15 genes of this family identified in the human genome along with a short description of their characteristics. It is a widely dispersed family with genes present on seven chromosomes in eight different subfamilies. All eukaryotes and some prokaryotes have

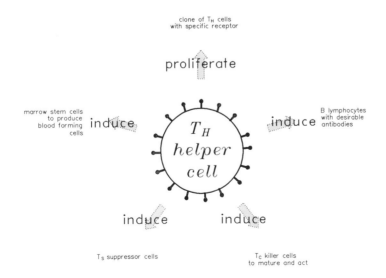

Fig. 9.8. *Some of the events following the activation of a T_H helper cell. These inductions are mediated by various agents such as interferons, lymphokines, cell growth factors, cytokines and so on. Complement components are also involved.*

TABLE 9.7. The P450 Genes Identified in the Human Genome

Symbol	Locus	Description
CYP2A	19q13.1–q13.2	Subfamily IIA, phenobarbitol-inducible
CYP2B	19q13.1–q13.2	Subfamily IIB, phenobarbitol-inducible
CYP2F	19q13.1–q13.2	Subfamily IIF
CYP11B1	8q21–q22	Subfamily XIB, polypeptide 1 11-beta hydroxylase
CYP11B2	8q21–q22	Subfamily XIB, polypeptide 2 11-beta hydroxylase
CYP21	6p21.3	Subfamily XXI, 21-hydroxylase
CYP21P	6p21.3	Subfamily XX1, psuedogene
CYP11A	On 15	Subfamily XIA, cholesterol side chain cleavage
CYP1	15q22–q24	Subfamily I, aromatic compound inducible
CYP19	15q21	Subfamily XIX, aromatization of androgens
CYP2E	On 10	Subfamily IIE, ethanol-inducible
CYP17	On 10	Subfamily XVII, steroid 17-alpha-hydroxylase
CYP2C	10q24.1–q24.3	Subfamily IIC, mephenytoin 4-hydroxylase
CYP2D	On 22	Subfamily IID, debrisoquine, sparteine metabolizing
CYP3	7q21.3–q22.1	Subfamily III, niphedipine oxidase

been found to have P450 enzymes, and altogether ten subfamilies have been identified. All mammals have eight of these subfamilies.

The amino acid sequence of a protein from anyone of the ten subfamilies identified is about 36% similar to the other nine subfamily members. Within a subfamily the similarity is about 68%. The prokaryote *Pseudomonas putida* has one P450 gene that encodes an enzyme with a cysteinyl-containing peptide with an amino acid sequence similar to all other cysteinyl peptides in eukaryotes, as shown in Figure 9.9. This is the region of the cytochrome molecule that associates with the iron in the heme. These similarities in sequence provide some evidence that the P450 family is constituted of homologous members. It may be an extremely ancient one going back 2 billion years, and before the origin of the eukaryote cell.

The P450 cytochromes have a myriad of monooxygenase activities ranging from oxidation of simple aliphatic compounds to oxidative deamination, to sulfoxide formation, and to hydroxylation of aromatic compounds, among others. The enzymes are not free in the cytosol but bound in the endoplasmic reticulum of particular cell types, especially in the livers of vertebrates. The liver enzymes catalyze the oxidation and detoxification of many different toxic substances. They are inducible, so their concentration is greatly increased. Doses of barbiturates and carcinogenic hydrocarbons like methylcholanthrene are especially active inducers.

The induction of the many different kinds of P450 enzymes is dependent on the nature of the toxic substances that reach the liver after being ingested. Some are absent in embryonic and newborn mammals but appear as active enzymes postpartum. Transcription of the genes appears to be in part under hormonal control. Phenobarbitol causes a marked transcriptional activation of the rat P450IIB subfamily. Transcriptional activation of P450 genes also occurs in the kidney, lung, spleen, and intestines of the mouse. In many cases the activation of the different genes is tissue-specific. Since some of the enzymes participate in the synthesis and degra-

a. *P 450 protein*

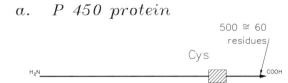

b. *comparison of amino acid sequences*

species	gene	distance from Cys, in amino acids
		−7 · · −4 · −2 · 0 · 2 · 4 · 6 · 8 · 10 · 13
Pseudomonas P 450 cam	CI	PHE GLY HIS GLY SER HIS LEU | CYS LEU GLY GLN HIS LEU ALA ARG ARG GLU ILE ILE VAL THR
yeast lan	LI	PHE GLY GLY GLY ARG HIS ARG | CYS ILE GLY GLU HIS PHE ALA TYR CYS GLN LEU GLY VAL LEU
fish P₁ 450	I	PHE GLY MET ASP LYS ARG ARG | CYS ILE GLY GLU ALA ILE GLY ARG ASN GLU VAL PHE LEU PHE
chicken P450 II C	II	PHE SER ALA GLY LYS ARG ILE | CYS ALA GLY GLU GLY LEU ALA ARG MET GLU ILE PHE LEU PHE
bovine P 450 scc	XI A	PHE GLY TRP GLY VAL ARG GLN | CYS VAL GLY ARG ARG ILE ALA GLU LEU GLU MET THR LEU PHE
porcine c 17	XVII	PHE GLY ALA GLY PRO ARG SER | CYS VAL GLY GLU MET LEU ALA ARG GLN GLY LEU PHE LEU PHE
rabbit form 2	II	PHE SER LEU GLY PRO ARG ILE | CYS LEU GLY GLU GLY ILE PHE ARG THR GLU LEU PHE PHE PHE
rat P 450 pcn 1	III	PHE GLY ASN GLY PRO ARG ASN | CYS ILE GLY MET ARG PHE PHE LEU MET ASN MET LYS LEU ALA
mouse P₃ 450	I	PHE GLY LEU GLY LYS ARG ARG | CYS ILE GLY GLU ILE PRO ALA LYS TRP GLU VAL PHE LEU PHE
human P 450 c 21	XXI	PHE GLY CYS GLY ALA ARG VAL | CYS LEU GLY GLU PRO VAL ALA ARG LEU GLU LEU PHE VAL LEU

Fig. 9.9. *Diagram of the linear P450 protein* (**a**) *and the approximate locations of the highly conserved cystein-containing peptide* (**b**). *Data from Nebert and Gonzales (1987).*

dation of steroids and the catabolism of fatty acids, they must be considered important agents for the maintenance of homeostasis. It is highly likely that there exist more P450 genes than we are presently aware of especially in animals. Animals ingest all kinds of toxic substances, most of them being natural plant products, and since the genes must be induced to be transcribed only in the presence of certain specific compounds, many of the latent genes may not be activated and easily identified.

4. The homeobox superfamily

The **homeobox** is a DNA sequence or domain about 180 bp long that is conserved in a great many, if not all, eukaryotic taxa ranging from *Saccharomyces* to man. Homeobox genes contain one or more of these **homeodomains** (HDs) and all are believed to encode binding proteins with amino acid sequences encoded by the HDs. These genes constitute a superfamily. The proteins they encode will be referred to in this discussion as HD proteins. Of course, they have many more amino acid residues besides the relatively limited HD-related sequences.

In 1950 Lewis described a complex of genes on chromosome 3R of *Drosophila melanogaster* when mutated led to the replacement of one body part by another. An example of this is shown in Figure 9.10. In this case mutations in the gene designated *bx* (bithorax) transforms the anterior portion of the third thoracic, or metathoracic segment (T3), to the anterior portion of the second thoracic, or mesothoracic, segment (T2). In the normal fly only T2 bears wings, as is true for the Diptera in general. T3 bears a pair of **halteres**. These are homologous to wings, but they do not serve as wings to keep the fly aloft. Rather they serve as balancers to keep the fly from yawing in flight by acting as gyroscopes. A *bx* mutation, which is usually a deletion (Hogness et al., 1985), may cause the halteres to develop into almost perfect wings. Thus one segmental structural element, the T3 haltere, is transformed into a homologous one, a T2 wing. This is called **homeosis**, or homoeosis (Bateson, 1894). Homeo is derived from the Greek and simply means similar.

By the late 1970s it began to become evident that these genes were involved as regulators in the development of the segmental body characteristics of *melanogaster* (Lewis, 1978). It was then that they were recognized as *homeotic* genes. In the 1980s they were found to be related in structure not only to one another, but to developmental regulators in other eukaryotes by virtue of possession of the 180bp homodomain.

The name homoeotic now applies to all genes that bear homeoboxes whether involved in segment transformation or not.

These genes are also found in animals that are not segmented, such as nematodes and echinoderms, so they should not be considered as regulating segmentation only. Actually some of the homeotic genes

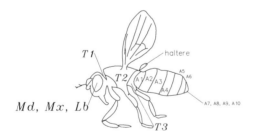

Fig. 9.10. *Diagram of a normal fruit fly, showing the body segments composing the thorax and abdomen. Each of the three thoracic segments, T1, T2, and T3, bears a pair of legs. T2 bears a pair of wings and T3 a pair of halteres. Genes in the* abd-A *affect the development of abdominal segments A1, A2, A3, and A4. The rest of abdominal segments are affected by mutations in the* Abd-B *region. The head segments are Md, Mx, and Lb.*

of *Drosophila* appear to be involved in neuronal differentiation, and the differentiation of cells along a dorsal–ventral as well as an anterior–posterior axis, as would be expected for segmentation (Akam, 1987; Beachy, 1990).

Drosophila *boxes*. Two distinct regions located on the *Melanogaster* chromosome 3R have genes that regulate certain phases of the flies' development. These are generally referred to as *BX-C* and *ANT-C*, acronyms for the Bithorax and Antennapedia complexes, respectively. Within each complex are genes that are structurally related. The *BX-C* is organized into three large functionally related regions designated as the *Ubx* (Ultrabithorax), *abdA* (abdominal-A), and *AbdB* (Abdominal-B) domains (Fig. 9.11). The *ANT-C*, which is about 11 cM from the *BX-C*, between it and the centromere at 84AB on the polytene map, includes at least seven genes (Fig. 9.12).

Mutations within the *Ubx* domain produce homeotic transformations in ectodermal derivatives such as thoracic segments, the nervous system, and mesodermal muscular derivatives (Beachy, 1990). This domain (diagrammed in Fig. 9.11) consists of six identifiable loci grouped into two regions, the *abx/bx* and the *bxd/pbx* regions, occupying about 110 kb. Table 9.8 lists some of the phenotypic effects of mutations at these loci. Within the *abx/bx* region is a transcription unit about 77 kb long indicated by the arrow in Figure 9.11. This transcribed region produces a family of 3.2- and 4.3-kb mRNAs, which are derived after hnRNA splicing from the regions of the arrow shown in black. A family of six proteins, each with an HD, is encoded by these mRNAs, which differ by virtue of alternative splicing. One of these has been shown to bind with high affinity to specific DNA sequences (Beachy et al., 1985). The others presumably act in the same general fashion for other sequences. The basic function of these *Ubx* products appears to be the regulation of transcription of specific genes. They have

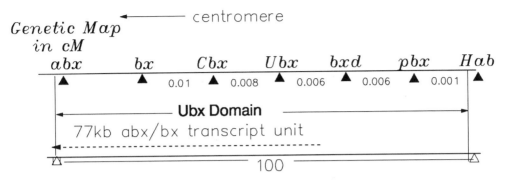

Fig. 9.11. *The BX-C gene cluster on chromosome 3R of D. melanogaster. The Ubx domain of the BX-C region of chromosome 3R is shown as the genetic map and the DNA map. The ab-dominal region is immediately to the right and not shown. The entire BX-C region is about 300 kb long. The genetic map shows the distances between some of the gene loci in centimorgans.*

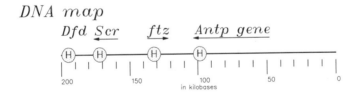

Fig. 9.12. *The ANT-C region on chromosome 3R shown as the DNA map. Only a part of the region is shown which includes the Dfd, Scr, ftz, and Antp genes. H indicates the positions of the DNA HDs. The four exons of the Antp gene encode the antennapedia protein, the HD of which has the amino acid sequence given at the bottom of the figure.*

TABLE 9.8. Some Phenotypic Effects of Mutations at the *BX-C* and *ANT-C* Loci

BX-C		ANT-C	
Mutation	Effect	Mutation	Effect
bx/bx	AT3 → AT2	*Antp/+*	T2 → T1
pbx/pbx	PT3 → PT2	*ftz/ftz*	Die in embryo; body segments lacking
Ubx/Ubx	Lethal	*Dfd/+*	Md and Mx segments abnormal
Ubx/+	AT3 → AT2	*Scr/+*	Lb and T1 segments affected
bxd/bxd	A1 → AT3		

A = Anterior; P = posterior; T2 = metathoracic segment; T3 = mesothoracic segment; A1 = first abdominal segment (see Fig. 9.11).

been shown to act as transcription factors in cultured *Drosophila* cells, yeast, and mammalian cells. What role(s) the encoded products of these transcribed regions play is not known, but they are involved in the processes of development.

Mutations in the *ANT-C* produce changes in the head and anterior thoracic segments (Kaufman et al., 1980; Scott et al., 1983; Gehring, 1987). Figure 9.12 shows the organization of the complex according to Gehring (1985). The loci of HDs and four genes of the complex are indicated on the DNA map. Like the *BX-C* genes the *ANT-C* are tightly linked, being less than 1 cM apart. The entire complex is not shown here; it probably occupies a space of about 300 kb, of which only about 200 kb is shown in Figure 9.12. The *Antp* region, as identified by mutations within it, is about 100-kb long (Scott et al., 1983) and consists of eight exons, four of which encode the 62 amino acid sequences constituting the HD of the protein. This sequence is given at the bottom of Figure 9.12. *Antp* is a complex gene that is identified with two major transcripts of 3.4 and 5.3 kb. Mutations within the *Antp* region are generally found to be deletions or inversions or both.

The nature and phenotypic effect of one type of inversion has already been described on page 219. The antennae of the fly are replaced by a pair of T2 legs. It has been concluded from this that the function of the *Antp*[+] gene is probably to ensure that these legs are in the proper T2 position. Deletions at the *Antp* locus in general cause a transformation of T2 toward T1 and head segments, but the more posterior T3 segment and the development of the nervous system in the posterior abdominal segments can also be affected. Transcripts of *Antp*[+] and its protein have been found to be mainly localized in the T2 and parts of the T1 and T3 segments (Hafen et al., 1983; Wirz et al., 1986). This supports the conclusion that the main function of the gene is to be involved with T2, but it is apparent that it also interacts with the genes of the *BX-C* system. Deletion of *Ubx* leads to the expression of the *Antp* gene in those posterior segments where *Ubx* is normally expressed. Thus it is possible that *Ubx*[+] represses the *Antp* gene in the posterior segments.

Mammalian and other vertebrate boxes. Figure 9.13 shows the chromosomal locations of 20 human homeoboxes isolated by Boncinelli et al. (1989). Typically, as in other eukaryotes in which they occur, the HDs are in clusters. For example, six HDs lie on mouse chromosome 6 in a 70-kb region, and five HDs have been located on chromosome 11 in a 50-

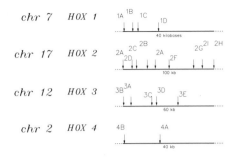

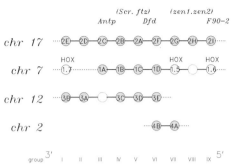

Fig. 9.13. *Homeobox (HD) sequences identified in the human genome. The four chromosomes involved have two or more boxes. Boxes on each chromosome are clustered in regions of less than 10^{-4} cM in total length. After Boncinelli et al. (1989).*

Fig. 9.14. *The 20 homeodomains (HDs) of the human domain arranged to show relationships in amino acid sequences. The three mouse HDs, HOX-1.5, HOX 1.6, and HOX 1.7 are included. The HDs that line up vertically are most closely related in structure. At the top of the figure certain of the Drosophila HDs are listed to show their relationship to the mammalian HDs. They too are arranged vertically to indicate the HDs of human and mouse to which they are most closely related. After Boncinelli et al., 1989).*

kb region (Martin, 1987). A great deal of similarity exists not only among the four HD types, HOX1, 2, 3, and 4, but in other HDs in vertebrates and *Drosophila*. Figure 9.14 shows schematically how the 20 human HDs are related to one another and to the HDs of mouse and *Drosophila*. The HDs at different loci in humans have been lined up in columns to indicate the similarity of amino acid sequences. Thus the HDs 2A and 1C are identical and the 2D and 3A differ by only one amino acid. The HDs of the *ANT-C* complex of *Drosophila* are indicated at the top of Figure 9.14 and positioned above the human HDs to which they are most similar in sequence.

Although the connection between the control of development and the HD containing proteins is not well established in vertebrates as it is in *Drosophila*, there is evidence that such a connection does exist in the former (Dressler and Gruss, 1988; Rossant and Joyner, 1989). Homeobox genes have been isolated from the mouse and shown to be actively expressed during an early stage of development leading up to the primitive streak stage (Gaunt et al., 1986). The gene *Mo-10* encodes an HD that is identical to *Antp* at 44 of 60 residues (McGinnis et al., 1984a; McGinnis, 1985). It appears to be involved in the formation

of the metameric pattern in the mouse. When an excess of synthetic homeobox mRNA for the *Xenopus* HD protein Xhox-1A is injected into *Xenopus* eggs, development of the embryo is disturbed. The morphogenesis of somites is chaotic, indicating that the role of Xhox-1A is to direct development of the normal segmented somite pattern (Harvey and Melton, 1988). The vertebrate growth hormone (GH) is actively expressed in the anterior pituitary. Its activation is under control of an HD protein, GHF-1, that is also specifically expressed in the anterior pituitary (Bodner et al., 1988; Karin et al., 1990). Apparently the GHF-1 binds to the GH promoter and initiates transcription of the GH gene. This is strong support for the role of HD proteins in the development of a vertebrate.

Degrees of HD homologies. Significant homologies are found in HDs in yeast and in the members of several invertebrate phyla including the Annelida, Echinodermata, Arthropoda, and all Chordata examined (McGinnis, 1985). They have also been found in *Caenorhabditis elegans*, a member of the Phylum Aschelminthes (Costa et al., 1988; Finney et al., 1988). Vertebrate sequences exhibit 70%–90% homology with *Drosophila* (Levine et al., 1984).

The *MAT* gene of yeast discussed in Chapter 5 (p. 242) encodes regulatory proteins one of which, α2, has been shown to be a DNA-binding protein (Johnson and Herskowitz, 1985). Both *MATα* and *MATa* encode proteins that have significant amino acid sequence homology to the HDs of the higher animals. Thus it is apparent that the 180-bp sequence is highly conserved and very ancient, perhaps going back to the beginnings of the eukaryotic line generally considered to be about 10^9 years ago. The degree of homology of the *Drosophila Antp*[+] gene HD and an HD of *Xenopus* is amazingly high—59 of 60 amino acid residues are identical (Mueller et al., 1984; De Robertis et al., 1985). It is generally assumed that the chordate and invertebrate lines separated about 500,000 years ago. Some homology exists between eukaryotic HDs and prokaryotic regulatory genes. The latter, like the eukaryotic proteins, have a helix–turn–helix recognition domain that binds directly to specific nucleotides in DNA. These considerations lead to the conclusion that proteins with HDs are probably master regulatory elements that act in the control of transcription of genes involved in various ways in the processes of development. They may start processes early in development that result in cascades of effects that lead to the adult state in the multicellular animals (Han et al., 1989). Homeoboxes have been found in plants where they play similar roles to those in animals (Hake, 1992). Since fungi also have an apparent homologue of HD, it is probable that HD-type genes are present in all eukaryotes (Kües and Casselton, 1992).

5. The collagen family

The collagens are a family of proteins found throughout the metazoans (re-

viewed by Bornstein and Sage, 1980). They have been most studied in the vertebrates, and 15 genes have been identified in humans that encode the various family members (Table 9.9). Actually there may be more than 20 collagen genes, since at least 23 collagen types have been identified. The primary function of collagens is supportive, to maintain the integrity of vertebrate organs. They are also involved in processes such as cell adhesion and movement and wound healing. About a third or more of the the total body protein of the average mammal is collagen. The various subunits, as listed in Table 9.9 are polypeptide chains all of which show some degree of homology in amino acid sequences. The molecular structure is that of a trimer composed of one, two, or three different subunit chains that are combined together as triple helices.

The factors involved in the transcription of these genes are complex and diversified (Ramirez and Di Liberto, 1990). The synthesis of the various collagens is regulated by a variety of specific cell-type enhancers and promoters. As is true for transcription in general, both *cis*-acting DNA and *trans*-acting binding proteins are involved in the regulation of expression of these genes. The different sets of genes are under different types of promoter control. The collagen I genes, *COL1A1* and *COL1A2* code for α1(I) and α2(I) subunits, respectively, which form a trimer with two αI and one α2 chains. The genes are on different chromosomes, but have similar enhancers and promoters, one of them being CCAAT, a common promoter type for other unrelated genes. On the other hand, the two genes, *COL4A1* and *COL4A2*, which encode the α1 and α2 subunits of collagen type IV, are in the same region of chromosome 13, arranged head to tail and separated by a 127-bp segment containing a binding site for the transcriptional factor SP1 (Poschl et al., 1988). This is a promoter that acts bidirectionally and has a structure unrelated to that of *COLI* genes.

TABLE 9.9. List of Collagens and Their Encoding Genes in the Human Genome

Type	Subunits	Gene	Locus
I	α1(I)	COL1A1	17q21.3–q22
	α2(I)	COL1A2	7q21.3–q22.1
II	α1(II)	COL2A1	12q14.3
III	α1(III)	COL3A1	2q31–q32.3
IV	α1(IV)	COL4A1	13q34
	α2(IV)	COL4A2	13q34
V	α1V	COL5A1	?
	α2V	COL5A2	2q14–q32
	α3V	COL5A3	?
VI	α1VI	COL6A1	2q22.3
	α2VI	COL6A2	2q22.3
	α3VI	COL6A3	2q37
VII	α(VII)	COL7A1	?
VIII	α1(VIII)	COL8A1	?
IX	α1(IX)	COL9A1	6q12–14
	α2(IX)	COL9A2	?
	α3(IX)	COL9A3	?
X	α1(X)	COL10A1	?
XI	α1(XI)	COL11A1	1p.24
	α2(XI)	COL11A2	6p21.3
	α3(XI)	COL11A3	12q14.3
XII	α1(XII)	COL12A1	?
XIII	α1(XIII)	COL13A1	10

6. Gene families in plants

Families of genes have been identified in higher plants especially those that are of economic importance. Gene family clusters are not common. Only two have been identified in the tomato *Lycopersicon esculentum* (Tanksley and Mutschler, 1990). Four esterase genes have been mapped to the same locus on chromosome 2, and two peroxidase genes are on the same chromosome at a different locus. Twelve other family groups of two or more are dispersed or identified only with respect to syntenic location. Most genes mapped in tomato as well as other plants are identified by their mutant morphological effects and therefore cannot be assigned a biochemical role. This is true of maize, for which more genes have been identified than for any other plant (Coe et al., 1990). It, too, has families, but the members are generally widely scattered. This may be the result of the

origin of these plant genomes more by polyploidy than simple gene duplication by other means. A relatively new genome derived from interspecific crosses leading to a hexaploid aneuploid (p. 269)—the cultivated wheat *Triticum aestivum*—has a complement consisting of three genomes A, B, and D. Each of these donates duplicate genes which may be considered paralogous family members (Milne and McIntosh, 1990; Hart and Gale, 1990). Duplicate genes have long been known in cultured plants whether they are sources of food or merely ornamentals. A somewhat similar genomic organization is found in *Xenopus*, an animal whose genome has relatively recently been derived from a polyploid state (Chapter 10).

As in animals, there are members of plant gene families that are turned on and off. One example is found in the seed storage families and is expressed in cotton seeds (Dure and Chan, 1984). Three families of genes are involved in the synthesis of storage proteins at different times of development of the embryo in the seed.

7. Structure and function of families

In this chapter and Chapters 5 and 6 a number of families have been described. Five families have been described in some detail: the globin, immune system, P450, homeobox, and collagen families. They all have in common a group of genes related in structure and function, but at the same time they all perform somewhat different roles in the lives of the organisms in which they function as important genetic elements. Furthermore they are all five (and this may be true of families in general) ancient, some going back a billion years or more.

Circumstantial evidence points to members of a single family being derived by the duplication of single ancestral genes. These may have been carried over in the evolution of the primordial pre-eukaryotes to the present eukaryotic types. Of the many significant differences between prokaryotic genophores and eukaryotic chromosomes is the fact that gene families hardly exist in prokaryotes, whereas eukaryotic genomes are loaded with family members. A reasonable explanation for this difference is that the state of differentiation in eukaryotes is far more complex than that found in prokaryotes. Therefore more genes are required for differentiation, which results not only in more complex individuals but a greater variety of different types of individuals, that is, species.

The most obvious of the many examples of the different members of a family participating at different periods of development are the α- and β-globin genes discussed in Chapter 6, p. 310. Why different periods in the life of the human embryo and fetus have different compositions of hemoglobin is not well understood. But disturbing this replacement of one kind of globin by another in time can cause serious disturbances in phenotype. This succession in the activation of related, different genes that encode products that function at different times in development is reminiscent of Haekel's "ontogeny recapitulates phylogeny" aphorism (Haeckel, 1866; Gould, 1977). It is well known that in the development of all members of the vertebrate subphylum, the embryos start out with very similar morphological traits, but as development proceeds the members of the different Classes, Orders, Genera, and species take on more and more of the traits that are characteristic of the taxa to which they belong. It is well within reason that these series of steps of activation of gene family members is an important factor in the processes that have led not only to a complex organization but to a greater variety of forms, since not all members of the related families in different taxa are identical. We no longer accept the premise of Haeckel that an organism in its development goes through the same processes it did in its evolution, but we cannot deny that over successive generations eukaryotes have accumulated more genes of a similar type by duplication. These duplicates evolve by mutation and respond to different activation factors in transcription at different times, thus leading to variations in phenotype. They assume different but related functions and may be active at different periods of development. It should be recognized that most of the proteins that have been isolated and sequenced have generally been from differentiated cells. Gene products from egg and early embryonic tissues for the most part have not, and these may in fact represent a far larger quota of protein types than are present in a fully differentiated adult. One example drawn from many is the distinction between fetal and adult forms of the muscle acetylcholine receptor of cattle (Mischina et al., 1986), and of course the embryonic and fetal globins of primates and other vertebrates.

A second aspect of family function, not necessarily unrelated to the first, involves the types of roles played by family members in protecting the organism against exigencies of the environment. The immune and P450 families are each in their own way examples of this. The immune system family of related genes is one of the most complex known to us. In addition to protecting against foreign invaders it also has members that are involved in cell adhesion and the triggering of cell division. The general functions appear to be cell surface recognition, but it is evident that the functions have diverged so that new functions have arisen over time.

The P450 system of genes bear some relationship in function to the immune system genes in as much as it reacts to exogenous influences that can be deleterious. But instead of reacting with cell surfaces its products are enzymes that catalyze the oxidation of various compounds both endogenous and exogenous. Exogenous compounds that are deleterious are generally destroyed in the liver. They actually induce the transcription of P450 enzymes that oxidatively inactivate them. Those that act on endogenous compounds such as steroids act to maintain homeostasis.

Regulatory processes leading to phenotypic change are basically of two different types: (1) transient changes in direct response to the environment that are generally reversible when the environmental stimuli are removed, abated, or modified and (2) the "permanent" type of change that occurs in the differentiation process of the developing eukaryote. In general prokaryotes have only the first type of regulation, while the eukaryotes have both as exemplified by the immune system and P450 families.

The homeobox family illustrates yet another aspect of family function—the regulation of development by encoding binding proteins that are involved in the regulation of development of transcription of proteins with important functions in differentiation (Dressler and Gruss, 1988). Every binding protein identified with this system has within it the 180-bp HD sequence, which is highly conserved. Although the functions of the products of the regulated genes are not identified, their importance cannot be overestimated. Once they are encoded, however, it would appear that they start a cascade of effects, culminating in a fully developed adult. It is the HD binding proteins that trigger the processes, and like the binding proteins of lambda phage that make the "switch" (Ptashne, 1986), they should be considered the candidates for master switches in the Superkingdom Eukaryota.

The members of the collagen family provide examples to illustrate another noteworthy feature of gene families, namely, that the different members may have quite different transcriptional controls. Thus, though all produce similar proteins, the encoding genes may be activated not only at different times but by quite different *trans*-acting proteins binding to different enhancer and promoter elements. This and other family examples point to the conclusion that the differentiation of a group of cells into a complex multicellular animal or plant is the result primarily of the actions of families of genes. Without families a metazoan would simply be a colony of identical cells.

B. Gene Clusters Involved in Sequential Reactions in Metabolism

The close linkage of genes involved in reaction sequences leading to the synthesis of a final product such as an amino acid is commonly found in bacteria such as *E. coli* and *Salmonella*, but it is not common in eukaryotes. The fungi, however, have more evidence of this type of gene organization than most other eukaryotes. Table 9.10 lists some examples found in *Neurospora*, yeast, and one case in *Drosophila*.

TABLE 9.10. Some Gene Clusters Involved in Sequential Reactions That Encode Multifunctional Enzyme Complexes

Gene cluster	Pathway	Organism
aro genes	Five genes in aromatic amino acid biosynthesis pathway	*Neurospora crassa*
qua genes	Seven genes in quinic acid utilization pathway	*Neurospora crassa*
ilv genes	Two genes in biosynthesis pathway leading to isoleucine and valine	*Neurospora crassa*
gal genes	Three genes in galactose utilization (LeLoir) pathway	*Saccharomyces cerevisiae*
rd genes	Three genes encoding three enzymes in pyrimidine biosynthesis pathway	*Drosophila melanogater*

A representative example of a group of tightly linked genes involved in a metabolic pathway is one leading to the biosynthesis of chorismic acid in *Neurospora crassa* (Fig. 9.15a). All together eight genes are known to be involved in this sequence of reactions. Of these, five are tightly linked in a cluster on chromosome 2 and encode enzymes that catalyze steps 2–6 (Fig. 9.15b,c). This cluster of *aro* genes encodes a multifunctional enzyme that catalyzes the steps 2–6 (Gaertner and Cole, 1977). Mutants have been isolated that lack individual enzyme activities of the multi-

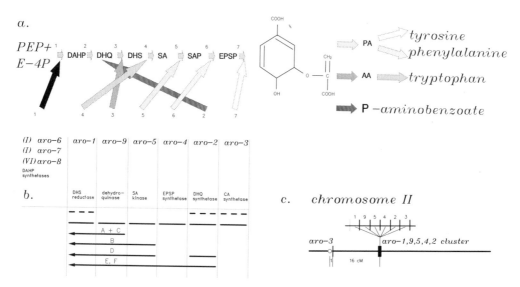

Fig. 9.15. a. *Reactions involved in the biosynthesis of chorismic acid in the polyaromatic amino acid pathway in Neurospora crassa.* **b.** *Complementation map for the five aro genes in the cluster (see p. 413).* **c.** *Genetic map of the six linked aro genes on chromosome II. PEP = phosphenolpyruroic acid; E-4-P = D erythrose-* *4-PO$_4$; DAHP = 3 deoxy-D-arabinose-heptulonic acid-7-PO$_4$; DHQ = 5-dehydroquinic acid; DHS = 5-dehydroshikimic acid; SA = shikimic acid; SAP = shikimic acid-5-PO$_4$; EPSP = 3-enolpyruvylshikimic acid-5-PO$_4$; CA = chorismic acid. Data from a review by Giles (1978).*

functional complex, and these have been used to construct the map shown in Figure 9.15c. Gene clusters such as this have also been found in two other fungi, *Ustilago* (Berlyn and Giles, 1972) and *Saccharomyces*.

The enzyme complex moves as a unit when centrifuged in a sucrose gradient. This is different from the behavior of enzymes produced by *E. coli* and *Salmonella*, which are encoded coordinately by operons in these prokaryotes that do not remain together when centrifuged in gradients. The polycistronic mRNAs encoded by the operons do not produce multifunctional complexes, but individual enzymes. Multifunctional enzyme complexes have been found, on the other hand, in many eukaryotes. While in some cases the multifunctional enzyme complex is encoded in total or part by clustered genes, this is not always the case. In *Neurospora* two genes involved in the pathway leading to the synthesis of isoleucine and valine are linked on chromosome 5 (Wagner et al., 1960; Kiritani, 1962). As shown in Figure 9.16, these two amino acids are synthesized jointly by a number of the same enzymes. This synthesis occurs in the mitochondria (Wagner and Bergquist, 1963) by an enzyme complex of about 400 kD incorporating five activities (Bergquist et al., 1974). However, only the genes encoding the reductoisomerase and dehydratase are linked. The others involved are on different chromosomes. The complex synthesizes valine from pyruvate and isoleucine from threonine plus pyruvate and is not active unless released from isolated mitochondria that are actively carrying out oxidative phosphorylation and respiring. The complex apparently has the five enzymes in the pathway illustrated in Figure 9.16.

Gene clusters are also found in *Drosophila melanogaster*. The rudimentary locus (*r*) in *melanogaster* is a complex on the X chromosome that encodes at least three enzyme activities in the pyrimidine pathway: aspartate transcarbamylase, carbamylphosphate synthetase and dihydroorotase (Rawls and Fristrom, 1975). These are apparently in a multienzyme complex (Jarry, 1976). Three genes on chromosome 2 of *melanogaster* are involved in purine metabolism: *ade2-1*, *ade3-1*, and *bur^{gva-1}* (Johnstone, 1985). Two of these, *ade2-1* and *ade3-1*, are closely linked at 18.4 and 20.0, respectively, on 2, but *bur^{gva2-1}* is at 55.8.

C. Conservation of Chromosome Structure

It is a remarkable fact, not often emphasized by those who study chromosomes, that eukaryotic chromosome structure is basically the same, ranging from fungi to all plants and animals. Centromeres and telomeres have conserved DNA

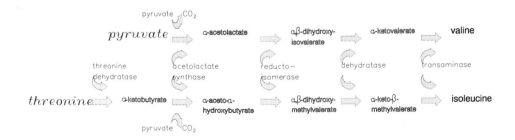

Fig. 9.16. *The isoleucine–valine pathway in Neurospora.*

structure; the structures of histones and other proteins, such as toposoimerases, associated with chromosomes are highly conserved; all chromosomes with the exception of those found in the Dinoflagellates have nucleosomes; the 10-nm and 30-nm fibers and the loops, folds, and coils are universally involved in the basic structure of chromosomes in the cycles that constitute the lives of cells; and the distribution of repetitive DNA sequences is not entirely random. Added to these characteristics, all of which differentiate chromosomes from genophores, are two others we discuss in this section, namely, the banding patterns exhibited by chromosomes after the application of certain stains and the conservation of linkage of certain genes.

Analyses of banding patterns and genetic maps of mammalian chromosomes give strong evidence that even those from different taxa separated as widely as the ordinal level have many regions with chromatin structural and gene linkage similarities. Indeed, conserved linkage patterns similar to those found in mammals have even been found in the bony fishes.

1. Banding patterns

G-banding patterns developed after condensation of the chromosomes in the prophase and metaphase stages of the mitotic cycle make it possible to do comparative banding pattern studies of some bony fishes and amphibia, and all reptiles, birds, and mammals (Bickmore and Sumner, 1989). [Apparently G bands are not developed in the chromosomes of the lower chordates, the Chondrichthyes, Agnatha and Protochrodates (Holmquist, 1989)]. As shown in Table 9.1, the Giemsa dark G bands and the Giemsa light R bands have several different characteristics, and for this reason it is assumed that the banding patterns reflect differences in functional chromosomal domains.

Whatever the banding patterns reflect, it is a fact that closely related species of animals and plants nearly always have related banding patterns. Figure 7.10 makes this clear for the chromosomes of humans, chimpanzee, gorilla, and orangutan. There are exceptions, however, a prominent one being the difference between these four primate species and the gibbon. In addition, there is little similarity in banding between the mouse and human chromosomes even though these two species have considerable similarities in patterns of linked genes. However, since an average R band in a metaphase chromosome may contain ten or more megabases of DNA, a considerable number of different genes may be linked within one of them, dispersed among the *SINES* repetitive sequences. Despite the exceptions, it has been found that in many cases sections of chromosomes from different species in different orders of mammals show definite homologies in banding patterns. Figure 9.17 shows, for example, that similar banding patterns can be traced in the cat and human genomes. Note that regions with similar banding patterns have the loci of similar genes.

2. Conservates

A conservate is defined here as a similar syntenic relationship of two or more genes in different taxa of plants or animals. It must be emphasized that synteny merely specifies that two or more syntenic genes are on the same chromosomes. They may occupy different arms and be so far apart that they segregate independently with 50% recombination, as described in Chapter 8. Furthermore closely linked syntenic genes may or may not be linked in the same order in different species even though synteny is conserved. In order to make these differences clear Nadeau (1989) has defined two types of conserved syntenic relationships: (1) **syntenic conservation**, in which the syntenic genes may be

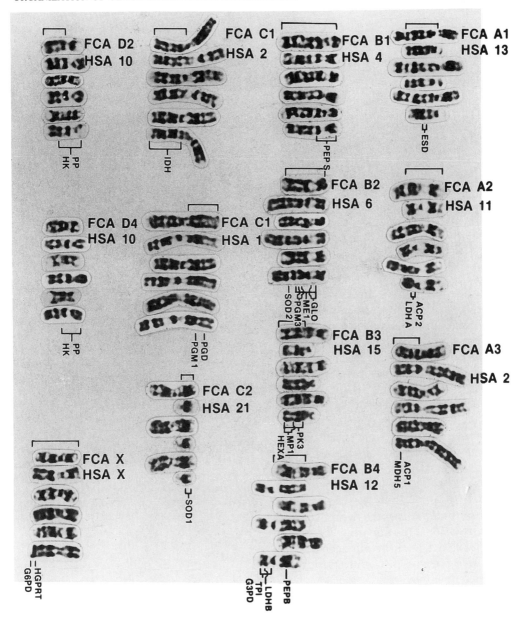

Fig. 9.17. *Comparison of banding patterns in cat and human genomes. FCA, cat; HSA, human chromosomes. Courtesy of Nash and O'Brien (1982).*

widely separated such as on different arms, or if closely associated on the same arm, linked in different order in two different species (Fig. 9.18a); and (2) **linkage conservation**, in which the genes are not only syntenic, but linked in the same order relative to one another and the centromere (Fig. 9.18b).

The Mammalian X. The most obvious case of linkage conservation is one that

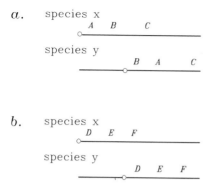

a.

species x

A   B   C

species y

B   A   C

b.

species x

D   E   F

species y

D   E   F

Fig. 9.18. *Conserved linkage types.* **a.** *Syntenic conservation.* **b.** *Linkage conservation. See text.*

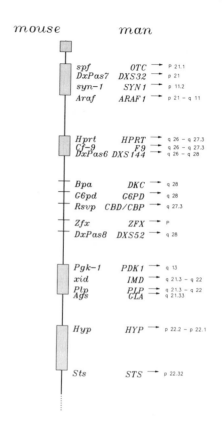

mouse   *man*

spf	OTC	→	P 21.1
DxPas7	DXS32	→	p 21
syn–1	SYN1	→	p 11.2
Araf	ARAF1	→	p 21 – q 11
Hprt	HPRT	→	q 26 – q 27.3
Cf–9	F9	→	q 26 – q 27.3
DxPas6	DXS144	→	q 26 – q 28
Bpa	DKC	→	q 28
G6pd	G6PD	→	q 28
Rsvp	CBD/CBP	→	q 27.3
Zfx	ZFX	→	P
DxPas8	DXS52	→	q 28
Pgk–1	PDK1	→	q 13
xid	IMD	→	q 21.3 – q 22
Plp	PLP	→	q 21.3 – q 22
Ags	GLA	→	q 21.33
Hyp	HYP	→	p 22.2 – p 22.1
Sts	STS	→	p 22.32

Fig. 9.19. *Comparative map for human and mouse X-linked homologous genes. The genes are mapped on the mouse X, and the chromosomal location of the human homologue is given after each mouse gene. Areas of conserved linkages are indicated by the heavy bars.*

has been recognized for at least 30 years; among the mammals X chromosomes are remarkably similar in gene content. This is known as **Ohno's Law**. The genes *G6PD*, *HPRT*, *GLA*, and *PGK* are invariably found on the X chromosome in all species in which homologues of these genes have been identified. The law holds for all sex-linked genes of eutherian mammals without exception up to 1989. The correspondence between the human and mouse Xs is exceptionally clear. The 42 genes syntenic on the human X, are also syntenic on the mouse X (Lalley et al., 1989). Linkage conservation is present (Fig. 9.19), but rearrangements have apparently disrupted whatever the original relationships may have been. All genes on the p arm of the metacentric human X are present in the single arm of the acrocentric mouse X (Fig. 9.19).

Ohno's Law is not completely obeyed when one compares eutherian to metatherian and protherian X chromosomes. In particular, all human sex-linked genes on the short Xp arm are on autosomes in marsupials and the duckbill platypus. Most of the human Xq genes are on the marsupial Xs, with at least one exception: *TBG* (thyroxin-binding globulin) is X-linked in humans and autosomal in kangaroos and platypus. The Xp arm genes of humans are not all silenced in females, and it is of

some interest that they are autosomal in kangaroos.

Between mouse and man. The most extensive linkage data for mammals is for humans and the laboratory mouse (Searle et al., 1989). Between these two there are at least 69 conserved linkage groups (Nadeau et al., 1991). There is even more conserved synteny. Figure 9.20 gives examples of autosomal synteny between man and mouse. The mouse map is the reference, and the chromosomal assignments for the human orthologous genes are shown

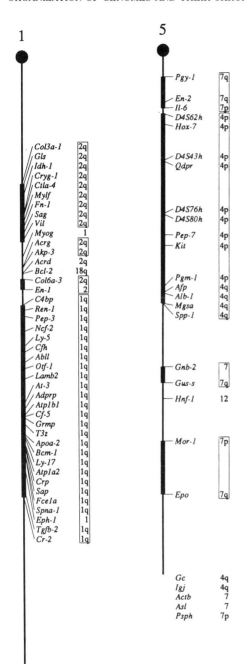

to the right for mouse chromosomes 1 and 5. Conserved linkage segments are highlighted. Two large segments of mouse 1 are conserved on the q arms of human 1 and 2. Chromosome 5 of mouse has considerably homology with human 4 and 7, but note that some segments are on human q and some on human p. The genes indicated at the bottoms of the mouse chromosomes are syntenic, but they have not yet been mapped to those chromosomes, although some have been to the human. The grid in Figure 9.21 shows how extensive the total known conservation of linkage between mouse and humans is.

Conserved segments in the mouse range from about 1 cM to more than 30 cM (Searle et al., 1989). Obviously a close homologous relationship exists between human and mouse genomes. They differ primarily in that their homologous segments are differently spatially organized among the chromosomes of their genomes.

Vertebrate conservates in general. Gene maps with sufficient linkage data to make comparisons between different species of Eutheria are available for members of the following orders: Primates (Creau-Goldberg et al., 1990), Rodentia (Stallings et al., 1990; Levan et al., 1990), Lagomorpha (Fox, 1990), Carnivora (O'Brien, 1990; Meera et al., 1990; Serov and Pack, 1990), Artiodactyla (Womack, 1987; Gallagher and Womack, 1992), and Perrisodactyla (Weitkamp and Sandburg, 1990).

Among the Primates extensive syntenic homology and considerable linkage homology is evident between the human chimpanzee, gorilla, and orangutan maps. The gibbon gene linkage map, as well as the banding pattern, shows that extensive rearrangements have occurred in this line

Fig. 9.20. *Linkage maps of mouse chromosomes 1 and 5 with loci of human homologous genes indicated. From Nadeau et al. (1991) with permission.*

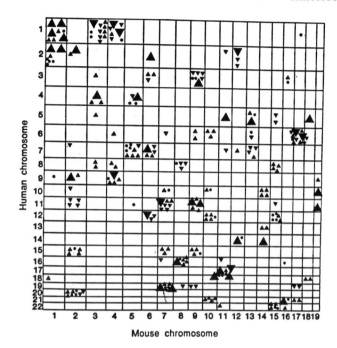

Fig. 9.21. *A comparison of human and mouse linkages of autosomal homologous genes. This grid shows the chromosomal loci of related mouse and human genes. The length of the horizontal side of each rectangle is proportional to the length of each indicated mouse chromosome. The lengths of each human chromosome are proportional to the vertical sides of the same rectangles. The loci of homologous genes in man and mouse are indicated by triangles. Triangles pointing down indicate p arm locations, and those pointing up, q arm locations on the human chromosomes. The circles indicate unknown arm locations. Large triangles refer to 5 loci. For every human chromosome at least two genes are also syntenic in the mouse. Reproduced from McKusick (1990), with permission of the publisher.*

of apes. However, some conservates have been maintained. As in the case of the human–mouse relationship, the gibbon–higher ape relationship demonstrates that considerable chromosome scrambling may occur without destroying conservates especially in related taxa. The Old Word (Catarrhini) monkeys and the New World (Platyrrhini) monkeys show considerable linkage homology with the apes and humans. The catarrhines—rhesus (*Macaca*), baboon (*Papio*), African green monkey (*Cercopithecus*)—and the platyrrhine capuchin (*Cebus*) and owl (*Aotus*) monkeys have many conservates in common. However, the owl monkey group, composed of eight different species, has 12 different karyotypes and the banding patterns show extensive rearrangement of chromosome elements as in the case of the gibbon (reviewed by O'Brien et al., 1988), but even so, conservates are maintained with all the other Primates (Lalley et al., 1989). Banding patterns similar to those of the majority of apes and monkeys are also apparent in the prosimians like the mouse lemur.

Linkage maps for mammals outside the Order Primate are not extensive except for the mouse, as already discussed. Banding patterns for about 250 carnivore species do show conservation of these patterns to a degree similar to that found in primates (Wurster-Hill and Centerwall, 1982). Linkage studies demonstrate conserved synteny comparable to that found among the primates for the cat (Nash and O'Brien, 1982) compared to human. The dog has scrambled chromosomes like the gibbon and owl monkey, but again conservates are apparent. It also has many small chromosomes ($n = 39$) making identification of linkage groups with chromo-

somes difficult. Linkage data for the cow are the most extensive among the non-primates except for the mouse (Womack, 1987, 1990). Syntenic and linkage homology is pronounced between cattle, human, and mouse. Panels a and b of Figure 9.22 illustrate detailed linkage relations between human chrosomes 9 and 12 and bovine linkage groups U18 and U3, respectively (Threadgill and Womack, 1990; Threadgill et al., 1990). Not only is there

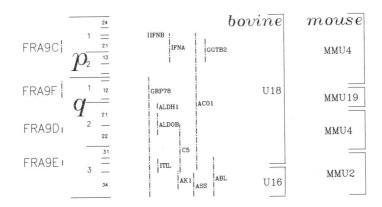

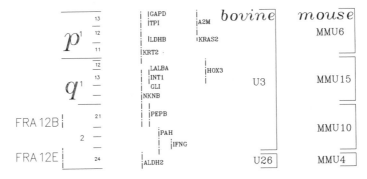

Fig. 9.22. a. *Comparison of chromosomal locations of genes in cattle linkage groups U18 and U16 with human 9 and mouse 4, 19, and 2. Common fragile sites on human 9 are indicated on the left.* **b.** *Comparison of chromosomal locations of genes in cattle linkage groups U3 and U26 with human 12 and mouse 6, 15, 10, and 4. Common fragile sites on human 12 are indicated on the left. From Threadgill and Womack and Threadgill et al. (1990) with permission.*

considerable homology for synteny of genes between these two sets of chromosomes, but the locations of fragile sites show some concordance. This leads to the possibility that fragile sites may demarcate conservates (O'Brien et al., 1988). The common conservate between human 12 and bovine U3 spans the region from pter of human 12 to q24. This region amounts to about 4% of the total average mammalian genome that has been conserved over a period of perhaps the 65 million years that separate the primate from the artiodactyl lines.

Linkage data outside the Class Mammalia are not presently extensive. However, some interesting similarities between distant kinships are emerging from data now available. Linkage maps are being constructed for salmonid (May and Johnson, 1990) and poeciliid (Morizot et al., 1990, 1991) fishes. The map for the sword tail, *Xiphophorus*, actually shows evidence for at least eight conservates common to the human map. Three of these are also found in the mouse genome and are extensively distributed throughout the vertebrates. Table 9.11 lists the vertebrates with one or more of these four conservates: *PKZ/MPI, GPI/PEPD, TKI/GAIK* and *TPI/GAPD*. The plus marks indicate their presence, the minus and question marks indicate an apparent nonsynteny, and blanks indicate no available data. The evolutionary distance between fish is at least 350 million to 400 million years.

TABLE 9.11. Examples of Synteny Among Vertebrates

Species	1 PKZ MPI	2 GPI PEPD	3 TP1 GAPD	4 TK1 GALK
Homo sapiens	+	+	+	+
Mus musculus (lab mouse)	+	+	+	+
Rattus norvegicus (rat)	−	+		+
Cricetulus griseus (hamster)	+	+		+
Felis catus (domestic cat)	+		+	
Canis familiaris (dog)	+		+	+
Oryctolagus cuniculus (rabbit)			+	
Sus scrofa domesticus (pig)	+			
Ovis aries (sheep)	−		+	
Bos taurus (cow)	−		+	
Mustela vison (mink)	?	+	+	+
Pan troglodytes (chimpanzee)	+		+	+
Gorilla gorilla			+	
Pongo pygmaeus (orangutan)	+		+	
Hylobates concolor (gibbon)	+		+	
Macaca mulatta (rhesus)			+	
Papio sp (baboons)	+	+		
Cercopithecus aethiops (African green monkey)	+		+	+
Cebus capucinus (capuchin monkey)	+			
Microcebus murinus (mouse lemur)	+	+	+	
Saguinus oedipus (marmoset)	+	+		
Aotus trivirgatus (owl monkey)	+			
Rana pipiens (leopard frog)		+		
Salmonid fishes		+		
Xiphophorus	+	+	+	

Organisms other than vertebrates. The banding patterns seen in the polytene chromosomes of the Diptera have been studied extensively in *Drosophila* sp and other genera such as *Sciara, Rhynchosciara, Simulium, Anopheles,* and *Chironomus.* These bands are not the same as the bands noted in mammalian chromosomes after staining, nor do they correspond to pachytene chromomeres. Nonetheless, they do provide band marks enabling the recognition of chromosomal aberrations. Conservation of banding patterns with blocks interrupted by inversions and translocations is evident in all species within these genera with the exception of a small number of species of *Drosophila.* The latter are in complexes of related species which show no banding differences within a complex. They are called **homosequential species complexes** (Carson, 1983). The homosequential Hawaiian *Drosophila* may show great differences in phenotype, but this is not common among *Drosophila* nor other Dipteran genera. With few exceptions found only in the genus *Drosophila,* different species have altered banding patterns resulting from chromosomal rearrangements. Generally it is possible to trace the changes in banding patterns because even in large complexes of 20 or more species the patterns remain conserved in blocks, just as described for conserved blocks of genes in vertebrates. Actually in many *Drosophila* species, groups with whole arms of chromosomes remain conserved to a high degree, and as previously described for mammals, chromosome numbers appear to be reduced by centric fusions of acrocentrics, or increased by dissociation to form acrocentrics from metacentrics (Sturtevant and Novitski, 1941).

Comparisons of linkage maps between different species of *Drosophila* do show a close linkage relationship. For example, *D. pseudoobscura* chromosomes 1(X) and 2 bear what appear to be many genes homologous to *D. melanogaster* chromosomes 1 and 2 (Anderson, 1989). A dissociation appears to have occurred to produce an $n = 5$ genome for *pseudoobscura* from the $n = 4$ genome of *melanogaster.* Chromosome 3 of *pseudoobscura* appears to bear genes homologous to those on 2 of *melanogaster.* On the other hand, the X of *pseudoobscura* is metacentric and its right arm bears apparent homologies to genes on 2 of *melanogaster.* A comparison of the linkage maps of *D. subobscura* and *melanogaster* also shows homologies between the chromosome arms of the two species (Pinsker and Sperlich 1984).

The Australian sheep blowfly (*Lucilio cuprinius*) has been analyzed genetically to a sufficient extent to make it possible to compare its linkage map with those of the house fly (*Musca domestica*) and *D. melanogaster* (Foster et al., 1981). There appear to be definite relationships—enough to warrant the assumption that the major linkage groups have survived largely intact during the evolution of these three different forms of higher Diptera from a common ancestor.

3. Significance of conservates

The significance, if any, of conservates is yet to be established. This question is addressed from the evolutionary point of view in Chapter 10. Here we consider certain findings that may provide reasons for some genes to remain linked over long periods of time.

1. The gene *CSFIR* (colony-stimulating factor 1), located at 5q33–q35, is closely linked to *PDGFR* (platelet-derived growth factor receptor) at 5q31–q32 in the human genome. In the mouse the two genes are linked on chromosome 18; therefore the two genes form a conservate. The coding regions between *PDGFR* and *CSFIR* on human chromosome 5 are contiguous. The genes lie head to tail and their amino

acid sequences indicate that they are homologous and may have arisen through duplication. The 3' end of *PDGF* is located less than 500 bp from the 5' untranslated exon of *CSFIR*. The possibility exists that both genes respond to the same promoter–enhancer sequence upstream from *PDGF*, since no such sequence has been identified between them (Roberts et al., 1988).

2. Genes within a cluster encompassing 15–250 kb may replicate during the same interval during S phase. Early-replicating genes are not found in late-replicating clusters. This coordinate timing of replication may indicate that contiguous genes in these clusters (also read conservates) act coordinately in encoding their products.

3. The genes *TK-1* (cytosolic thymidine kinase) and *GALK* (galactokinase) are linked in all mammals in which they have been assigned loci (Table 9.11). Chinese hamster cells in culture that are selected for absence of galactokinase activity also show a reduction in thymidine kinase activity and vice versa (Wagner et al., 1985). It has not been found possible to obtain cell lines that are completely deficient in both. Cells in culture from two unrelated humans homozygous for galactokinase deficiency (*GALK⁻/GALK⁻*) showed definite negative alterations in thymidine kinase levels (Schoen et al., 1984; Okajima et al., 1987). It is not known whether these two genes are coregulated, but in any case there does seem to be some kind of relationship between the two.

4. The genes *MT* (metallothionein) and *TAT* (tyrosine amino transferase) are linked in both humans and hamster. When *MT* is amplified in cultured cells in the presence of Cd^{3+}, *TAT* is coordinately amplified (Stallings, unpublished).

5. The *ABO* alleles, the gene *AK1* (adenylate kinase-1), and the oncogene *ABL* (Abelson) are linked on human chromosome 9 at q34. Several patients with blood type A have been reported to change blood type when they contracted hemolytic leukemia. In addition they also developed low levels of adenylate kinase while other enzyme levels remained normal (Kahn et al., 1971; Race and Sanger, 1975).

6. The genes *TPI* and *GADP* encode the enzymes triosephosphate isomerase and glyceraldehyde-3-phosphate dehydrogenase, respectively. Both are key enzymes in the glycolytic pathway; the isomerase converts dihydroxyacetone phosphate to glyceraldehyde-3-phosphate, which in turn is oxidized to 3-phosphoglycerol phosphate in the presence of glyceraldehyde-phosphate dehydrogenase and NAD^+. These are two of the most important steps in the glycolytic pathway. They result in the formation of 2 ATPs per molecule of glucose. If the isomerase were not present, only one ATP per glucose would be formed. Is it necessary that the two genes be closely linked for glycolysis to maintain maximum efficiency?

These isolated instances point to the possibility of types of relations between genes apparently unrelated in function that have been previously undetected and unimagined. It is probable that the more that is learned about the genes in conservates, the more the significance of these entities in the functioning of chromosomes will be appreciated.

D. *Cis–Trans* Effects and Complementation

The classical definition of an allele of a gene is that it occupies the same locus as its sister alleles in an allelic series and no crossovers should be expected between sister alleles. However, in 1940 Oliver demonstrated that *D. melanogaster* females, heterozygous for the lozenge alleles *lz³* and *lzᵍ*, and having the lozenge phenotype, produced *lz⁺* progeny. Further genetic analysis of this locus proved that the *lz* gene is about 0.1 cM in length and consists of four sites $lz^{BS} - lz^k - lz^{46} - lz^g$ that are separable by crossing over (Green and Green, 1956; Green, 1961, 1990). Despite the fact that they occupy separate loci, flies heterozygous for some of the different alleles may have mutant phenotypes similar to those that are homozygous. Thus, flies that are *trans:* $lz^{46} + / + lz^g$ are mutant in phenotype. On the other hand, when these "alleles" are *cis:* $lz^{46}\ lz^g/ + +$, the phenotype is wild-type, $+ +$. This condition is called *pseudoallelism* (Lewis, 1950) and lz^{46} and lz^g are **pseudoalleles**. Each has a separate locus (Figure 9.23) and a number of "true" sister alleles. A large number of pseudoallelic loci have been found in *Drosophila*, but presently the term pseudoallelism is not general use. Aggregates of pseudoalleles are also frequently referred to as complex loci with separate "genes." It should be noted that the dictionary definition of locus is location, and that its original usage in genetics (Morgan et al., 1915) was just that, the position of a gene on the genetic map. However, locus is now used more or less synonymously with gene. Some authors even use the term *gene locus* to identify a gene. If locus and gene are synonymous then gene locus can mean either gene-gene or locus-locus.

The *BX-C* complex (discussed on p. 394) has "genes" that were originally called pseudoalleles because of the *cis-trans* effects they showed in certain combinations. When two of them are paired in *cis*, the phenotype is normal except when one of them is dominant. Otherwise, when *trans*, nearly all combinations produce a mutant phenotype. Table 9.12 lists some examples of these effects. On p. 394 we used the example of *cis* and *trans* to describe complementation effects as a means of deciding whether two genes are allelic or not. If they complemented in *trans*, they were considered nonallelic and separate genes. By that definition the genes in the *BX-C* are not separate genes, since (except for *bx* and *phx*, and *bx* and *bxd*) they do not complement. This has been described as the *cis–trans position effect* or the *cis–vection effect* (Lewis, 1954, 1955). Lozenge and BX-C are by this *cis–trans* test considered complex loci.

Extensive complementation studies have been carried out with the ascomcycete fungi such as the yeasts *Aspergillus* and *Neurospora*, among others. One example of this is the gene cluster involved in the aromatic amino acid pathway in *Neurospora* (described on p. 402 and in Fig. 9.15). As shown in Figure 9.15b, there is complementation of various kinds in this complex. To understand this it is necessary to know that the vegetative phase of *Neurospora* is haploid (it has the haplontic cycle shown in Fig. 5.14), and therefore cannot form heterozygous nuclei except in the short-lived diploid zygote. However, it does form heterokaryons in which nuclei of diverse genotype can exist within the same mycelium. Mutants lacking the individual enzymes in the aromatic pathway have been tested for complementation by synthesizing heterokaryons with nuclei from two different mutant strains. A heterokaryon with, for example, an *aro-1⁻* nucleus and an *aro-9⁻* nucleus can be written as $aro\text{-}1^- aro\text{-}9^+/aro\text{-}1^+ aro\text{-}9^-$ and

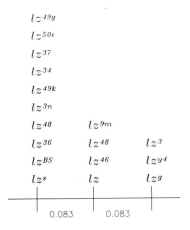

Fig. 9.23. *The three lozenge loci according to Green and Green (1956).*

TABLE 9.12. Phenotypic Effects of *BX-C* Genes in *Trans* Arrangement

bx³ + / + bxd	0
bx³ + / + pbx	0
bxd + / + pbx	PT3 → PT2
bx³ + / + Ubx	AT3 → AT2
UBX + / + bxd	PT3 → PT2 and A1 → AT3
Ubx + / + pbx	PT3 → PT2
Cbx + / + bxd	PT2 → PT3

considered as an approximation to a heterozygote. If the heterokaryon grows without being supplemented with chorismic acid, complementation has occurred between these two mutant strains. Mutants for all five of the clustered genes shown in Figure 9.15 have been tested and complemented. This is indicated by the solid lines in the second row of Panel b. By this complementation test all five are single genes, and most of the single-gene mutants derived from them produce a multienzyme complex of the expected molecular weight of 230,000. The inactive enzymes in the different enzyme complexes are presumably the result of missense mutations that inactivate them. On the other hand, other types of mutants, called **pleiotropic**, have been isolated that are designated by the letters A–F in Figure 9.15b. These do not complement with one another in the same pattern as the single-gene mutants indicated by the arrows in the figure (Case and Giles, 1971). One group, labeled E, F, lacks all five enzyme activities while other groups have only one (D), two (B) or three (A & C) activities, and always at much reduced levels. They produce enzyme aggregates of molecular weights of about 60,000–85,000 and are believed to be the result of nonsense rather than missense mutations. This indicates the possibility that a single mRNA encoding the complete 5-enzyme aggregate is ordinarily transcribed unless there are chain-terminating codons resulting from base pair deletions or duplications chang-

ing the reading frame. Complementation between two different gene encoding units, as discussed above, is to be expected and easily understood if the active forms of the genes are present in heterozygotes in the *trans* as well as *cis* arrangements.

What is not so easily understood is that complementation may also occur between two different mutant alleles of the same gene. This is called **allelic complementation**. It can result in a nonmutant phenotype if the mutations are in different parts of the same gene. The *aro-1, aro-2,* and *aro-3* genes of the aromatic pathway are shown by the top row of Figure 9.15b to have mutant alleles that in certain heterokaryotic combinations produce enzyme activity expected only from the active wild-type allele. Figure 9.24 illustrates this by a complementation map for mutant alleles of *aro-2*. The lines that overlap show that heterokaryons made from those mu-

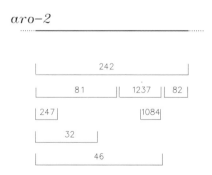

Fig. 9.24. *Interallelic complementation map for the aro-2 gene of Neurospora crassa.*

tants do not complement. The other combinations do. Thus, mutant 242 does not complement with any of the other seven. Mutants 81, 1237, and 82 complement with one another, mutant 46 complements only with 82, and so forth. Ordinarily the activity of the enzyme produced by the heterokaryon is not the same as that produced by the wild-type allele. The explanation for this, which is supported by experimental evidence, is that the enzyme involved is multimeric and each mutant allele produces a polypeptide with a different defect. If we consider the enzyme to be dimeric, then the mutant genes above will encode proteins with a single polypeptide type: P_1P_1 or P_2P_2 in homokaryons. But the heterokaryon will allow the formation of a third or hybrid type, P_1P_2, which unlike P_1P_1 and P_2P_2 may be active (Fincham, 1966). Thus complementation between two mutant genes does not necessarily prove that they are alleles of two different genes.

III. WHAT IS A GENETIC UNIT?

This question has been asked ever since the beginnings of this century. The orthodox opinion has been that the gene is *the* unit of inheritance because it is a demonstrable unit of function. This concept has been supported physically by the demonstration that at least many genes encode specific RNA and polypeptide molecules. Some critics, especially Goldschmidt (1955), have maintained that genes do not exist as such. But they have neglected to recognize that different levels of genetic organization exist. These range from single genes that contribute a specific product to groups of genes, linked and even unlinked, that act together in the widest sense as haplotypes. The physical gene *is* a unit of selection, but it must also be recognized that there are other units of function and hence selection that are of a higher order

than single genes. These have been described as *supergenes* (Darlington and Mather, 1949). Possible examples of *supergenes* have been described in this section and chapter as complexes, families, and conservates. The MHC can reasonably be called a supergene. Its combination of closely linked separate genes encoding separate polypeptides is conserved presumably by selection. However, the genes may not necessarily be related in function as in the MHC; they may be coadapted to cooperatively encode products to produce an adaptive characteristic. Conservates, by this definition, can be considered supergenes and hence genetic units. Complementation tests also indicate that clusters and complexes should also fall into the category of supergenes. The *aro* genes of *Neurospora* clustered on chromosome 2 function as a unit, as do the *BX-C* genes in *Drosophila*. This discussion is continued in the next chapter after some additional facts that bear on this question are presented.

IV. SUMMARY

1. The genetic activity of chromosomal DNA is to a large degree dependent on the way it is packaged. The protein scaffold is apparently an important element in this packaging in the mitotic phase of the cell cycle. At this time most genes are turned off. The packaging of chromatin at the time genes are active in the interphase is not understood, as discussed in Chapter 7.

2. The distribution of some of the repeated sequences of DNA is not random. It is possible that only a minor part of them may be random in their distribution.

3. The fractionation of DNA fragments by equilibrium centrifugation reveals that birds and mammals have a higher GC content in their DNA than the lower vertebrates. Upon centrifugation, the fragments separate into distinct, discrete

aggregates with similar buoyant densities, called isochores. Most of the genes in the nuclear DNA of warm-blooded vertebrates are localized in the ischores with the highest GC content.

4. The *SINES* family repeats are found in highest concentration in the GC-rich isochores, whereas the *LINES* repeats appear to be more highly concentrated in the GC-poor regions.

5. If genes are defined as DNA sequences containing regulatory plus transcribed regions of DNA, one type of calculation indicates that the human genome may contain at least 20,000 structural genes. Estimates that the human genome may contain 50,000–100,000 are also current, but no one really knows how many there are.

6. Families of genes occur in the eukaryotes. A family contains genes that show a considerable degree of homology with respect to sequence structure and general function. They appear to have arisen by duplication from primordial mother genes.

7. Families operate in important ways in eukaryotes through functioning by the members at different times during development, and by providing an extensive amount of variability to protect the organism against environmental stress. It is probable that it would not have been possible for complex multicellular organisms such as higher animals and plants to arise and continue to evolve without the participation of family genes.

8. Gene clusters controlling metabolic pathways are found in some eukaryotes especially the fungi. In some cases they are found to be involved in the encoding of multifunctional enzymes and can conceivably be considered as single multifunctional genes.

9. A high degree of conservation of chromosome structure exists among all eukaryotes. In cases in which banding patterns can be developed with the appropriate stains, definite patterns of homology can be discerned among many mammalian species in different orders. Comparisons of the gene maps of vertebrates ranging from bony fishes to mammals reveal that many groups of homologous genes have been conserved in their linkage. These conserved groups do not necessarily have genes related in function, although family members may be present. To distinguish them from families these conserved linked groups are called conservates. The maps for human and mouse show that about 48 conservates exist between the two.

10. *Cis–trans* effects and complementation can be used to define genes. Some closely linked genes operate as a functional unit in development and metabolic pathways. These frequently show *cis*-vection effects in heterozygotes.

10 The Evolution of Genomes and Their Chromosomes

Would to God your horizon may broaden everyday! The people who bond themselves to systems are those who are unable to encompass the whole truth and try to catch it by the tail; a system is like the tail of truth, but truth is like a lizard; it leaves its tail in your fingers and runs away knowing full well it will grow a new one in a twinkling.

—Ivan Turgenev in a letter
           to Leo Tolstoy (1856)

The very nature of the subject of evolution lends itself to the development of a multiplicity of different systems of thought. The efforts to explain the mechanisms functioning in the evolution of new forms, or explain how the present forms happen to be as they are, have led to a great variety of interpretations of the huge mass of data that have accumulated since the beginning of the last century. Of the many theories that have been advanced Charles Darwin's theory of natural selection and Sewell Wright's shifting balance theory are the two systems of thought that we use in this chapter to focus on the evolution of eukaryotic genomes and their chromosomes. Mainly we try to develop an approach to an explanation of why these genomes and chromosomes are as they are. This is one way of developing an understanding of how they function. Our dialectic is that the evolution of the eukaryotes is a genomic rather than a purely genic phenomenon. Certainly genes as such are involved, but only as parts of the genome.

I. ORGANIC EVOLUTION IS A CONSEQUENCE OF CHANGES IN DNA

New DNA genotypes may produce new phenotypes to be acted upon by environmental factors. From these interactions new forms arise, some of which survive and produce offspring for successive generations and in the process may produce new species phyletically or cladistically. Phyletic or anagenic evolution is change that occurs in a single line of descent without branching as in cladistic evolution. In the former case the mother form becomes a new form while in the latter a mother form or species gives rise to a new species while it itself may remain untransformed. Actually, a given lineage may involve both forms so it is not always easy to separate one type from another. What happens to these genotypic alterations in time measured in generations is the essence of the evolutionary processes.

Lewontin (1974) has stated that *we know virtually nothing about the genetic changes that occur in species formation.* This may be an overstatement, but it certainly does not exaggerate the problems one faces in attempting to make a coherent statement about the evolutionary processes. One should begin by recognizing that it is not a *single* mechanism that is involved, but a number of basic mechanisms, some of which we can identify now, but no doubt others that we still cannot. Thus we use the phrase "evolutionary processes" rather than the singular "evolutionary process."

A. DNA Changes Occur by Chance

Chance genotypic changes are the basic events providing the materials for evolutionary processes. As mentioned in Chapter 5 (p. 185), there are at least seven types of events that produce genomic alterations. In this chapter we concentrate on the role played by alterations in the organization of genomes by chromosomal rearrangements, changes in ploidy, and repetitive sequences, duplication, and meiotic recombination, while at the same time not ignoring the importance of other factors. Changes resulting from all of these are in one way or another subject to natural selection, which acts as a sieve casting out genotypic changes that produce phenotypes that do not survive.

1. DNA changes are subject to natural selection

A great deal of the theory of evolutionary mechanics has in the past dealt with Darwinian selection acting upon one or very few genes. This imposes a limit in order to keep the mathematics relatively simple even though most evolutionists recognize that the approach is simplistic (Franklin and Lewontin, 1970). Wright recognized this at least as early as the 1920s, when he began the development of his shifting balance theory. We begin our discussion by first describing this theory.

2. The shifting balance theory incorporates both chance and selection

The **shifting balance theory** of evolution as developed by Wright (1931, 1980, 1982) has as its core premise that a distinct difference exists between what he has described as **genic selection** and **organismic selection**. By this he meant that the gene, although a unit of action, is not

the only unit of selection; instead the individual organisms or breeding populations are the units upon which natural selection predominantly operates. To put it in more specific genetic terms, it is the genome that is acted upon, not the individual genes *per se*. His argument in support of this thesis is based upon the fact that genes operate not alone, but in concert, and there is not a one-to-one reaction between gene and specific character. Hence, a gene's expression and its survival in a population, or selective value, is contingent upon the state of the rest of the genome that it occupies. Figure 10.1 illustrates the differences between Wright's concept of a gene's relation to the phenotype (a), and the concept held by geneticists such as Haldane, who pictured the relation to be a one-to-one, gene-to-unit character (b). The condition expressed in Panel a is closer to the actual situation even in prokaryotes than that expressed in Panel b, which is far too simplistic.

A mutation of a specific gene may result in a specific phenotypic change even with pleiotropy, but the resultant specific new phenotype is still the product of the interactions of many genes. This, of course, does not mean that a deleterious mutant allele will not be selected against, all other things being equal.

Wright (1980) has put this concept into a paragraph worth quoting:

> A person considering evolution in terms of Mendelian heredity for the first time is likely to think that it is merely a question of how the best alleles in all loci, relative to a given ecological niche, come to be assembled. After a moment's consideration, it is fairly obvious, even on the simplistic assumption of a one-to-one relation between gene and unit character, that the selective value of any one of the unit characters will depend on the others with which it is combined (p. 827).

Furthermore, in bisexually reproducing eukaryotes it is not *that* phenotype that is inherited because "an individual does

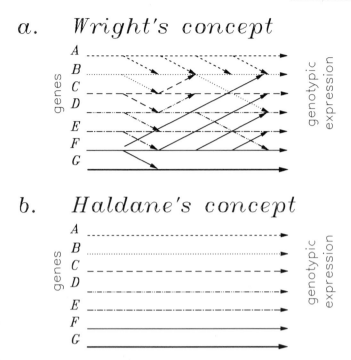

Fig. 10.1. *Diagramatic representation of Wright's concept* **(a)** *versus Haldane's* **(b)** *regarding the action of genes. In* **(a)** *all primary gene products (see Fig. 6.1) interact with other gene products. Every intersection of lines represents an interaction. The final effect is that a characteristic is the product of the action of many genes. In* **(b)** *there is a direct one-to-one relation between gene and characteristic.*

not transmit his array of coadapted characters as a unit, combinations are broken up so rapidly in terms of the evolutionary process (except for almost completely linked genes), that it is only the average effect of each allelic difference that counts."

These considerations, based in large part on his experiences in the analysis of guinea pig coat color inheritance, and of domestic animal-breeding programs, led Wright to the development of the **shifting balance theory**. The word "balance" as used by Wright refers not only to balance in interactions between the alleles of different genes that are coadapted to produce a particular phenotype, but a balance between mutation and gene flow into and out of a population, and a balance between selection and randomness, or **genetic drift**.

Mutation comprises all the various types of genotypic variations resulting from the alterations of DNA of genomes referred to in this and previous chapters. In addition to gene mutation and chromosomal rearrangements, a major source of variation is recombination by crossing-over and independent assortment as described in Chapter 4, p. 150. A balance does in fact exist between the mutation rate, the recombination rate, and viability. An extremely high rate of change can be expected to be selected against. Gene flow refers to the flow of new mutant forms to various other demes. Selection is the differential and nonrandom reproduction of genotypes as a consequence of the superior fitness of their phenotypes. Genetic drift, also called the Sewall Wright effect, is a random process of changes in gene (allele) frequencies expected in small populations where certain alleles, or even chromosomal aberrations, may be lost or fixed by chance without the necessary participation of natural selection. Selection is directed change; drift is random, undirected change (Wright, 1948). Random changes are not, of course, exempt from natural selection. In large populations they may become fixed, if they are fit or selectively neutral, but in small isolated populations they may by chance replace more fit genotypes originally present in the genome.

An essential part of Wright's theory—which has given it the name "shifting balance"—proposes that a well-adapted genotypic state may shift to a better one even though it has to pass through less fit states. Thus in a population with a variety of semi-isolated subpopulations or demes a superior adaptive genotype may arise by chance that can spread or diffuse throughout the population. This type of shift has in the past been considered unlikely because recombination should inevitably reduce the superior adaptive state to a lesser one (Haldane, 1960). More recently, however, experimental support for the shifting balance theory has been provided by Wade and Goodnight (1991). It must now be considered to be one of the important possible modes of evolution.

Wright's shifting balance theory has been criticized as being posited as an alternative to natural selection, but this is a misconception. The theory basically states that the joint action of all factors involved in the evolutionary processes must be taken into consideration including both drift *and* natural selection. Also it should be recognized that other theories such as those developed by Fisher (1930) and Haldane (1932) are, as pointed out by Wright (1988) himself, also valid. They simply address different aspects of the evolutionary processes as we envision them. When considering total genome evolution, however, it

would seem that Wright's approach is the most appropriate one.

II. VIABLE OFFSPRING HAVE GENOMES CONSISTING OF ARRAYS OF COADAPTED ALLELES

All genotypes that produce viable phenotypes capable of producing viable offspring are constituted of a constellation of genes whose alleles act together harmoniously to produce a fit phenotype. Such a gene array is said to be **coadapted**.

A. Selection May Keep Some Coadapted Alleles Together as Supergenes

Some coadapted genes may be linked together to form **supergenes** such as the MHC. The supergene concept, as advanced by Darlington and Mather (1949), was popular in the 1950s and 1960s but interest in it lagged after that, probably because the gene's function as an encoder of specific RNAs and polypeptides drew the most attention from researchers. In this discussion we regard the supergene as an important genetic entity and a conceptual guide to understanding the functioning of genomes.

1. Selection may form supergenes

Fisher (1930) early pointed out that if two genes, each with two alleles, were on the same chromosome and the four linked types: AB, ab, aB, and Ab had differential selective advantages such that A was advantageous in the presence of B, but not with b, and a not with B, natural selection would favor close linkage between A and B. The selection process would favor haplotypes AB and ab, because heterozygotes, AB/ab, would result in the disadvantageous combinations Ab and aB. Tight linkage between A and B would thus be favored. It follows that groups of more than two genes whose linkage is advantageous can evolve to form a supergene (Bodmer and Parsons, 1962). A supergene is an assemblage of linked genes formed because fitness is greater with reduced recombination (Turner, 1967a).

Selection for linkage between two genes is related to a phenomenon called **linkage disequilibrium**. This is a tendency for alleles at different loci to segregate together whether they are physically linked or not. The alleles A and B may occur together in gametes in excess: more often than one would predict from the frequency of A and B in a population with AB, Ab, aB, and ab genotypes. Thus, if two genes are asyntenic, then from AB/ab individuals in a population one should expect Ab, ab, Ab, and aB gametes in equal numbers, or if linked closer than ~15 cM, in a ratio depending on map distances only. Tight physical linkage helps to maintain linkage disequilibrium, if selection favors it. Or, if by chance, two widely separated genes are brought together in physically close linkage by translocation or inversion, linkage disequilibrium will be enhanced.

Thus linkage disequilibrium preserves the linkages of a pair of genes presumably by natural selection and the resultant *gametic* excess of certain combinations (Lewontin, 1988). The hypothesis has been advanced that linkage disequilibrium is important in maintaining specific allelic arrays of closely linked genes (Franklin and Lewontin, 1970). It has been maintained that tests of this hypothesis in natural populations of *Drosophila* have not supported it (reviewed by Kimura, 1983). However, these results do not negate the possibility that positive selection for close linkage between certain genes exists despite their allelic state, and supergenes may arise by this means. Supergenes like the MHC have arisen most probably by gene duplication rather than by rearrangements, but their unit genes may be maintained in close linkage by linkage disequilibrium.

2. Genomes may congeal

If coadaptive linkage is advantageous, why then does not the genome congeal into a single linkage group? Since all genes act together in concert, should not the most favorable grouping be maintained by the genome congealing with reduced recombination between the linked elements (Turner, 1967b)? Genome congealing in fact does occur. We have already described the forms of the plant *Oenothera*, which because of their reciprocal translocations in balanced heterozygotes effectively have only one or two linkage groups (p. 206). The Indian muntjac has a haploid number of 3 or 4 instead of the $n = 23$ found in the Chinese muntjac. Members of the deer family Cervidae, to which the muntjacs belong, generally have even higher haploid numbers than 23. Two animals, the nematode *Parascaris equorum univalens* and the Australian ant *Myrmecia pilosula*, have but one chromosome haploid (Crosland and Crozier, 1986). This ant is of particular interest because it is a member of a sibling species group with members having diploid numbers of 9, 10, 16, 24, 30, 31, and 32. Some of these sibling species are morphologically indistinguishable from *pilosula* and one another. This is in agreement with the general observation that chromosome number differences may have scant effect on morphology within a group of related species. This is also true of the muntjacs, and butterflies in the genus *Lyceaena*. Plants in the genus *Brachycome* (Compositae) have a wide range of chromosome numbers ranging from $n = 2$ to $n = 45$ (Smith-White et al., 1970). The basic primitive number is inferred to be 9. Most species have multiples of nine chromosomes and are polyploids. Those with two chromosomes presumably have congealed genomes (Smith-White and Carter, 1970). Another composite, *Haplopappus gracilis*, also has $n = 2$ (Jackson, 1973). As in the animal groups, however, examples of similar low chromosome numbers indicating congealed genomes are rare. In these two plant genera as well as in the ant genus *Myrmecia* the condition of **dysploidy** is prevalent. Dysploidy is a frequent occurrence in angiosperms and insects in which the basic chromosome numbers within a related group differ by one chromosome such as 5, 6, 7, etc.

Although genome congealing is expected to reduce recombination, it does not necessarily reduce it to zero. Crossing-over rates vary widely among the eukaryotes.

3. Selection against complete congealing of the genome also appears to occur

The complete congealing of the genome is extremely rare among eukaryotes whether animal or plant. Although it appears that natural selection may favor close linkage between certain genes, it is also apparent that there are certain limits to this tendency. Most plant and animal species have haploid chromosome numbers above $n = 10$, and all, with very few exceptions, have crossing-over between homologues. It has been shown that theoretically the reduction of recombination for a genome can be expected to reduce fitness especially in face of a changing, fluctuating environment (Turner, 1967b; Maynard Smith, 1977). A balance exists between too much recombination and too little. The maintenance of a restriction, but not suppression, in recombination within supergenes such as the *MHC* or within the α- and β-*globin* complexes is apparently important. Recombination between them and other genetic units is not only tolerable, but may be beneficial, and therefore a factor opposing congealing.

B. Are Conservates Coadapted Arrays?

The conservates (discussed in Chapter 9, p. 404) may have selective advantages and hence be maintained by selection pressure. The genes could have been brought together by rearrangements and their linkage conserved following the process of formation of supergenes described above. By this reasoning they may be described as supergenes. One can visualize that the active units of a conservate form advantageous haplotypes much as different haplotypes are formed in the *MHC*. Furthermore, breaks within conservates may eventuate in nonpermissive rearrangements. This is an idea that was apparently first developed by the corn geneticist Brink (1932), who almost 60 years ago in a paper entitled "Are Chromosomes Aggregates of Groups of Physiologically Interdependent Genes?" made the following statement:

> It might be assumed that the chromosome in its essential make-up consists not of genes which are entirely distinct from each other in function, but of aggregates of groups of genes which are physiologically interdependent. On this hypothesis it is supposed that propinquity of genes within a group is essential to normal gene action. On this view translocations involving breaks between groups of genes would not alter the genotype. If, however, the chromosome is broken in such a way as to separate the members of a gene group, more or less profound changes in the physiological properties of the complex would follow.

The discussion of nonpermissive and permissive rearrangement in Chapter 7 (p. 345) is certainly germane to Brink's observation. Estimates for the spontaneous rates of reciprocal translocations have been made for a variety of animals and they fall between 10^{-4} and 10^{-3} per gamete per generation (Lande, 1979). These estimates are based on the observation that inversions and translocation heterozygotes and homozygosity rarely produce phenotypic effects. They are selected against when heterozygotes, as pointed out in Chapter 5, but they may have normal fitness when homozygous and can then become fixed. Actually translocations are rarely detected *within* species, but there is a high frequency of such differences between closely related ones. White (1978) has stated that "over 90 percent (and perhaps over 98 percent) of all speciation events are accompanied by karyotypic changes, and that in the majority of these cases the structural chromosomal rearrangements have played a primary role in initiating divergence." Wilson et al. (1974b, 1975, 1977), and Bush et al. (1977) have marshaled a considerable amount of data from mammals indicating that the rate of organismic changes may be related to the rate of karyotypic changes. Their results would seem to support White's conclusion.

But are two closely related species different because they have different structural rearrangements? Perhaps, but the chance fixation of a translocation or inversion in a small population by drift can lead to partial reproductive isolation with demes of the same species not carrying the aberration. This isolation may lead to the formation of a new species, not because the aberration is the basic reason for the phenotypic differences between the new form and its sibling species, but because it initiates the train of events leading to the speciation event. Once the sexual isolation occurs, other types of mutational events lead to divergence and significant differences between the new form and the stem from which it arose. Of course, the possibility still exists that the aberrations are not entirely neutral in the formation of specific new phenotypic differences.

C. Chromosome Rearrangements May Create as Well as Preserve Linkage of Coadapted Alleles

Inversions are more frequently observed as polymorphisms in wild populations than are translocations. Paracentric inversions are common in different species of *Drosophila*, while pericentric ones are less common (Dobzhansky, 1970; Pat-

terson and Stone, 1952). In most other eukaryotes pericentrics are more likely to be detected because they may move the position of the centromere, whereas paracentrics do not, although they may occur quite commonly.

Dobzhansky and associates have made an extensive study of the inversion detected in natural populations of *Drosophila pseudoobscura* and *D. persimilis* (Dobzhansky, 1970; Wright, 1978). A common salivary band sequence of the third chromosome was designated as "standard" (ST). Seventeen different types of these were found in various different populations of *D. pseudoobscura*, a species found in most of the western United States and in Mexico, Central America, and Columbia. When the frequencies of the different types of inversions found in different parts of *pseudoobscura's* range were compared, it became obvious that while adjacent regions had similar patterns, regions distantly apart showed great differences, as can be seen from data in Table 10.1. Particularly notable are the differences in patterns of distribution between the northern and southern part of the range.

Not only are there significant geographic differences in the frequencies of occurrence of these inversion patterns, but particular inversions show seasonal frequency changes as shown in Figure 10.2 for some populations of *pseudoobscura*. It is quite clear from these data that the frequency of the standard arrangement, ST, begins to decline in the spring when the fly population begins to increase. These flies in this region of California go dormant in the summer dry season and become active again in late winter and March.

Chromosomal polymorphisms in *Drosophila subobscura* were detected and analyzed several years after this species, endemic to Europe, North Africa, and Asia north of the Himalayas, was introduced to South America. Seven Chilean populations were studied over the period from

TABLE 10.1. Percentages of Eight Different Types of Third Chromosome Inversions Found in Populations of *Drosophila pseudoobscura* in Different Parts of Its Range

Location	Inversion type							
	PP	AR	CU	SC	ST	OL	CH	TL
British Columbia	2	41	0	5	37	1	8	6
California N. Sierras	0	46	0	4	31	4	8	7
California Coast Range	0	18	0	15	48	1	13	5
S. California	0	25	0	1	47	0	22	5
E. California	0	62	0	0	24	0	13	1
Arizona, Utah, Nevada	2	88	0	0	3	0	7	0
New Mexico, Colorado, Wyoming	25	65	0	0	2	1	2	2
Texas	71	17	0	0	0	3	0	8
Baja California, Mexico	0	26	0	5	58	0	11	0
Chihuahua, Mexico	27	8	0	2	0	1	61	1
Nuevo Leon, Mexico	9	3	0	3	0	47	0	35
SE Mexico	0	0	51	1	0	6	1	34
SW Mexico	4	0	31	41	0	4	0	18
Guatemala	0	0	9	65	0	0	0	25
Colombia	0	0	0	35	0	0	0	65

Data from Dobzhansky and Epling (1944) and Dobzhansky (1963).

1981 to 1986 (Prevosti et al. 1990). Twenty different inversion types were identified on the basis of polytene chromosome analyses. The old world populations of *subobscura* show a definite correlation between specific inversion types and latitude, and the Chilean populations distributed from south latitudes of 32°02' to 45°35' show for all but four of the inversions the same patterns of distribution. Thus an adaptation to latitude in the Palaearctic appears also to occur in South America. The pattern has evolved very quickly in South America and indicates a rapid response to selection, much more than is usually assumed to be possible.

Experiments conducted in the laboratory with flies breeding in population cages have yielded results in agreement with field findings. (A population cage is a large box about 50 cm × 30 cm × 15 cm in which a population of *Drosophila* can be maintained over a period of a number of generations so that changes in frequencies of genotypes that might occur can be observed and measured by taking samples

at intervals.) The *pseudoobscura* inversions *ST*, *AR*, and *CH* were studied for changes in frequency in population cages maintained at two different temperatures: 16.5°C and 25°C. Populations were started out with certain frequencies and then followed for several generations for possible changes in frequencies. Changes at 16.5°C were slight but flies reared at 25°C showed significant frequency change of the three types of inversions relative to one another (Wright and Dobzhansky, 1946). It was also found that the flies heterozygous for inversions were fitter in terms of selective values than the homozygotes (Dobzhansky, 1947).

III. THE NEED TO PRESERVE CERTAIN LINKED ARRAYS MAY LIMIT THE NUMBER OF CHROMOSOMES IN A GENOME

If eukaryotic genomes had no chromosomes but only genes, each with its own centromere, the processes of meiosis and

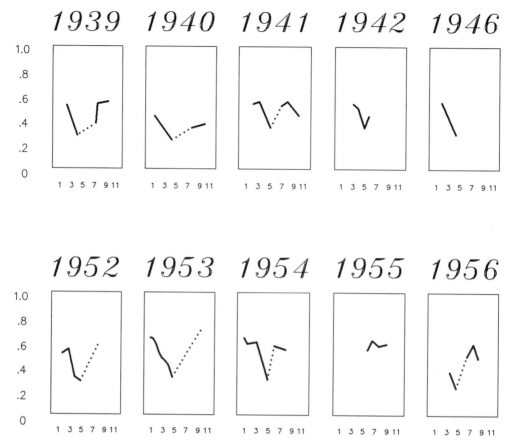

Fig. 10.2. *Monthly frequencies of the ST (standard arrangement, defined by Dobzhansky, 1970) of Drosophila pseudoobscura of Pinon Flats, California, between 1939 and 1956. dotted line indicates determination skipped one or more months. Data from Wright (1977).*

mitosis would likely be subject to a great amount of nondisjunction intolerable in any organism with a large number of genes. All eukaryotes have limits to permissible aneuploidy. Hence one should expect a limit to the number of chromosomes per genome. A balance between extremes in chromosome number is to be expected. The data in Table 5.8 make it clear that the ranges in number are extensive. If, however, the ferns and the flowering plants are set aside because their high numbers are mostly the result of extensive polyploidy, the average haploid number, ex-

cept for certain Lepidoptera, falls somewhere in the range of 20–30. The mammals have been extensively karyotyped and found to have a range of 3–52, but a mode about at 24 (Imai and Crozier, 1980).

If supergenes of whatever type, be they families such as MHC or globins or conservates as defined on p. 404, lose their advantage or become inactive when disassembled by rearrangements, then it is possible that the upper limit to chromosome numbers per genome may be set in part by the number of these conserved units. The mammal with the largest known number of chromosomes, the rodent *Tympanoctomys barrier* ($n = 52$), may be near or at that limit for mammals (Contraras et al., 1990). And it may not be a coincidence that the number of linkage conservates shared by man and mouse probably is about 50–60 if one includes the undispersed subfamilies such as the MHC.

The profound effects of breaks described by Brink should not be expected if the breaks occur between supergenes, and we do know that many inversions and translocations are tolerated in plants and animals. Although breaks within supergenes may be nearly always selected against, it is possible that some may lead in very rare instances to rearrangement homozygotes that have the ability to survive and reproduce in new niches. Those that do survive would be expected to be the result of breaks that occurred between the functional parts of genes in the spacer regions. As pointed out in Chapter 5 translocations and inversions are strongly selected against, and such an event would be expected to be extremely rare. Wright (1941) has estimated that if one assumes a selective value of 0.5 in plants heterozygous for a reciprocal translocation, with sexual reproduction the chance of fixation is 10^{-3} with a breeding population of 10, 2×10^{-6} in a population of 20, and 3×10^{-14} in a population of 50. As for animals in which the elimination of unbalanced combinations does not occur in a gametophyte, the chance of fixation is about 4×10^{-6} in a population of 50. These estimates are for viable homologous translocations. Therefore the chance that a rare homologous viable translocation involving a break in a supergene would become fixed is even less.

Nonetheless some of these rare occurrences may have become fixed. One possible example is to be found in the α- and β-globin gene families of the amphibia. Two family groups are found in *Xenopus*: One group encodes larval and the other adult globins. The larval α- and β-globins are tightly linked in one cluster as are the adult. The genes in the two groups are paralogous, having probably arisen as a result of tetraploidization. All birds and all mammals have the α-globin genes on a different chromosome than the β-globin genes. Since the α- and β-globins are almost certainly members of the same ancestral family, it is logical to assume that the linked relationship is the ancestral one. The question then arises as to whether genomic changes such as these were involved in the origin of birds and mammals from their ancestral tetrapod. One would like to know the situation in the reptiles from which we know the birds and mammals evolved.

The paleontological evidence strongly supports the hypothesis that the anamniote Amphibia and the Amniota (reptiles, birds, and mammals) arose independently from an ancestral tetrapod during the Paleozoic. The Amniota are descended from an ancestral form, the Anthracosauria, and not from the Amphibia (Panchen and Smithson, 1988). The molecular sequence data derived from present forms clearly indicate the close relationship of birds, mammals, and crocodilians; the Amphibia are most closely related to the fishes (Bishop and Friday, 1988). Both paleontological and molecular data indicate the separation of the amphibious anamniote and terrestrial amniote lines occurred about 350 YBP in the Carboniferous. It would be of interest to know the linkage relationships of the α- and β-hemoglobin families in the reptiles and the fishes. If the fish have the two linked like the Amphibia, and the reptiles have them unlinked, was the separation an adaptation to a terrestrial lifestyle? Questions such as these are germane to inquiries about the nature of genomes and their evolution. The comparative anatomy of gross structure of organisms has told us a great deal about them, and even more knowledge will come from studying the comparative anatomy of their genomes. Comparisons make evident the generality of biological phenomena, their possible origins from common precursors, and frequently to a better understanding of their functions.

IV. MULTIGENE FAMILIES ARE ESSENTIAL ELEMENTS OF EUKARYOTIC GENOMES

There seems to be general agreement among evolutionists that the complexity of form and function of present-day plants and animals has evolved accompanied by the accumulation of a great amount of genetic information stored in multigene families. This complexity is inferred to be dependent upon the multiplicity of these families and their members as described in Chapter 9. Eukaryotic genomes have been built by duplication and differentiation of the basic gene types found in prokaryotes that enable them to perform the fundamental life processes of simple cells. A multicellular organism carries out the same basic heterotrophic metabolic activities as does a heterotrophic bacterium.

It is becoming increasingly apparent as more and more information is accumulated about the sequences of amino acids in proteins, the base sequences in repetitive DNA, and the structural elements in the RNAs involved in protein synthesis that a relatively small number of fundamental molecular arrays or domains have given rise to the majority of existing structures (Doolittle, 1979; Field et al., 1988). It is as if a set of building blocks of DNA sequences exist that have been and are assembled in a variety of different ways to produce all the variety we now witness in the living world of prokaryotes and eukaryotes. Among the eukaryotes the genic exons may have been shuffled to produce the great variety of proteins in the bios, and perhaps only 1,000–7,000 basic exon

subunits are needed to form all proteins (Gilbert, 1978; Ohta, 1989; Dorit et al., 1990). The immunoglobin superfamily is one example of a set of genes encoding proteins with a variety of functions sharing a relatively small number of polypeptide types or domains (Table 9.6 and Fig. 9.3). One can in fact extend this beyond the immunoglobulin superfamily because similar structural relationships found among the immunoglobulin proteins can be extended to trypsinogens, plasminogens, thrombins, haptoglobins, orosmomucoids, and so forth. For example, it has been estimated that at least 133 genes or gene families are similar and possibly homologous in *Drosophila* and vertebrates (Merriam et al., 1991). Just as an alphabet of only 26 letters can form virtually an unlimited number of words, so can a limited set of sequences formed from four nucleotides go into the making of the seemingly unlimited forms of life.

A. Gene Families Arise Initially by Duplication of Primordial Genes

A comparison of the structures of some of the enzymes of bacteria with their eukaryotic analogues reveals sufficient homology of amino acid sequences to justify the assumption that the eukaryotic genes encoding them have evolved from primordial prokaryotic genes. Some of the enzymes involved in basic metabolic functions such as glyceraldehyde-3-phosphate dehydrogenase, triose phosphate isomerase, and lactate dehydrogenase have structures that indicate a common origin in the distant past (Martin and Cerff, 1986; Doolittle et al., 1989). However, there is also good evidence that bacteria have obtained eukaryotic enzyme-encoding genes by horizontal transfer (Doolittle et al., 1990).

The principal mode of extensive gene duplication appears to be unequal crossing-over between either sister or nonsister chromatids (reviewed by Tartof, 1988). The first example of unequal crossing-over was described by Sturtevant (1925) in *Drosophila melanogaster*. He showed that a dominant, sex-linked mutation designated Bar (*B*) was the result of unequal crossing-over. Bar is viable when homologous and reduces the number of facets in the compound eye to about half the normal number. Flies carrying two B^+ genes on one chromosome are phenotypically mutant Bar and have a direct tandem duplication of seven polytene bands covering the region 16A1-7 on the polytene chromosome map as shown in Fig. 5.8. The crossovers are nearly all the result of exchange between nonsister chromatids at the four-strand stage. Unequal sister strand exchanges of this type are infrequent both in *Drosophila* and yeast (Jackson and Fink, 1985). However, sister strand exchanges do occur at high frequency at certain loci such as the bobbed locus (*bb*) of *Drosophila*, which contains the rDNA genes.

Unequal crossing-over has been demonstrated in mice (Harbers et al., 1986) and humans (Collins and Weissman, 1984; Jeffreys et al., 1988). It is almost certainly a general phenomenon in eukaryotes. Since misalignment of pairing homologues would be expected in regions with repeated copies, one should expect unequal crossovers to occur mostly in those regions. However, it also occurs in regions with unique sequences. The reason for this appears to be

that, at least in some cases, transposable elements are involved. The rate of spontaneous gene duplication in *D. melanogaster* has been measured for the maroon-like (*ma-1*) and rosy (*ry*) genes (Shapura and Finnerty, 1986). The rates are approximately 2.7×10^{-6} for *ma-1* and 1.7×10^{-4} for *ry*. Unexpectedly, duplication also occurs in males, which do not have measurable meiotic crossing-over. This may be the result of mitotic crossing-over in 2N germ line cells, or sister strand exchanges.

Unequal crossing-over is by no means the only possible mode of gene duplication leading to formation of families. Table 10.2 lists some of the other possibilities currently considered to be likely prospects.

B. Gene Families Evolve by Changes in Duplicates

After duplication of a particular gene, diversification may occur to form a family. Otherwise an amplified set of identical genes with the same function will exist. Most family members have related but not identical functions. Even in families like the α- and β-globin genes, which encode globins that act in oxygen transport, each member functions in its own way in time and space. Some family members in fact do not even seem to have similar functions even though they encode proteins with

---

TABLE 10.2. **Suggested Possible Mechanisms of Gene Family Formation**

1. Unequal crossing-over between homologous nonsister or sister strands
2. Gene conversion
3. Transpositions
4. RNA-mediated exchanges
5. Duplication of whole genomes that results in polyploidy, leading to the formation of paralogous gene family members
6. Duplication of individual chromosomes, leading to a tolerated aneuploidy

similar amino acid sequences. This leads one to the realization that this process of diversification must have been an extremely important one in the evolution of eukaryotes. But by what process do they evolve? Initially they arise by duplication and then are followed by such events as are listed in Table 10.2, leading to diversification (reviewed by Maeda and Smithies, 1986).

To consider this diversification we turn first to the prokaryotes, organisms that, since most of their genes are singlets, we generally think of as having few or no gene families. This, however, is probably not true. Consider the following question. How did the genes arise that encode the enzymes that catalyze the reactions leading up to the synthesis of an essential amino acid such as methionine? The steps that lead from the beginning products to methionine involve at least 15 enzymes in *E. coli*, starting with inorganic NH_3 and SO_4^{2-} and a carbon source such as glucose. That such a system arose by chance all at once to supply methionine to the first organisms is inconceivable. Horowitz (1965) has proposed that when the first organisms arose in the oceans all the constituents necessary to build and maintain them were present, having been synthesized from simple inorganic sources by the action of heat and solar radiation in the presence of a reducing atmosphere. Methionine was present to start with to be used in the synthesis of proteins as were the other essential amino acids. The critical period was reached when the methionine supply began to be deficient and an enzyme was needed to convert the immediate precursor to methionine. This precursor was already present in the primordial soup, as were all the other precursors. One has to assume that a gene arose to encode this enzyme, and that the RNA transcriptional and translational systems necessary to synthesize polypeptides were already in existence. Methionine was then synthesized from this precursor, but over time it

too became deficient in the soup. A precursor to it is present, but cannot be metabolized. To overcome this difficulty it is posited that a duplication of the newly arisen gene occurred by unequal crossing-over. With two genes, each having the same function, it is possible that one could mutate by chance to a state enabling it to encode an enzyme that catalyzes a different reaction. That new reaction could be converting the second prior precursor to the one immediately before methionine. By this backward procedure a series of reaction steps leading back to the initial products was created by the operation of natural selection on viable mutant forms. If this hypothesis is correct, the enzymes involved in the sequence of reactions should have structural relations, as should the genes that encode them. This has been found to be the case for two enzymes in *E. coli* that have a similar structure. These two are cystathionine-γ-synthetase and β-cystathionase. They catalyze the last two reactions leading to homocysteine the immediate precursor to methionine by methylation of the end sulfhydral group (Belfaiza et al., 1986). If this is true for the other enzymes in the pathway, then a family exists that originated by gene duplication. And one can assume that many other families exist in *E. coli* and other heterotrophic prokaryotes for analogous reasons. These probably are similar families in the heterotrophic fungi such as yeast and *Neurospora* (discussed on pp. 401 and 402) and to a much more limited extent in the higher eukaryotes.

Although prokaryotes may be considered to have families that have a common descent, it must be appreciated that the type of families they have are fundamentally different from the vast majority of eukaryotic gene families described in Chapter 9. Most prokaryotic and fungal gene families function serially in consecutive metabolic reactions. Most gene families of higher eukaryotes are groups of genes that encode related macromolecules that do not act serially in a chain of metabolic reactions.

C. Eukaryotic Gene Families Can Be Placed in Three Categories

The known multigene families of the eukaryotes can be functionally categorized as either terminal RNA encoders of 18S–28S RNAs, 5S RNAs, and URNAs, or encoders of proteins such as antibodies, hemoglobins, collagens, enzymes, and so forth. A third category of DNA family sequences with indeterminate functions exists in the form of repetitive sequences such as satellites found in the heterochromatin, and the *SINES* and *LINES* found in mammalian euchromatin. The telomeric repeats can also be included in this latter category. Whether these repetitive sequences should be called genes is a matter of judgment. Table 10.3 lists some of the characteristics of these different types of families. Added to these characteristics is one general one that applies to all primary types: A family may show little variance between members within a single species,

but usually shows considerable unexpected differences from equivalent families in related species. As diagrammed in Figure 10.3, a gene family with five clustered genes may show little intraspecific allelic variation, but large unexpected interspecific differences generally exist for that same family. The explanation for this has been a challenge for evolutionists who have yet to provide definitive answers.

Various names have been proposed to describe this phenomenon of relative homogeneity of family structure within a species. Descriptive terms such as concerted, coincidental, and co- or species-specific evolution have been suggested. In this discussion we use the term **concerted evolution**, and define it as the tendency of a family of genes to evolve in unison, or to become homogenized (Arnheim, 1983; Dover, 1986; Ohta, 1989). Homogenization may maintain all gene members with identical structure, or maintain their allelic differences (Fig. 10.4). There are several facets to this enigmatic situation, among them the maintenance of ~100% homology among members of a very large family.

TABLE 10.3. Three Types of Eukaryotic Multigene Families

Type	Characteristics
	I. Terminal RNA encoders
A. 18S–28S	Exist in multiples of 10–600 + . Usually completely sequence-identical within a species.
B. 5S	Exist in multiples of 100–1200 + . Usually completely sequence-identical within a species.
C. tRNA	Exist in multiples of 4 or 5 to several hundred for each type of tRNA.
	II. Protein encoders
	Many different families. Members of each family may exist as singlets or be amplified. Each member type has less than 100% sequence identity to other member types; they may be only 30%–70% similar within a family.
	III. Indeterminate repetitive
	Many different *sequences* families. Occur in multiples of several thousand to millions. Sequence identities may be ~100%, as in the case of telomeric repeats, to less than 70% as in the case of *Alu*'s and *Kpn*'s.

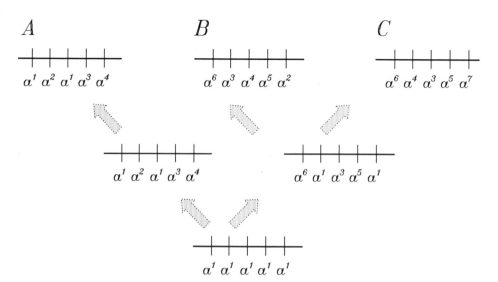

Fig. 10.3. *Relation between homologous gene families within and between species. The primordial family is indicated at the bottom with all members identical. Divergence in three species derived from this family is indicated in* **A, B,** *and* **C.**

1. Maintaining sequence homology in rRNA families

Consider, for example, the terminal RNA encoding genes. Within a given species of eukaryote several hundred to several thousand members of a family may exist, all of them identical, or with an extremely high degree of homogeneity, approaching 100%. The 18S and 28S ribosomal RNA genes of *Xenopus laevis* constitute a multigene family of about 450 repeating units located in the nucleolar organizer region (Brown et al., 1971). There are nontranscribed spacers between each of the transcribed units. All spacers are constituted of repeated homologous units, but are heterogeneous in length. A related species, *Xenopus muelleri*, has 18S and 28S genes that appear to be virtually identical to *X. leavis* genes, but the spacer regions have significant differences in sequence. What maintains the homogeneity of the transcribed rDNA sequences in both these species? Certainly mutations must be occurring in these genes just as they do in all other types of genes.

Concerted evolution of their many rDNA genes is general among the eukaryotes. It does not appear probable that natural selection operates in the usual sense to maintain the common identity of the individual genes in these gene clusters. If one gene in a cluster of several hundred mutated to a deleterious or neutral condition, its elimination by natural selection would be improbable, since the many hundreds of normal genes would presumably mask its presence in carrying out their normal functions. Selection would conceivably only apply if molecular drift toward dysfunction applied to all simultaneously. Logically some mechanism other than selection would appear to maintain the common individual identity of these genes, and natural selection operates directly only on the cluster as a whole (Williams, 1990). Thus the cluster may be considered as a supergene acting as a unit of selection.

Since spontaneous base pair mutational changes must occur, some correction mechanism must be operating to maintain the homogeneity of clusters of this

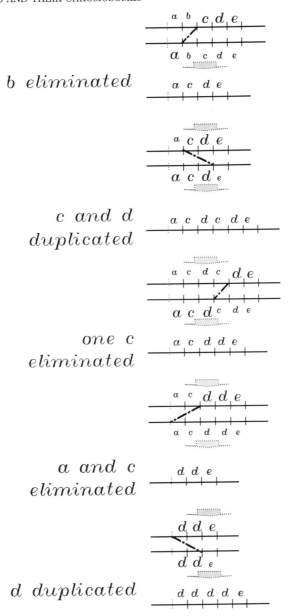

b eliminated

c and d duplicated

one c eliminated

a and c eliminated

d duplicated

Fig. 10.4. *Fixation of one member of a cluster of genes in a multigene family. See text.*

type. Numerous proposals have been made to explain how this homogeneity might be maintained. Some of them are as follows:

1. A mechanism such as gene conversion (Chapter 4, p. 160), which corrects by substituting a "normal" allele for a mutant one, has been invoked to explain a number of instances in which homogeneity of structure has been conserved. A problem with it is that the conversion can go either way, mutant or normal. A bias

toward the normal may exist because of some selective mechanism operating during the resolution of the heteroduplex during the crossover process (see p. 160). Some experimental findings support such a possibility, which is discussed in the next section. In addition to the nonreciprocal transfer of genetic material from one allele to another on homologous chromosomes, the term gene conversion is also used to describe the transfer of DNA sequences from one gene to a related one elsewhere in the genome on a homologous or nonhomologous chromosome. This a form of ectopic recombination.

2. It is proposed that gene amplification in which one gene is tandemly replicated many times over (Chapter 5, p. 248) could be a mechanism for maintaining a high degree of identity. This has documented support in the amplification of rDNA genes, which is known to occur in the oocytes of animal cells, some of it by the rolling circle method. This does not explain how mutated genes are purged from the cluster, however.

3. Transposition of a genetic element (described in Chapter 5) has been cited as a possible mode of diffusion of an element to form a family of similar sequence genetic units (Campbell, 1983). This is certainly a reasonable proposal, and the SINES and LINES families are examples, among others.

4. Homologous but unequal crossing-over can theoretically result in homogenization such that one member of a set of genes with different alleles replaces all other members of the set (Smith, 1973; Perelson and Bell, 1977; Ohta, 1989). Monte Carlo computer simulations show that after large numbers of crossovers within a family one member may become fixed. Figure 10.4 illustrates this. Mispairing by one repeat with one crossover can result in the gene's increasing by one, decreasing by one, or not changing in number if not involved in the crossover.

2. Evolution of protein-encoding gene families

In contrast to families with reiterated identical elements maintained in linear clusters such as the rDNA genes, the families of protein-encoding genes generally have heterogeneity in structure and function and may be dispersed throughout the genome. Otherwise they are transiently amplified clusters.

DNA and protein sequence studies lead to the conclusion that protein-encoding multigene families have not evolved by neat and precise tandem duplications. This is in part because crossing-over and the other mechanisms listed in Table 10.2 are not neat and precise processes. The generation of sequence variants by unequal crossing-over is well established (Maeda and Smithies, 1986). One example among many is the origin of proline-rich genes (PR) by unequal crossing-over in humans (Lyons et al., 1988). The PR family consists of six tandemly linked genes on chromosome 12p. The genes are not all identical in structure, but have different insertions and deletions, the origins of which are best explained by homologous but unequal crossing-over. The entire cluster has a length of ~600 kb and the members have length differences mainly resulting from different numbers of tandem repeats in third exons. These are all functioning genes and presumably each one performs a specific role.

As described in Chapter 4 and this chapter, gene conversion can result in aberrant segregation ratios in which one allele of a pair replaces another during recombination. The encoded sequences in high-cytosine (Hc) late-acting chorion gene families of Bombyx mori have about 91% sequence similarity with one another except for the several hundred base pair regions at their 3′ ends. These regions are not conserved and are variable in length. The nucleotide differences within the highly conserved regions were not found to be the result of independent mutations within the gene family (Xiang et al., 1988). All genes in the family have similar sequences; their sequences have become "homogenized" (Dover, 1982, 1986; Dover and Tautz, 1986). The 91% homology was maintained by the spread of one type of change through the population of family members. Thus it is probable that mechanisms other than selection for equivalent independent mutations are involved. The most likely explanation based on the available data is that the homologies are maintained by gene conversion.

Gene conversion appears to have been a factor in maintaining a degree of sequence homogeneity in a rat P450 subfamily (Matsunaga et al., 1990). Certain of the exon sequences of the four clustered genes in the family have only about 80% sequence similarities, but other regions have a high degree of sequence conservation that is apparently maintained by gene conversion. These regions code for the essential noncovalently bound heme active sites of the P450 enzymes.

On the other hand, gene conversion can also theoretically lead to diversification of family members. Gene conversion in the chicken immunoglobulin family apparently contributes to some of the wide variety of different λ genes altered during cell ontogeny (Reynaud et al., 1987; Thompson and Neiman, 1987). What is particularly interesting in this case is that somatic conversion is involved in the diversification in addition to V-J joining that occurs in the immunoglobulin gene clusters, and that some sequences derived from a pseudogene pool participate.

The untidiness involved in the origin of some families is revealed in the structure of the human α-amylase family, the members of which are clustered on the short arm of chromosome 1 at 1p21. Five active genes and one pseudogene have been identified (Fig. 10.5a). Two of these genes,

AMY2A and *AMY2B*, encode pancreatic amylase, and three, *AMY1A*, *AMY1B*, and *AMY1C*, the amylase salivary enzymes of the parotid glands (Samuelson et al., 1990; Groot et al., 1990). The entire cluster spans 200–300 kb depending on the haplotype, and both the salivary and pancreatic genes are closely related in sequence, the major difference being that *AMY1* has one more exon than the *AMY2* gene (Horii et al., 1987). In addition to the *AMY* pseudogene, *AMYP1*, there is also present in the cluster copies of the γ-actin pseudogenes, *ACTGP3*, that are associated with both *AMY2* and *AMY1* types (Fig. 10.5b). The promoter and nontranslated exons of the *AMY1* genes are derived from the γ-actin pseudogene, and all genes except *AMY2B* have retroviral LTRs associated with them. The family is actually a chimera that incorporates, in addition to the active amylase gene sequences, remnants of γ-actin and *AMY1* pseudogenes, endogenous retroviral sequences with LTRs, and *L1* long interspersed repeat elements. Each *AMY1* gene is associated with an intact retrovirus designated *ERVA1* (Fig. 10.5a), and its transcription start is within the γ-actin *ACTGP3* pseudogene sequence.

The mouse amylase gene family does not contain either retroviral or γ-actin sequences, which indicates that these elements were introduced into the human line after the separation of primate and mouse lines. Either that or the mouse line has had its amylase genes expunged of the actin and retroviral elements. DNA isolated from human, chimpanzee, gorilla, orangutan, and a macaque monkey hybridized with γ-actin and amylase probes, indicating that the pseudogene was introduced prior to the separation of the ape and old world monkey lines, and at about the same time that the retroviral elements were integrated.

Aside from the integration of the foreign elements into the amylase complex, it is evident from the detailed sequence

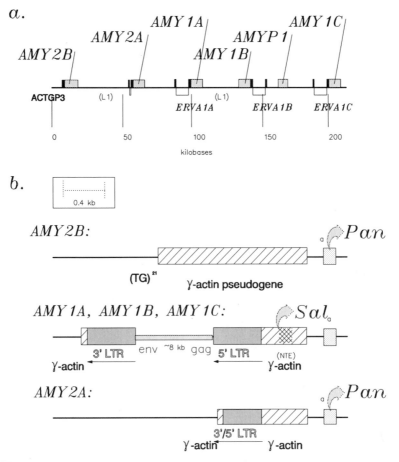

Fig. 10.5. *The human amylase gene complex.* **a,b.** *See text. Reproduced from Samuelson et al. (1990), with permission of the publisher.*

analyses of Groot et al. (1990) that the present-day family has evolved by consecutive inter- and intrachromosomal unequal, homologous crossovers plus inversions. The result is a hodgepodge that lends itself to the formation of numerous different haplotypes with 3–12 functional amylase genes (Groot et al., 1989). Each haplotype analyzed has had both Type 1 and Type 2 genes so that the amylase secretions of both pancreas and parotid are present.

The amylase cluster is by no means the only example of excess complexity within families. Other examples already de-

scribed in previous chapters are the β- and α-globin families (p. 307) and the immunoglobulin families (p. 285).

3. Gene families arise from genotypic changes that have selective value

We have described a hypothetical selective process by which the earliest organisms may have developed metabolic pathways by using the heterotrophic prokaryotes as an example. These prokaryotic families are fundamentally different from the typical eukaryotic family; but, can families such as those described for the higher eukaryotes in Chapter 9 also be expected to arise by natural selection of chance genotypic alterations? These families must have evolved by chance mutation of duplicate genes to new gene forms that encode new proteins with related but not identical functions. How has this come about to produce the higher, more complex forms from the so-called lower? Sufficient data are now available to make an approach to answering this question.

The P450 gene superfamily provides some insight into the selective mechanisms that may be involved (Nelson and Strobel, 1987; Gonzalez et al., 1990). The animal drug-metabolizing enzymes encoded by this superfamily of genes are described in Table 9.7. It is estimated that the mammals have between 60 and 200 functional P450 genes and each gene encodes a functionally unique enzyme structurally related to the other family members. Most of the genes in the different subfamilies (CYP1, etc., as listed in Table

9.7) are induced to transcribe by different families of drugs. Some are also developmentally regulated and some are tissue- and sex-specific.

More than 110 of the proteins encoded by P450 genes have been sequenced in 15 eukaryotes and four prokaryotes. Some of the substrates acted upon by these enzymes are plant products that are toxic to animals. In accord with the structure of other families the data show that intraspecific variation in structure among the enzymes is low, while differentiation between even closely related species in unexpectedly high.

A phylogenetic tree (Fig. 10.6) has been constructed from the amino acid sequence data of 34 of the proteins (Nelson and Strobel, 1987). Three evolutionary events appear to be indicated by the tree: (1) The divergence of bacterial and eukaryotic genes is set at about 1.4×10^9 years ago, which is the approximate date of the earliest eukaryotic microfossils; (2) the sep-

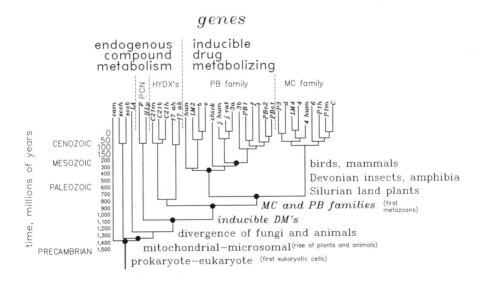

Fig. 10.6. *The phylogenetic tree deduced for some P450 genes from the amino acid sequences of their products. Data from Nelson and Strobel (1987).*

aration of the plant and animal lines of descent appears to have been about 1.2×10^9 years ago; (3) about 0.4×10^9 years ago a great increase of new genes occurred particularly in the *CYP2* family. This is about the time that the animals began to inhabit the land and feed upon the many different forms of terrestrial plants whose ancestors had made the sea–land transition in the late Silurian or early Devonian. It is well established that land plants in particular produce a wide variety of products that are toxic to animals. A logical assumption is that they make these substances to protect themselves from herbivorous animals. This leads to the possibility that the animal *CYP* family arose by selection of random variants that provided a protective response. The data in Figure 10.6 show that the genes encoding the inducible animal drug-metabolizing enzymes arose after plants and animals separated, and the large *PB* family arose in the Paleozoic when eukaryotes began to move on land.

A similar situation may have been the reason for the development of the immunoglobulin superfamily. Animals respond to the changes in their external environment through their nervous systems, but also by protective proteins of various sorts. The immunoglobulin superfamily may have arisen after the separation of plant and animal lines about 750 MYBP (Bodmer et al., 1986). This was followed at 400 MYBP by the MC and PB parts of the P450 family with the appearance of the first metazoans. Both families therefore may have arisen as an adaptation to the exigencies of the external environment.

An explanation of this sort does not immediately strike one as being the causation behind the origin of families like the globins, collagens, and the many others that are involved in the functioning and structure of complex animals and plants, but it is possible that families such as these arose primarily for endogenous rather than

exogenous reasons, and secondarily as an adaptive response to the external environment.

What then can be the factors involved in the evolutionary origins of these kinds of families? We can probably best begin to address this question by considering the roles played by the present-day families in determining what an organism is, as well as how it copes with its environment. These considerations have been dealt with in previous chapters, but we summarize and amplify them here.

It is important to note that families of structural genes are not limited by time or space in their actions in the developing embryo and adult. They have a tempo to their activity in the developing embryo, such as, for example, the cascade of activity found for the globins in the human embryo and fetus. Their activities are also spatially confined—thus, the globins are active in blood cells to become erythrocytes, and the collagen genes are active in different tissues and organs of the embryo and adult. Furthermore, the ubiquitous members of the homeotic gene families not only act at different times in different tissues, but they assume quite different roles in different organisms. Thus family members can evolve in different directions, giving genomes great plasticity.

A second point to be made about families is that a great many more probably exist, particularly in the higher eukaryotes, than we have cognizance of presently. This possibility is supported by estimates of mRNA types detected in various organisms. The assumption is that the number of detectable mRNAs should be a measure of the number of transcribing genes. One should expect a lesser number of these in the lower eukaryotes than in the higher, more complex multicellular forms. The number of genes encoding for basic metabolic functions (the so-called housekeeping genes) should be about the same for all eukaryotes, and therefore it is not unreasonable to assume that most of them above that number ($\sim 3 \times 10^3$) should be family members. Yeast has been reported to have about 3,000–4,000 mRNAs (Hereford and Rosbash, 1977) and the phycomycete fungus *Achlya* about 2,000–3,000 (Timberlake et al., 1977). Presumably these low numbers for these simple fungi are a reflection of the absence of large numbers of families. These estimates are to be compared with those for the number of different mRNAs detected at various stages of development of some insects, echinoderms, and *Xenopus laevis*, which have been reported to be between (8–27) $\times 10^3$ (Hough-Evans et al., 1980). The tobacco plant has been reported to have about 25,000 mRNAs (Kamalay and Goldberg, 1980). These figures fall within the range of the estimates made for total number of encoding genes and indicate that perhaps 20,000 of the genes in the human genome are family members. However, this may be a considerable underestimate. Van Ness et al. (1979) have reported that mouse brain tissue has both poly(A)$^+$ and poly(A)$^-$ mRNAs. The complexity determinations indicated that a total of 170 $\times$ 10^3 different sequences 1.5 kb in length are present in the total RNA. Of these 78 $\times 10^3$ are nonpolyadenylated [poly(A)$^-$]. The two different populations of mRNAs showed no homology. The polyadenylated fraction transcribed cDNAs. These findings are a further support to the conclusion that no one knows how many genes there are in the human genome, that there may be many more families of genes than present data would indicate, and that the estimate previously given that 82% of the genes are family members may be much too low. Animals with large brains may have ten times more genes, most of which may be family members, than higher plants. Thus estimates for total gene numbers in the range of $(50-100) \times 10^3$ in the large mammalian genomes may not be out of order.

Certainly data such as these may be used to support the hypothesis that the $\sim$50% unique DNA fraction in the human genome that does not appear to be transcribed can in fact be transcribed and the transcripts translated under certain conditions. Also it should be borne in mind that some RNAs although not used to translate polypeptides or even function in translation as rRNAs and so on may carry out regulatory roles. RNAs appear to carry out such a role in the development of the early insect embryo (Kalthoff, 1983; Elbetieha and Kalthoff, 1988).

Impressive evidence exists that complexity of structure and function, including behavior of phenotype, is accompanied by complexity of the genotype and hence genomic structure. The evidence that life evolved from its beginnings with increase in complexity is well established. We posit that this increase was the result of adaptations to fill the various niches available in the earth's environment. Filling these many niches was made possible by the development of myriad families of genes. Eukaryotic evolution is therefore basically gene family evolution.

The beginnings of the diversification of genes in families may have begun about 570 MYBP in what has been called the **Cambrian Explosion** (Erwin et al., 1987). It was during the Cambrian that the primitive animal phyla and multicellular plants began to appear in the sea. Many of these were the ancestors of the present-day plants and animals. And it is possible that it was in the Cambrian that the present-day type of genomes with multiple chromosomes began to appear to accommodate the multiplicity of family members. This would follow from the observation that there appears to be selection against congealing of the genome.

That variance in family members within a species is less than that between species is a fact that must be considered in thinking about their evolution. It has become evident with the advent of molecular genetics that the processes involved in the origin of genotypic alterations are of greater magnitude than previously imagined. In addition to gene mutation, recombination, and chromosomal rearrangements, the processes of transposition (Erwin and Valentine, 1984), gene conversion, unequal crossing-over, and RNA-mediated exchanges via retroviruses (to mention only a few) must now be reckoned with. Furthermore, exchanges between chromosomes are not primarily limited to homologues but may occur just as frequently between nonhomologues by such means as transposition and RNA-mediated exchanges in addition to translocations. Molecular processes such as these have undoubtedly been important in the evolution of families, and they are currently receiving considerable attention by both the theoreticians and experimentalists (Slatkin, 1986; Walsh, 1986; Tachida, 1987; Nagylaki, 1988; Ohta, 1989). The name **molecular drive** has been assigned by Dover (1982) to describe the effects of these processes on the origin and evolution of families. "*Molecular drive is the process by which mutations are spread through a family (homogenization) and through a population (fixation) as a consequence of a variety of mechanisms of nonreciprocal DNA transfer within and between chromosomes*" (Dover, 1986). Both random processes and directional processes are involved. The random events may include equal and unequal crossing-over at both the somatic and germ line level, unbiased gene conversion, and transposition; the directional processes might include biased gene conversion, and duplications including amplification. Even random processes can lead to a concerted evolution in which chance determines which family members prosper. The directional processes can accelerate the tendency toward homogeniety once isolation is established and the winners are selected. The phenomenon of concerted evolution may well be one of the fundamental factors in the origin of new species.

Genotypic changes within a species are certainly a most important factor contributing to its evolution, but not the only one. Intraspecific genetic exchanges or unilateral contributions through viral participation must also be considered to be a factor. Unilateral contributions can be extreme. For example, good evidence exists that a *GAPDH* gene carried in the genome of *E. coli* encodes a glyceraldehyde-3-phosphate dehydrogenase that is more similar in amino acid sequence to the eukaryotic enzyme than the prokaryotic one. The bacterium also has a *GAPDH* gene that encodes a prokaryotic-type enzyme. Thus we have an example of a "naturally occurring horizontal gene transfer from a eukaryote to a prokaryote" (Doolittle et al., 1990). Findings such as these are certainly broadening our horizons about the evolutionary processes. Sylvanen (1986) has marshaled evidence that he considers supports the thesis that the transfer of genes from one species to another may be a major factor in evolution. The most likely mode of transmission probably would be viruses such as the retroviruses and would include transposable elements. Also the interspecific crosses that occur among the flowering plants in particular with the establishment of allopolyploids must be considered in this connection.

V. UNIQUE DNA SEQUENCES WITH INDETERMINATE FUNCTIONS ALSO EVOLVE

Single-copy DNA constitutes a large part of the DNA of higher eukaryotes. As shown in Table 2.3, ~60% of the human genome is unique. A part of this DNA, perhaps 3%–5% encodes proteins that perform identifiable functions as enzymes, structural elements, and regulatory elements.

These are the members of gene families for the most part. The rest have no identified function as yet, but they may also include families of related sequences. On average ~70% of animal genomic and ~30% of plant genomic DNA is unique (Schmidkte and Epplen, 1980). Certain exceptions exist among the animals; the anamniotic selachians, teleosts, and amphibians have an average of about 30% unique DNA (Morescalchi and Olmo, 1982). This is in contrast to the amniotes: The birds and mammals have a unique DNA fraction averaging in the 70% range, while the reptiles form an intermediate class between them and the anamniotes with a unique fraction between 45% and 68%.

Studies of the conservation of these "nongenic" unique sequences have indicated that they may be evolving rapidly compared to the identified genes that are for the most part related in DNA sequence structure among related species. A single-stranded probe for a protein-encoding gene from one species generally hybridizes with DNA of related forms, and the proteins with similar functions of related species nearly always show close amino acid sequence compositions. However, at least in some taxa, a fraction of the unique DNA shows differences divergent enough that it does not hybridize with DNA of a closely related species.

Unique fractions of DNA (containing 1–10 homologous sequences) from five different species of the *Drosophila virilis* group have been shown to have 30%–50% divergence between species (Riede et al., 1983). In the *melanogaster* group *D. simulans*, and *D. melanogaster* are morphologically very similar, but are sexually isolated. A fraction of about 20% of the DNA from *simulans* does not hybridize with *melanogaster* DNA (Werman et al, 1990). This nonhybridizing fraction is a mixture of repeats and single-copy DNA sequences with the single-copy sequences probably predominant. The differences between

these sequences in the two species appear to be the result of base substitutions, although small rearrangements and deletions may also be involved. Thus a significant fraction of the DNA of two closely related species can have significant differences in nucleotide sequences. One concludes from this that portions of the *simulans* and *melanogaster* genomes are evolving rapidly.

DNA–DNA hybridization experiments with material from humans and the catarrhine primates—chimpanzee, gorilla, orangutan, the gibbons, and baboons—have shown clearly that the unique sequence components have diverged considerably in this group of primates (Caccone and Powell, 1989). However, that small fraction of the unique part of the genome that is transcribed and translated is for the most part highly conserved. In general the functionally identified parts of the genomes of related taxa encode proteins that have a high degree of amino acid sequence similarity. The amino acid data for the nine types of polypeptide chains from human and chimpanzee sources show only a 0.4% difference between them and for human and gorilla the difference is 0.7% (reviewed by Goodman et al., 1989, 1990).

The data presently available clearly indicate that the functionally identified fraction of the DNA of related taxa is conserved in sequence to a high degree. A good deal of the functionally unidentified element may still be conserved, and not be detected by the relatively crude technique of hybridization. We are dealing, after all, with ~60% or more of a mammalian genome, or ~1.8 × 10^9 bp. Furthermore, the possibility that the relatively large changes in the unidentified unique fraction between taxa may constitute the important genetic differences between them cannot be ignored. It is also possible that a large part of the functionally unidentified repetitive and unique sequences may play an important role in allowing for reshuffling the genome by rearrangements within and between chromosomes.

The fact that only a small percentage of the total DNA per genome in the higher eukaryotes is identified as to function has led many observers to the conclusion that the bulk of the unique fraction and the repetitive elements of genomic nuclear DNA are unimportant and carried as excess baggage. This may be true but it also amounts to stating that if one does not understand something, it can't be important. Some may indeed be excess baggage, but as discussed below excess does not necessarily mean nonfunction.

VI. MIDDLE REPETITIVE SEQUENCES MAY ALSO EVOLVE RAPIDLY

Families of middle repetitive sequences (MRS DNA) show evidence for extensive, rapid evolutionary change both in sequence and repeat number. The size of these sequences may range from dinucleotide doublets with two base pairs per strand to >10 kb, and the different types may be either confined to centromeric regions or interspersed among unique sequences in the euchromatin (as described in Chapter 2). Those located in the euchromatin in particular frequently show a disposition to move about in the genome as transposing elements (as described in Chapter 5). Besides varying considerably in structure between and even within species they may show great variation in the amounts of the different types.

A. CpG and GpT Repeats

The dinucleotide repeat CpG (CG/GC) is found in both animals and plants. Methylation occurs specifically in the cytosines of these doublets. When cytosine is methylated, deamination occurs at a high frequency and the cytosine is changed to thy-

mine (Bird, 1986). For this reason the dinucleotide CpG is found much less frequently than expected in DNA on the basis of chance. There are, however, CpG-rich islands where the dinucleotide occurs at frequencies similar to what would be expected by chance (Bird, 1987). These "islands" are regions 500–1,000 bp long and they frequently occur upstream from the transcribed regions of genes. The cytosines in these islands remain unmethylated and thus are not converted to thymine. Presumably there is selection against methylation because genes that are methylated may become inactive. It is thought that interisland regions are deficient in CpG because the cytosines were methylated and converted to thymine. Thus C-to-T transitions result in either TpG or CpA (Bird, 1980). It has been suggested that this type of CpG loss is responsible for the production of pseudogenes such as the one in the α-globin family (Bird, 1987).

The vascular plants (angiosperms, gymnosperms, and pteridophytes) have the trinucleotides CpNpGp in which the cytosines are heavily methylated along with the methylated CpGs (Belanger and Hepburn, 1990). The nonvascular plants, on the other hand, appear to limited to methylated cytosines in the trinucleotides.

In addition to CpG the doublet GpT is interspersed in the genomes of all eukaryotes. It occurs in clusters in which $(GpT)_n = 15–30$. In the human genome clusters occur about every 30 kb. The functions of these dinucleotide repeats are unknown, but the possibility that they do have a function should not be ignored, since their occurrence is widespread among the eukaryotes.

B. MRS in Centric Regions

These MRS are frequently called satellites and are ubiquitous among the eukaryotes with the exception of the fungi. Their structure and disposition have already been described in Chapter 2 in some detail. In summary in can be stated that (1) centric MRS types are generally not found in the euchromatic arms of chromosomes; (2) specific types may be concentrated in centric regions of specific chromosomes of a genome, while others may be found in centric regions of all members of a genome; (3) encoding genes are sparsely represented in the areas occupied by them; (4) they may vary greatly in amount among closely related taxa, and some chromosomes in the same genome may have little of them while others may have most of the genomic total; (5) crossing-over within the centric regions occupied by them is strongly reduced; and (6) genes normally located in the euchromatin have their regulatory patterns altered when placed in juxtaposition to these sequences in the heterochromatin. The role of centric MRS in chromosome function is not known, and it has been suggested that their repetitive presence is maintained by the reduction in crossing-over in these regions they occupy (Charlesworth et al., 1986). Unequal crossing-over in these regions of highly repetitive DNA would be expected to both increase and decrease the amount. If a certain concentration of them is required, then this would be best maintained by the absence of crossing-over. It is of some interest in this connection that those fungi such as yeast and *Neurospora* that have simple centromeres and no measurable centric MRS do not show depression of crossover rates in the centric regions (Mortimer and Schild, 1980; Fincham et al., 1979).

C. MRS in Euchromatic Regions

All eukaryotes have MR nontranscribing **spacer DNA**, so called because it occupies spaces between transcribed genic regions primarily in the euchromatin. The fungi have little of this interspersed spacer, but plants and animals ordinarily have 25%–30% of their nuclear DNA in this form. Examples of some of the different types of euchromatic MRS have been described and discussed at some length in Chapters 1, 2, and 5. Here we summarize some of these characteristics and consider the possible evolutionary roles that some of the different types may play in the genomes they occupy.

Like the centric MRS the euchromatic MRS occur in families. Their disposition in the euchromatin may not be random. Some are associated with specific genes, and some are concentrated in R bands, and others are in G bands. An MRS family may have less than 100 or more than 10^6 members. The members are more complex in structure than the simple GpC and GpT couplets, and they may have structures similar to known, functional genes in the genomes that they occupy. Some are transposable and are associated with retroviruses. Some contain open reading frames and are transcribed. The transcripts in some cases are translated. Members of many of the families appear to be evolving in concert; as a general rule, however, they are evolving more rapidly than the "true" multigene families. The central question about these families that we address here is Do they have a function? If so their evolution should be partly under the control of natural selection.

The clusters of rDNA genes in all eukaryotes have interspersed nontranscribed regions. In the slime mold *Physarum polycephalum* the nontranscribed spacers between the transcribed regions are 23 kb in length and contain DNA replication origins (Ferris, 1985). Each spacer consists of reiterations of different DNA sequences only 1,200 bp in length and these sequences seem to be undergoing concerted evolution. In *Drosophila mercatorum* and *D. hydei* the 5-kb noncoding rDNA spacers also appear to be undergoing concerted evolution, and in addition show a strong phenotypic and adaptive effect

(Templeton et al., 1989). The effects are significant, depending on the altitude at which the different natural populations are found.

The genomes of plants show great interspecific differences in DNA content. These differences may amount to 1,000-fold and are mostly attributable to variations in MRS amounts rather than in the numbers of translated genes (Flavell et al., 1974; Flavell, 1986). A study made of the DNA of the genomes of two species of lily, *Lilium henryi* and *L. longiflorum*, with genome sizes of 32×10^9 and 34×10^9 bp, respectively, reveals that both have an abundant retrotransposon repeat identified as *del* (Joseph et al., 1990). In the genome of *henryi*, *del* is present about 13,000 times, but it is three times more common in *longiflorum*. This dispersed MRS is 9.35 kb long, which means that it contributes $3 \times 13,000 \times 9.350 = 0.364 \times 10^9$ bp to the *longiflorum* genome. This is a significant fraction of the 2×10^9 bp difference in the complexity of the two genomes. This is but one example among many to be found, especially among the angiosperms, of great differences in the number of specific MRS to be found between related taxa. The question that confronts is; if *del* or related types of MRS do have a function, why are they present in such vastly different numbers? DNA amount differences such as these lead to more DNA per chromosome rather than more chromosomes per genome, as discussed in the next section.

The *SINES* and *LINES* (short and long interspersed sequences) found in the various mammalian taxa have probably been more extensively investigated than any other group of MRS. In humans and related primates these types of repeats are represented by the *Alu*'s and *L1*'s respectively. The *SINES Alu*'s (described in Chapter 5) are clustered principally in the R bands of human chromosomes (Chen and Manuelidis, 1989). They are apparently

derived from 7SLRNA genes. A cognate repeat, *B1*, found in rodents is also a 7SLRNA derivative. In addition some *SINES* are derived from tRNA genes. All of these probable derivatives vary only about 4%–15% from consensus sequences (reviewed by Deininger and Daniels, 1986). A 350-bp *SINES* family member, designated as *C*, found in the rabbit has >200,000 members interspersed in the genome. Some of them are inserted within genes and function as polyadenylation signals (Krane and Hardison, 1990).

The *LINE* family *L1* is found in most mammalian species and is apparently derived from a protein-encoding gene transcribed by RNA POLII (Singer and Scowronski, 1985). Related elements are also found in trypanosomes, corn, *Neurospora*, and *Drosophila* (Di Nocera and Sakaki, 1990). Human chromosomes have them primarily clustered in the dark G bands in contrast to the R-band-clustered *SINES* (Chen and Manuelidis, 1989). Members of the *L1* family have large ORFs the sequences of which are highly conserved intraspecifically. They therefore appear to be under strong selection pressure and subject to concerted evolution. In this respect they resemble protein-encoding families. One of the ORFs has definite sequence homology to RNA-dependent DNA polymerases (Hattori et al., 1986). Unlike *Alu*'s, *L1*'s have been shown to be capable of transposition. Like *Alu*'s they may have one or more function(s). For example, the *Alu* sequences surrounding the human α_1-acid glycoprotein genes may have been involved in gene duplication and gene conversions (Merritt et al., 1990).

Like the *del* sequence in *Lilium* sp., both *Alu*'s and *L1*'s may show wide divergence in numbers in closely related taxa. In humans and higher apes *Alu*'s range in number per genome from about 300,000 to 900,000, and the *L1*'s from ~50,000 to 100,000 (Hwu et al., 1986). Within this primate group both repeat types show significant sequence homogeneity. Concerted evolution is evident, but accompanied by deletion and insertion events.

VII. GENOMES HAVE EVOLVED BY INCREASE IN BOTH DNA AMOUNT AND CHROMOSOME NUMBER

A. The Amount of DNA per Genome Can Increase Without an Increase in Chromosome Numbers

When the amount of DNA per genome is compared among the different species within some taxa a very wide range of values is found. This has been found to be true for some grasses, echinoderms, insects, and fungi (Sparrow and Nauman, 1976). And it has been noted in Chapter 5, (p. 248) that the salamanders of the genus *Plethodon* have members with significant difference in the amounts of DNA per genome but with equal numbers of chromosomes. Figure 10.7 shows the distribution of the amounts of DNA per genome in picograms for 80 species of grasses. The increase in the amount of DNA per genome appears to be associated with an increase in chromosome size, and not with increase in chromosome number.

This phenomenon has been called **cryptopolyploidy** to distinguish it from true polyploidy. The data are plotted on a semilog scale and clearly show a doubling pattern of the DNA. Some groups, such as the vertebrates, also show considerable differences in the amounts of DNA per genome, but this is generally accompanied by an increase in the number of chromosomes. Also, members of the vertebrate subclass Eutheria have an almost uniform amount of DNA per genome, but a wide distribution of chromosome numbers.

The evolutionary significance of cryptopolyploidy of the type described here is obscure. It appears to be associated primarily with differences in the amounts of repetitive DNA per genome, and not with the numbers of functional genes.

B. Changes in Ploidy Have Probably Been of Great Importance in Genomic Evolution

Polyploidy of the various types described in Chapter 5 is generally widespread among recent day plant taxa and is relatively rare among animals such as the vertebrates. But even in groups such as the present-day mammals there are signs pointing to its participation in the

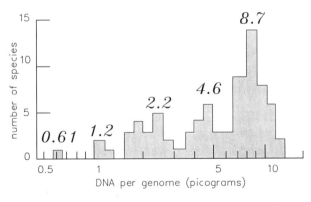

Fig. 10.7. *The distribution of amounts of DNA/genome for 80 species of grasses. Reproduced from Sparrow and Nauman (1976), with permission of the publisher.*

past toward the development of their present-day genomes. Many paralogous gene family members may have arisen originally by an increase in ploidy.

One of the few groups of vertebrates in which polyploidy seems to be well established at least among certain genera are the frogs (Anura). The genus *Xenopus* is a good example because it has species with polyploid series with chromosomes and nuclear DNA contents in the proportions of 2:4:8:12 (Kobel and du Pasquier, 1986; Graf, 1989). The only known diploid species is *X. tropicalis* with $n = 10$. Polyploid species are *epitropicalis* ($n = 20$), *laevis* ($n = 18$), *vestitus* ($n = 36$), and *ruwenzoriensis* ($n = 54$). The multiplier is either 10 (for $n = 10$ and 20) or 9 (for 18, 36, and 54). Hence besides the diploid, there also exist tetraploid, octaploid, and dodecaploid species. Interestingly, however, all *Xenopus* species of whatever ploidy generally assemble only bivalents at meiosis. They are therefore comparable to plant allopolyploids that function as diploids, and in apparent agreement with the origin of plant allopolyploids, they appear to be the result of interspecific hybridizations in the past.

Despite the fact that these polyploid frogs have extra sets of homologous chromosomes the number of active genes of each type as expressed in the diploid does not increase proportionately with the increase in number of chromosome sets (Kobel and du Pasquier, 1986). The MHC of *Xenopus* is well developed and similar to the MHC of mammals; yet all species, whether polyploid or not, have only one functioning MHC complex with the exception of *ruwenzoriensis*, which is thought to be of very recent polyploid origin. Other families, such as the globin and vitellogenin, and singlets, such as albumin and cytosolic MDH, have only two or three times more active members, even in the dodecaploid species. All species of whatever ploidy demonstrated only one nucleolus cytologically.

It is probable that the genes that became inactive in polyploids, leaving only one active member in a polyploid set, have been expunged in descendent taxa. The active remainders were the ancestors of today's singlets in animals, like mammals, that have an apparent polyploid ancestry. A detailed study of the inheritance of 29 genes with electrophoretically identifiable allelic protein products has been made with the tetraploid *X. laevis* (Graf, 1989). Duplicate loci were found for *ALB*, *ADH*, *PEP*, *GPD*, and *GPI*, but such others as *FH* and *TRF* were present as singlets. All showed mendelian segregation patterns in crosses. Paralogous associations were found, such as *ALB-2* and *ADH-1*, linked on chromosome 1, and *ALB-1* and *ADH-2*, linked on chromosome 2.

Duplicate loci that remain active in the tetraploid species may show tissue or developmental differences in activity. The duplicate malate dehydrogenase genes in the tetraploid species *X. muelleri* and *X. laevis* show differential tissue-specific expression. This finding supports the proposition that family members may arise by genome duplication as well as by gene duplication, as discussed in the previous sections.

Although polyploidy has in the past not been considered as important a factor in animal genomic evolution as it has in plant genomic evolution, it is becoming more and more apparent that even present-day vertebrates such as fishes, amphibians, reptiles, and mammals, which are functional diploids, may have had polyploid ancestors (Schultz, 1980; Bogart, 1980). Clear evidence exists for ancestral tetraploidy in the salmonid fish (Johnson et al., 1987). Some species in this taxon have been clearly derived from tetraploids, as in the case of *Xenopus*, and are undergoing diploidization. Others are true diploids but have the duplicate gene patterns, indicating a polyploid origin. Recently obtained linkage data for mammals are also pointing to possible polyploid origins of their genomes. Table 10.4 lists some conserved paralogous segments in mouse and human. The most logical explanation for the origin of these duplicate genes is that mammals have polyploid ancestors. These may be very ancient conditions that originated 400 MYBP or earlier and have been maintained since.

VIII. HAVE GENOMES EVOLVED BY CHANGES IN BOTH TRANSCRIBED AND NONTRANSCRIBED REGIONS?

It is generally assumed that if a specific DNA sequence is conserved over a period of time, it is highly probable that selection must be operating to conserve it, and that it is also probable that the sequence has a function. These assumptions follow logically from the reasonable premise that if something works, the natural tendency will be for it to be preserved. If they are brought to bear on the evidence presented in this and the preceding chapters about the characteristics of genomic DNA in connection with the phylogeny of various taxa, one is led to the conclusion that genomes have evolved by changes in *some* regions of their nontranscribed DNA as well in their transcribed DNA. The main regions of conserved nontranscribed gene sequence DNA are those upstream from the cap sites, which are of a regulatory nature, as discussed in Chapter 6. Others are the spacers between some genes, particularly those that occur between the rDNA genes that appear in some cases to evolve as a homogeneous families. In addition some intervening sequences such as *Alu*'s and *L1*'s in the primates have at least some parts conserved. This is especially true of the ORF's in the *Alu*'s, some of which are transcribed. Widely unrelated taxa generally have unrelated nontranscribed MRS, but closely related taxa may have some that are clearly homologous in structure.

Many ambiguities exist in the relations between these transcribed and nontranscribed entities, but one thing appears clear: The transcribed sequences, especially those that encode mRNAs, are far more conserved over time than the nontranscribed. Figure 10.8a shows the percentage sequence divergence of the amino acid sequences for the α-globins of seven different vertebrates from human α-globin plotted against divergence from a common ancestor in MY. The slope of the line indicates that amino acid substitutions due to nucleotide substitutions have occurred in this polypeptide at the rate of about $\sim 0.1\%$ per 106 Y. On the other hand, sequence change data for nontranscribed sequences in a number of vertebrates show rapid changes on the order of 0.37% per 10^6Y.

This rapid rate of structural changes in non-protein-coding DNA sequences is accepted by many evolutionists as strong evidence that it has no function. While this may be true for some MRS, it is certainly no evidence that it is true for all. It appears that one reason for assuming no function is that many MRS sequence family members vary tremendously in number per genome. Some examples have been noted in this and previous chapters such as in the euchromatin of flowering plants and amphibia and for the centric MRS in the kangaroo rats. This assumption posits that if something varies greatly in amount, it can't be important: a conclusion of dubious validity at best. A sequence may be functionally significant, but less or more of it may make little if any phenotypic difference so long as some is present (Zuckerkandl et al., 1989).

It has been shown that different classes of *Alu* repeats exist in the human genome (Willard et al., 1987; Jurka and Smith, 1988). Although most of the repeats may be considered analogous to pseudogenes, some may actually be functional genes with functions presently unknown. Support for

TABLE 10.4. Conserved Paralogous Segments in Mouse and Human Genomes

Human	Mouse	Gene	Human	Mouse	Gene
19q12–q13.2	7	APOE, Apolipoprotein E	1q21–q23	1	APOA2, Apolipoprotein A
19q12–q13.2	7	ATP1A3, Na, K-ATPase	1q21–q23	1	ATP1A2, Na, K-ATPase
19q12–q13.2	7	PEPD, peptidase D	1q25–or42	1	Peptidase C
19q13.1–q13.2	7	TGFB, tumor growth factor	1q41	1	TGFB2, tumor, Growth factor
12p12.1	6	KRAS2, Kirsten rat sarcoma	17q21q22	11	KRT1, Type 1 cytokeratin
12q12–q13	15	Hox3, Homebox3	17q21q22	11	HOX2, Homeobox2
12q12–q13	15	INT1, Mammary tumor onc	17	11	Mammary tumor

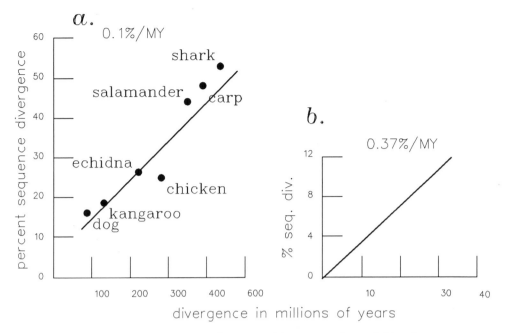

Fig. 10.8. *Rates of divergence of amino acid sequence in α-globins for seven vertebrates* (**a**) *and for some MRS* (**b**). *See text.*

this conclusion rests in the observation that a "source sequence" exists that appears to be evolving at only one tenth the rate considered to be the **neutral mutation rate** in *Alu* sequences (Zuckerkandl et al., 1989).

Some base substitutions do occur in encoding genes without causing an amino acid substitution. These are called **synonymous base substitutions**; they are substitutions that change a codon to a synonymous codon and hence no amino acid substitution occurs. Synonymous substitutions probably occur at a rate similar to the rate that nonsynonymous ones do, but presumably because they are not selected against they are **selectively neutral**. Also some base substitutions result in amino acid substitutions that appear to be selectively neutral. These observations have led to the **neutral allele** or **neutral mutation–random drift hypothesis** of Kimura (1983). The occurrence of alleles of a specific gene encoding different **allozymes** that have no apparent effect on the selec-

tive value of the phenotype is common and well documented. (The term allozyme is defined here as designating an altered polypeptide that may or may not have an enzymatic function.) Members of all natural populations appear to be heterozygous for many of their genes, and are frequently polymorphic for two or more alleles, some of which encode allozymes that appear to be selectively neutral.

It follows from this that not all conserved sequences have originated through the operation of natural selection. Ample evidence exists to show that molecular changes such as the nucleotide base substitutions resulting in point mutations go on in all sequences at a fairly constant rate. The rates are not the same for all sequences; each sequence type, however, changes at an approximately linear rate when number of substitutions found in related taxa are plotted against divergence times from a common ancestor, as shown in Figure 10.8. This linear rate of exchange that appears to be constant over long periods of time is referred to as the **molecular clock**.

One of the more interesting phenomena to become apparent after the molecular data including the clock began to become available was the fact that in the Primates in particular **organismic** evolution seems to occur more rapidly than molecular evolution (Wilson et al., 1974a,b). By organismic is meant anatomical and behavioral characteristics. The higher apes and *Homo sapiens* differ very little in the structure of their identified proteins, but do in certain aspects of their anatomy and behavior. Observations such as this have led to the proposal that the differences are the result of changes in the genome other than in the genes encoding the types of proteins common to all the different species. The differences may be in the regulation of the temporal and spatial activity of orthologous genes in the different species rather in the structure of their translated products. If this be the case, then the regulatory genes and the upstream binding sites for their binding protein products, as well as possible regulatory roles of nontranscribed unique and MRS sequences, should be considered. Major roles may be played by these elements in the speciation processes (Wilson, 1975).

IX. SUMMARY, CONCLUSIONS, AND AFTERTHOUGHTS

A. Origins of Eukaryotic Genomes

Unless one is willing to admit that eukaryotic genomes probably arose from the simpler genomes of the prokaryotic type by the addition of supplementary genetic material, one is left with no logical explanation for their origin, since prokaryotic remains appear in the fossil record well before the evidence for eukaryotic remains. There is enough circumstantial support to legitimate as a reasonable theory the hypothesis that the first eukaryotic cells arose about 1.4×10^9 YA with a synergism between aerobic free-living prokaryotes able to use oxygen as a hydrogen sink, and anaerobic protonucleated cells with only a hexose glycolytic pathway capable of producing a limited amount of ATP. The three carbon end products of the nucleated protoeukaryote's glycolytic pathway were used by the prokaryotic partner to provide additional ATP with the production of CO_2 and H_2O. Photosynthetic bacteria contributed the additional ability to utilize light energy to produce carbohydrates to complete the carbon cycle.

If such was the case, it follows logically that the prokaryoyic infiltrations of presumably protoeukaryotic nucleated cells evolved to the present-day types of cells, in which the cytoplasmic mitochondria and chloroplasts are the residues of the original invaders. (See Dyer and Obar, 1985, for extensive literature support.) The animal and fungal lines evolved with mitochondria only, while the green plant lines had both.

Present-day eukaryotic genomes are therefore theoretically derived from at least three sources: the two types of bacterial genophores (the heterotrophic and autotrophic photosynthetic) and the chromosomes (?) of the original anaerobic protoeukaryotic heterotrophic host. Since present-day mitochondria and chloroplasts are not free-living, but are mostly reduced to the catabolic reactions of the citric acid cycle, and the electron transport system in the mitochondria, and the anabolic reactions of photosynthesis in the chloroplasts, it is probable that they have lost to the chromosomes of the nuclei those genes that made it possible to maintain a free-living existence. Protein synthesis goes on in both types of organelles. They have their own transcriptional and translational systems distinct from the nuclear and cytosolic systems, but this synthesis is mostly limited to certain subunits of the enzymes that are involved in the electron transport pathway in the mitochondria and to the chromoproteins involved in photosynthesis within the plastids. Other anabolic processes, such as the synthesis of isoleucine and valine in the fungi within the mitochondria, are carried out in the presence of enzymes encoded by chromosomal genes and the mRNAs translated in the cytosol. There appears to be continual passage of polypeptides from their sites of synthesis in the cytosol into the mitochondria and plastids.

In this view the chromosomes of the eukaryotes arose as chimeras possessing just short of 100% of the genetic material necessary to make a functional organism. The essential functions of oxidative metabolism and photosynthesis have been maintained in the highly hydrophobic environments of the inner membranes of mitochondria and chloroplasts. The segregation of these functions in these special environments appears to be necessary for these activities. Otherwise these organelles might have long since been lost.

Once established the primordial composite chromosomes evolved endogenously by the various processes discussed in this and the preceding chapters—the main ones perhaps being duplication by unequal crossing-over, amplification, polyploidy, and rearrangements by translocations, inversions, and transpositions. Also, exogenous exchange of DNA between members of unrelated taxa by transformations of nonsexual types, including by viruses of various kinds, must now be considered.

B. Repetitive DNA

The fact that most eukaryotes, especially those in the multicellular plant and animal taxa, have a large segment of their DNA in the repetitive category has led to the conclusion by some observers that this repetitive fraction is unimportant in genomic function. They arrive at this conclusion primarily because many of the repeated sequences that have been identified can vary greatly in amount between closely related taxa with no apparent significant phenotypic effects. This is especially true of the angiosperms, which may have genomic DNA contents ranging from 10^8 to 10^{11} bp even between members of the same family. Those species with DNA contents per genome in the range of 10^8 bp have 10 to more than 100 times less repetitive DNA than close relatives yet function quite effectively. But repetitive DNA is never completely absent, and may perform important essential functions even at low concentrations. For example, the repeat sequences such as TTAGGG in the telomeric regions can vary widely in number, yet it is highly probable that they are essential for chromosome replication. Theoretically their complete absence should be lethal. While the dosage of functioning encoding genes appears to be exceedingly important, the dosage of these repeating entities seems not to be, provided they are present at certain minimal levels. They are not important from the standpoint of aneuploidy, but may otherwise have important functions particularly in the replication and organization of the chromatin during the interphase of the cell cycle.

C. Euploidy and Aneuploidy

The above considerations lead to the conclusion that two fundamental categories of eukaryotic DNA exist: (1) that which is rigorously maintained at constant dosage, the **CD fraction**, and (2) that which is not and constitutes the the variable dosage, or **VD fraction**. Both categories perform necessary functions. The CD fraction includes the identified "genes" of various sorts that are maintained within a framework of *dosage balance* in activity with one another. The interesting fact about this balance is that it is *chromosomal* rather than simply *genic*. What is meant by this is that the dosage of an encoding gene may be amplified *on* a chromosome horizontally without drastic phenotypic effects. But changing dosage vertically by aneuploid hyperploidy or hypoploidy can have significant phenotypic effects, which in most cases can be lethal.

Strong selective pressure must therefore operate to maintain the CD fraction balanced between chromosomes. This is especially true of the autosomes. On the other hand, selection pressure need only maintain the VD fraction dosage within certain broad limits, which may range from extremely small dosages (say one to a few hundred) to as much as an organism may be able to tolerate economically.

References

INTRODUCTION

Altmann R (1889) Uber Nukleinsauren. Arch Anat Physiol Lpz Physiol, Abt. 524.

Anderson E (1925) Crossing over in the case of attached X chromosomes in *Drosophila melanogaster*. Genetics 10:403.

Arber W, Linn S (1969) DNA modification and restriction. Annu Rev Biochem 38:467.

Arrighi FE, Hsu TC (1971) Localization of heterochromatin in human chromosomes. Cytogenetics 10:81.

Auerbach L (1874) "Organologische Studien," 1 und 2 Heft. "Zur Charakteristik und Lebensgeschichte der Zellkerne." Breslau: E. Morganstern.

Avery OT, McCloed CM, McCarty M (1944) Transformation of pneumococcal types induced by a desoxyribonucleic acid fraction isolated from Pneumococcus Type III. J Exp Biol Med 79:137.

Baer KE von (1828) "De ovi mammalium et hominis genesi. Epistola Acad Imp Scientiarum Petropolitanum." Leipzig: Leopold Voss. An English translation of this important work appears in Meyer AM (1956) "Human Generation." Stanford, CA: Stanford University Press.

Balbiani EG (1881) Sur la structure du noyau des cellules salivaires chez les larves de *Chironomus*. Zool Anz 4:637, 662.

Bateson W (1902) "Mendel's Principles of Heredity, a Defence." Cambridge: Cambridge University Press.

Bateson W (1909) "Mendel's Principles of Heredity." Cambridge: Cambridge University Press.

Bateson W (1916) The mechanism of Mendelian heredity (a review). Science 44:536.

Beadle GW, Emerson S (1935) Further studies of crossing over in attached-X chromosomes of *Drosophila melanogaster*. Genetics 20:192.

Beadle GW, Tatum EL (1941) Genetic control of biochemical reactions in Neurospora. Proc Natl Acad Sci USA 27:499.

Beneden E van (1883) Recherches sur la maturation de l'oeuf, et la fecondation *Ascaris megalocephela*. Arch Biol 4:265.

Beneden E van, Neyt A (1887) Nouvelles recherches sur la fecondation et la division mitosique chez l'Ascaride megalocephale. Bull Acad R Med Belg Ser 3 14:215.

Boveri T (1888) Zellen-Studien. II. Die Befruchtung und Teilung des Eies von *Ascaris megalocephala*. Jena Z Naturwiss 22:285.

Boveri T (1902) "Das Problem der Befruchtung." Jena: Gustav Fischer.

Boveri T (1909) Die Blastomerenkerne von *Ascaris megalocephala* und die Theorie der Chromosomenindividualitat. Arch Zellforsch 3:181.

Britten R, Kohne D (1966) Nucelotide sequence repetition in DNA. Carnegie Inst Washington Yearb 65:78.

Brenner S et al (1961) An unstable intermediate carrying information from genes to ribosomes for protein synthesis. Nature 190:576.

Bridges CB (1916). Nondisjunction as proof of the chromosome theory of heredity. Genetics 1:1.

Bridges CB (1925). Sex in relation to chromosomes and genes. Am Nat 59:127.

Butenandt A et al (1940) Kynurenin als Augenpigmentbildung ausloesendes Agens bei Insekten. Naturwissenschaft 28:63.

Caspersson TO (1950) "Cell Growth and Cell Function." New York: Norton.

Chargaff E (1951) The composition of the deoxyribonucleic acid of salmon sperm. J Biol Chem 192:223.

Correns C (1900) G. Mendel's Regel uber das Verhalten der Nachkommenschaft der Rassen-bastarde. Ber Dtsch Bot Ges 18:158.

Creighton HB, McClintock (1931) A correlation of cytological and genetical crossing over in *Zea mays*. Proc Natl Acad Sci USA 17:485.

Cremer T (1985) "Von der Zellenlehre zur Chromosomentheorie." Berlin: Springer-Verlag.

Cuenot L (1902) La loi de Mendel et l'heredite de la pigmentation chez les souris. Arch Zool Exp Gen Ser 3 10:27.

Cuenot L (1903) L'Heredite de la pigmentation chez les souris. Arch Zool Exp Gen Ser 4 1(3): Notes Rev, 33.

Delage Y "La Structure du protoplasma et les theories sur l'heredite et les grands problemes de la biologie generale." Paris: C. Reinwald.

Demerec M (1940) Genetic behavior of euchromatic segments inserted into heterochromatin. Genetics 25:618.

Drake S (1978) "Galileo at Work. His Scientific Biography." Chicago: University of Chicago Press.

Ephrussi B (1942) Chemistry of the "eye color hormones" of Drosophila. Q Rev Biol 17:327.

Emerson RA, East EM (1913) The inheritance of quantitative characters in maize. Bull Agric Exp Sta Nebr 4:1.

Farmer JB, Moore JES (1904) On the maiotic phase (reduction division) in animals and plants. Q J Microsc Sci NS 48:489.

Feulgen R (1914) Uber die Kohlen hydatgruppe in der Nucleisaure. Z Phys Chem 92:154.

Feulgen R, Rossenbeck H (1924) Mikroskopisch-chemischer Nachweis von Typus der Thymonucleisaure und die darauf beruhende elektive Farbung von Zellkern in Microskopischen Preparaten. Z Phys Chem 137:203.

Flemming W (1882) "Zellsubstanz, Kern- und Zelltheilung." Leipzig, Jena: Gustav Fischer.

Fol H (1879) Recherches sur la fecondation et le commencement de l'henogenie chez divers animaux. Mem Soc Phys Hist Nat Geneve, vol 26.

Fol H (1891) La quadrille des centres. Arch Sci Phys Nat (II Per), Bd 25 Zitiert nach Boveri (1908).

Garrod AE (1902) The incidence of alkaptonuria: A study in chemical individuality. Lancet 2:1616.

Garrod AE (1909) Inborn Errors in Metabolism. "Reprinted in Harris H, ed (1963): Garrod's Inborn Errors of Metabolism." Oxford Monographs on Medical Genetics. London: Oxford University Press.

Gould SG (1977) "Ontogeny and Phylogeny." Cambridge, MA: Harvard University Press.

Greenstein, JP (1943) Friedrich Miescher, 1844–1895. Founder of nuclear chemistry. Sci Monthly 57:523.

Hayes W (1968) "The Genetics of Bacteria and Their Viruses." New York: J Wiley.

Heitz E, Bauer H (1933) Beweise fur die Chromensomennatur die Kernschleifen in den Knauelkernen von *Bibio hortulanus*. Z Zellforsch Mikrosc Anat 17:67.

Hertwig O (1878) Beitrage zur Kenntnis der Bildung, Befruchtung und Theilung des thierischen Eies. III Teil Morphol Jahrb 4:156.

Hertwig O (1884) "Das Problem der Befruchtung und der Isotropie des Eies. Eine Theorie der Vererbung." Jena: Gustav Fischer.

Hertwig O (1895) "The Cell. Outlines of General Anatomy and Physiology," Campbell M (trans), Campbell H (ed). New York: Swan Sonnenschein, Macmillan.

Hoagland MB et al (1958) A soluble ribonucleic acid intermediate in protein synthesis. J Biol Chem 231:241.

Horowitz NH (1948) The one gene–one enzyme hypothesis. Genetics 33:612.

Horowitz NH (1991) Fifty years ago: The Neurospora revolution. Genetics 127:631.

Horowitz NH, Leupold U (1951): Some recent studies bearing on the one gene–one enzyme hypothesis. Cold Spring Harbor Symp Quant Biol 16:65.

Hsu TC (1979) "Human and Mammalian Cytogenetics. An Historical Perspective." Heidelberg: Springer-Verlag.

Hughes A (1959) "A History of Cytology." London: Abelard-Schuman.

Ingram VM (1957) Gene mutations in human haemoglobin: The chemical difference between normal and sickle cell haemoglobin. Nature 180:326.

Janssens FA (1909) Spermatogenese dans la batraciens. V. La théorie de la chiasmatypie, nouvelles interpretation des cineses de maturation. La Cellule 25:387.

Johannsen W (1909) "Elemente der exackten Erblikeitslehre." Jena: Gustav Fischer.

Koelliker Von RA (1841) "Beitraeger zur Kenntnis der Geschlechsverhaeltnisse und der Samenfluessigkeit wirbelloser Thiere nebst einem Versuche uber das Wesen und die Bedeutung der sogennanten Samenthiere." Berlin.

Kossel A (1882) Zur Chemie der Zellkerns. Hoppe-Seyler's Z Physiol Chem 7:7.

Lejeune J et al (1959) Etude des chromosomes somatiques de neuf enfants mongoliens. C R Acad Sci Ser D 248:1721.

Lindegren CC (1949) Chromosome maps of Saccharomyces. Proceedings of the Eighth International Congress on Genetics. Hereditas, Suppl 338.

Longley AE (1924) Chromosomes in maize and maize relatives. J Agric Res 28:673.

Mayr E (1982) "The Growth of Biological Thought. Diversity, Evolution and Inheritance." Cambridge, MA: Harvard University Press.

McClintock B (1929). Chromosome morphology in *Zea mays*. Science 69:629.

McClung CE (1901) Notes on the accessory chromosome. Anat Anz 20:220.

McClung CE (1902) The accessory chromosome, sex determinant? Biol Bull 3:43.

Meselson M, Yuan R (1968) DNA restriction enzyme from *E. coli*. Nature 217:1110.

Meyen FJF (1837–1839) "Neues System der Pflanzen-Physiologie." Bd 1–3, Berlin: Haude und Spenersche Buchhandlung.

Miescher F (1871) Chemische Zuzammensetzung der Eiterzelle. In: Medizinisch-chemische Untersuchungen (Hoppe-Seyler F, Hrsg) S 441.

Montgomery TH (1901) A study of the germ cells of the Metazoa. Trans Am Phil Soc 20:154.

Morgan TH (1910) The method of inheritance of two sex-limited characters in the same animal. Proc Soc Exp Biol Med 8:17.

Morgan TH (1911) An attempt to analyze the constitution of the chromosomes on the basis of sex-limited inheritance in Drosophila. J Exp Zool 11:365.

Morgan TH et al (1915) "The Mechanism of Mendelian Heredity." New York: Henry Holt.

Muller HJ (1927) Artificial transmutation of the gene. Science 46:84.

Naegeli E (1884) "Mechanisch-physiologische Theorie der Abstammungslehre." Munich, Leipzig: R. Oldenbourg.

Nirenberg MW, Matthaei HJ (1961) The dependence of cell-free protein synthesis in *E. coli* upon naturally occurring or synthetic polyribonucleotides. Proc Natl Acad Sci USA 47:1588.

O'Brien SJ (1990) "Genetic Maps" (5th ed). Cold Spring Harbor, NY: Cold Spring Harbor Press.

Olby RC (1966) "Origins of Mendelism." London: Constable.

Overton E (1893) On the reduction of the chromosomes in the nuclei of plants. Ann Bot 7:139.

Painter TS (1933) A new method for the study of chromosome rearrangements and the plotting of chromosome maps. Science 78:585.

Pauling L et al (1949) Sickle cell anemia, a molecular disease. Science, 110:543.

Portugal FH, Cohen JS (1977) "A Century of DNA." Cambridge, MA: MIT Press.

Rabl C (1885) Uber Zellthei. Morphol Jahrb 10:214.

Randolph LF (1928) Chromosome numbers in *Zea mays* L. Cornell University Exp Sta Mem 117:1.

Reichert K (1847) Beitrag zur Entwicklungsgeschichte der Samenkorperchen bei der Nematoden. Arch Anat (Muller's Arch) 14:88.

Remak R (1852) Uber extracellulare Entstehung thierischer Zellen und uber Vermehrung derselben durch Theilung. Arch Anat Physiol Wissenschaft Medizin (Muller's Arch) 19:47.

Remak R (1855) "Untersuchen uber die Entwicklung der Wirbelthiere." G Reimer. Berlin:

Rhoades MM (1984) The early years of maize genetics. Annu Rev Genet 18:1.

Schleiden MJ (1838) Beitrage zur Phytogenesis. Arch Anat Physiol Wissenschaft Medizin (Mullers Archiv) 5:137.

Schneider A (1873) Untersuchen uber Platyhelminthen. 14. Jahresber Oberhessischen Ges Natur- und Heilkd Giessen. 14:69.

Schroedinger E (1944) "What Is Life?" New York: Macmillan.

Schwann T (1839) "Mikroskopische Untersuchungen uber die Ubereinstimmung in der Struktur und dem Wachsthum der Thiere und Pflanzen." Berlin: Verlag der Sanderschen Buchhandlung (GE Reimer).

Stern C (1931) Zytologisch-genetische Untersuchungen als Beweise fur die Morganische Theorie des Faktorenaustauschs. Biol Zentralbl 46:547.

Stevens NM (1905) Studies in spermatogenesis with special reference to the "accessory chromosome." Carnegie Inst Washington Publ 36:359.

Strasburger E (1884) "Neue Untersuchungen uber den Befruchtungsvorgang bei den Phanerogamen als Grundlage fur eine Theorie der Zeugung." Jena: Gustav Fischer.

Strasburger E (1988) "Uber Kern und Zelltheilung im Pflanzenbereich, nebst einem Anhang uber Befruchtung." Jena: Gustav Fischer.

Sturtevant AH (1913). The linear arrangements of six sex-linked factors in Drosophila as shown by their mode of association. J Exp Zool 14:43.

Sturtevant AH (1925) The effects of unequal crossing over at the bar locus in Drosophila. Genetics 26:517.

Sturtevant AH (1965) "A History of Genetics." New York: Harper and Row.

Sutton WS (1902) On the morphology of the chromosome group in Brachystola magna. Biol Bull 4:24.

Sutton WS (1903) The chromosomes in heredity. Biol Bull 4:231.

Tijo JH, Lean A (1956) The chromosome number of man. Hereditas 42:1.

Tschermak E von (1900) Uber kunstlicher Kreuzung bei Pisa sativum. Z Landwirtsch Versuchswes Dtsch-Osterr 3:465.

Virchow R (1858). "Die Cellularpathologie." Available In English translation by F Chance (1863). Republished by Dover Publications in 1971.

Vries H De (1900) Sur la loi de disjonction des hybrides. C R Acad Sci Ser D 130:845.

Vries H De (1889) "Intracellular Pangenesis." Jena: Gustav Fischer. Translation by CS Gager (1910). Chicago: Open Court.

Wagner RP (1989) On the origins of the gene—enzyme hypothesis. J Hered 80:503.

Waldeyer W (1888) Uber Karyokinese und ihre Beziehungen zu den Befruchtungsvorgangen Arch Mikrosc Anat 32:1 English translation by WB Benham (1889) in Q J Microsc Sci NS 30:159.

Watson JD, Crick FHC (1953) A structure for deoxyribonucleic acid. Nature 171:737.

Watson JD, Crick FHC (1953) Genetic implications for the structure of deoxyribonucleic acid. Nature 171:964.

Weiss MC, Green H (1967) Human—mouse hybrid cell lines containing partial complements of human chromosomes and functioning human genes. Proc Natl Acad Sci USA 58:1104.

Weismann A (1892) "Das Keimplasma: Eine Theorie der Vererbung." Jena: Gustav Fischer.

Wilson EB (1895) "An Atlas of the Fertilization and Karyokinesis of the Ovum." New York: Macmillan

Wilson EB (1896) "The Cell in Development and Inheritance" (2nd ed, 1900; 3rd ed, 1925). New York: Macmillan.

Wright S (1917) Color inheritance in mammals. J Hered 8:224.

Yanofsky C et al (1964) On the colinearity of gene structure and protein structure. Proc Natl Acad Sci USA 51:266.

CHAPTER 1

Athey BD et al (1990) The diameters of frozen-hydrated chromatin fibers increase with DNA linker length: Evidence in support of variable diameter models for chromatin. J Cell Biol 111:795.

Boulikas T et al (1980) Points of contact between H1 and the histone octamer. Proc Natl Acad Sci USA 77:127.

Bradbury EM (1992) Reversible histone modification and the chromosome cycle. BioEssays 14:9.

Bradbury EM, Baldwin JP (1986) Neutron scatter studies of chromatin structure. In, Pifat-Mrzljak G (ed): "Supramolecular Structure and Function." Heidelberg: Springer-Verlag.

Bradbury EM, Matthews HR (1980) Chromatin structure, histone modifications and the cell cycle. NATO Advanced Study Inst Ser, Ser A Life Sci 38:411.

Britten RJ, Kohne DE (1966) Nucleotide sequence repetition in DNA. Carnegie Inst Washington Yearb 65:78.

Britten RJ, Kohne DE (1968) Repeated sequences in DNA. Science 161:529.

Burlingame RW et al (1985) Crystallographic structure of the octameric histone core of the nucleosome at a resolution of 3.3 Å. Science 228:546.

Comings DE, Okada TA (1973) Some aspects of chromosome structure in eukaryotes. Cold Spring Harbor Symp Quant Biol 38:145.

D'Anna JA, Isenberg I (1974) A histone cross-complexing pattern. Biochemistry 13:4992.

Davidson EH et al (1973) General interspersion of repetitive with non-repetitive sequence elements in the DNA of Xenopus. J Mol Biol 77:1.

Earnshaw WC, Heck MMS (1985) Localization of topoisomerase II in mitotic chromosomes. J Cell Biol 100:1716.

Earnshaw WC et al (1985) Topoisomerase II is a structure component of mitotic chromosome scaffolds. J Cell Biol 100:1706.

Elgin SCR et al (1973) A prologue to the study of the nonhistone chromosome proteins. Cold Spring Harbor Symp Quant Biol 38:821.

Felsenfeld G, McGhee JD (1986) Structure of the 30nm chromatin fiber. Cell 44:375.

Gasser SM, Laemmli UK (1986a) The organization of chromatin loops: Characterization of the scaffold attachment site. EMBO J 5:511.

Gasser S, Laemmli UK (1986b) Cohabitation of scaffold binding regions with upstream-enhancer elements of three developmentally regulated genes of D. melanogaster. Cell 46:521.

Gasser SM et al (1986) Metaphase chromosome structure: Involvement of topoisomerase two. J Mol Biol 188:613.

Germond GE et al (1975) Folding of the DNA double helix in chromatin-line structures from simian virus 40. Proc Natl Acad Sci USA 72:1843.

Graever GN, Wang JC (1988) Supercoiling of intracellular DNA can occur in eukaryotic cells. Cell 55:849.

Haapala O (1984) Chromatic macrocoiling and chromosome compaction. Hereditas 100:17.

Hershka A (1983) Ubiquitin: Roles in protein modification and breakdown. Cell 34:11.

Hewish DR, Burgoyne LA (1973) Chromatin substructure. The digestion of chromatin DNA at regulatory spaced sites by a nuclear deoxyribonulcease. Biochem Biophys Res Commun 52:504.

Hozier J et al (1977) The chromosome's fiber: Evidence for an ordered superstructure of chromosomes. Chromosoma 62:301.

Isenberg I (1979) Histones. Annu Rev Biochem 48:159.

Jaworski A et al (1987) Left-handed DNA in vivo. Science 238:773.

Kavenoff R et al (1973) On the nature of chromsome-sized DNA molecules. Cold Spring Harbor Symp Quant Biol 38:1.

Klug A et al (1985) Crystallographic structure of the octamer histone core of the nucleosome. Science 229:1109.

Labhart PT et al (1982). Involvement of higher order chromatin structures in metaphase chromosome organization. Cell 30:115.

Laemmli UK et al (1977). Metaphase chromosome structure: The role of the nonhistone proteins. Cold Spring Harbor Symp Quant Biol 42:51.

Lewin, B (1987) "Genes." Third Edition. New York: Wiley.

Lewis CD, Laemmli UK (1982) Higher order metaphase chromosome structure: Evidence for metalloprotein interactions. Cell 29:171.

Manuelidis L (1990) A vie of interphase chromosomes. Science 250:1533.

Manuelidis L, Chen TL (1990) A unified model of eukaryotic chromosomes. Cytometry 11:8.

Maxam A, Gilbert W (1977) A new method for sequencing DNA. Proc Natl Acad Sci USA 74:560.

Maxam AM, Gilbert W (1980) Sequencing end labeled DNA with base-specific chemical cleavages. Methods Enzymol 65:499.

McGhee JD et al (1983) Higher order structure of chromatin: orientation of nucleosomes within the 30nm chromatin solenoid is independent of species and spacer length. Cell 33:831.

Messing J (1981) M13mp2 and derivatives. In Walter AG (ed): "Recombinant DNA: 3rd Cleveland Symposium on Macromolecules." Amsterdam: Elsevier.

Mirkovitch J et al (1984). Organization of the higher level chromatin loop: Specific DNA attachment sites on nuclear scaffold. Cell 39:223.

Morse RH, Cantor CR (1985) Nucleosome core particles suppress the thermal untwisting of core DNA and adjacent linker DNA. Proc Natl Acad Sci USA 82:4653.

Moyzis RK et al (1981a). An alternative view of mammalian DNA sequence organization. I. Repetitive sequence interspersion in Syrian hamster DNA: A model system. J Mol Biol 153:841.

Moyzis RK et al (1981b) An alternative view of mammalian sequence organization. II. Short repetitive sequences are organized into scrambled tandom clusters in Syrian hamster DNA. J Mol Biol 153:871.

Nieto MA, Palacian E (1988) Structural changes of nucleosomal particles and isolated core-histone octamers induced by chemical modification of lysine residues. Biochemistry 27:5635.

Okada TA, Comings DE (1979) Higher order structure of chromosomes. Chromosoma 72:1.

Paulson JR, Laemmli UK (1977) The structure of histone-depleted metaphase chromosomes. Cell 12:817.

Pettijohn DE (1988) Histone-like proteins and bacterial chromosome structure. J Biol Chem 262:12793.

Pettijohn DE et al (1984) Nuclear proteins that become part of the mitotic apparatus: A role in nuclear assembly. J Cell Science Suppl 1:187.

Rattner JB, Lin CC (1985) Radial loops and helical coils coexist in metaphase chromosomes. Cell 42:291.

Rich A et al (1984) The chemistry and biology of left-handed Z-DNA. Annu Rev Biochem 53:791.

Richmond TJ et al (1984) Structure of the nucelosome core particle at 7Å resolution. Nature 311:532.

Sambrook J et al (1989) Molecular Cloning: A Laboratory Manual 2nd Edition. Cold Spring Harbor, NY: Cold Spring Harbor Press.

Sanger F et al (1977) DNA sequencing with chain-terminating inhibitors. Proc Natl Acad Sci USA 74:5463.

Sedat J, Manuelidis L (1977) A direct approach to the structure of eukaryotic chromosomes. Cold Spring Harbor Symp Quant Biol 42:331.

Singer MT (1982) Highly repeated sequences in mammalian genomes. Int Rev Cytol 76:67.

Spiker S, Isenberg I (1978) Evolutionary conservation of histone–histone binding sites. Evidence from interkingdom complex formation. Cold Spring Harbor Symp Quant Biol 42:157.

Stack SM, Clark CR (1974) Chromosome polarization and nuclear rotation in *Allium cepa*. Cytologia 39:553.

Subirana JA et al (1985) The layered organization of nucleosomes in 30nm chromatin fibers. Chromosoma 91:377.

Taniguchi T, Takayama S (1986) High-order structure of metaphase chromosomes: Evidence for a multiple coiling model. Chromosoma 93:511.

Thoma F, Koller T (1977) Influence of histone H1 on chromatin structure. Cell 12:101.

Timberlake WE (1978) Low repetitive DNA content in Aspergillus nidulans. Science 202:973.

Vinograd J et al (1968) Early and late helix-coil transitions in closed circular DNA. The number of superhelical turns in polyoma DNA. J Mol Biol 33:173.

Von Holt C et al (1979) More histone structures FEBS Lett 100:201.

Waldeyer W (1888) Ueber Karyokinese und ihre Beziehungen zu Befruchtungs-Vorgaengen. Arch Mikrosc Anat 32:1.

Wang JC (1985) DNA topoisomerases. Annu Rev Biochem 54:665.

Watson JD, Crick FCH (1953) A structure for deoxyribose nucleic acid. Nature 171:737.

White JH et al (1988) Helical repeat and linking number of surface-wrapped DNA. Science 241:323.

Widom J, Klug A (1985) Structure of the 300 Å chromatin filament: X-ray diffraction from oriented samples. Cell 43:207.

Winnaker E (1987) "From Genes to Clones: Introduction to Gene Technology." Weinheim, Germany: VCH Verlagsgesellschaft.

Wray, WI, Stubblefield TE (1970) A new method for the rapid isolation of chromosomes. Exp Cell Res 59:469–478.

Yanisch-Perron C et al (1985) Improved M13 phage cloning vectors and host strains: Nucleotide sequences of M13mp18 and puc19 vectors. Gene 33:103.

Zatsepina OV et al (1983) Chromonema and chromomere. Structural units of mitotic and interphase chromosomes. Chromosoma 88:91.

CHAPTER 2

Arndt-Jovin DJ et al (1983) Left-handed Z-DNA in bands of acid-fixed DNA polytene chromosomes. Proc Natl Acad Sci USA 80:4344.

Austerberry CF et al (1984) Specific DNA rearrangements in synchronously developing nuclei of *Tetrahymena*. Proc Natl Acad Sci USA 81:7383.

Babu KA, Verma RS (1985) Structural and functional aspects of nucleolar organizer regions (NORs) of human chromosomes. Int Rev Cytol 94:151.

Bahr GF, Larsen PM (1974) Structural "bands" in human chromosomes. Adv Cell Mol Biol 3:192.

Barnes SR et al (1985) The organization, nucleotide sequence, and chromosomal distribution of a satellite DNA from *Allium cepa*. Chromosoma 92:185.

Beerman S (1959) Chromatin-diminution bei Copepoden. Chromosoma 10:504.

Beerman S (1977) The diminution of heterochromatic chromosome segments in Cyclops (Crustacea, Copepoda). Chromosoma 28:297.

Beerman S (1984) Circular and linear structures in chromatin diminution in cyclops. Chromosoma 89:321.

Biessmann H et al (1990) Addition of telomere-associated HeT DNA sequences "heals" broken chromosome ends in Drosophila. Cell 61:663.

Blackburn EH, Karrer KM (1986) Genomic reorganization in ciliated protozoans. Annu Rev Genet 20:501.

Blackburn EH, Szostak JW (1984) The molecular structure of centromeres and telomeres. Ann Rev Biochem 53:163.

Blackburn EH et al (1983) DNA termini in ciliate macronuclei. Cold Spring Harbor Symp Quant Biol 47:1195.

Bloom KS, Carbon J (1982) Yeast centromere DNA is a unique and highly ordered structure in chromosomes and small microchromosomes. Cell 29:305.

Bloom KS et al (1983) Structural analysis and sequence organization of yeast centromeres. Cold Spring Harbor Symp Quant Biol 47:1175.

Boveri T (1899) Die Entwicklung von Ascaris megalocephela mit besonderer Ruecksicht auf die Kernverhaeltnisse. In Festschrift fuer C. Von Kupffer.

Braselton JP (1981) The ultrastructure of meiotic kinetochores of *Luzula*. Chromosoma 82:143.

Brenner S et al (1981) Kinetochore structure, duplication, and distribution in mammalian cells: Analysis by human autoantibodies from scleroderma patients. J Cell Biol 91:95.

Brinkley BR (1985) Microtubule organizing centers. Annu Rev Cell Biol 1:145.

Brown DD, Dawid I (1968) Specific gene amplification in oocytes. Science 160:272.

Brown SW (1966) Heterochromatin. Science 151:417.

Brutlag DL (1980) Molecular arrangement and evolution of heterochromatic DNA. Annu Rev Genet 14:121.

Brutlag DL et al (1978) DNA sequence organization in Drosophila heterochromatin. Cold Spring Harbor Symp Quant Biol 42:1137.

Bull JJ (1983) "Evolution of Sex Determining Mechanisms." Menlo Park, California: Benjamin/Cummings.

Burkholder GD (1988) The analysis of chromosome organization by experimental manipulation. In Proceedings of the 18th Stadler Genetics Symposium on Chromosomes, Structure and function: Impact of New Concepts. New York: Plenum.

Burkholder GD, Duczek LL (1982) The effect of chromosome banding techniques on the proteins of isolated chromosomes. Chromosoma 87:425.

Callan HG (1955) Recent work on the structure of cell nuclei. In: "Symposium On the Fine Structure of Cells." IUBS Publ Series B. 21:89.

Callan HG (1957) Recent work on the structure of cell nuclei. (Symposium on the Fine Structure of Cells.) IUBS Publ Series B 21:89.

Callan HG (1963) The nature of lampbrush chromosomes. Int Rev Cytol 72:1.

Callan HG (1986) "Lambrush Chromosomes." Berlin: Springer-Verlag.

Callan HG, Lloyd L (1960) Lampbrush chromosomes. In Walker PMB (ed): "New Approaches in Cell Biology." New York: Academic Press.

Carbon J (1984) Yeast Centromeres: Structure and Function. Cell 37:351.

Carlson WR (1978) The B chromosome of corn. Annu Rev Genet: 16:5.

Clarke L, Carbon J (1983) Genomic substitutions of centromeres in *Saccharomyces cerevisiae*. Nature 305:23.

Clarke L, Carbon J (1985). The structure and function of yeast centromeres. Annu Rev Genet 19:29.

Comings DE (1972) The structure and function of chromatin. Hum Genet 3:237.

Comings DE (1978) Mechanisms of chromosome banding and implications for chromosome structure. Annu Rev Genet 12:25.

Comings DE (1980). Arrangement of chromatin in the nucleus. Hum Genet 53:131.

Comings DE, Okada TA (1972) Holocentric chromosomes in *Oncopeltus*: Kinetochore plates are present in mitosis but absent in meiosis. Chromosoma 37:177.

Cordeiro-Stone M, Lee CS (1976) Studies on the satellite DNA's of Drosophila nasutoides: Buoyant densities, melting temperatures, reassociation rates and localizations in polytene chromosomes. Mol Biol 104:1.

Corneo G et al (1971) Renaturation properties and localization in heterochromatin of human satellite DNAs. Biochim Biophys Acta 247:528.

Darzynkiewicz Z, Crissman HA (1990) "Methods in Cell Biology," vol 33: "Flow Cytometry." San Diego: Academic Press.

Davidson EH et al (1973) General interspersion of repetitive with non-repetitive sequence elements in the DNA of *Xenopus*. J Mol Biol 77:1.

Dawid IB, Wellauer PK (1974) Ribosomal RNA and related sequences in *Drosophila melanogaster*. Cold Spring Harbor Symp Quant Biol 42:1185.

Deininger PL, Schmid CW (1976) An electron microscope study of the DNA sequence organization of the human genome. J Mol Biol 106:773.

Devine EA et al (1985) Chromosomal localization of several families of repetitive sequences by in situ hybridization. Am J Hum Genet 37:114.

Dodge JD (1985) The chromosomes of dinoflagellates. Int Rev Cytol 94:5.

Earnshaw WC, Rothfield N (1985) Identification of a family of human centromere proteins using autoimmune sera from patients with scleroderma. Chromosoma 91:313.

Emery HS, Weiner AM (1981) An irregular satellite sequence is found at the termini of

the linear extrachromosomal rDNA of *Dictyostelium discoideum*. Cell 26:411.

Flavell RB (1980) The molecular characterization and organization of plant chromosomal DNA sequences. Annu Rev Plant Physiol 31:569.

Flavell RB (1986). Repetitive DNA and chromosome evolution in plants Philos Trans R Soc London Ser B 312:227.

Gall JG (1968) Differential synthesis of the genes for ribosomal RNA during amphibian oogenesis. Proc Natl Acad Sci USA 60:553.

Gall JG, Atherton DD (1974) Satellite DNA sequences in *Drosophila virilis*. J Mol Biol 85:633.

Gall JG, Callan HG (1962) Tritiated uridine incorporation in lampbrush chromosomes. Proc Natl Acad Sci USA 48:562.

Gatti M, Pimpinelli S (1983) A gene controlling condensation of heterochromatin in *Drosophila melanogaster*. Science 221:83.

Goday C, Pimpinelli S (1984) Chromosome organization and heterochromatin elimination in *Parascaris*. Science 224:411.

Godward MBE (1985) The kinetochore. Int Rev Cytol 94:77.

Gorovsky MA (1980) Genome organization and reorganization in *Tetrahymena*. Annu Rev Genet 14:203.

Gottschling DE, Cech TR (1984) Chromatin structure of the molecular ends of *Oxytricha* macronuclear DNA: Phased nucleosomes and a telomeric complex. Cell 38:501.

Grady DL et al (1992) Highly conserved DNA sequences are present at human centromeres. Proc Natl Acad Sci USA 89:1695.

Haapala O (1984a) Chromatid macrocoiling and chromosome compaction. Hereditas 100:17.

Haapala O (1984b) Metaphase and chromomere banding are distinct entities of chromosome structure. Hereditas 100:75.

Hamkalo BA (1985) Visualizing transcription in chromosomes. Trends Genet 1:255.

Hatch FT, Mazrimas JA (1974) Fractionation and characteristics of satellite DNAs of the kangaroo rat (dipodymis ordii). Nucleic Acids Res. 1:559.

Heitz E (1928) Das heterochromatin der moose. Jahrb Wissenschaft Bot 69:762.

Hennig W (1985) Y chromosome function and spermatogenesis in *Drosophila hydei*. Adv Genet 23:179.

Hennig W (1986) Heterochromatin in germ line restricted DNA. In "Results and Problems in Cell Differentiation in Germ Line-Soma Differentiation." Hennig W (ed.): Berlin: Springer-Verlag.

Herrick G et al (1985) Mobile elements bounded by C4A4 telomeric repeats in *Oxytricha fallox*. Cell 43:759.

Herzog M, Soyer MO (1982) Dinoflagellate DNA. Cell Biol 27:151.

Herzog M, Soyer MO (1983) The native structure of dinoflagellate chromosomes and their stabilization by Ca^{++} and Mg^{++} cations. Eur J Cell Biol 30:33.

Herzog M et al (1984) Ultrastructural and biochemical nuclear aspects of eukaryotic classification. Origins of Life 13:205.

Hilliker AJ et al (1980) Cytogenetic analysis of the chromosomal region immediately adjacent to the rosy locus in *Drosophila melanogaster*. Genetics 95:95.

Holmquist G (1989) Evolution of chromosome bands: Molecular ecology of noncoding DNA. J Mol Evol 28:469.

Holmquist G et al (1982) Characterization of Giemsa dark- and light-band DNA. Cell 31:121.

Huberman JA et al (1986) Centromeric DNA from Saccharomyces uvarum is functional in Saccharomyces cerevisiae. Chromosoma 94:162.

Hughes-Schrader S (1948) Cytology of coccids (Coccoidea–Homoptera). Adv Genet 2:127.

Hughes-Schrader S, Ris H (1941) The diffuse spindle attachments of coccids verified by the mitotic behavior of induced chromosome fragments. J Exp Zool 87:429.

Jabs EW, Persico MG (1987) Characterization of the human centromeric regions of specific chromosomes by means of alphoid DNA sequences. Am J Hum Genet 41:374.

Jhanwar SC, Chaganti RSK (1981) Pachytene chromomere maps of Chinese hamster autosomes. Cytogenet Cell Genet 31:70.

Jhanwar SC et al (1982) Mid-pachytene chromomere maps of human autosomes. Cytogenet Cell Genet 33:240.

Johnson EM (1980) A family of inverted repeat sequences and specific single-strand gaps at the termini of the *Physarum* rDNA palindrome. Cell 24:875.

Jones JDG, Flavell RB (1982a) The mapping of highly repeated DNA families and their relationship to C-bands in chromosomes of *Secale cereale*. Chromosoma 86:595.

Jones JDG, Flavell RB (1982b) The structure, amount and chromosomal localization of defined repeated DNA sequences in species of the Genus *Secale*. Chromosoma 86:613.

Jones JDG, Flavell RB (1983) Chromosomal structure and arrangement of repeated DNA sequences in the telomeric heterochromatin of *Secale cereale* and its relatives. Cold Spring Harbor Symp Quant Biol 47:1209.

Korenberg JR, Rykowski MC (1988) Human genome organization: Alu, lines, and the molecular structure of metaphase chromosome bands. Cell 53:391.

Koshland DE et al (1988) Poleward chromosome movement driven by microtubule depolymerization in vitro. Nature 331:499.

Kubai DF (1975) The evolution of the mitotic spindle. Int Rev Cytol 43:167.

Kuhn EM, Therman E (1979) Chromosome breakage and rejoining of sister chromatids in Bloom's syndrome. Chromosoma 73:275.

Labhart P et al (1982) Model for the structure of the active nucleolar chromatin. Cold Spring Harbor Symp Quant Biol 47:557.

Lee CS (1981) Restriction enzyme analysis of a highly diverged satellite DNA from *Drosophila nasutoides*. Chromosoma 83:367.

Leutwiler LS, Hough-Evans BR (1984) The DNA of *Arabidopsis thaliana*. Mol Gen Genet 194:15.

Lima DE, Feria A (1962) Metabolic DNA in *Tipula oleracea*. Chromosoma 13:47.

Long EO, Dawid IB (1980) Repeated genes in eukaryotes. Annu Rev Biochem 49:727.

Marchand GE, Holm DG (1988a) Genetic analysis of the heterochromatin of chromosome 3 in *Drosophila melanogaster*. I. Products of compound-autosome detachment. Genetics 120:503.

Marchand GE, Holm DG (1988b) Genetic analysis of the heterochromatin of chromosome 3 in *Drosophila melanogaster*. II. Vital loci identified through EMS mutagenesis. Genetics 120:519.

Masumoto HH et al (1989) A human centromere antigen (CENP-B) interacts with a short specific sequence in alphoid DNA, a human centromeric sequence. J Cell Biol 109:1963.

McClintock B (1951) Chromosome organization and genic expression. Cold Spring Harbor Symp Quant Biol 16:13.

Merry DE et al (1985) Anti-kinetochore antibodies: Use as probes for inactive centromeres. Am J Hum Genet 37:425.

Meyerwitz EM et al (1985). The 68C glue puff of *Drosophila*. Cold Spring Harbor Symp Quant Biol 50:347.

Meyerwitz EM et al (1987) How *Drosophila* larvae make glue: Control of SGS-3 gene expression. Trends Genet 3:288.

Meyne J et al (1989) Labeling of human centromeres using an alphoid DNA consensus sequence application to the scoring of chromosomal aberrations. Mutat Res 226d:75.

Meyne J et al (1990) Distribution of nontelomeric sites of the (TTAGGG)n telomeric sequence in vertebrate chromosomes. Chromosoma 99:3–10.

Miklos GLG (1985). Localized highly repetitive DNA sequences in vertebrate and invertebrate genomes. In MacIntyre RJ (ed): "Molecular and Evolutionary Genetics." New York: Plenum.

Miller OL, Hamkalo BA (1972) Visualization of RNA synthesis on chromosomes. Int Rev Cytol 33:1.

Mitchell AR et al (1986) Molecular hybridization to meiotic chromosomes in man reveals sequence arrangement on the no. 9 chromosome and provides clues to the nature of "parameres." Cytogenet Cell Genet 41:89.

Mitchison TJ (1988) Microtubule dynamics and kinetochore function in mitosis. Annu Rev Cell Biol 4:527.

Mitchison T et al (1986) Sites of microtubule assembly and disassembly in the mitotic spindle. Cell 45:515.

Moyzis RK et al (1981a) An alternative view of mammalian DNA sequence organization. I. Repetitive sequence interspersion in Syrian hamster DNA: A model system. J Mol Biol 153:841.

Moyzis RK et al (1981b) An alternative view of mammalian sequence organization. II. Short repetitive sequences are organized into scrambled tandem clusters in Syrian hamster DNA. J Mol Biol 153:871.

Moyzis RK et al (1987) Human chromosome-specific repetitive DNA sequences: Novel markers for genetic analysis. Chromosoma 95:375.

Moyzis RK et al (1988) A highly conserved repetitive DNA sequence (TTAGGG)n present at the telomeres of human chromosomes. Proc Natl Acad Sci USA 85:6622.

Moyzis RK et al (1989) The distribution of interspersed repetitive DNA sequences in the human genome. Genomics 4:273.

Muller HJ (1938) The remaking of chromosomes. The Collecting Net 8:182.

Mullins JI, Blumenfeld M (1979) Satellite Ic: A possible link between the Satellite DNAs of *Drosophila virilis* and *Drosophila melanogaster*. Cell 17:615.

Nagl W (1978) "Endopolyploidy and Polyteny in Differentiation and Evolution." Amsterdam. Elsevier/North Holland.

Narayan RKJ (1991) Molecular organization of the plant genome: Its relation to structure, recombination and evolution of chromosomes. J Genet 70:43.

Nordheim A et al (1981) Antibodies to left-handed Z-DNA bind to interband regions of *Drosophila* polytene chromosomes. Nature 294:417.

Olmsted JB (1986) Microtubule associated proteins. Annu Rev Cell Biol 2:421.

Pardue ML, Hennig W (1991) Heterochromatin: Junk or collectors item? Chromosoma 100:3.

Patterson JT, Stone WS (1952) "Evolution in the genus *Drosophila*." New York: Macmillan.

Pavan C, Da Cunha AB (1969) Chromosomal activities in *Rhynchosciara* and other Sciaridae. Annu Rev Genet 3:425.

Peacock WJ et al (1973) The organization of highly repeated DNA sequences in *Drosophila melanogaster* chromosomes. Cold Spring Harbor Symp Quant Biol 38:405.

Pech M et al (1979a) Patchwork structure of bovine satellite DNA. Cell 18:883.

Pech M et al (1979b) Nucleotide sequence of a highly repetitive component of rat DNA. Nucleic Acids Res 7:417.

Peterson JB, Ris H (1976) Electron microscopic study of the spindle and chromosome movement in the yeast *Saccharomyces*. J Cell Sci 22:219.

Pimpinelli S, Goday C (1989) Unusual kinetochores and chromatin diminution in *Parascaris*. Trends Genet 5:310

Pimpinelli S et al (1985) On biological functions mapping to the heterochromatin of *Drosophila melanogaster*. Genetics 109:701.

Procunier JD, Dunn RJ (1978) Genetic and molecular organization of the 5S locus and mutants in *D. melanogaster*. Cell 15:1087.

Procunier JD, Tartof KD (1978) A genetic locus having trans and contiguous cis-functions that controls the disproportionate replication of RNA genes in *Drosophila melanogaster*. Genetics 88:67.

Prosser J et al (1986) Sequence relationships of three human satellite DNAs. J Mol Biol 187:145.

Rattner JB (1986) Organization within the mammalian kinetochore. Chromosoma 93:515.

Rattner JB (1987) The organization of the mammalian kinetochore: A scanning electron microscope study. Chromosoma 95:175.

Reeder RH (1984) Enhancers and ribosomal gene spacers. Cell 38:349.

Rich A et al (1984) The chemistry and biology of left-handed Z-DNA. Annu Rev Biochem 53:791.

Rieder CL (1982) The formation, structure, and composition of the mammalian kinetochore and kinetochore fiber. Int Rev Cytol 79:1.

Ris H, Witt PL (1981) Structure of the mammalian kinetochore. Chromosoma 82:153.

Risely MS et al (1986) Changes in DNA topology during spermatogenesis. Chromosoma 94:217.

Rizzo PJ, Nooden LD (1974) Dinoflagellates. Biochim Biophys Acta 349:402.

Schmid CW, Deininger PL (1975) Sequence organization of the human genome. Cell 6:345.

Schneider E et al (1973) Loss of heteropyknosis of the constitutive heterochromatin in specifically activated cells of the thyroid gland of *Microtis agrestis*. Chromosoma 41:167.

Schrader F (1953) "Mitosis" (2nd ed) New York: Columbia University Press.

Schweizer D (1981) Counterstain-enhanced chromosome banding. Human Genet 57:1.

Sigee DC, Kearns LP (1982) Differential retention of proteins and bound divalent cations in dinoflagellate chromatin fixed under varied conditions: An X-ray microanalytical study. Cytobios 33:51.

Singer MT (1982) Highly repeated sequences in mammalian genomes. Int Rev Cytol 76:67.

Skinner DM (1967) Satellite DNAs in the crab *Garcinus lateralis* and *Cancer pagurus*. Proc Natl Acad Sci USA 58:103.

Skinner D (1972) Satellite DNA's. Bioscience 27:790.

Sorenson JC (1984) The structure and expression of nuclear genes in higher plants. Adv Genet 22:109.

Spector DL (1984) Dinoflagellate nuclei. In Spector DL (ed). "Dinoflagellates." New York: Academic Press.

Stack SM, Clark CR (1974) Chromosome polarization and nuclear rotation in *Allium cepa* roots. Cytologia 39:553.

Stebbins GL (1950) "Variation and Evolution in Plants." New York: Columbia University Press.

Sturtevant AH, Novitski E (1941). The homologies of the chromosome elements of Drosophila. Genetics 26:517.

Sullivan KF, Glass CA (1991) CENP-B is a highly conserved mammalian centromere protein with homology to the helix–loop–helix family of proteins. Chromosoma 100:360.

Sumner AT (1986) Electron microscopy of the parameres formed by the centromeric heterochromatin of human chromosome 9 at pachytene. Chromosoma 94:199.

Szostak JW (1984) Replication and resolution of telomeres in yeast. Cold Spring Harbor Symp Quant Biol 47:1187.

Szostak JW, Blackburn EH (1982) Cloning yeast telomeres on linear plasmid vectors. Cell 29:245.

Taggart RT et al (1985) Variable numbers of pepsinogen genes are located in the centromeric region of human chromosome 11 and determine the high frequency electrophoretic polymorphism. Proc Natl Acad Sci USA 82:6240.

Therman E (1986) "Human Chromosomes" (2nd ed). New York: Springer-Verlag.

Traverse KH, Pardue ML (1989) Studies of He–T DNA sequences in the pericentric regions of *Drosophila* chromosomes. Chromosoma 97:261.

Triemer RE, Fritz L (1984) Cell cycle and mitosis in dinoflagellates. In Spector DL (ed): "Dinoflagellates." New York: Academic Press.

Vaarama A (1954) Cytological observations on pleurozium scheberi with special reference to centromere evolution. Annu Bot Soc Zool-Bot Fenn Vanamo 28:1.

Valdivia MM, Brinkley BR (1985) Factionation and characterization of the kinetochore from mammalian metaphase chromosomes. J Cell Biol 101:1124.

Vig BK (1984) Sequence of centromere separation: Orderly separation of multicentric chromosomes in mouse L cells. Chromosoma 90:39.

Vig BK, Paweletz N (1988) Sequence of centromere separation: Generation of unstable multicentric chromosomes in a rat cell line. Chromosoma 96:275.

Vinograd J et al (1968) Early and late helix-coil transitions in closed circular DNA. The number of superhelical turns in polyoma DNA. J Mol Biol 33:173.

Walmsley RM, Peters D (1985) Genetics of chromosome length in yeast. Proc Natl Acad Sci USA 82:506.

Wang TS-F (1991) Eukaryotic DNA polymerases. Ann Rev Biochem 60:513.

Waye JS, Willard HF (1989) Chromosome specificity of satellite DNAs: Short- and long-range organization of a diverged dimeric subset of human alpha satellite from chromosome 3. Chromosoma 97:475.

Wheeler L, Altenberg LC (1977) Hoechst 33258 banding of Drosophila nasutoides chromosomes. Chromosoma 62:351.

Wheeler L et al (1978) Localization of Drosophila nasutoides satellite DNA's in metaphase chromosomes. Chromosoma 70:41.

Willard HF (1985) Chromosome-specific organization of human alpha satellite DNA. Am J Hum Genet 36:524.

Willard HF, Waye JS (1987) Heirarchical order in chromosome specific human alpha satellite DNA. Trends Genet 3:192.

Wray R, Stubblefield E (1970) A new method for the rapid isolation of chromosomes. Exp Cell Res 59:469.

Wurster-Hill DH et al (1986) Banded chromosome studies and B chromosomes in wild-caught raccoon dogs, Nyctereutes procyonides viverrinus. Cytogenet Cell Genet 42:85.

Wurster-Hill DH et al (1988) Fragile sites, telomeric DNA sequences, B chromosomes, and DNA content in raccoon dogs, Nyctereutes procyonides, with comparative notes on foxes, coyote, wolf, and raccoon. Cytogenet Cell Genet 49:278.

Yao M-C et al (1987) A conserved nucleotide sequence at the sites of developmentally regulated chromosomal breakage in Tetrahymena. Cell 48:779.

Young BS et al (1983) Teleomere regions in Drosophila share complex DNA sequences with pericentric heterochromatin. Cell 34:85.

Yunis JJ (1981) Mid-prophase human chromosomes: The attainment of 2,000 bands. Hum Genet 56:293.

Yunis JJ, Prakash O (1982) The origin of man: A pictorial legacy. Science 215:1525.

Zatsepina OV et al (1983) Chromonema and chromomere. Structural units of mitotic and interphase chromosomes. Chromosoma 88:91.

Zinkowski RP et al (1986) Characterization of kinetochores in multicentric chromosomes. Chromosoma 94:243.

Zinkowski RP et al (1991) The centromere–kinetochore complex: A repeat subunit model. J Cell Biol 113:1091.

CHAPTER 3

Alberts BM (1984) The DNA enzymology of protein machines. Cold Spring Harbor Symp Quant Biol 49:1.

Ashburner M (1990) Puffs, genes, and hormones revisited. Cell 61:1

Ashburner M, Berendes HD (1978) In Ashburner M, Wright TRF (eds): "The Genetics and Biology of Drosophila." New York: Academic Press.

Baril EF et al (1988) A multiprotein DNA alpha complex from HeLa cells: Interaction with other proteins in DNA replication. In Kelly T, Stillman B (eds): "Eukaryotic DNA Replication," vol 6: "Cancer Cells." Cold Spring Harbor, NY: Cold Spring Harbor Laboratory Press.

Barlow PW, Sherman MI (1974) Cytological studies on the organization of DNA in giant trophoblast nuclei of the mouse and rat. Chromosoma 47:119.

Berrios M et al (1985) In situ localization of topoisomerase II, a major polypeptide component of the Drosophila nuclear matrix fraction. Proc Natl Acad Sci USA 82:4142.

Blackburn EH (1990) Telomeres: Structure and synthesis. J Biol Chem 265:5919.

Blow JJ, Laskey RA (1986) Initiation of replication in nuclei and purified DNA by a cell free extract of *Xenopus* eggs. Cell 47:577.

Blumenfeld M, Forrest HS (1972) Differential under-replication of satellite DNAs during *Drosophila* development. Nature New Biol 239:170.

Blumenthal AB et al (1974) The units of DNA replication in *Drosophila melanogaster* chromosomes. Cold Spring Harbor Symp Quant Biol 38:205.

Bonner WM et al (1988) Qualitative and kinetic characterization of soluble histone pools: Linkage between protein and DNA synthesis during the cell cycle. In Kelly T, Stillman B (eds): "Eukaryotic DNA Replication," vol 6: "Cancer Cells." Cold Spring Harbor, New York: Cold Spring Harbor Laboratory Press.

Bramhill D, Kornberg A (1988) A model for initiation at origins of DNA replication. Cell 54:915.

Brill SJS et al (1987) Need for DNA topoisomerase activity as a swivel for DNA replication and for transcription of ribosomal RNA. Nature 326:414.

Bryant PE (1988) Use of restriction endonucleases to study relationships between DNA double-strand breaks, chromosomal aberrations and other end-points in mammalian cells. Int Rev Radiat Biol 54:869.

Budd M et al (1988) Yeast polymerase and ARS-binding proteins. In Kelly T, Stillman B (eds): "Eukaryotic DNA Replication," vol 6: "Cancer Cells." Cold Spring Harbor, NY: Cold Spring Harbor Laboratory Press.

Bunnell BH et al (1990) Elevated expression of a p58 protein kinase leads to changes in the CHO cycle. Proc Natl Acad Sci USA 87:7467.

Calza RE et al (1984) Changes in gene position are accompanied by a change in time of replication. Cell 36:689.

Campbell J (1986) Eukaryotic DNA replication. Annu Rev Biochem 55:733.

Chang LMS et al (1982) Evolutionary conservation of DNA polymerase beta structure. Proc Natl Acad Sci USA 79:758.

Clayton DA (1982) Replication of animal mitochondrial DNA. Cell 28:693.

Cohen MM, Levy HP (1989) Chromosome instability syndromes. Adv Hum Genet 18:43.

Cook PR, Lang J (1984) The spatial organization of sequences involved in initiation and termination of eukaryotic DNA replication. Nucleic Acids Res 12:1069.

Cooney CA et al (1984) 5-Methyldeoxycytidine in the *Physarum* minichromosome containing the ribosomal RNA genes. Nucleic Acids Res 12:1501.

Cooney CA et al (1988) Methylation is coordinated on the putative replication origins of *Physarum* ribosomal DNA. J Mol Biol 204:889.

Crissman HA (1985) Correlated measurements of DNA, RNA and protein in individual cells by flow cytometry. Science 228:1321.

Crissman HA et al (1985) Normal and perturbed Chinese hamster ovary cells: Correlation of DNA, RNA and protein content by flow cytometry. J Cell Biol 101:141.

Cross F et al (1989) Simple and complex life cycles. Annu Rev Cell Biol 5:341.

Daar I et al (1991) The ras oncoprotein and M-phase activity. Science 253:74.

D'Anna JA, Prentice DA (1983) Chromatin structural changes in synchronized cells blocked in early S phase by sequential use of isoleucine deprivation and hydroxyurea blockade. Biochemistry 22:5631.

D'Anna JA, Tobey RA (1984) Changes in histone H1 content and chromatin structure of cells blocked in early S phase by 5-fluorodeoxyuridine and aphidicolin. Biochemistry. 23:5024.

D'Anna JA et al (1985a) G1 and S-phase syntheses of histones H1 and H10 in mitotically selected CHO cells: Utilization of high-performance liquid chromatography. Biochemistry 24:2005.

D'Anna JA et al (1985b) Time-dependent changes in H1 content, H1 turnover, DNA elongation, and the survival of cells blocked in early S phase by hydroxyurea, aphidicolin or 5-fluorodeoxyuridine. Biochemistry 24:5020.

DaCunha AB (1972) Chromosome activities and differentiation. Chromosomes Today 3:296.

DaCunha AB et al (1969) Studies on cytology and differentiation in Sciaridae. II. Redundancy in salivary gland cells of Hybisciara Fragilis (Diptera, Sciaridae). Genetics 61(Supple):335.

Dillworth SM (1987) Two complexes that contain histones are required for nucleosome assembly in vitro: Role of nucleoplasmin and N1 in *Xenopus* egg extracts. Cell 51:1009.

Dodson M et al (1985) Specialized nucleoprotein structures at the origin of replication of bacteriophage lambda: Complexes with lambda O protein and with lambda O, lambda P and *Escherichia coli* DnaB proteins. Proc Natl Acad Sci USA 82:4678.

Doree M et al (1989) M phase-promoting factor: Its identification as the M phase-specific H1 kinase and its activation by dephosphorylation. J Cell Sci Suppl 12:39.

Downey KM et al (1988) Proposed roles for DNA polymerase alpha and delta at the replication fork. In Kelly T, Stillman B (eds): "Eukaryotic DNA Replication," vol 6: "Cancer Cells." Cold Spring Harbor, NY: Cold Spring Harbor Laboratory Press.

Dunphy WG, Newport JW (1988) Unraveling of mitotic control mechanisms. Cell 55:925.

Echols H, Goodman MF (1991) Fidelity mechanisms in DNA replication. Annu Rev Biochem 60:477.

Emerit I, Cerutti P (1981) Clastogenic activity from Bloom Syndrome fibroblast cultures. Proc Natl Acad Sci USA 78:1868.

Endow SA, Gall JG (1975) Differential replication of satellite DNA in polyploid tissues of *Drosophila virilis*. Chromosoma 50:175.

Fairman MP et al (1988) Characterization of cellular proteins required for SV 40 replication in vitro. In Kelly T, Stillman B (eds): "Eukaryotic DNA Replication," vol 6: "Cancer Cells." Cold Spring Harbor, NY: Cold Spring Harbor Laboratory Press.

Friedberg EC (1985) "DNA Repair." New York: WH Freeman.

Friedberg EC (1988) Deoxyribonucleic acid repair in the yeast *Saccharomyces cerevisiae*. Microbiol Rev 52:70.

Friedberg E. (1991) Eukaryotic DNA repair: Glimpses through the yeast *Saccharomyces cerevisiae*. BioEssays 13:295.

Fuller RS et al (1984) The DnaA protein complex with chromosomal replication origin (ORI C) and other DNA sites. Cell 38:889.

Gage LP (1974) Polyploidization of the silk gland of *Bombyx mori*. J Mol Biol 86:97.

Gautier J et al (1990) Cyclin is a component of maturation-promoting factor from *Xenopus*. Cell 60:487.

Ghiara JB et al (1991) A cyclin B homolog in *S. cerevisiae*: Chronic activation of the cdc28 protein kinase by cyclin prevents exit from mitosis. Cell 65:163.

Goldman MA et al (1984) Replication timing of genes and middle repetitive sequences. Science 224:686.

Goto T, Wang JC (1984) Yeast DNA polymerase II is encoded by a single-copy essential gene. Cell 36:1073.

Goto T, Wang JC (1985) Cloning of yeast TOPI, the gene encoding topoisomerase I, and the construction of mutants defective in both DNA topoisomerases I and II. Proc Natl Acad Sci USA 82:7178.

Gregory P et al (1985). Alterations in the time of X chromosome replication induced by 5-Azacytidine in a patient with 48,XXXY/47,XXY. Cytogenet Cell Genet 39:234.

Greider CW, Blackburn EH (1985) Identification of a specific telomere terminal transferase activity in tetrahymena extracts. Cell 42:405.

Greider CW, Blackburn EH (1987) The telomere terminal transferase of Tetrahymena is a ribonucleoprotein enzyme with two kinds of primer specificity. Cell 51:887.

Groppi VE, Coffino P (1980) G1 and S phase mammalian cells synthesize histones at equivalent rates. Cell 21:195.

Hammond MF, Laird CD (1985a) Control of DNA replication and spatial distribution of defined DNA sequences in salivary gland cells of *Drosophila melanogaster*. Chromosoma 91:279.

Hammond MP, Laird CD (1985b) Chromosome structure and DNA replication in nurse and follicle cells of *Drosophila melanogaster*. Chromosoma 91:267.

Harland RM, Laskey RA (1980) Regulated replication of DNA microinjected in eggs of *Xenopus laevis*. Cell 21:761.

Hartwell LH, Weinert TA (1989). Checkpoints: Controls that ensure the order of cell cycle events. Science 246:629.

Hartwell LH et al (1974) Genetic control of the cell division cycle in yeast. Science 183:46.

Hatton KS et al (1988a) Temporal order of replication of multigene families reflects chromosomal location and transcriptional activity. In Kelly T, Stillman B (eds): "Eukaryotic DNA Replication," vol 6: "Cancer Cells." Cold

Spring Harbor, NY: Cold Spring Harbor Laboratory Press.

Hatton KS et al (1988b) Replication program of active and inactive multigene families in mammalian cells. Mol Cell Biol 8:2149–2158.

Henderson E et al (1987) Telomeric DNA oligonucleotides form novel intramolecular structures containing guanine–guanine base pairs. Cell 51:899.

Holmquist G (1987) Role of replication time in the control of tissue-specific gene expression Am J Hum Genet 40:151.

Holmquist GP (1989) Evolution of chromosome bands: Molecular ecology of noncoding DNA. J Mol Evol 28:469.

Holmquist G et al (1982) Characterization of Giemsa Dark-and light band DNA. Cell 31:121.

Horosh I et al (1989) Purification and characterization of the RAD3 ATPase/DNA helicase from *Saccharomyces cerevisiae*. J Biol Chem 264:20532.

Hotta Y, Stern H (1984) The organization of DNA segments undergoing repair synthesis during pachytene. Chromosoma 89:127.

Hotta Y et al (1966) Synthesis of DNA during meiosis. Proc Natl Acad Sci USA 56:1184.

Hotta Y et al (1985) Meiosis specific transcripts of a DNA component replicated during chromosome pairing: Homology across the phylogenetic spectrum. Cell 40:785.

Huberman JA (1987) Eukaryotic DNA replication: A complex picture partially clarified. Cell 48:7.

Hubscher U et al (1981) Evidence that a high molecular weight replicative DNA polymerase is conserved during evolution. Proc Natl Acad Sci USA 78:6771.

Ikeda J-E et al (1981) Replication of adenovirus DNA–protein complex with purified proteins. Proc Natl Acad Sci USA 76:884.

Inoué S (1981) Cell division and the mitotic cycle. J Cell Biol 91:131s.

Jackson DA, Cook PR (1986) Replication occurs at a nucleoskeleton. EMBO J 5:1403.

Jackson V, Chalkley R (1985). Histone synthesis and deposition in the G1 and S phases of hepatoma tissue culture cells. Biochemistry 24:6921.

Joenge H et al (1989) Oxygen toxicity and chromosomal breakage in ataxia telangiectasia. Carcinogenesis 8:341.

Johnson LM et al (1985) Isolation of the gene encoding yeast DNA polymerase I. Cell 43:369.

Kaguni LS et al (1988) Structural and catalytic features of the mitochondrial DNA polymerase from *Drosophila melanogaster* embryos. In Kelly T, Stillman B (eds): "Eukaryotic DNA Replication," vol 6: "Cancer Cells." Cold Spring Harbor, NY: Cold Spring Harbor Laboratory Press.

Kelly TJ (1988) SV40 DNA replication. J Biol Chem 263:17889.

Kishimoto T et al (1984) Induction of starfish oocyte maturation by maturation-promoting factor of mouse and surf clam oocytes. J Exp Zool 231:293.

Kleinschmidt JA et al (1985) Co-existence of two different types of soluble histone complex in nuclei of *Xenopus laevis* oocytes. J Biol Chem 260:1166.

Kornberg A (1980) "DNA Replication." San Francisco: WH Freeman.

Kornberg A (1982) "DNA Replication, 1982 Supplement." San Francisco: WH Freeman.

Kornberg A (1988) DNA replication. J Biol Chem 263:1.

Kraemer KH et al (1987) Xeroderma pigmentosum. Cutaneous, ocular and neurologic abnormalities in 830 published cases. Arch Dermatol 123:241.

Laird CD (1973) DNA of Drosophila chromosomes. Annu Rev Genet 7:177.

Laird CD et al (1974) Organization and transcription of DNA in chromosomes and mitochondria of *Drosophila*. Cold Spring Harbor Symp Quant Biol 38:311.

Lamb MM, Laird CD (1987) Three euchromatic DNA sequences under-replicated in polytene chromosomes of *Drosophila* are localized in constrictions and ectopic fibers. Chromosoma 95:227.

Laskey RA et al (1989) S phase of the cell cycle. Science 246:609.

Lee M, Nurse P (1988) Cell cycle control genes in fission yeast and mammalian cells. Trends Genet 4:287.

Lima de Faria A, Jaworska H (1968) Late DNA synthesis in Heterochromatin. Nature 217:138.

Loeb LA, Preston BD (1986) Mutagenesis by apurinic/apyrimidinic sites. Annu Rev Genet 20:201.

Lohka MJ et al (1988) Purification of maturation-promoting factor: An intracellular regulator of early mitotic events. Proc Natl Acad Sci USA 85:3009.

Matsushima H et al (1991) Colony stimulating factor 1 regulates novel cyclins during the G1 phase of the cell cycle. Cell 65:701.

Mazia D (1961) Mitosis and the physiology of cell division. In Brachet J, Mirsky AE (eds): "The Cell," vol 3. New York: Academic Press.

Mazia D (1987) The chromosome cycle and the centrosome cycle in the mitotic cycle. Int Rev Cytol 100:49.

McCarroll RM, Fangman WL (1988) Time of replication of yeast centromeres and telomeres. Cell 54:505.

McIntosh RJ, Koonce MP (1989) Mitosis. Science 246:622.

Meyerwitz EM et al (1985) The 68C glue puff of *Drosophila*. Cold Spring Harbor Symp Quant Biol 50:347.

Meyerwitz EM et al (1987) How *Drosophila* larvae make glue: Control of SGS-3 gene expression. Trends Genet 3:288.

Meyne J et al (1990) Distribution of nontelomeric sites of the (TTAGGG)n telomeric sequence in vertebrate chromosomes. Chromosoma 99:3.

Millerd A, Whitfeld PR (1973) DNA and RNA synthesis during the expansion phase of cotyledon development in *Vicia faba* L. Plant Physiol 51:1005.

Morin GB (1989) The human telomere terminal transferase enzyme is a ribonucloprotein that synthesizes TTAGGG repeats. Cell 59:521.

Muller HJ (1947) Pigrim Trust Lecture: The gene. Proc R Soc Ser B 134:1.

Murray AW, Szostak JW (1985) Chromosome segregation in mitosis and meiosis. Annu Rev Cell Biol 1:289.

Nagl W (1978) "Endopolyploidy and Polyteny in differentiation and Evolution." Amsterdam: North-Holland.

Newlon CS (1988) Yeast chromosome replication and segregation. Microbiol Rev 52:568.

Nicklas RB, Jaqua RA (1965) X chromosome DNA replication: Developmental shift from synchrony to asynchrony. Science 147:1041.

O'Farrell PH et al (1989). Directing cell division during development. Science 246:635.

Osley A (1991) The regulation of histone synthesis in the cell cycle. Annu Rev Biochem 60:827.

Painter TS, Biesele JJ (1966) Endomitosis and polyribosome formation. Proc Natl Acad Sci USA 56:1920.

Painter TS, Reindorp EC (1939) Endomitosis in the nurse cells of the ovary of *Drosophila melanogaster*. Chromosoma 1:267.

Pardee AB (1989) G1 events and regulation of cell proliferation. Science 246:603.

Pavan C, DaCunha AB (1969) Chromosomal activities in *Rhynchsciara* and other Sciaridae. Annu Rev Genet 3:425.

Pettijohn DE, Price CM (1988) Nuclear proteins of the mitotic apparatus. In Adolph KW (ed): "Chromosomes," vol 3. Boca Raton, FL: CRC Press.

Pettijohn DE et al (1984) Nuclear proteins that become part of the mitotic apparatus: A role in nuclear assembly. J Cell Sci Suppl 1:187.

Plevani P et al (1985) Polypeptide structure of DNA primase from a yeast DNA. J Biol Chem 260:7102.

Plevani P et al (1988) Biochemical and genetic characterization of the yeast DNA polymerase–primase complex. In Kelly T, Stillman B (eds): "Eukaryotic DNA Replication," vol 6: "Cancer Cells." Cold Spring Harbor, NY: Cold Spring Harbor Laboratory Press.

Radman M, Wagner R (1986) Mismatch repair in *Escherichia coli*. Annu Rev Genet 20:523.

Reddy GPV, Pardee AB (1980) Multienzyme complex for metabolic channeling in mammalian DNA replication. Proc Natl Acad Sci USA 77:3312.

Ripley LS (1990) Frameshift mutation: Determinants of specificity. Annu Rev Genet 24:189.

Roberge M et al (1990) The topoisomerase II inhibitor VM-26 induces marked changes in histone H1 kinase activity, histones H1 and H3 phosphorylation and chromosome condensation in G2 phase and mitotic BHK cells. J Cell Biol 111:1753.

Roufa DJ (1978) Replication of a mammalian genome: The role of de novo protein biosynthesis during S phase. Cell 13:129.

Rudkin GT, Tartof KD (1973) Repetitive DNA in polytene chromosomes of *Drosophila melanogaster*. Cold Spring Harbor Symp Quant Biol 38:397.

Schieste RH, Prakash L (1988) Rad1, an excision repair gene of *Saccharomyces cerevisiae* is also involved in recombination. Mol Cell Biol 8:3619.

Schieste RH, Prakash L (1990) RAD10, an excision repair gene of *Saccharomyces cerevisiae* is involved in the RAD1 pathway of recombination. Mol Cell Biol 10:2485.

Schmidt M (1980) Two phases of DNA replication in human cells. Chromosoma 76:101.

Shaham M et al (1980) A diffusable clastogenic factor in ataxia telangiectasia. Cytogenet Cell Genet 27:155.

Shippen-Lentz D, Blackburn EH (1990) Functional evidence for an RNA template in telomerase. Science 247:544.

Simpson RT et al (1985) Chromatin reconstituted from tandemly repeated cloned DNA fragments and core histones: A model system for study of higher order structure. Cell 42:799.

Spear BB (1977) Differential replication of DNA sequences in *Drosophila* chromosomes. Am Zool 17:695.

Stahl H et al (1988) DNA unwinding function of the SV40 large tumor antigen. In Kelly T, Stillman B (eds): "Eukaryotic DNA Replication," vol 6: "Cancer Cells." Cold Spring Harbor, NY: Cold Spring Harbor Laboratory Press.

Stein A, Bina M (1984) A model chromatin assembly system. Factors affecting nucleosome spacing. J Mol Biol 178:341.

Stinchcomb DT el al (1981) DNA sequences that allow the replication and segregation of yeast chromosomes. ICN-UCLA Symposium. Mol Biol Cell Biol 22:473.

Sundin O, Varshafsky A (1981) Arrest of segregation leads to accumulation of richly intertwined catenated dimers: Dissection of the final stages of SV40 DNA replication. Cell 25:659.

Sung P et al (1987) The RAD3 gene of *Saccharomyces cerevisiae* encodes a DNA-dependent ATPase. Proc Natl Acad Sci USA 84:6045.

Surana U et al (1991) The role of cdc28 and cyclins during mitosis in the budding yeast *S. cerevisiae*. Cell 65:145.

Svaren J, Chalkley R (1990) The structure and assembly of active chromatin. Trends Genet 6:52.

Takahashi M (1989) A fractal model of chromosomes and chromosomal DNA replication. J Theor Biol 141:117.

Takanari H (1985) Studies of endoreduplication V: A Three dimensional scheme for diplo- and quadruple chromosomes and a model for DNA replication. Cytogenet Cell Genet 39:188.

Tsurimoto T, Stillman B (1991) Replication factors required for SV40 DNA replication in vivo. J Biol Chem 266:1961.

Umek RM, Kowalski D (1988) The ease of DNA unwinding as a determinant of initiation at yeast replication origins. Cell 52:559.

Umek RM et al (1989) New beginnings in studies of eukaryotic replication origins. Biochim Biophys Acta 1007:1.

Van der Vliet PC et al (1988) Interaction of cellular proteins with the adenovirus origin of DNA replication. In Kelly T, Stillman B (eds): "Eukaryotic DNA Replication," vol 6: "Cancer Cells." Cold Spring Harbor, NY: Cold Spring Harbor Laboratory Press.

Van Houten B (1990) Nucleotide excision repair in *Escherichia coli*. Microbiol Rev 54:18.

Vogt VM, Braun R (1977) The replication of ribosomal DNA in *Physarum polycephalum*. Eur J Biochem 80:557S.

von Sonntag C (1987) "The Chemical Basis of Radiation Biology." London: Taylor and Francis.

Walbot V, Dure LS (1976) Developmental biochemistry of cotton seed embryogenesis and germination. VII: Characterization of the cotton genome. J Mol Biol 101:503.

Wang T S-F. (1991): Eukaryotic DNA polymerases. Annu Rev Biochem 60:513.

Ward JF (1975) Molecular mechanisms of radiation-induced damage to nucleic acids. Adv Radiat Biol 5:181.

Ward JF (1990) The yield of double-strand breaks produced intracellularly by ionizing radiation: A review. Int J Radiat Biol 57:1141.

Wold MS et al (1988) Cellular proteins required for SV40 DNA replication in vitro. In Kelly T, Stillman B (Eds): "Eukaryotic DNA Replication," vol 6: "Cancer Cells." Cold Spring Harbor, NY: Cold Spring Harbor Laboratory Press.

Wu RS, and Bonner WM (1981) Separation of basal histone synthesis from S-phase histone synthesis in dividing cells. Cell 27:321.

Wu RS et al (1983) Fate of newly synthesized histone in G1 and G0 cells FEBS Lett 162:161.

Yunis JJ, Prakash O (1982). The origin of man: A pictorial legacy. Science 215:1525.

Zahler AM, Prescott DM (1988) Telomere terminal transferase activity in the hypotrichous ciliate *Oxytricha nova* and a model for the replication of the ends of DNA molecules. Nucleic Acids Res 16:6953.

Zahler AM, Prescott DM (1989) DNA primase and the replication of the telomeres in *Oxytricha nova*. Nucleic Acids Res 17:6299.

Zakian VA (1989) Structure and function of telomeres. Annu Rev Genet 23:579.

Zakian VA et al (1990). How does the end begin? Trends Genet 6:12.

CHAPTER 4

Adzuma K, Mizuuchi K (1989) Interaction of proteins located at a distance along DNA: Mechanism of target immunity in the mu DNA stand–transfer reaction. Cell 57:41.

Albini SM, Jones GH (1987) Synaptonemal complex spreading in *Allium cepa* and *A. fistulosum*. I. The initiation and sequence of pairing. Chromosoma 95:324.

Anderson LK, Stack SM (1988) Nodules associated with axial cores and synaptonemal complexes during zygotene in *Psilotum nudum*. Chromosoma 97:96.

Anderson LK, Stack SM, Sherman JD (1988) Spreading synaptonemal complexes from *Zea mays*. I. No synaptic adjustment of inversion loops during pachytene. Chromosoma 96:295.

Baker BS, Carpenter ATC, Esposito MS, Esposito RE, Sandler L (1976) The genetic control of meiosis. Annu Rev Genet 10:53.

Bateson W, Saunders ER, Punnett RC (1905) Experimental studies in the physiology of heredity. Reports to the Evolution Committee. R Soc London 2:80.

Bennett MD (1982) Nucleotypic basis of the spatial ordering of chromosomes in eukaryotes and the implications of the order for genomic evolution and phenotypic variation. In "Genome Evolution." Systematics Association special volume 20. Dover GA, Flavell RB (eds): New York: Academic Press, p 239.

Bernstein H, Hopf FA, Michod RE (1988) Is meiotic recombination an adaptation for repairing DNA, producing genetic variation, or both? In Michod RE, Levin BR (eds): "The Evolution of Sex." Sunderland, MA: Sinnauer Associates.

Bojko M (1989) Two kinds of "recombination nodules" in *Nuerospora crassa*. Genome 32:309.

Callan HG (1972) Replication of DNA in the chromosomes of eukaryotes. Proc R Soc London Ser B 181:19.

Callow RS (1984) Comments on Bennett's model of somatic chromosome disposition. Heredity 54:171.

Carpenter ATC (1979) Recombination nodules and synaptonemal complex in recombination-defective females of *Drosophila melanogaster*. Chromosoma 75:259.

Carpenter ATC (1984) Meiotic roles of crossing-over and gene conversion. Cold Spring Harbor Symp Quant Biol 49:23.

Carpenter ATC (1987) Gene conversion, recombination nodules, and the initiation of meiotic synapsis. BioEssays 6:232.

Carpenter ATC (1991) Distributive segregation: Motors in the polar wind? Cell 64:885.

Cavalier-Smith T (1988). Origin of the cell nucleus. BioEssays 9:72.

Charlesworth B (1991) When to be diploid. Nature 351:273.

Chovnick A, Ballantyne GH, Baillie DL, Holm DG (1970). Gene conversion in higher organisms: Half-tetrad analysis of recombination within the rosy cistron of *Drosophila melanogaster*. Genetics 66:315.

Cleveland LR (1947) The origin and evolution of meiosis. Science 105:287.

Comings DE, Harris DC, Okada TA, Holmquist GP (1977) Nuclear proteins. IV. Deficiency of non-histone proteins and condensed chromatin of *Drosophila virilis* and mouse. Exp Cell Res 105:349.

Curtis D, Clark SH, Chovnick A, Bender W (1989) Molecular analysis of recombination events in *Drosophila*. Genetics 122:653.

Darlington CD (1932) "Recent Advances in Cytology." Philadelphia: P. Blakiston's Son and Co.

Dawson DS, Murray AW, Szostak JW (1986) An alternative pathway for meiotic chromosome segregation in yeast. Science 234:713.

Dresser ME (1987) The synaptonemal complex and meiosis: An immunocytochemical approach. In Moens PB (ed): "Meiosis." New York: Academic Press.

Egel-Mitani M, Olson LM, Egel R (1982) Meiosis in *Aspergillus nidulans*: Another example for lacking synaptonemal complexes in the absence of crossover interference. Hereditas 97:179.

Esposito MS (1978) Evidence that mitotic recombination occurs at the two-strand stage. Proc Natl Acad Sci USA 75:4436–4440.

Fabre F, Boulet A, Roman H (1984) Gene conversion at different points in the mitotic cycle of *Saccharomyces cerevisiae*. Mol Gen Genet 195:139.

Fogel S, Mortimer R, Lusnak K, Tavares F (1979) Meiotic gene conversion: A signal of the basic recombination event in yeast. Cold Spring Harbor Symp Quant Biol 43:1325.

Gillies CB (1984) The synaptonemal complex in plants. CRC Crit Rev Plant Sci 2:81.

Gillies CB (1988) Telomeric C bands and synaptonemal complex formation in rye (*Secale cereale*). Genome 30(Suppl 1):131.

Goldstein P, Triantaphyllou AC (1982) The synaptonemal complexes of *Meloidogyne*: Relationship of structure and evolution of parthenogenesis. Chromosoma 87:117.

Golin J, Esposito M (1981) Mitotic recombination: Mismatch correction and replication resolution of Holliday structures formed at the two strand stage in *Saccharomyces*. Mol Gen Genet 183:252.

Grell RJ (1976) Distributive pairing. In Novitski E, Ashburner M (eds): "Genetics and Biology of Drosophila," vol Ia. New York: Academic Press, p 435.

Haber JE, Rogers DT, McCusker JH (1980) Homothallic conversions of yeast mating-type genes occur by intrachromosomal recombination. Cell 22:277.

Hasenkampf CA (1984) Synaptonemal complex formation in pollen mother cells of *Tradescantia*. Chromosoma 90:275.

Hawley RS (1980) Chromosomal sites necessary for normal levels of meiotic recombination in *Drosophila melanogaster*. I. Evidence for and mapping of the sites. Genetics 94:625.

Henderson SA (1988) Four effects of elevated temperature on chiasma formation in the locust *Schistocerca gregaria*. Heredity 60:387.

Hickey DA, Rose MR (1988) The role of gene transfer in the evolution of eukaryotic sex. In Michod RE, Levin BR (eds): "The Evolution of Sex." Sunderland, MA: Sinnauer Associates.

Hotta Y, Stern H (1974) DNA scission and repair in Lilium. Chromosoma 46:279.

Jinks-Robertson S, Petes TD (1986) Chromosomal translocations generated by high-frequency meiotic recombination between repeated yeast genes. Genetics 114:731.

John B (1976) Myths and mechanisms of meiosis. Chromosoma 54:295.

John B, Miklos GLG (1979) Functional aspects of satellite DNA and heterochromatin. Int Rev Cytol 58:1.

Käfer E (1961) The process of spontaneous recombination in vegetative nuclei of *Aspergillus nidulans*. Genetics 46:1581.

Kitani Y, Olive LS, El-Ani AS (1962) Genetics of *Sordaria fimicola*. V. Aberrant segregation at the *g* locus. Am J Bot 49:697.

Klar AJS, McIndoo J, Strathern JN, Hicks JB (1980) Evidence for a physical interaction between the transposed and substituted sequences during the mating type gene transposition in yeast. Cell 22:291.

Klingwell B, Rattner JB (1986) Sister chromatids are associated together at discrete points in metaphase chromosomes. J Cell Biol 103:492.

Kmiec E, Holloman W (1984) Synapsis promoted by *Ustilago* rec 1 protein. Cell 36:593.

Kondrashov AS, Crow JF (1991) Haploidy or diploidy: Which is better? Nature 351:314.

Kubai DF (1982) Meiosis in *Sciara coprophila*: Structure of the spindle and chromosome behavior during the first meiotic division. J Cell Biol 93:655.

Lindegren CC (1953) Gene conversion in Saccharomyces. J Genet 51:625.

Lica LM, Narayanswami S, Hamkalo BA (1986) Mouse satellite DNA, centromere structure and sister chromatid pairing. J Cell Biol 103:1145.

Lichten M, Borts R, Haber JE (1987) Meiotic gene conversion and crossing over between dispersed homologous sequences occurs frequently in *Saccharomyces cerevisiae*. Genetics 115:233.

Loidl J (1991) Coming to grips with a complex matter. A multidisciplinary approach to the synaptonemal complex. Chromosoma 100:289.

Lu BC (1967) Meiosis in *Coprinus lagopus*: A comparative study with light and electron microscopy. J Cell Sci 2:529.

Lu BC (1970) Genetic recombination in *Coprinus II*. Its relations to the synaptonemal complexes. J Cell Sci 6:669.

Lu BC, Raju NB. (1970) Meiosis in Coprinus II. Chromosome pairing and the lampbrush diplotene stage of meiotic prophase. Chromosoma 29:305.

Maguire MP (1978) Evidence for separate genetic control of crossing over and chiasma maintenance in maize. Chromosoma 65:173.

Maguire MP (1980) Adaptive advantage for chiasma interference: A novel suggestion. Heredity 45:127.

Maguire MP (1988a) Crossover site determination and interference. J Theor Biol 134:565.

Maguire MP (1988b) Interactive meiotic systems. In Gustafson JP, Appels R (eds): "Chromosome Structure and Function. Impact of New Concepts." New York: Plenum.

Maguire MP (1990). Sister chromatid cohesiveness: Vital function, obscure mechanism. Biochem Cell Biol 68:1231.

Maguire MP (1992) The evolution of meiosis. J Theor Biol 154:43.

Maizels N (1989) Might gene conversion be the mechanisms of somatic hypermutation of mammalian immunoglobulin genes? Trends Genet 5:4.

Margulis L, Sagan D, Olendzenski L (1985) What is sex? In Halverson HO, Monray A (eds): "The Origin and Evolution of Sex." New York: Alan R Liss.

Mendel G (1866) Versuche uber Pflanzenhybriden. Verh Naturforsch-Vereins Brunn 4:3.

Moens PB, Church K (1979) The distribution of synaptonemal complex material in metaphase I bivalents of *Locusta* and *Chloealtis* (Orthoptera: Acrididae). Chromosoma 73:247.

Morgan TH (1911) An attempt to analyze the constitution of the chromosomes on the basis of sex-limited inheritance in Drosophila. J Exp Zool 11:365.

Morgan TH, Lynch CJ (1912) The linkage of two factors in Drosophila that are not sexlinked. Biol Bull 23:174.

Mortimer R, Fogel S (1974) Genetical interference and gene conversion. In Grell R (ed): "Mechanisms in Recombination." New York: Plenum, p 263.

Moses MJ, Poorman PA, Roderick TH, Davisson MT (1982) Synaptonemal complex analysis of mouse chromosomal rearrangements. IV. Synapsis and synaptic adjustment in two paracentric inversions. Chromosoma 84:457.

Murray AW, Szostak JW (1985) Chromosome segregation in mitosis and meiosis. Annu Rev Cell Biol 1:289.

Nicklas RB (1983) Measurements of the force produced by the mitotic spindle in anaphase. J Cell Biol 97:542.

Nicklas RB (1988) Chromosomes and kinetochores do more in mitosis than previously thought. In Gustafson JP, Appels R (eds): "Chromosome Structure and Function. Impact of New Concepts." New York: Plenum.

Nicklas RB, Kubai DF (1985) Microtubules, chromosome movement, and reorientation after chromosomes are detached from the spindle by micromanipulation. Chromosoma 92:313.

Nicklas RB, Stachly CO (1967) Chromosome micromanipulation. I. The mechanics of chromosome attachment to the spindle. Chromosoma 21:1.

Olsen LW, Eden U, Egel-Mitani M, Egel R (1978) Asynaptic meiosis in fission yeast? Hereditas 89:189.

Orr-Weaver TL, Sozstak JW (1985) Fungal recombination. Microbiol Rev 49:33.

Parvinen M, Söderström KO (1976) Chromosome rotation and formation of synapsis. Nature 269:534.

Perrot V., Richerd S., Valéro M (1991) Transition from haploidy to diploidy. Nature 315:315.

Powers PA, Smithies O (1986) Short gene conversions in the human fetal globin region: A by-product of chromosome pairing during meiosis? Genetics 112:343.

Rasmussen SW (1977) Meiosis in *Bombyx mori* females. Philos Trans R Soc Lond B 277:343.

Reynaud CA, Anquez U, Grimal H, Weill JC (1987) A hyperconversion mechanism generates the chicken light chain preimmune repertoire. Cell 48:379.

Rockmill B, Roeder GS (1988) RED I: A yeast gene required for the segregation of chromosomes during the reductional division of meiosis. Proc Natl Acad Sci USA 85:6057.

Roeder GS, Steward SE (1988) Mitotic recombination in yeast. Trends Genet 4:263.

Roman H (1956) Studies of gene mutation in yeast. Cold Spring Harbor Symp Quant Biol 21:175.

Sen D, Gilbert W (1988) Formation of parallel four-stranded complexes by guanine-rich motifs in DNA and its implications for meiosis. Nature 334:364.

Slighton JL, Blechel AE, Smithies O (1980) Human γ^G- and γ^A-globin genes: Complete nucleotide sequences suggest that DNA can be exchanged between these duplicated genes. Cell 21:627.

Smithies O, Powers PA (1986) Gene conversions and their relationship to homologous pairing. Philos Trans R Soc Lond B 312:291.

Snow R (1979) Maximum likelihood estimation of linkage and interference from tetrad data. Genetics 92:231.

Stack S, Anderson L (1986) Two-dimensional spreads of synaptonemal complexes from solanaceous plants. III. Recombination nodules and crossing over in *Lycopersicum esculentum* (tomato). Chromosoma 95:253.

Stern H, Hotta Y (1987) The biochemistry of meiosis. 1987. In Moens PB (ed): "Meiosis." New York: Academic Press.

Sturtevant AH (1913a) The linear arrangement of sex-linked factors in Drosophila as shown by their mode of association. J Exp Zool 14.43.

Sturtevant AH (1913b) A third group of linked genes in *Drosophila ampelophila*. Science 37:990.

Symington L, Fogarty L, Kolodner R (1983) Genetic recombination of homologous plasmids catalyzed by cell-free extracts of *Saccharomyces cerevisiae*. Cell 35:805.

Thompson CB, Neimann PE (1987) Somatic diversification of the chicken immunoglobulin light chain gene is limited to the rearranged variable gene segment. Cell 48:369.

Vilardell J, Coll MD, Querol E, Eqozau J (1989) Histone electrophoretic pattern in the characterization of synaptonemal complexes. Cell Mol Biol 35:207.

Walters MS (1970) Evidence on the time of chromosome pairing from the preleptotene spiral stage in *Lilium longiflorum* "Croft." Chromosoma 29:375.

Weismann A (1892) Das Keimplasma. Eine Theorie der Vererbung. Jena: Gustav Fischer.

Westergaard M, von Wettstein D (1972) The synaptonemal complex. Annu Rev Genet 6:71.

Wettstein D von, Rasmussen SW, Holm PB (1984) The synaptonemal complex in genetic segregation. Annu Rev Genet 18:331.

Winkler H (1930) Die Konversion der Gene. Jena: Gustav Fischer.

Wise D (1988) The diversity of mitosis: The value of evolutionary experiments. Biochem Cell Biol 66:515.

Zickler D, Sage J (1981) Synaptonemal complexes with modified lateral elements in *Sordaria humana*: Development of and relation to the "recombination nodules." Chromosoma 84:305.

CHAPTER 5

Alitalo K et al (1983) Homogeneously staining chromosomal regions contain amplified copies of an abundantly expressed cellular oncogene (c-myc) in malignant neuroendocrine cells from a human colon carcinoma. Proc Natl Acad Sci USA 80:1707.

Aragonicillo C et al (1978) Influence of homologous chromosomes on gene-dosage effects in allohexaploid wheat (*Triticum aestivum* L). Proc Natl Acad Sci USA 75:1446.

Astaurov BL (1969) Experimental polyploidy in animals. Annu Rev Genet 3:99.

Averett JE (1980) Polyploidy in plant taxa: Summary. In Lewis WH (ed): "Polyploidy." New York: Plenum.

Avery AG et al (1959) "Blakeslee: The Genus *Datura*." New York: Ronald Press.

Babcock EB (1947) The genus *Crepis*. Univ Calif Berkeley Publ Bot, vols 21 and 22.

Bains W (1986) The multiple origins of human Alu sequences. J Mol Evol 23:189.

Baker B et al (1986) Transposition of the maize controlling element "activator" in maize. Proc Natl Acad Sci USA 83:4844.

Baker BS, Belote JM (1983) Sex determination and dosage compensation in *Drosophila melanogaster*. Annu Rev Genet 17:345.

Baltimore D (1970) RNA-dependent DNA polymerase in virions of RNA tumor viruses. Nature 226:1209.

Baltimore D (1985) Retroviruses and retrotransposons: The role of reverse transcription in shaping the eukaryotic genome. Cell 40:481.

Baverstock PR (1982) A sex-linked enzyme in birds—Z-chromosome conservation but no dosage compensation. Nature 296:763.

Beerman S (1984) Circular and linear structures in chromatin diminution in Cyclops. Chromosoma 89:321.

Bell G (1988) Uniformity and diversity in the evolution of sex. In Michod RE, Levin BR (eds): "The Evolution of Sex: An Examination of Current Ideas." Sunderland, MA: Sinauer.

Belote JM, Lucchesi JC (1980) Control of X chromosome transcription in the maleless gene in Drosophila. Nature 285:573.

Benazzi M (1957) Considerazioni sulla evoluzione cromosomica negli animali. Boll Zool 24:373.

Berger R et al (1985) Report of the Committee on Chromosomal Rearrangements in Neoplasia and Fragile Sites. (In: Human Gene Mapping 8. Helsinki Conference (1985). Eighth International Workshop on Human Gene Mapping.) Cytogenet Cell Genet 40:490.

Bickham JW (1984) Patterns and modes of chromosome evolution in reptiles. In Sharma AK, Sharma A (eds): "Chromosomes in Evolution of Eukaryotic Groups." Boca Raton, FL: CRC Press.

Bickham JW et al (1985) Diploid–triploid mosaicism: An unusual phenomenon in side-necked turtles (Platemys platycephala). Science 227:1591.

Bingham PM, Judd BH (1981) A copy of the copia transposable element is very tightly linked to the w^a allele at the white locus of D. melanogaster. Cell 25:705.

Bishop JM (1983) Cellular oncogenes and retroviruses. Annu Rev Biochem 52:301.

Bishop JM (1985) Viral oncogenes. Cell 42:23.

Bishop JM (1986) Oncogenes as hormone receptors. Nature 321:112.

Bishop JM (1987) The molecular basis of cancer. Science 235:305.

Blackburn E (1987) Developmentally controlled DNA excision in Ciliates. Trends Genet 3:235.

Blackburn EH, Karrer KM (1986) Genomic reorganization in ciliated protozoans. Annu Rev Genet 20:501.

Blakeslee AF et al (1922) A haploid mutant in the jimson weed, Datura stramonium. Science 55:646.

Boeke JD et al (1985) TY elements transpose through an RNA intermediate. Cell 40:491.

Bogart JP (1980) Evolutionary implications of polyploidy in amphibians and reptiles. In Lewis WH (ed): "Polyploidy." New York: Plenum.

Bogart JP, Tandy M (1976) Polyploid amphibians: Three more diploid–tetraploid cryptic species of frogs. Science 193:334.

Borst P (1986) Discontinuous transcription and antigenic variation in Trypanosomes. Annu Rev Biochem 55:701.

Borst P, Greaves DR (1987) Programmed rearrangements altering gene expression. Science 235:658.

Boveri J (1910) Die Potenzen der Ascaris-Blastomeren. In "Festschrift R. Hertwig," vol 3, p 131.

Boveri T (1887) Uber Differenzierung der Zellkerne wahrend der Forschung des Eies von Ascaris megalocephala. Anat Anz 2:688.

Bowman JT, Simmons JR (1973) Gene modulation in Drosophila: Dosage compensation of Pgd^+ and Zw^+ genes. Biochem Genet 10:319.

Brennard J et al (1982) Cloned DNA sequences of the hypoxanthine/guanine phosphoribosyl gene from a mouse neuroblastoma cell line found to have amplified genome sequences. Proc Natl Acad Sci USA 79:1950.

Bridges CB (1921) Triploid intersexes in Drosophila melanogaster. Science 54:252.

Brinkley BR et al (1984) Compound kinetochores of the Indian muntjac: Evolution by linear fusion of unit kinetochores. Chromosoma 91:1.

Brinkley BR et al (1985) The kinetochore of mammalian chromosomes: Structure and function in normal mitosis and aneuploidy. In Dellarco VL et al (eds): "Aneuploidy." New York: Plenum, p 243.

Britten RJ et al (1988) Sources and evolution of human Alu repeated sequences. Proc Natl Acad Sci USA 85:4470.

Brown AL (1984) On the origin of the Alu family of repeated sequences. Nature 312:106.

Brown DD, Dawid I (1968) Specific gene amplification in oocytes. Science 160:272.

Brown SW, Chandra HS (1977) Chromosome imprinting and the differential regulation of homologous chromosomes. In Prescott DM (ed): "Cell Biology: A Comprehensive Treatise." New York: Academic Press.

Brown WT et al (1988) Multilocus analysis of the fragile X syndrome. Hum Genet 78:210.

Bruere AN et al (1974) The significance of G-bands and C-bands of three different Robertsonian translocations of domestic sheep. Cytogenet Cell Genet 13:479.

Bruere AN et al (1978) Centric fusion polymorphism in Romney marsh sheep of England. J Hered 69:8.

Bull JJ (1983) "Evolution of Sex Determining Mechanisms." Menlo Park, CA: Benjamin/Cummings.

Burgoyne PS et al (1986) Spermatogenic failure in male mice lacking H-Y antigen. Nature 320:170.

Burnham CR (1956) Chromosomal interchanges in plants. Bot Rev 22:419.

Butner KA, Lo CW (1986) High frequency DNA rearrangements associated with mouse centromeric satellite DNA. J Mol Biol 187:547.

Campuzano S et al (1986) Excess function hairy-wing mutations caused by gypsy and copia insertions within structural genes of the acheate-scute locus of Drosophila. Cell 44:303.

Capanna E (1985) Karyotypic variability and chromosome transilience in rodents: The case of the genus Mus. In Luckett WP, Hartenberger J-L (eds): "Evolutionary Relationships Among Rodents: A Multi-disciplinary Analysis." New York: Plenum.

Carr DH, Gideon M (1977) Population cytogenetics of human abortuses. In Hook SB, Porter IH (eds): "Population Cytogenetics." New York: Academic Press.

Carson H (1983) Chromosomal sequences and interisland colonizations in Hawaiian Drosophila. Genetics 103:465.

Carson PS (1972) Dosage compensation. Mol Gen Genet 114:273.

Catcheside DJ (1936) Origin, nature and breeding behavior of Oenothera lamarckiana trisomics. J Genet 33:1.

Cattanach BM et al (1982) Male, female and intersex development in mice of identical chromosome constitution. Nature 300:445.

Chatterjee RN (1985) X chromosomal organization and dosage compensation. Chromosoma 91:259.

Chin DJ (1982) Appearance of crystalloid endoplasmic reticulum in compaction-resistant Chinese hamster cells with a 500-fold increase in 3-hydroxy-3-methylglutaryl-coenzyme A reductase. Proc Natl Acad Sci USA 79:1184.

Ciferri O et al (1969) Dosage compensation. Genetics 61:567.

Clare J, Farabaugh P (1985) Nucleotide sequence of a yeast TY element: Evidence for an unusual mechanism of gene expression. Proc Natl Acad Sci USA 82:2829.

Cleland RE (1932) Further data bearing upon circle formation in Oenothera, its cause and genetical effect. Genetics 17:572.

Cleland RE (1972) "Oenothera: Cytogenetics and Evolution." New York: Academic Press.

Cleland RE, Blakeslee AF (1931) Segmental interchange, the basis of chromosomal attachments in Oenothera. Cytologia 2:175.

Close RL (1984) Rates of chromosome loss during development in different tissues of the bandicoots Perameles nasuta and Isoodon macrourus (Marsupialia: Peramelidae). Austr J Biol Sci 37:53.

Coen ES, Carpenter R (1986) Transposable elements in Antirrhinum majus: Generation of genetic diversity. Trends Genet 2:292.

Cowell JK (1982) Double minutes and homogeneously staining regions: Gene amplification in mammalian cells. Annu Rev Genet 16:21.

Craig NL (1990) P element transposition. Cell 62:399.

Crawford BD et al (1985) Coordinate amplification of metallothionein I and II genes in calcium resistant Chinese hamster cells: Implications for mechanisms regulating metallothein gene expression. Mol Cell Biol 5:320.

Crouse HV (1960) The controlling element in sex chromosome behavior in Sciara imprinting. Genetics 45:1429.

Crouse HV (1979) X heterochromatin subdivision and cytogenetic analysis in Sciara coprophila (Diptera, Sciaridae). Chromosoma 74:219.

Crouse HV et al (1971) L-chromosome inheritance and the problem of chromosome imprinting in *Sciara* (Sciaridae, Dystera). Chromosoma 34:324.

Daniels GE, Deininger RL (1985) Integration site preferences of the Alu family and similar repetitive DNA sequences. Nucleic Acids Res 13:8939.

Darlington CD (1956) "Chromosome Botany." London: Allen and Unwin.

Davies PL, Hew CL (1990) Biochemistry of fish antifreeze proteins. FASEB J 4:2460.

Davis BM (1911) Genetical studies on *Oenothera*. II. Some hybrids of Oenothera biennis and O. grandiflora that resemble Oenothera lamarckiana. Am Nat 45:193.

Davis BM (1916) *Oenothera lamarckiana*, hybrid of *O. franciscana* bartlett × *O. biennis* linnaeus. Am Nat 50:688.

Dawley RM et al (1985) Triploid progeny of pumpkinseed × green sunfish hybrids. J Hered 76:251.

Deka N et al (1986) Repetitive human DNA sequences. II. Properties of a transposon-like human element. Cold Spring Harbor Symp Quant Biol 51:471.

Deumling B, Clermont L (1989) Changes in DNA content and chromosomal size during cell culture and plant regeneration of *Scilla siberica*: Selective chromatin diminution in response to environmental conditions. Chromosoma 97:439.

Devlin RH et al (1982) Autosomal dosage compensation in *Drosophila melanogaster* strains trisomic for the left arm of chromosome 2. Proc Natl Acad Sci USA 79:1200.

Devlin RH et al (1984) Dosage compensation is transcriptionally regulated in autosomal trisomies of *Drosophila*. Chromosoma 91:65.

Devlin RH et al (1988) The influence of whole-arm trisomy on gene expression in *Drosophila*. Genetics 118:87.

Di Nocera PP, Sakaki Y (1990) Lines: A superfamily of retrotransposable ubiquitous DNA elements. Trends Genet 6:29.

Doolittle WF, Sapienza C (1980) Selfish genes, the phenotype paradigm and genome evolution. Nature 284:601.

Doolittle WF (1982) Selfish DNA after fourteen months. In Dover GA, Flavell RB (eds): "Genome Evolution." London: Academic Press.

Dooner HK, Nelson OE (1977a) Controlling element-induced alterations in UDP glucose: Flavenoid glucosyltransferase, the en-zyme specified by the bronze locus in maize. Proc Natl Acad Sci USA 74:5623.

Dooner HK, Nelson OE (1977b) Genetic control of UDP glucose: Flavonol 3-O-glycosyl transferase in the endosperm of maize. Biochem Genet 15:509.

Dooner HK, Nelson OE (1979) Interaction among *C, R* and *Vp* in the control of the *Bz* glucosyltransferase during endosperm development in maize. Genetics 91:309.

Dooner HK et al (1985) A molecular genetic analysis of insertions in the bronze locus in maize. Mol Gen Genet 200:240.

Doring H-P, Starlinger P (1986) Molecular genetics of transposable elements in plants. Annu Rev Genet 20:175.

Dyer A et al (1970) Aneuploidy: A redefinition. Notes R Bot Garden Edinburgh 30:177.

Elder FFB (1980) Tandem fusion, centric fusion, and chromosomal evolution in the cotton rats, genus *Sigmodon*. Cytogenet Cell Genet 26:199.

Emerson SH (1936) Trisomic derivatives of *Oenothera lamarckiana*. Genetics 21:200.

Emerson SH, Sturtevant AH (1931) Genetical and cytological studies in *Oenothera*. III. The translocation hypothesis. Z Indukt Abstamm Vererbungs 59:395.

Emori Y et al (1985) The nucleotide sequences of copia and copia-related RNA in *Drosophila* virus-like particles. Nature 315:773.

Endoh H, Okada N (1986) Total DNA transcription in vitro: A procedure to detect highly repetitive and transcribable sequences with tRNA-like structures. Proc Natl Acad Sci USA 83:251.

Endow SA, Atwood KC (1988) Magnification: Gene amplification by an inducible system of sister chromatid exchange. Trends Genet 4:348.

Engels WR (1982) P elements in *Drosophila melanogaster*. In Berg DE, Howe MM (eds): "Mobile DNA." Washington, DC: American Society for Microbiology.

Engels WR (1983) The P family of transposable elements in *Drosophila*. Annu Rev Genet 17:315.

Engels WR (1989) P elements in Drosophila melanogaster. In: Berg DE, Howe MM (eds): "Mobile DNA." Washington, DC: Amer Soc for Microbiol.

Engels WR, Preston CR (1984) Formation of chromosome rearrangements by P factors in *Drosophila*. Genetics 107:657.

Engels WR et al (1987) Somatic effects of P element activity in *Drosophila melanogaster*: Pupal lethality. Genetics 117:745.

Epstein CJ (1985) Mouse monosomics and trisomics as experimental systems for studying mammalian aneuploidy. Trends Genet 1:129.

Epstein CJ (1988) Mechanisms of the effects of aneuploidy in mammals. Annu Rev Genet 22:51.

Evans HJ (1962) Chromosome aberrations induced by ionizing radiation. Int Rev Cytol 13:221.

Evans HJ (1974) Effects of radiation on mammalian chromosomes. In German J (ed): "Chromosomes and Cancer." New York: John Wiley.

Evans HJ (1983) Effects on chromosomes of carcinogenic rays and chemicals. In German J (ed): "Chromosome Mutation and Neoplasia." New York: Alan R. Liss.

Feddersen RM, Van Ness BG (1985) Double recombination of a single immunoglobulin κ-chain allele: Implications for the mechanism of rearrangement. Proc Natl Acad Sci USA 82:4793.

Federoff NV (1983) Controlling elements in maize. In Shapiro JA (ed): "Mobile Genetic Elements." New York: Academic Press.

Federoff (1989) Maize transposable elements and development. Cell 56:181.

Field LM et al (1983) Molecular evidence that insecticide resistance in peach potato aphids Myzue-persicae Sulz results from amplification of an esterase gene. Biochem J 251:309.

Finch RA (1983) Tissue-specific elimination of alternative whole parental genomes in one barley hybrid. Chromosoma 88:386.

Fincham JRS, Sastry GRK (1974) Controlling elements in maize. Annu Rev Genet 8:15.

Fink R et al (1986) The mechanism and consequences of retrotransposition. Trends Genet 2:118.

Finnegan DJ (1985) Transposable elements in eukaryotes. Int Rev Cytol 93:281.

Finnegan DJ (1989) Eukaryotic transposable elements and genome evolution. Trends Genet 5:103.

Finnegan DJ, Fawcett DH (1986) Transposable elements in *Drosophila melanogaster*. In Maclean N (ed): "Oxford Survey of Eukaryotic Genes," vol 3. Oxford: Oxford University Press.

Flavell AJ, Ish-Horowicz D (1981) Extrachromosomal circular copies of the eukaryotic transposable element copia in cultured *Drosophila cells*. Nature 292:591.

Fogel S, Welch JW (1982) Tandem gene amplification mediates copper resistance in yeast. Proc Natl Acad Sci USA 79:5342.

Ford M, Fried M (1986) Large inverted duplications are associated with gene amplification. Cell 45:425.

Fossey A, Liebenberg H (1987) Cytotaxonomic studies in *Themeda triandra* forsk. Part II: Aneuploidy in a diploid population. S Afr Tydskr Plantk 53:362.

Fourney RM et al (1984) Accumulation of winter flounder antifreeze RNA after hypophysectomy. Gen Comp Endocrinol 54:392.

Fredga K et al (1976) Fertile XX and XY-type females in the wood lemming *Myopus schisticolor*. Nature 261:225.

Galili G et al (1986) Gene-dosage compensation of endosperm proteins in hexaploid wheat *Triticum aestivum*. Proc Natl Acad Sci USA 83:6524.

Gardiner PR et al (1987) Identification and isolation of a variant surface glycoprotein from *Trypanosoma vivax*. Science 235:774.

Garfinckle DJ et al (1985) TY element transposition: Reverse transcriptase and virus-like particles. Cell 42:507.

Gartler SM, Andina RJ (1981) Mammalian X-chromosome inactivation. Adv Hum Genet 7:99.

Gartler SM, Riggs AD (1983) Mammalian X-chromosome inactivation. Annu Rev Genet 17:155.

Garvey EP, Santi DV (1986) Stable amplified DNA in drug resistant *Leishmania* exists as chromosomal circles. Science 233:535.

Geitler L (1938) Weitere cytogenetische Untersuchungen an naturlichen Populationen von *Paris quadrifolia*. Z Indukt Abstam Vererbungs 75:161.

Gerbi SA (1986) Unusual chromosome movements in Sciarid flies. In Hennig W (ed): "Germ Line–Soma Differentiation." Berlin: Springer-Verlag.

German J (1972) Genes which increase chromosomal instability in somatic cells and predispose to cancer. Prog Med Genet 8:61.

German J (1983) Patterns of neoplasia associated with chromosome-breakage syndromes. In German J (ed): "Chrosome Mutation and Neoplasia." New York: Alan R. Liss.

Geyer-Dusznska I (1966) Genetic factors in oogenesis and spermatogenesis in Cecidomyidae. Chromosomes Today 1:174.

Ghiselin MT (1988) The evolution of sex: A history of competing points of view. In Michod RE, Levin BR (eds): "The Evolution of Sex: An Examination of Current Ideas." Sunderland, MA: Sinauer.

Gierl A et al (1988) In Nelson O (ed): "Plant Transposable Elements." New York: Plenum.

Gierl A et al (1989) Maize transposable elements. Annu Rev Genet 23:71.

Gill BS (1987). Chromosome banding methods, standard chromosome band nomenclature, and applications in cytogenetic analysis. In "Wheat and Wheat Improvement." Agronomy (Monogr No. 13, 2nd ed).

Givol D et al (1981) Diversity of germ line immunoglobulin VH genes. Nature 292:424.

Goday C, Pimpinelli S (1984) Chromosome organization and heterochromatin elimination in Parascaris. Science 224:411.

Goday C, Pimpinelli S (1986) Cytological analysis of chromosomes in the two species, *Parascaris univalens* and *P. equorum*. Chromosoma 94:1.

Goday C, Pimpinelli S (1989) Centromere organization in meiotic chromosomes of *Parascaris univalens*. Chromosoma 98:160.

Goday C et al (1985) Centromere ultrastructure in germ-line chromosomes of *Parascaris*. Chromosoma 91:121.

Goldblatt P (1980) Polyploidy in angiosperms: Monocotyledons. In Lewis WH (ed): "Polyploidy." New York: Plenum.

Grant V (1971) "Plant Speciation." New York: Columbia University Press.

Greenblatt IM (1984) A chromosome replication pattern deduced from pericarp phenotypes resulting from movements of the transposable elements, modulator, in maize. Genetics 108:471.

Gullotta F et al (1981) Descriptive neuropathology of chromosomal disorders in man. Hum Genet 57:344.

Hahn P et al (1986) Chromosomal changes without DNA overproduction in hydroxyurea-treated mammalian cells: Implications for gene amplification. Cancer Res 46:607.

Hall JG (1990) Genomic imprinting: Review and relevance to human disease. Am J Hum Genet 46:857.

Hamkalo B et al (1985) Ultrastructural features of minute chromosomes in a methotrexate-resistant mouse 3T3 cell line. Proc Natl Acad Sci USA 82:1126.

Hamlin JL et al (1984) DNA sequence amplification in mammalian cells. Int Rev Cytol 90:31.

Hannah IC, Nelson OE (1976) Characterization of ADP-glucose pyrophosphorylase from shrunken-2 and brittle-2 mutants of maize. Biochem Genet 14:547.

Hassold TJ (1986) Chromosome abnormalities in human reproductive wastage. Trends Genet 2:105.

Hayman DL, Martin PG (1969) Cytogenetics of marsupials. In Benirschke K (ed): "Comparative Mammalian Cytogenetics." Berlin: Springer.

Hecht F (1988) Enigmatic fragile sites on human chromosomes. Trends Genet 4:121.

Heiter PA et al (1981) Human immunoglobulin kappa light-chain genes are deleted or rearranged in lambda-producing B cells. Nature 290:368.

Helentjaris T et al (1988) Identification of the genomic locations of duplicate nucleotide sequences in maize by analysis of restriction fragment length polymorphisms. Genetics 118:353.

Heppich S et al (1982) Premeiotic chromosome doubling after genome elimination during spermatogenesis of the species hybrid Rana esculenta. Theor Appl Genet 61:101.

Hew CL et al (1988) Multiple genes provide the basis for antifreeze protein diversity and dosage in the ocean pout, Macrozoarces americanus. J Biol Chem 263:12049.

Hicks J et al (1979) Transposable mating type genes in *Saccharomyces cerevisiae*. Nature 282:478.

Hicks JB et al (1977) The cassette model of mating type interconversion. In Bukhari A et al (eds): "DNA Insertion Elements, Plasmids and Episomes." Cold Spring Harbor, NY: Cold Spring Harbor Laboratory, p 457.

Highton R, Larson A (1979) The genetic relationships of the salamanders of the genus *Plethodon*. System Zool 28:579.

Hill AB, Schimke RT (1985) Increased gene amplification in L5I78Y mouse lymphoma cells with hydroxyurea-induced chromosomal aberrations. Cancer Res 45:5050.

Honjo T (1985) Origin of immune diversity: Genetic variation and selection. Annu Rev Biochem 54:803.

Hook EB et al (1989) The natural history of cytogenetically abnormal fetuses detected at midtrimester amniocentesis which are not terminated electively: New data and estimates of the excess and relative risk of late fetal death associated with 47, +21 and some other abnormal karyotypes. Am J Hum Genet 45:855.

Hsu TC (1973) Longitudinal differentiation of chromosomes. Annu Rev Genet 7:153.

Hughes-Schrader S, Monahan DF (1966) Hermaphroditism in *Icerya zeteki* Cockerell and the mechanism of gonial reduction in icerine coccids. Chromosoma 20:15.

Hunkerpiller T, Hood L (1986) The growing immunoglobulin gene superfamily. Nature 323:15.

Hyrien O, Buttin G (1986) Gene amplification in insects. Trends Genet 2:275.

Irifune K (1990) Karyomorphological study on speciation of the aster *Ageratoides* subsp. amplexifolius complex in Japan. J Sci Hiroshima Univ 23:163.

Jaffe E, Laird C (1986) Dosage compensation in *Drosophila*. Trends Genet 2:316.

Jenness DD et al (1983) Binding of alpha-factor pheromone to yeast a cells: Chemical and genetic evidence for a alpha-factor receptor. Cell 35:521.

Jinks-Robertson S, Petes TD (1985) High frequency meiotic gene conversion between repeated genes on nonhomologous chromosomes in yeast. Proc Natl Acad Sci USA 82:3350.

John B, Freeman M (1975) Causes and consequences of Robertsonian exchange. Chromosoma 52:124.

Johnson MS, Turner JRG (1979) Absence of dosage compensation for a sex linked enzyme in butterflies (*Heliconius*). Heredity 43:71.

Johnstone SA, Hopper JE (1982) Isolation of the yeast regulatory gene GAL 4 and analysis of its dosage effects. Proc Natl Acad Sci USA 79:6971.

Junakovic N et al (1984) Genomic distribution of copia-like elements in laboratory stocks of *Drosophila melanogaster*. Chromosoma 90:378.

Jurka J, Smith T (1988) A fundamental division in the Alu family of repeated sequences. Proc Natl Acad Sci USA 85:4775.

Kafatos FC et al (1985) Studies on the developmentally regulated expression and amplification of chorion genes. Cold Spring Harbor Symp Quant Biol 50:537.

Kafayan L et al (1985) Localization of sequences resulting in Drosophila chorion gene amplification and expression. Cold Spring Harbor Symp Quant Biol 50:527

Kafer E (1977) Meiotic and mitotic recombination in *Aspergillus* and its chromosomal aberrations. Adv Genet 19:33.

Kallman KD (1962) Population genetics of the gynogenetic teleost, *Mollenisia formosa* (Girard). Evolution 16:497.

Kanalis JJ, Suttle C (1984) Amplification of the UMP synthetase gene and overproduction in pyrazofurin-resistant rat hepatoma cells. J Biol Chem 259:1848.

Karess RE, Rubin GM (1984) Analysis of Transposable element functions in *Drosophila*. Cell 38:135.

Kaufman BP (1946) Organization of the chromosome. I. Break distribution and chromosome recombination in *Drosophila melanogaster*. J Exp Zool 102:293.

Kazazian HH Jr et al (1988) Haemophilia A resulting from de novo insertion of L1 sequences represents a novel mechanism for mutation in man. Nature 332:164.

Khandellwal S (1990) Chromosome evolution in the genus *Ophioglossum* L. Bot J Linnean Soc 102:105.

Kimber G, Sears ER (1987) Evolution in the genus *Triticum* and the origin of cultivated wheat. In "Wheat and Wheat Improvement." Agronomy (Monogr No. 13, 2nd ed).

Kimber J, Riley R (1963) Haploid angiosperms. Bot Rev 29:480.

Klar AJS, Fogel S (1977) The action of homothallism genes in *Saccharomyces cerevisiae* diploids during vegetative growth and the equivalence of hma and HMalpha loci functions. Genetics 85:407.

Klar AJS, Fogel S (1979) Activation of mating type genes by transposition in *Saccharomyces cerevisiae*. Proc Natl Acad Sci USA 76:4539.

Klar AJS, Strathern JN (1984) Resolution of recombination intermediates generated during yeast mating type switching. Nature 310:744.

Klein HL, Petes TD (1981) Intrachromosomal gene conversion in yeast. Nature 289:144.

Kohno S et al (1986) Chromosome elimination in the Japanese hagfish, *Eptatretus burgeri* (Agnatha, Cyclostomata). Chromosoma 41:209.

Korenberg JR, Rykowski MC (1988) Human genome organization: Alu, lines, and the molecular structure of metaphase chromosome bands. Cell 53:391.

Koropatnick J et al (1985) Acute treatment of mice with cadmium salts results in amplification of the metallothionein-1 gene in mice. Nucleic Acid Res 13:5423.

Kraemer PM et al (1983) Spontaneous neoplastic evolution of Chinese hamster cells in culture: Multistep progression of phenotype. Cancer Res 43:4822.

Kuhn EM, Therman E (1979) Chromosome breakage and rejoining of sister chromatids in Bloom's syndrome. Chromosoma 73:275.

Kyhos DW (1965) The independent aneuploid origin of two species of *Chaenactis* (Compositae) from a common ancestor. Evolution 19:26.

Laird C et al (1987) Fragile sites in human chromosomes as regions of late-replicating DNA. Trends Genet 3:274.

Lande R (1979) Effective deme size during long term evolution estimated from rates of chromosomal rearrangement. Evolution 33:234.

Lao W, Chen S (1986) HsaI: A restriction enzyme from a human being. Sci Sin Ser B 29:947.

Lawrence WJC, Scott-Moncrieff R (1935) The genetics and chemistry of flower colour in *Dahlia*: A new theory of specific pigmentation. J Genet 30:155.

Lee CS (1975) A possible role of repetitious DNA in recombinatory joining during chromosome rearrangement in *Drosophila melanogaster*. Genetics 79:467.

Lee W-H (1984) Expression and amplification of the N-myc gene in primary retinoblastoma. Nature 309:458.

Leipoldt M, Schmidtke J (1982) Gene expression in phylogenetically polyploid organisms. In Dover GA, Flavell RB (eds): "Genome Evolution." London: Academic Press.

Lewis S et al (1985) DNA elements are asymmetrically joined during the site-specific recombination of kappa immunoglobulin genes. Science 228:677.

Lewis WH (1980) "Polyploidy: Biological Significance." New York: Plenum.

Li WY et al (1982) Nucleotide sequence of 7S RNA. J Biol Chem 257:5136.

Liming S et al (1980) Comparative genetic studies on the red muntjac, Chinese muntjac, and their F1 hybrids. Cytogenet Cell Genet 26:22.

Lindgren V (1981) The location of chromosome breaks in Bloom's syndrome. Cytogenet Cell Genet 29:99.

Lindsley DL et al (1972) Segmental aneuploidy and the genetic gross structure of the *Drosophila* genome. Genetics 71:157.

Liu C-P et al (1980) Mapping of heavy chain genes for mouse immunoglobulins M and D. Science 209:1348.

Looney JE, Hamlin JL (1987) Isolation of the amplified dihydrofolate reductase domain from methotrexate-resistant Chinese hamster ovary cells. Mol Cell Biol 7:569.

Love A, Kapoor BM (1966) An allopolyploid *Ophioglossum*. Nucleus 9:132.

Love A, Kapoor BM (1967) The highest plant chromosome number in Europe. Svensk Bot Tidskrift 61:29.

Lowe CH et al (1970) Chromosomes and evolution of the species group of Cnemidophorus. Syst Zool 19:28.

Lucchesi JC (1983) The relationship between gene dosage, gene expression and sex in *Drosophila*. Dev Genet 3:275.

Lucchesi JC, Rawls JM (1973) Regulation of gene function: A comparison of enzyme activity levels in relation to gene dosage in diploids and triploids of *Drosophila melanogaster*. Biochem Genet 9:41.

Luykx P, Syren RM (1979) The cytogenetics of *Incisitermes schwarzi* and other Florida termites. Sociobiology 4:191.

Lyon MF (1961) Gene action in the X-chromosome of the mouse (*Mus musculus*). Nature 190:372.

Lyon MF (1986) X chromosomes and dosage compensation. Nature 320:313.

Lyon MF (1988) The William Allan Memorial Award Address: X-chromosome inactivation and the location and expression of X-linked genes. Am J Hum Genet 42:8.

MacGregor HC (1982) Big chromosomes and speciation among amphibia. In Dover GA, Flavell RB (eds): "Genome Evolution." New York: Academic Press.

MacGregor HC, Uzzell TM (1964) Gynogenesis in salamanders related to Ambystoma jeffersonianum. Science 143:1043.

MacKay TFC (1985) Transposable element-induced response to artificial selection in Drosophila melanogaster. Genetics 111:351.

Mangelsdorf PC, Fraps GS (1931) A direct quantitative relationship between vitamin A in corn and the number of genes for yellow pigmentation. Science 73:241.

Martin R et al (1987) Variation in the frequency and type of sperm chromosome abnormalities among normal men. Hum Genet 77:108.

Martinez-Cruzado JC et al (1988) Evolution of the chorion locus n Drosophila I. Genetics 119:663.

Matthey R (1970) Nouvelles donne sur cytogenetique et la speciation des Leggada (Mammakia-Rodentia-Muridae). Experientia 26:102.

Mattei MG et al (1981) Structural anomalies of the X chromosome and inactivation center. Hum Genet 56:401.

Mattei MG et al (1982) X-autosome translocations: Cytogenetic characteristics and their consequences. Hum Genet 61:295.

May CE, Appels A (1987) The molecular genetics of wheat: Toward an understanding of 16 billion base pairs of DNA. In "Wheat and Wheat Improvement." Agronomy (Monogr No. 13, 2nd ed).

McCarrey JR, Thomas K (1987) Human testis-specific PKG gene lacks introns and possesses characteristics of a processed gene. Nature 326:501.

McClaren A, Monk A (1982) Fertile females produced by inactivation of an X chromosome of "sex-reversed" mice. Nature 300:445.

McClintock B (1951) Chromosome organization and genic expression. Cold Spring Harbor Symp Quant Biol 16:13.

McClintock B (1956) Mutations in maize. Carnegie Inst Washington Yearb 55:323.

McClintock B (1984) The significance of responses of the genome to challenge. Science 226:792.

McClure MA et al (1988) Sequence comparison of retroviral proteins: Relative rates of change and general phylogeny. Proc Natl Acad Sci USA 85:2469.

McIntosh RA, Cusick JE (1987) Linkage map of hexaploid wheat. In "Wheat and Wheat Improvement." Agronomy (Monogr No. 13, 2nd ed).

Menzel MY et al (1986) Characteristics of duplications and deficiencies from chromosome translocations in Gossypium hirsutum. J Hered 77:189.

Merriam RW, Ris H (1954) Size and DNA content of nuclei in various tissues of male, female and worker honeybees. Chromosoma 6:522.

Metz CW (1938) Chromosome behavior, inheritance and sex determination in Sciara. Am Nat 72:485.

Meyer BJ (1988) Primary events in C. elegans sex determination and dosage compensation. Trends Genet 4:337.

Meyer BJ, Casson LP (1986) Caenorhabditis elegans compensates for the difference in X chromosome dosage between the sexes by regulating transcript levels. Cell 47:871.

Migeon BR et al (1986) Complete reactivation of X-chromosomes from human chorionic villi with a switch to early DNA replication. Proc Natl Acad Sci USA 83:2182.

Miller DW, Miller LK (1982) A virus mutant with an insertion of a copia-like transposable element. Nature 299:562.

Milne DL, McIntosh RA (1990) Triticum aestivum (common wheat). In O'Brien SJ (ed): "Genetic Maps," 5th ed. Cold Spring Harbor, NY: Cold Spring Harbor Laboratory Press.

Mizuno S et al (1976) Interspecific "common" repetitive DNA sequences in salamanders of the genus Plethodon. Chromosoma 58:1.

Mohandas T et al (1982) Genetic evidence for the inactivation of a human autosomal locus attached to an inactive X chromosome. Am J Hum Genet 34:811.

Moorehead PS (1976) A closer look at chromosomal inversions. Am J Hum Genet 28:294.

Morton NE et al (1982) Cytogenetic surveillance of spontaneous abortions. Cytogenet Cell Genet 32:232.

Mount SM, Rubin GM (1985) Complete nucleotide sequence of the *Drosophila* transposable elements copia: Homology between copia and retroviral proteins. Mol Cell Biol 5:1630.

Moyzis RK et al (1989) The distribution of interspersed repetitive DNA sequences in the human genome. Genomics 4:273.

Muldal S (1952) The chromosomes of the earthworms. I. The evolution of polyploidy. Heredity 6:55.

Muller F et al (1982) Nucleotide sequence of satellite DNA contained in the eliminated genome of *Ascaris lumbricoides*. Nucleic Acids Res 10:7493.

Muller HJ (1925) Why is polyploidy rarer in animals than in plants? Am Nat 59:346.

Muller HJ (1932) Further studies on the nature and causes of gene mutations. Proc Sixth Int Congr Genet 1:231.

Muller HJ (1950) Evidence for the precision of genetic adaptation. Harvey Lect 43:165.

Muntzing A (1949) Accessory chromosomes in *Secale* and *Poa*. Hereditas Suppl 35:402.

Nakagome Y et al (1983) Distribution of break points in human structural rearrangements. Am J Hum Genet 35:288.

Nakai Y, Kohno S (1987) Elimination of the largest chromosome pair during differentiation into somatic cells in the hagfish, *Myxine garmani* (Cyclostomata, Agnatha). Cytogenet Cell Genet 45:80.

Nasmyth KA (1982). The regulation of yeast mating type chromatin structure by SIR: An action at a distance affecting both transcription and transposition. Cell 30:567.

Nasmyth K, Shore D (1987) Transcriptional regulation in the yeast life cycle. Science 237:1162.

Nasmyth KA, Tatchell K (1980) The structure of the transposable yeast mating type locus. Cell 19:753.

Natarajan AT, Obe G (1984) Molecular mechanisms involved in the production of chromosomal aberrations. III. Restriction endonucleases. Chromosoma 90:120.

Nedospasov SA et al (1986) Tandem arrangement of genes coding for tumor necrosis factor (TNF-alpha) and lymphotoxin (TNF-beta)

in the human genome. Cold Spring Harbor Symp Quant Biol 51:611.

Nelson OE, Klein AS (1984) Characterization of an SPM-controlled bronze-mutable allele in maize. Genetics 106:769.

Nevers P, Saedler H (1977) Transposable genetic elements as agents of gene instability and chromosomal rearrangements. Nature 268:109.

Nevers P et al (1985) Plant transposable elements. Adv Bot Res 12:102.

O'Brien SJ, Seuanez HN (1988) Mammalian genome organization: An evolutionary view. Annu Rev Genet 22:323.

O'Hare K (1987) Chromosome plasticity and transposable elements in *Drosophila*. Trends Genet 3:87.

O'Hare K, Rubin GM (1983) Structure of P transposable elements and their sites of insertion and excision in the *Drosophila melanogaster* genome. Cell 34:25.

Ohno S (1970) "Evolution by Gene Duplication." New York: Springer-Verlag.

Orgel LE, Crick F (1980). Selfish DNA. Nature 288:645.

Orr NH et al (1986) The genome of frog erythrocytes displays centuplicate replications. Proc Natl Acad Sci USA 83:1369.

Orr-Weaver TL, Spradling AC (1986) *Drosophila* chorion gene amplification requires an upstream region regulating S18 transcription. Mol Cell Biol 6:4624.

Page DC et al (1987) The sex determining region of the human Y chromosome encodes a finger protein. Cell 51:1091.

Pan W-C, Blackburn EH (1981) Single extrachromosomal ribosomal RNA gene copies are synthesized during amplification of the rDNA in Tetrahymena. Cell 23:459.

Patterson JT, Stone WS (1952) "Evolution in the genus Drosophila." New York: Macmillan.

Paulson E et al (1985) Transposon-like element in human DNA. Nature 316:359.

Perkins DD, Barry EG (1977). The cytogenetics of *Neurospora*. Adv Genet 19:134.

Piechaczyk M et al (1984) Unusual abundance of vertebrate 3-phosphate dehydrogenase pseudogenes. Nature 312:469.

Pimpinelli S, Goday C (1989) Unusual kinetochores and chromatin diminution in Parascaris. Trends Genet 5:310.

Pohlman R et al (1984) The nucleotide sequence of the maize controlling element activator. Cell 37:635.

Prasad J et al (1981) Transcription in X-chromosomal segmental aneuploids of Drosophila melanogaster and regulation of dosage compensation. Genet Res 38:103.

Rasch EM et al (1970) Cytogenetic studies of Poecili (Pisces). II. Triploidy and DNA levels in naturally occurring populations associated with the gynogenetic teleost Poecilia formosa (Girard) Chromosoma 31:18.

Renner O (1949) Die 15-chromosomigen Mutanten der Oenothera lamarckiana und ihre Verwandten. Z Indukt Abstamm Vererbungs 83:1.

Rhoades MM (1951) Duplicate genes in maize. Am Nat 85:105.

Rickards GK (1983) Orientation behavior of chromosome multiples of interchange (reciprocal translocation) heterozygotes. Am Rev Genet 17:443.

Rieger R (1966) Alternative interpetation Versuche zur Entstehung chromosomaler Structurumbauten: Bruch-Reunions und Austach-Hypothese. Biol Zentralbl 85:29.

Rio DC (1990) Molecular mechanisms regulating Drosophila P element transposition. Annu Rev Genet 24:543.

Risler H, Kempter E (1962) Die Haploidie der Männchen und die Honigbiene (Apis mellifera) und die Wiederherstellung der Diploidie der Drohnen. Z Zellforsch 12:351.

Roberts DB, Evans-Roberts S (1979) The X-linked alpha-chain gene of Drosophila LSP-1 does not show dosage compensation. Nature 280:691.

Robertson HM et al (1988) A stable genomic source of P element transposase in Drosophila melanogaster. Genetics 118:461.

Robertson W (1916) Chromosome studies. I. Taxonomic relationships shown in the chromosomes of Tettigidae and Acrididae. V-shaped chromosomes and their significance in Acrididae, Locutididae and Gryllidae: Chromosomes and variation. J Morphol 27:179.

Roeder S, Fink GR (1983) In Shapiro JA (ed): "Mobile Genetic Elements." New York: Academic Press.

Roehrdanz RL et al (1977) Lack of dosage compensation for an autosomal gene relocated to the X chromosome in Drosophila melanogaster. Genetics 85:489.

Rogers J (1985) The origin and evolution of retroposons. Int Rev Cytol 93:187.

Rogers J (1986) The origin of retroposons. Nature 319:725.

Roiha H et al (1988) P element insertions and rearrangements at the singed locus of Drosophila melanogaster. Genetics 119:75.

Rose M, Winston F (1984) Identification of a Ty insertion within the coding sequence of the S. cerevisiae URA3 gene. Mol Gen Genet 193:557.

Rosenfeld RG et al (1979) Sexual and somatic determination of the human Y chromosome: Studies in a 46,XYp phenotypic female. Am J Hum Genet 31:458.

Rous P (1911) A sarcoma of the fowl transmissable by an agent separable from tumor cells. J Exp Med 13:397.

Rudkin GT, Tartof KD (1973) Repetitive DNA in polytene chromosomes of Drosophila melanogaster. Cold Spring Harbor Symp Quant Biol 38:397.

Sager R, Kitchin R (1975) Selective silencing of eukaryotic DNA. Science 189:426.

Sakai RK, Mahmaod F (1985) Homozygous chromosomal aberrations in Anopheles stephani. J Hered 76:231.

Sakaki Y et al (1986) The LINE 1 family of primates may encode a reverse transcriptase-like protein. Cold Spring Harbor Symp Quant Biol 51:465.

Sandmeyer SB et al (1990) Integration specificity of retrotransposons and retroviruses. Annu Rev Genet 24:291.

Sawada I et al (1985) Evolution of Alu family repeats since the divergence of human and chimpanzee. J Mol Evol 22:316.

Schimke RT (1984) Gene amplification in cultured animal cells. Cell 37:705.

Schimke RT (1988) Gene amplification in cultured cells. J Biol Chem 263:5989.

Schimke RT et al (1986) Overreplication and recombination of DNA in higher eukaryotes: Potential consequences and biological implications. Proc Natl Acad Sci USA 83:2157.

Schimke RT et al (1988) Enhancement of gene amplification by perturbation of DNA synthesis in cultured mammalian cells. In "Eukaryotic DNA Replication," vol 6: "Cancer

Cells." Cold Spring Harbor, NY: Cold Spring Harbor Laboratory.

Schmid M et al (1985) Chromosome banding in Amphibia IX. The polyploid karyotypes of Odontophyrnus americanus and Ceratophrys ornata (Anura, Leptodactylidae). Chromosoma 91:172.

Schneuwly S et al (1987a) Molecular analysis of the dominant homeotic antennapedia phenotype. EMBO J 6:201.

Schneuwly S et al (1987b) Redesigning the body plan of *Drosophila* by ectopic expression of the homoeotic gene antennapedia. Nature 325:816.

Schroecksnadel H et al (1982) Komplette Triploidie (69XXX) mit einer Ubenlebensdauer von 7 Monaten. Wochenschr 94:309.

Schroeder TM, German J (1974) Bloom's syndrome and Fanconi's anemia: Demonstration of two distinctive patterns of chromosomal disruption and rearrangement. Humangenetik 25:299.

Schultz RJ (1980) Role of polyploidy in the evolution of fishes. In Lewis WH (ed): "Polyploidy." New York: Plenum.

Schwab M et al (1983) A cellular oncogene (C-KI-ras) is amplified, overexpressed, and located within karyotypic abnormalities in mouse adrencortical tumour cells. Nature 303:497.

Schwartz D (1960) Electrophoretic and immunochemical studies with endosperm proteins of maize mutants. Genetics 45:1419.

Schwartz D, Dennis ES (1986) Transposase activity of the AC controlling element in maize is regulated by its degree of methylation. Mol Gen Genet 205:476.

Scott GK et al (1985) Antifreeze protein genes are tandemly linked and clustered in the genome of the winter flounder. Proc Natl Acad Sci USA 82:2013.

Scott GK et al (1988) Differential amplification of antifreeze protein genes in the Pleuronectinae. J Mol Evol 27:29.

Sears ER (1952) Homeologous chromosomes in *Triticum aestivum*. Genetics 37:624.

Sears ER (1976) Genetic control of chromosome pairing in wheat. Annu Rev Genet 10:31.

Selsing E, Storb U (1981) Somatic mutation of immunoglobulin light-chain variable-region genes. Cell 25:47.

Sessions SK (1982) Cytogenetics of diploid and triploid salamanders of the *Ambystoma jeffersonianum* complex. Chromosoma 84:599.

Shale DM (1986) Engineering herbicide tolerance in transgenic plants. Science 233:478.

Shull GH (1928) *Oenothera* cytology in relation to genetics. Am Nat 62:97.

Skowronski J, Singer MF (1986) The abundant LINE-1 family of repeated DNA sequences in mammals: Genes and pseudogenes. Cold Spring Harbor Symp Quant Biol 51:457.

Snyder M, Doolittle WF (1988) P elements in *Drosophila*: Selection at many levels. Trends Genet 4:147.

Spradling AC, Mahowald AP (1980) Amplification of genes for chorion proteins during oogenesis in *Drosophila melanogaster*. Proc Natl Acad Sci USA 77:1096.

Soares MB et al (1985) RNA-mediated gene duplication: The rat preproinsulin I gene is a functional retroposon. Mol Cell Biol 5:2090.

Stanley HP et al (1984) Meiotic chromatin diminution in a vertebrate, the Holocephalan fish *Hydrolagus colliei* (Chondrichthyes, Holocephali). Tissue Cell 16:203.

Stanley SM (1975) Clades versus clones in evolution: Why we have sex. Science 190:382.

Stark GR, Wahl GM (1984) Gene amplification. Annu Rev Biochem 53:447.

Stebbins GL (1950) "Variation and Evolution in Plants." New York: Columbia University Press.

Stein JP et al (1983) Tissue-specific expression of a chicken calmodulin pseudogene lacking intervening sequences. Proc Natl Acad Sci USA 80:6485.

Sunchen G et al (1984) Ty-mediated gene expression of the LYS2 and H1S4 genes of *Saccharomyces cerevisiae* is controlled by the same SPT genes. Proc Natl Acad Sci USA 81:2431.

Sutherland GR (1982) Heritable fragile sites on human chromosomes. VIII. Preliminary population cytogenetic data on the folic-acid-sensitive fragile sites. Am J Hum Genet 34:452.

Sutherland GR, Mattei JF (1987) Report of the Committee on Cytogenetic Markers. (In Human Gene Mapping 9.) Cytogenet Cell Genet 46(Nos. 1–4).

Temin HM (1989) Retrovirus variation and evolution. Genome 31:17.

Temin HM, Mizutani S (1970) Viral-dependent DNA polymerase. Nature 226:12111.

Therman E, Sarto GE (1983) Inactivation center on the human X-chromosome. In Sandberg AA (ed): "Cytogenetics of the Mammalian X Chromosome," Part A. New York: Alan R. Liss.

Thomas HM, Pickering RA (1983) Chromosome elimination in *Hordeum vulgare* × H. *bulbosum* hybrids. 1. Comparison of stable and unstable amphidiploids. Theor Appl Genet 66:135.

Tlsty TD et al (1984) UV radiation facilitates methotrexate resistance and amplification of the dihydroxyfolate reductase gene in cultured 3T6 mouse cells. Mol Cell Biol 4:1050.

Tobler H (1986) The differentiation of germ and somatic cell lines in nematodes. In Hennig W (ed): "Results and Problems in Cell Differentiation," vol 13: "Germ Line–Soma Differentiation." Berlin: Springer-Verlag.

Tonegawa S (1983) Somatic generation of antibody diversity. Nature 302:575.

Triantaphyllou AC (1984) Chromosomes in evolution of nematodes. In Sharma AK, Sharma A (eds): "Chromosomes in Evolution of Eukaryotic Groups," vol 2. Boca Raton, FL: CRC Press.

Ullu E, Tschudi C (1984) Alu sequences are processed 7SL RNA genes. Nature 312:171.

Uzzell TM, Goldblatt SM (1967) Serum proteins of salamanders of the *Ambystoma jeffersonianum* complex, and the origin of the triploid species of this group. Evolution 21:345.

Varmus H (1988) Retroviruses. Science 240:1427.

Varshavsky A (1981) Phorbol ester dramatically increases incidence of methotrexate-resistant mouse cells: Possible mechanisms and relevance to tumor promotion. Cell 25:561.

Vijayalaxmi et al (1983) Bloom's syndrome: New evidence for an increased mutation frequency in vivo. Science 221:851.

Vine DT et al (1976) Inversion homozygosity of chromosome No. 9 in a highly inbred kindred. Am J Hum Genet 28:203.

Virkki N (1984) Chromosomes in evolution of Coleoptera. In Sharma AK, Sharma A (eds): "Chromosomes in Evolution of Eukaryotic Groups," vol 2. Boca Raton, FL: CRC Press.

Voiculescu I et al (1986) Familial pericentric inversion of chromosome 12. Hum Genet 72:320.

Vries H De (1901) "Die Mutationstheorie," vol 1. Leipzig: Viet.

Vries H De (1903) "Die Mutationstheorie," vol 2. Leipzig: Viet.

Vries De H (1909) On triple hybrids. Bot Gaz 47:1.

Wagner M (1986) A consideration of the origin of processed pseudogenes. Trends Genet 2:134.

Wagner WH, Wagner FS (1980) Polyploidy in pteridophytes. In Lewis WH (ed): "Polyploidy." New York: Plenum.

Wahl GM (1984) Effect of chromosomal position on amplification of transfected genes in animal cells. Nature 307:516.

Walker TG (1984) Chromosomes and evolution in pteridophytes. In Sharma AK, Sharma A (eds): "Chromosomes in Evolution of Eukaryotic Groups." Boca Raton, FL: CRC Press.

Walter P, Blobel G (1983) Disassembly and reconstitution of signal recognition particles. Cell 34:525.

Wayne RK et al (1987a) Chromosomal evolution of the Canidae: I. Species with high diploid numbers. Cytogenet Cell Genet 44:123.

Wayne RK et al (1987b) Chromosome evolution of the Canidae: II. Divergence from the primitive carnivore karyotype. Cytogenet Cell Genet 44:134.

Weiner AM et al (1986) Non viral retroposons: Genes, pseudogenes, and transposable elements generated by the reverse flow of genetic information. Annu Rev Biochem 55:631.

Weismann A (1892) "Das Keimplasma. Eine Theorie der Vererbung. Jena: Gustav Fischer.

Wendel JF et al (1986) Duplicated chromosome segments in *Zia mays*. I. Further evidence from hexokinase enzymes. Theor Appl Genet 72:178.

Wessler SR (1988) Phenotypic diversity mediated by the maize transposable elements AC and SPM. Science 242:399.

Wett MJD de (1980). Origin of polyploids. In Lewis WH (ed): "Polyploidy." New York: Plenum.

White MJD (1946) The cytology of the *Cecidomyidae* (Diptera) II. The chromosome cycle and anomalous spermatogenesis of *Miastor*. J Morph 79:323.

Wharton LT (1943) Analysis of the metaphase chromosome morphology within the genus *Drosophila*. In "Studies in the genetics of Drosophila," vol 3: "The *Drosophilidae* of the Southwest." University of Texas publication No. 4313.

White MJD (1957) Cytogenetics of the grasshopper *Morata scurra*: II. Heterotic systems and their interaction, with a statistical appendix by B Griffing. Aust J Zool 5:305.

White MJD (1969) Chromosomal rearrangements and speciation. Annu Rev Genet 3:75.

White MJD (1973) "Animal Cytology and Evolution," 3rd ed. New York: Cambridge University Press.

Willard C et al (1987) Existence of at least three distinct Alu subfamilies. J Mol Evol 26:180.

Williams GC (1966) "Adaptation and Natural Selection." Princeton, NJ: Princeton University Press.

Williams RO et al (1979) Genomic rearrangements correlated with antigenic variation in *Trypanosoma brucei*. Nature 282:847.

Wilson EB (1925) "The Cell in Development and Heredity." New York: Macmillan.

Wintersberger U, Wintersberger E (1987) RNA makes DNA: A speculative view of the evolution of DNA replication mechanisms. Trends Genet 3:198.

Wramsby HJ et al (1987) Chromosomal analysis of human oocytes recovered from pre-ovulation follicles in stimulated cycles. N Engl J Med 316:121.

Wright S (1977) "Evolution and the Genetics of Populations," vol 3: "Experimental Results and Evolutionary Deductions." Chicago: University of Chicago Press.

Wurster-Hill DH, Centerwall WR (1982) The interrelationships of chromosome banding patterns in Canids, Mustelids, Hyena and Felids. Cytogenet Cell Genet 34:178.

Wurster-Hill DH, Gray CW (1973) Giemsa banding patterns in the chromosomes of twelve species of cats (Felidae). Cytogenet Cell Genet 12:377.

Wurster-Hill DH, Gray CW (1975) The interrelationships of banding patterns of Procyonids, Viverrids and Felids. Cytogenet Cell Genet 15:306.

Yosida TH, Amano K (1965) Autosomal polymorphisms in laboratory bred and wild Norway rats, *Rattus norvegicus* found in Misima. Chromosoma 16:658.

Young AP, Ringold GM (1983) Mouse 3T6 cells that overproduce glutamine synthetase. J Biol Chem 258:11260.

Young MW, Schwartz HE (1981) Nomadic gene families in Drosophila. Cold Spring Harbor Symp Quant Biol 45:629.

Young T et al (1982) In Hollaender A et al (eds): "Genetic Engineering of Microorganisms for Chemicals." New York: Plenum.

Yunis JJ, Prakash O (1982) The origin of man: A pictorial legacy. Science 215:1525.

Zachar Z et al (1985) A detailed developmental and structural study of the transcriptional effects of insertion of the copia transposon into the white locus of *Drosophila melanogaster*. Genetics 111:495.

CHAPTER 6

Adelman JP et al (1987) Two mammalian genes transcribed from opposite strands of the same DNA locus. Science 235:1514.

Allison LA et al (1985) Extensive homology among the largest subunits of eukaryotic and prokaryotic polymerases. Cell 42:599.

Belling J (1933) Crossing-over and gene rearrangement in flowering plants. Genetics 18:388.

Benzer S (1961) On the topography of the genetic fine structure. Proc Natl Acad Sci USA 47:403.

Biggs J et al (1985) Structure of the eukaryotic transcription apparatus: Features of the gene for the largest subunit of Drosophila RNA polymerase II. Cell 42:611.

Birnstiel ML et al (1985) Transcription termination and 3' processing: The end is in site! Cell 41:349.

Bogenhagen DF, Brown DD (1981) Nucleotide sequences in Xenopus 5SDNA required for transcription termination. Cell 24:261.

Brenda P, Davidson R (1971) J Cell Physiol 78:209.

Brill SJ, Sternglanz R (1988) Transcription-dependent DNA supercoiling in yeast DNA topoisomerase mutants. Cell 54:403.

Busch SJ, Sassone-Corsi P (1990) Dimers, leucine zippers and DNA-binding domains. Trends Genet 6:36.

Cabrera CV et al (1987) Phenocopies induced with antisense RNA identify the wingless gene. Cell 50:659.

Cedar H (1984) DNA methylation and gene expression. In Razin A, Cedar J, Riggs AD (eds): "DNA Methylation." New York: Springer-Verlag.

Cedar H (1988) DNA methylation and gene activity. Cell 53:3.

Cereghini S et al (1983) Structure and function of the promoter-enhancer region of polyomer and SV40. Cold Spring Harbor Symp Quant Biol 47:935.

Chada K et al (1985) Specific expression of a foreign beta-globin gene in erythroid cells of transgenic mice. Nature 314:377.

Chada K et al (1986) An embryonic pattern of expression of a fetal globin gene in transgenic mice. Nature 319:685.

Chapman VM (1986) X-chromosome regulation in oncogenes and early mammalian development. In Rossan J, Pedersen RA (eds): "Experimental Approaches to Mammalian Embryonic Development." Cambridge: Cambridge University Press.

Citron BA, Donelson JE (1984) Sequence of the Saccharomyces GAL region and its transcription in vivo. J Bacteriol 158:269.

Clegg JB (1987) Can the product of the omega gene be a real globin? Nature 329:465.

Collins FS et al (1984) G-gamma beta-hereditary persistence of fetal hemoglobin: Cosmid cloning and identification of specific mutation 5' to the G-gamma gene. Proc Natl Acad Sci USA 81:4894.

Conklin KF, Groudine M (1984) Chromatin structure and gene expression. In Razin A et al (eds): "DNA Methylation." New York: Springer-Verlag.

Constantini F et al (1985) Developmental regulation of human globin genes in transgenic mice. Cold Spring Harbor Symp Quant Biol 50:361.

Constantini F et al (1986) Correction of murine beta-thalessemia by gene transfer into the germ line. Science 233:1192.

Cooney CA, Bradbury EM (1990) DNA methylation and chromosome organization in eukaryotes. In Strauss RL, Wilson NG (eds): "The Eukaryotic Nucleus: Molecular Structure and Macromolecular Assemblies." Caldwell, New Jersey: Telford Press.

Cozzarelli NR et al (1983) Purified RNA polymerase III accurately and efficiently terminates transcription of 5SRNA genes. Cell 34:829.

Darlington GJ et al (1982) Expression of human hepatic genes in somatic cell hybrids. Somatic Cell Genet 8:403.

Davidson EJ, Britten RJ (1979) Regulation of gene expression: Possible role of repetitive sequences. Science 204:1052.

Davidson R et al (1966) Regulation of pigment synthesis in mammalian cells as studied by somatic cell hybridization. Proc Natl Acad Sci USA 56:1437.

DePamphilis ML (1988) Transcriptional elements as components of eukaryotic origins of DNA replication. Cell 52:635.

Donahue T et al (1983) A short nucleotide sequence required for regulation of HIS4 by the general control system of yeast. Cell 32:89.

Dynan WS (1989) Understanding the molecular mechanisms by which methylation influences gene expression. Trends Genet 5:35.

Dynan WS, Tjian R (1985) Control of eukaryotic messenger RNA synthesis by sequence specific DNA binding proteins. Nature 316:774.

Fink GR (1986) Translational control of transcription in eukaryotes. Cell 45:155.

Frels WI, Chapman VM (1980) Expression of the maternally derived X chromosome in the mural trophoblast of the mouse. J Embryol Exp Morph 56:179.

Futcher B (1988) Supercoiling and transcription, or vice versa? Trends Genet 4:271.

Gelinas R et al (1985) G to A substitution in the distal CCAAT box of the Agamma-globin gene in Greek hereditary persistence of fetal hemoglobin. Nature 313:323.

Giglioni B et al (1984) A molecular study of a family with Greek hereditary persistence of fetal globin and beta-thalassemia. EMBO J 3:2641.

Gilmour D et al (1986) Topoisomerase I interacts with transcribed regions in Drosophila cells. Cell 44:401.

Graever GN, Wang JC (1988) Supercoiling of intracellular DNA can occur in eukaryotic cells. Cell 55:849.

Graftstrom RH et al (1984) DNA methylation: DNA replication and repair. In Razin A, Cedar H, Riggs AD (eds): "DNA Methylation." New York: Springer-Verlag.

Grant SG, Chapman VM (1988) Mechanisms of X chromosome regulation. Annu Rev Genet 22:199.

Green PJ et al (1986) The role of antisense RNA in gene regulation. Annu Rev Biochem 55:569.

Grondine M et al (1983) Human fetal to adult switching: changes in chromatin structure of the beta-globin gene locus. Proc Natl Acad Sci USA 80:7551.

Gross D, Garrard WT (1988) Nuclease hypersensitive sites in chromatin. Annu Rev Biochem 57:159.

Grummt I et al (1985) Transcription of mouse rDNA terminates downstream of the 3′ end of 28S RNA and involves interaction of factors with repeated sequences in the 3′ spacer. Cell 43:801.

Grummt I et al (1986) A repeated 18 pair sequence motif in the mouse rDNA spacer mediates binding of a nuclear factor and transcription termination. Cell 45:837.

Guarante L (1987) Regulatory proteins in yeast. Annu Rev Genet 21:425.

Henikoff S et al (1986) Gene within a gene: Nested Drosophila genes encode unrelated proteins on opposite strands. Cell 44:33.

Hope IA, Struhl K (1985) GCN protein, synthesized in vitro, binds HIS3 regulatory sequences: Implications for general control of amino acid biosynthetic genes in yeast. Cell 43:177.

Hosbach HA et al (1983) The Xenopus laevis globin gene family: Chromosomal arrangement and gene structure. Cell 32:45.

Hsu S-L et al (1988) Structure and expression of the human theta globin gene. Nature 331:94.

Ingram VM (1959) The genetic control of protein structure. In Sutton HE (ed): "Genetics. Genetic Information and the Control of Protein Structure and Function. Transactions of the First Conference." Josiah Macy Jr Foundation.

Jalouzot B et al (1985) Replication timing of the H4 histone genes in Physarum polycephalum. Proc Natl Acad Sci USA 82:6475.

Johannsen W (1909) Elementen der exakten Erblichkeitslehre. Jena: Gustav Fischer.

Johnson PF, McKnight AL (1989) Eukaryotic transcriptional regulatory elements. Annu Rev Biochem 58:799.

Karlsson S, Nienhuis AW (1985) Developmental regulation of human globin genes. Annu Rev Biochem 54:1071.

Lyon MF (1986) X chromosomes and dosage compensation. Nature 320:313.

Lyon MF (1988) The William Allan Memorial Award Address: X-chromosome inactivation and the location and expression of X-linked genes. Am J Hum Genet 42:8.

Magram J et al (1985) Developmental regulation of a cloned adult beta-globin gene in transgenic mice. Nature 315:338.

Maniatis T et al (1987) Regulation of inducible and tissue-specific gene expression. Science 236:1237.

Maxson R et al (1983) Distinct organizations and patterns of expression of early and late histone gene sets in the sea urchin. Nature 301:120.

McKnight SL (1983) Constitutive transcriptional control signals of the herpes simplex virus tk gene. Cold Spring Harbor Symp Quant Biol 47:945.

Melton DA (1985) Injected anti-sense RNAs specifically block messenger RNA translation in vivo. Proc Natl Acad Sci USA 82:144.

Mitchell PJ, Tjian R (1989) Transcriptional regulation in mammalian cells by sequence-specific DNA binding proteins. Science 245:371.

Moffat AS (1991) Making sense of antisense. Science 253:510.

Mohandas T et al (1981) Reactivation of an inactive human X chromosome: Evidence for X-inactivation by DNA methylation. Science 211:393.

Mueller PP, Hinnebusch AG (1986) Multiple upstream AUG codons mediate translational control of GCN4. Cell 45:209.

Murphy S et al (1987) The in vitro transcription of the 7SK RNA gene by RNA polymerase III is dependent only on the presence of an upstream promoter. Cell 51:81.

Myers RM et al (1986) Fine structure genetic analysis of a beta-globin promoter. Science 232:613.

Old RW, Woodland HR (1984) Histone genes: Not so simple after all. Cell 38:624.

Orkin SH, Nathan DG (1981) The molecular genetics of thalassemia. Adv Hum Genet 11:233.

Orkin SH et al (1985) Thalessemia due to a mutation in the cleavage-polyadenylation signal in the human beta-globin gene. EMBO J 4:583.

Otto SP, Walbot V (1990) DNA methylation in eukaryotes: Kinetics of demethylation and de nova methylation during the life cycle. Genetics 124:429.

Parker CS, Topol J (1984) A Drosophila RNA polymerase II transcription factor contains a promoter-region specific DNA binding activity. Cell 36:357.

Platt T (1986) Transcription termination and the regulation of gene expression. Annu Rev Biochem 55:339.

Posakony J et al (1985) Identification of DNA sequences required for the regulation of Drosophila alcohol dehydrogenase. Cold Spring Harbor Symp Quant Biol 50:515.

Ptashne MA (1986a) "A Genetic Switch, Gene Control and Phage Lambda." Palo Alto, CA: Blackwell Scientific and Cell Press.

Ptashne M (1986b) Gene regulation by proteins acting nearby and at a distance. Nature 322:697.

Rae PMM, Steele RE (1979) Absence of cytosine methylation at CCGG and GCGC sites in the gamma DNA coding regions and intervening sequences of Drosophila and the gamma DNA of other higher insects. Nucleic Acids Res 6:2987.

Razin A, Riggs AD (1980) DNA methylation and gene function. Science 210:604.

Razin A et al (1986) Replacement of 5-methylcytosine by cytosine: A possible mechanism for transient DNA demethylation during differentiation. Proc Natl Acad Sci USA 83:2827.

Rich A (1984) Left-handed Z-DNA and methylation of D(CpG) sequences. In Razin A et al (eds): "DNA Methylation." New York: Springer-Verlag.

Saluz HP et al (1991) Studying DNA modifications and DNA-protein interactions in vivo. Trends Genet 7:207.

Sanford T et al (1983) RNA polymerase II from wild type and alpha-amanitin-resistant strains of Caenorhabditis elegans. J Biol Chem 258:12804.

Seguin C et al (1984) Competition for cellular factors that activate metallothionein gene for transcription. Nature 312:781.

Sentenac A, Hall BD (1982) Yeast nuclear RNA polymerases and their role in transcriptions. In Stratherm JN et al (eds): "Molecular Biology of the Yeast Saccharomyces: Metabolism and Gene Expression." Cold Spring Harbor, NY: Cold Spring Habor Laboratory Press.

Serfing E et al (1985) Enhancers and eukaryotic gene transcription. Trends Genet 1:224.

Sharman GB (1971) Late DNA replication in the paternally derived X chromosome of female kangaroos. Nature 230.

Sollner-Webb B, Tower SJ (1986) Transcription of cloned eukaryotic ribosomal RNA genes. Annu Rev Biochem 55:801.

Stein GS et al (1984) "Histone Genes Structure, Organization and Regulation." New York: John Wiley.

St John TP, Davis RW (1981) The organization and transcription of the galactose gene cluster in Saccharomyces. J Mol Biol 152:285.

Struhl K (1989) Molecular mechanisms of transcriptional regulation in yeast. Annu Rev Biochem 58:1051.

Struhl K et al (1985) Constitutive and coordinately regulated transcription of yeast genes: Promoter elements, positive and negative regulatory sites and DNA binding proteins. Cold Spring Harbor Symp Quant Biol 50:489.

Thoma F (1991) Structural changes in nucleosomes during transcription: Strip, split or flip? Trends Genet 7:175.

Tuan D et al (1985) The "B-like-globin" gene domain in human erythroid cells. Proc Natl Acad Sci USA 82:6384.

Urieli-Shoval S et al (1982) The absence of detectable methylated bases in Drosophila melanogaster. FEBS Lett 146:148.

Voelker RA et al (1985) Genetic and molecular variation in the RpII215 region of Drosophila melanogaster. Mol Gen Genet 201:437.

Wasylyk B, Chambon P (1983) Potentiator effect of the SV40 72bp repeat on initiation of transcription from heterologous promoter elements. Cold Spring Harbor Symp Quant Biol 47:921.

Wilson EB (1896) "The Cell in Development and Inheritance." New York: Macmillan.

Zweidler A (1984) Core histone variants of the mouse: Primary structure and differential expression. In Stein GS, Stein JL, Marzluff WF (eds): "Histone Genes." New York: John Wiley.

CHAPTER 7

Avivi L, Feldman M (1980) Arrangement of chromosomes in the interphase nucleus of plants. Hum Genet 55:281.

Avivi L, Feldman M (1982a) An ordered arrangement of chromosomes in the somatic nucleus of common wheat *Triticum aestivum*. I. Spatial relationships between chromosomes of the same genome. Chromosoma 86:1.

Avivi L, Feldman M (1982b) An ordered arrangement of chromosomes in the somatic nucleus of common wheat (*Triticum aestivum*). II. Spatial relationships between chromosomes of different genomes. Chromosoma 86:17.

Baker WK (1968) Position effect variegation. Adv Genet 14:133.

Bennett MD (1982) Nucleotypic basis of the spatial ordering of chromosomes in eukaryotes and the implications of the order for genome evolution and phenotypic evolution. In Dover GA, Flavell RB (eds): "Genome Evolution." London: Academic Press.

Berrios M et al (1985) In situ localization of topoisomerase II, a major polypeptide component of the *Drosophila* nuclear matrix fraction. Proc Natl Acad Sci USA 82:4142.

Biggin MD et al (1988) Zests encodes a sequence-specific transcription factor that activates the Ultrabithorax promoter in vitro. Cell 53:713.

Bjorkroth B et al (1988) Structure of the chromatin axis during transcription. Chromosoma 96:333.

Blobel G (1985) Gene gating: A hypothesis. Proc Natl Acad Sci USA 82:8527.

Boehm T, Rabbitts TH (1989) The human T cell receptor genes are targets for chromosomal abnormalities in T cell tumors. FASEB J 3:2344.

Bourgeois CA et al (1987) Evidence for the existence of nucleolar skeleton attached to the pore complex-lamina in human fibroblasts. Chromosoma 95:315.

Bouvier D et al (1982) RNA is responsible for the three-dimensional organization of nuclear matrix proteins in HeLa cells. Biol Cell 43:143.

Boveri T (1887) Uber Differenzierung der Zellkerne Wahrend der Forschung des Eies von Ascaris Megalocephala. Anat Anz 2:688.

Brosseau GE (1964) Evidence that heterochromatin does not suppress V-type position effect. Genetics 50:237.

Burkholder GD (1988) The analysis of chromosome organization by experimental manipulation. In "Proceedings of the 18th Stadler Genetics Symposium on Chromosomes, Structure and Function: Impact of New Concepts." New York: Plenum.

Calza RE et al (1984) Changes in gene position are accompanied by a change in time of replication. Cell 36:689.

Catcheside DJ (1936) Origin, nature and breeding behavior of *Oenothera lamarckiana* trisomics. J Genet 33:1.

Catcheside DG (1947) The P-locus position effect in *Oenothera*. J Genet 48:31.

Cermeno MC, Lacadeno JR (1983) Spatial arrangement analysis of wheat and rye genomes in Triticale interphase nuclei by gamma-radiation induced chromosomal interchange. Heredity 51:377.

Chada K et al (1985) Specific expression of a foreign beta-globin gene in erythroid cells of transgenic mice. Nature 314:377.

Chada K et al (1986) An embryonic pattern of expression of a fetal globin gene in transgenic mice. Nature 319:685.

Church K, Moens PB (1976) Centromere behavior during interphase and meiotic prophase in *Allium fistulosum* from 3-D EM reconstruction. Chromosoma 56:549.

Cockerill PN, Garrard WT (1986) Chromosomal loop anchorage of the kappa immunoglobulin gene occurs next to the enhancer in a region containing topoisomerase II sites. Cell 44:273.

Cole PA, Sutton E (1941) The absorption of ultraviolet radiation by bands of the salivary gland chromosomes of *Drosophila melanogaster*. Cold Spring Harbor Symp Quant Biol 9:66.

Comings DE (1980) Arrangement of chromatin in the nucleus. Hum Genet 53:131.

Constantini F, Lacy E (1981) Introduction of a rabbit beta-globin gene into the mouse germ line. Nature 294:92.

Cook PR (1988) The nucleoskeleton: Artefact, passive framework or active site? J Cell Science 90:1.

Cram LS et al (1983) Spontaneous neoplastic evolution of Chinese hamster cells in culture: Multistep progression of karyotype. Cancer Res 43:4828.

Cremer T et al (1982) Rabl's model of the interphase chromosome arrangement tested in Chinese hamster cells by premature chromosome condensation and laser-UV-microbeam experiments. Hum Genet 60:46.

Croce CM (1987) Role of chromosome translocations in human neoplasia. Cell 49:155.

Croce CM et al (1985) Coexpression of translocated and normal C-myc oncogenes in hybrids between Daudi and lymphoblastoid cells. Science 227:1235.

Denny CT et al (1986) Common mechanism of chromosome inversion in B- and T-cell tumors: Relevance to lymphoid development. Science 234:197.

Dobrovic A et al (1991) Molecular analysis of the Philadelphia chromosome. Chromosoma 100:479.

Dorn R et al (1986) Suppressor mutation of position effect variegation in Drosophila melanogaster affecting chromatin properties. Chromosoma 93:398.

Earnshaw WC et al (1985) Topoisomerase II is a structure component of mitotic chromosome scaffolds. J Cell Biol 100:1706.

Fearon ER et al (1990) Identification of a chromosome 18q gene that is altered in colorectal cancers. Science 247:49.

Feldherr CM et al (1984) Movement of a karyophilic protein through the nuclear pores of oocytes. J Cell Biol 99:2216.

Feldman M, Avivi L (1973) The pattern of chromosomal arrangement in nuclei of common wheat and its genetic control. In "Proceedings of the Fourth International Wheat Genetics Symposium."

Fleischmann G et al (1984) Drosophila DNA topoisomerase I is associated with transcriptionally active regions of genome. Proc Natl Acad Sci USA 81:6958.

Foe VE, Alberts BM (1985) Reversible chromosome condensation induced in Drosophila embryos by anoxia: Visualization of interphase nuclear organization. J Cell Biol 100:1623.

Gartler SM, Riggs AD (1983) Mammalian X-chromosome inactivation. Annu Rev Genet 17:155.

Gasser S, Laemmli UK (1986a) The organization of chromatin loops: Characterization of a scaffold attachment site. EMBO J 5:511.

Gasser S, Laemmli UK (1986b) Cohabitation of scaffold binding regions with upstream/-enhancer elements of three developmentally regulated genes of D. melanogaster. Cell 46:521.

Gehring WJ et al (1984) Varigated position effect. EMBO 3:2077.

German J (1972) Genes which increase chromosomal instability in somatic cells and predispose to cancer. Prog Med Genet 8:61.

German J (1983) Patterns of neoplasia associated with the chromosome-breakage syndromes. In German J (ed): "Chromosome Mutation and Neoplasia." New York: Alan R Liss.

Gilmour D et al (1986) Topisomerase I interacts with transcribed regions in Drosophila cells. Cell 44:401.

Grosveld F et al (1987) Position-independent, high-level expression of the human beta-globin gene in transgenic mice. Cell 51:975.

Hadlaczky GY et al (1986) Direct evidence for the non-random localization of mammalian chromosomes in the interphase nucleus. Exp Cell Res 167:1.

Haluska FG et al (1986) The t(8:14) translocation occurring in B-cell malignancies results from mistakes in V-D-J joining. Nature 324:158.

Hancock R, Boulikas T (1982) Functional organization of the nucleus. Int Rev Cytol 79:165.

Hancock R, Hughes ME (1982) Organisation of DNA in the interphase nucleus. Biol Cell 44:201.

Hartmann-Goldstein IJ (1967) On the relationships between heterochromatization and variegation in Drosophila with special reference to temperature-sensitive periods. Genet Res 10:143.

Henikoff S (1979) Position effects and variegation enhancers in an autosomal region of Drosophila melanogaster. Genetics 93:105.

Henikoff S (1981) Position effect variegation and chromosome structure of a heat shock puff in *Drosophila*. Chromosoma 83:381.

Hertwig O (1895) "The Cell," Campbell M (trans). New York: Macmillan.

Heslop-Harrison JS, Bennet MD (1983) The positions of centromeres in the somatic metaphase plate of grasses, J Cell Sci 64:163.

Heslop-Harrison JS, Bennet MD (1984) Chromosome order—Possible implications for development. J Embryol Exp Morphol 83(suppl):51.

Hilliker AJ, Trusis-Coulter SN (1987) Analysis of the functional significance of linkage group conservation in *Drosophila*. Genetics 117:233.

Hilliker AJ, Appels R (1989) The arrangement of interphase chromosomes: Structural and functional aspects. Exp Cell Res 185:297.

Hinton T et al (1952) Changing of the gene order and number in natural populations. Evolution 6:19.

Hochstrasser M, Sedat JW (1987a) Three-dimensional organization of *Drosophila melanogaster* interphase nuclei. I. Tissue-specific aspects of polytene nuclear architecture. J Cell Biol 104:1455.

Hochstrasser M, Sedat JW (1987b) Three-dimensional organization of *Drosophila melanogaster* interphase nuclei. II. Chromosomal spatial organization and gene regulation. J Cell Biol 104:1471.

Huberman JA (1987) Eukaryotic DNA replication: A complex picture partially clarified. Cell 48:7.

Huijser P et al. (1987) Poly(dC-dA/dG-dT) repeats in the *Drosophila* genome: A key function for dosage compensation and position effects? Chromosoma 95:209.

Jack JW, Judd BH (1979) Allelic pairing and gene regulation: A model for the zeste–white interaction in *Drosophila melanogaster*. Proc Natl Acad Sci USA 76:1368.

Jackson DA, Cook PR (1985) Transcription occurs at a nucleoskeleton. EMBO J 4:919.

Jaenisch R, Jahner D (1984) Methylation, expression and chromosomal position of genes in mammals. Biochim Biophys Acta 782:1.

Jaenisch R et al (1981) Chromosomal position and activation of retroviral genomes inserted into the germ line of mice. Cell 24:519.

James TC, Elgin SCR (1986) Identification of a nonhistone chromosomal protein associated with heterochromatin in *Drosophila melanogaster*. Mol Cell Biol 6:3862.

Judd BH (1955) Direct proof of a variegated position effect at the white locus in Drosophila melanogaster. Genetics 40:739.

Judd BH (1988) Transvection: Alleic cross talk. Cell 53:841.

Kornher JS, Kauffman SA (1986) Variegated expression of the SGS-4 locus in *Drosophila melanogaster*. Chromosoma 94:205.

Kraemer PM et al (1983) Spontaneous neoplastic evolution of Chinese hamster cells in culture: Multistep progression of phenotype. Cancer Res 43:4822.

Lacy E et al (1983) A foreign beta-globin gene in transgenic mice: integration at abnormal chromosomal positions and expression in inappropriate tissues. Cell 34:343.

Lebeau MM, Rowley JD (1984) Heritable fragile sites in cancer. Nature 308:607.

Lebrowski JS, Laemmli UK (1982a) Evidence for two levels of DNA folding in histone-depleted HeLa interphase nuclei. J Mol Biol 156:309.

Lebrowski JS, Laemmli UK (1982b) Nonhistone proteins and long-range organization of HeLa interphase DNA. J Mol Biol 156:325.

Levis R et al (1985) Effects of genomic position in the expression of transduced copies of the white gene of *Drosophila*. Science 229:558.

Lewin R (1981) Do chromosomes cross-talk? Science 214:1334.

Lewis EB (1950) The phenomenon of position effect. Adv Genet 3:75.

Lewis EB (1954) The theory and application of a new method of detecting chromosomal rearrangements in *Drosophila melanogaster*. Amer Naturalist 88:225.

Lichter P et al (1988) Delineation of individual human chromosomes in metaphase and interphase cells by in situ suppression hybridization using recombinant DNA libraries. Hum Genet 80:224.

Lifton RP et al (1978) *Drosophila* histone genes. Cold Spring Harbor Symp Quant Biol 42:1047.

Lindsley DL, Grell EH (1968) Genetic variations of *Drosophila melanogaster*. Carnegie Inst Washington Publ No. 627.

Lindsley DL et al (1960) Sex-linked recessive lethals in *Drosophila* whose expression is suppressed by the Y chromosome. Genetics 45:1649.

Locke J et al (1988) Dosage-dependent modifiers of position effect variegation in *Drosophila* and a mass action model that explains their effect. Genetics 120:181.

Maguire MP (1967) Premeiotic mitosis in maize: Evidence for pairing of homologues. Caryologia 25:17.

Mansukhani A et al (1988) Nucleotide sequence and structural analysis of the zeste locus of *Drosophila melanogaster*. Mol Gen Genet 211:121.

Manuelidis L (1984) Different CNS cell types display distinct and non-random arrangements of satellite DNA sequences. Proc Natl Acad Sci USA 181:3123.

Manuelidis L (1985) Individual interphase chromosome domains revealed by in situ hybridization. Hum Genet 71:288.

Manuelidis L, Borden J (1988) Reproducible compartmentalization of individual chromosome domains in human CNS cells revealed by in situ hybridization and three-dimensional reconstruction. Chromosoma 96:397.

Manuelidis L, Chen TL (1990) A unified model of eukaryotic chromosomes. Cytometry 11:8.

Mathog D (1990) Transvection in the ultrabithorax domain of the bithorax complex of *Drosophila melanogaster*. Genetics 125:371.

McKnight GS et al (1983) Expression of the chicken transferrin gene in transgenic mice. Cell 34:335.

Miklos GLG et al (1988) Microcloning reveals a high frequency of repetitive sequences characteristic of chromosome 4 and the beta-heterochromatin of *Drosophila melanogaster*. Proc Natl Acad Sci USA 85:2051.

Mirkovitch J et al (1984) Organization of the higher level chromatin loop: Specific DNA attachment sites on nuclear scaffold. Cell 39:223.

Mirkovitch J et al (1987) Relation of chromosome structure and gene expression. Philos Trans R Soc Lond B 317:563.

Moore GD et al (1979) Histone gene deficiencies and position effect variegation in *Drosophila*. Nature 282:312.

Muller HJ (1930) Types of visible variations induced by X-rays in *Drosophila*. J Genet 22:299.

Nelson WG et al (1986) The role of the nuclear matrix in the organization and function of DNA. Annu Rev Biophys Chem 15:457.

Nickerson JA et al (1989) Chromatin architecture and nuclear RNA. Proc Natl Acad Sci USA 86:177.

Oud JL et al (1989) 3-D arrangement of anaphase chromosomes of crepis capillaris root tip cells as studied by confocal scanning laser microscopy. J Cell Sci 92:329.

Palmiter RD (1982) Differential regulation of metallothionein-thymidine kinase fusion genes in transgenic mice and their offspring. Cell 29:710.

Pardoll DM et al (1980) A fixed site of DNA replication in eukaryotic cells. Cell 19:527.

Pathak S (1986) Specific chromosomal anomalies in human cancer. The Cancer Bull 38:129.

Pathak S (1989) Genetic susceptibility, somatic mosaicisms and predisposition to human cancer. Anticancer Res 9:17.

Pienta KJ, Coffey DS (1984) A structural analysis of the role of the nuclear matrix and DNA loops in the organization of the nucleus and chromosome. J Cell Sci Suppl 1:123.

Pienta KJ et al (1989) Cancer as a disease of DNA organization and dynamic cell structure. Cancer Res 49:2525.

Pirrotta V et al (1987) Structure and sequence of the Drosophila zeste gene. EMBO J 6:791.

Pirrotta V et al (1988) Developmental expression of the *Drosophila zeste* gene and localization of *zeste* protein on polytene chromosomes. Genes Dev 2:1839.

Puck T et al (1991) Confocal microscopy of genome exposure in normal, cancer and reverse-transformed cells. Somatic Cell Genet 17:489.

Rabbitts TH et al (1988) Chromosomal abnormalities in lymphoid tumors: Mechanism and role in tumour pathogenesis. Trends Genet 4:300.

Rabl C (1885) Uber Zellteilung. Morphol Jahrb 10:214.

Rawlins DJ, Shaw PJ (1988) Three dimensional organisation of chromosomes in *Crepis capillaris*. J Cell Sci 91:401.

Roberge M et al (1988) Chromosomal loop/nuclear matrix organization of transcriptionally active and inactive RNA polymerases in HeLa nuclei. J Mol Biol 201:545.

Roberts WM et al (1988) Tandem linkage of human *CSF-1* receptor (C-fms) and *PDGF* receptor genes. Cell 55:655.

Rowley JD (1980) Chromosome abnormalities in cancer. Cytogenet Cell Genet 2:175.

Rowley JD (1982) Identification of constant chromosome regions involved in human hematologic malignant disease. Science 218:749.

Rushlow CA, Chovnick A (1984) Heterochromatic position effect at the rosy locus of *Drosophila melanogaster*. Cytological, genetic and biochemical characterization. Genetics 108:589.

Rushlow CA et al (1984) Studies on the mechanism of heterochromatic position effect at the rosy locus of *Drosophila melanogaster*. Genetics 108:603.

Rutledge JC et al (1986) A balanced translocation in mice with a neurological defect. Science 231:395.

Sager R (1989) Tumor suppressor genes: The puzzle and the promise. Science 246:1406.

Sander M, Hsieh T (1985) Drosophila topoisomerase II double-strand DNA cleavage: Analysis of DNA sequence homology at the cleavage site. Nucleic Acids Res 13:1057.

Sankoff D et al (1990) Genomic divergence through gene rearrangement. Methods Enzymol 183:428.

Schultz J (1936) Variegation in *Drosophila* and the inert chromosome regions. Proc Natl Acad Sci USA 22:27.

Schwarzacher-Robinson T et al (1987) Genotypic control of centromere positions of parental genomes in Hordeum × Secale hybrid metaphases. J Cell Science 87:291.

Segall A et al (1988) Rearrangement of the bacterial chromosome: Forbidden inversions. Science 241:1314.

Selby PB (1979) Radiation-induced dominant skeletal mutations in mice: Mutation rate, characteristics and usefulness in estimating genetic hazard to humans from radiation. In Okada S et al. (eds): "Radiation Research." Tokyo: Toppan Printing Co.

Shotten DM (1988) Review: Video-enhanced light microscopy and its application in cell biology. J Cell Sci 89:129.

Sinclair DA et al (1983) Genes which suppress position-effect variegation in *Drosophila*

melanogaster are clustered. Mol Gen Genet 191:326.

Slobodyanuk SY, Serov OL (1983) Variations in the expression of the gene *PGD* due to the effect of chromosomal rearrangements in *Drosophila melanogaster*. Mol Gen Genet 191:372.

Slobodyanyuk SY, Serov OL (1987) Stage and organ specificity of the degree of variegation of the 6-phosphogluconate dehydrogenase gene in *Drosophila melanogaster*. Mol Gen Genet 208:329.

Smith HC, Berezney R (1980) DNA polymerase is highly bound to the nucleus matrix of actively replicating liver. Biochem Biophys Res Commun 97:1541.

Spector DL (1990) Higher order nuclear organization: Three dimensional distribution of small nuclear ribonucleoprotein particles. Proc Natl Acad Sci USA 87:147.

Sperling K (1982) Cell and chromosome cycles. In Rao PN et al (eds): "Premature Chromosome Condensation." New York: Academic Press.

Sperling K, Ludtke EK (1981) Arrangement of prematurely condensed chromosomes in cultured cells and lymphocytes of the Indian muntjac. Chromosoma 83:541.

Spofford J (1976) Position effect variegation in *Drosophila*. In Ashburner M, Novitski E (eds): "Genetics and Biology of Drosophila," vol 1C. New York: Academic Press.

Tartof KD et al (1984) A structural basis for variegating position effects. Cell 37:869.

Tartof KD et al (1989) Towards and understanding of position effect variegation. Dev Genet 10:162.

Travers A, Klug A (1987a) DNA wrapping and writhing. Nature 327:280.

Travers AA, Klug A (1987b) The bending of DNA in nucleosomes in its wider implications. Philos Trans R Soc Lond B 317:537.

Verheijen R et al (1988) The nuclear matrix: Structure and composition. J Cell Sci 90:11.

Vogelstein B et al (1980) Supercoiled loops and eukaryotic DNA replication. Cell 22:79.

Wagenaar EB (1969) End-to-end chromosome attachments in mitotic interphase and their possible significance to meiotic chromosome pairing. Chromosoma 26:410.

Wahl GM et al (1984) Effect of chromosomal position on amplification of transfected genes in animal cells. Nature 307:516.

Wu C-T, Goldberg ML (1989) The *Drosophila* zeste gene and transvection. Trends Genet 5:189.

Yunis JJ (1983) The chromosomal basis of human neoplasia. Science 221:227.

Yunis JJ et al (1987) Fragile sites are targets of diverse mutagens and carcinogens. Oncogene 1:59.

Zelesco PA, Graves JAM (1986) Chromosome segregation from cell hybrids. IV. Movement and position of segregant set chromosomes in early phase interspecific cell hybrids. J Cell Sci 89:49.

Zhimulev IF et al (1989) Position-effect variegation and intercalary heterochromatin: A comparative study. Chromosoma 98:378.

CHAPTER 8

Barski G, Sorieul S, Cornefert F (1960) Production dans des cultures in vitro de deux souches cellulaires en association de cellules de caractere "hybride." C R Acad Sci Ser D 251:1825.

Bell J, Haldane JBS (1937) The linkage between the genes for colour-blindness and haemophilia in man. Proc R Soc 123B:119.

Bernatzky R, Tanksley SD (1986) Toward a saturated linkage map in tomato based on isozymes and random cDNA sequences. Genetics 112:887.

Botstein D, White RL, Skolnick M, Davis RW (1980) Construction of a genetic linkage map in man using restriction fragment length polymorphisms. Am J Hum Genet 32:314.

Brenner S, Livak KJ (1989) DNA fingerprinting by sampled sequencing. Proc Natl Acad Sci USA 86:8902.

Burke DT, Carle GF, Olson MV (1987) Cloning of large segments of exogenous DNA into yeast by means of artificial chromosome vectors. Science 236:806.

Burmeister M, Monaco AP, Gillard EF, van Ommen G, Affara NA, Ferguson-Smith MA, Kunkel LM, Lehrach H (1988) A 10-megabase physical map of human Xp21, including the Duchenne muscular dystrophy gene. Genomics 2:189.

Coe EH, Hoisington DA, Neuffer MG (1990) Linkage map of corn. In O'Brien SJ (ed): "Genetic Maps." Cold Spring Harbor, NY: Cold Spring Harbor Laboratory Press.

Coulson A, Sulston J, Brenner S, Karn J (1986) Toward a physical map of the genome of the nematode Caenorhabditis elegans. Proc Natl Acad Sci USA 83:7821.

Coulson A, Waterston R, Kiff J, Sulston J, Kohara Y (1988) Genome linking with yeast artificial chromosomes. Nature 335:184.

Cox DR, Pritchard CA, Uglum E, Casher D, Kobori J, Myers RM (1989) Segregation of the Huntington's disease region of human chromosome 4 in a somatic cell hybrid. Genomics 4:397.

Creagan RP, Chen S, Ruddle FH (1975) Genetic analysis of the cell surface. Proc Natl Acad Sci USA 72:2237.

Crow JF, Dove WF (1990) Mapping functions. Genetics 125:669.

Davisson MT et al (1990) Locus map of the mouse. In O'Brien SJ (ed): "Genetic Maps." Cold Spring Harbor, NY: Cold Spring Harbor Laboratory Press.

Donahue RP, Bias WB, Renwick JH, McKusick VA (1968) Probable assignment of the Duffy blood group to chromosome 1 in man. Proc Natl Acad Sci USA 61:949.

Edgley ML, Riddle DL (1990) The nematode *Caenorhabditis elegans*. In O'Brien SJ (ed): "Genetic Maps." Cold Spring Harbor, NY: Cold Spring Harbor Laboratory Press.

Ephrussi B, Scaletta LJ, Stencheuer MA, Yoshida MC (1963) Hybridization of somatic cells in vitro. In Harris R (ed): "Cytogenetics of Cells in Culture." New York: Academic Press, p 13.

Evans GA, Lewis KA (1989) Physical mapping of complex genomes by cosmid multiplex analysis. Proc Natl Acad Sci USA 86:5030.

Felsenstein J (1979) A mathematically tractable family of mapping functions with different amounts of interference. Genetics 91:769.

Goss SJ, Harris H (1977) Gene transfer by means of cell fusion. J Cell Sci 25:17.

Guyer MF (1910) Accessory chromosomes in man. Biol Bull 19(4).

Haldane JBS (1919) The combination of linkage values and the calculation of distance between the loci of linked factors. J Genet 8:299.

Haldane JBS, Smith CAB (1947) A new estimate of the linkage between the genes for colour blindness and hemophilia in man. Ann Eugen 14:10.

Hamada H, Petrino M, Kakunaga (1982) A novel repeated element with Z-DNA forming potential is widely found in evolutionarily diverse eukaryotic genomes. Proc Natl Acad Sci USA 79:6465.

Harper ME, Saunders GF (1981) Localization of single copy DNA sequences on G-banded chromosomes by in situ hybridization. Chromosoma 83:431.

Harris H, Watkins JF (1965) Hybrid cells derived from mouse and man: Artificial heterokaryons of mammalian cells from different species. Nature 205:640.

Hsu TC (1979) "Human and Mammalian Cytogenics." New York: Springer-Verlag.

Jeffreys AJ, Wilson V, Thein SL (1985) Hypervariable "minisatellite" regions in human DNA. Nature 314:67.

Kohara Y, Akiyama K, Isono K (1987) The physical map of the whole E. coli chromosome: Application of a new strategy for rapid analysis and sorting of a large genomic library. Cell 50:495.

Kosambi DD (1944) The estimation of map distance from recombination values. Ann Eugen 12:172.

Kusano T, Chen S, Ruddle FH (1971) A new reduced human-mouse somatic cell hybrid. Proc Natl Acad Sci USA 68:82.

Lander ES, Waterman MS (1988) Genomic mapping by fingerprinting random clones: A mathematical analysis. Genomics 2:231.

Leary J, Brigati D, Ward DC (1983) Rapid and sensitive clorimetric method for visualizing biotin-labeled DNA probes hybridized to DNA immobilized on nitrocellulose: Bioblots. Proc Natl Acad Sci USA 80:4045.

Lewin R (1986) Shifting sentiments over sequencing the human genome. Science 233:620.

Lichter P, Tang C, Call K, Hermonson G, Evans G, Housman D, Ward D (1990) High-resolution mapping of human chromosome II by in situ hybridization with cosmid clones. Science 247:64.

Lindsley DL, Grell EH (1972) Genetic variations of Drosophila melanogaster. Carnegie Inst Washington Publ No. 627.

Littlefield JW (1964) Selection of hybrids from matings of fibroblasts in vitro and their presumed recombinants. Science 145:709.

Lyon MF (1990) L. C. Dunn and mouse genetic mapping. Genetics 125:231.

Morgan TH (1910) The method of inheritance of two sex-limited characters in the same animal. Proc Soc Exp Biol Med 8:17.

Morgan TH, Lynch CJ (1912) The linkage of two factors in Drosophila that are not sex-linked. Biol Bull 23:174.

Mortimer RK, Schild D (1987) Genetic map of Saccharomyces cerivisiae. In O'Brien SJ (ed): "Genetic Maps." Cold Spring Harbor, NY: Cold Spring Harbor Laboratory Press.

Morton NE, Lindsten J, Iselius L, Yee S (1982) Data and theory for a related chiasma map of man. Hum Genet 62:266.

Moyzis RK, Albright KL, Bartholdi MF, Cram LS, Deaven LL, Hildebrand CE, Joste N, Longmire JL, Meyne J, Schwarzacher-Robinson T (1987) Human chromosome specific repetitive DNA sequences: Novel markers for genetic analysis. Chromosoma 95:375.

Olson MV, Dutchik JE, Graham MY, Brodeur GM, Helms C, Frank M, MacCollin M, Scheinman R, Frank T (1986) Random clone strategy for genomic restriction mapping in yeast. Proc Natl Acad Sci USA 83:7826.

Orkin SH (1986) Reverse genetics and human disease. Cell 78:845.

Ott J (1985) "Analysis of Human Genetic Linkage." Baltimore: Johns Hopkins University Press.

Pardue M-L, Gall J (1970) Chromosomal localization of mouse satellite DNA. Science 168:1356.

Poustka AM, Lehrach H, Williamson R, Bates G (1988) A long range physical map encompassing the cystic fibrosis locus and its closely linked genetic markers. Genomics 2:337.

Renwick JH, Lawler SD (1963) Probable linkage between a congenital cataract locus and the Duffy blood group locus. Ann Hum Genet 27:67.

Rommens JM et al (1989) Identification of the cystic fibrosis gene: Chromosome walking and jumping. Science 245:1059.

Schwartz DC, Cantor CR (1984) Separation of yeast chromosome sized DNAs by pulsed field gradient electrophoresis. Cell 37:67.

Southern EM (1975) Detection of specific sequences among DNA separated by gel electrophoresis. J Mol Biol 98:503.

Stahl F (1989) The linkage map of T4. Genetics 123:245.

Stallings RL, Siciliano MJ (1981) Confirmational provisional, and/or regional assignment of 15 enzyme loci onto Chinese hamster autosomes 1, 2 and 7. Somat Cell Genet 7:683.

Stallings RL, Torney DC, Hildebrand CE, Longmire JL, Deaven LL, Jett JH, Doggett NA, Moyzis RK (1990) Physical mapping of human chromosomes by repetitive sequence fingerprinting. Proc Natl Acad Sci USA 87:6218.

Stallings RL, Ford AF, Nelson D, Torney D, Hildebrand CE, Moyzis RK (1991) Evolution and distribution of (GT)n repetitive sequences in mammalian genomes. Genomics 10:807.

Tartof KD, Bishop C, Jones M, Hobbs CA, Locke J (1989) Towards an understanding of position effect variegation. Dev Genet 10:162.

Weber JL, May PE (1989) Abundant class of human DNA polymorphisms which can be typed using polymerase chain reaction. Am J Hum Genet 44:388.

Weiss MC, Green H (1967) Human–mouse hybrid cell lines containing partial complements of human chromosomes and functional human genes. Proc Natl Acad Sci USA 58:1104.

Westphal EM, Burmeister M, Wienkler TF, Lehrach H, Bender K, Scherer G (1987) Tyrosine amino transferase and chymotrypsinogen B are linked to haptoglobin on human chromosome 16q: Comparison of genetic and physical distances. Genomics 1:313.

White R, Lalouel JM (1988) Sets of linked genetic markers for human chromosomes. Annu Rev Genet 22:259.

Wilson EB (1911) The sex chromosomes. Arch Mikrosc Anat 77:249.

CHAPTER 9

Akam M (1987) The molecular basis for metameric pattern in the Drosophila embryo. Development 101:1.

Anderson WW (1989) Linkage map of the fruit fly Drosophila pseudoobscura. In O'Brien SJ (ed): "Genetic Maps," 5th ed, book 3. Cold Spring Harbor, NY: Cold Spring Harbor Laboratory Press.

Bateson W (1894) "Materials for the study of variation." London: Macmillan.

Beachy PA (1990) A molecular view of the ultrabithorax homeotic gene of Drosophila. Trends Genet 6:46.

Beachy PA et al (1985) Segmental distribution of bithorax complex proteins during Drosophila development. Nature 313:545.

Bergquist A et al (1974) A pyruvate-valine enzyme complex that is dependent upon the metabolic state of the mitochondria. Proc Natl Acad Sci USA 71:4352.

Berlyn MB, Giles NH (1972) Studies of aromatic biosynthesis and catabolic enzymes in ustilago maydis and in mutants of ustilago violacea. Genet Res 19:261.

Bernardi B, Bernardi G (1990) Compositional patterns in the nuclear genome of cold-blooded vertebrates. J Mol Evol 31:265.

Bernardi G, Bernardi B (1986) Compositional constraints and genome evolution. J Mol Evol 24:1.

Bernardi G (1989) The isochore organization of the human genome. Annu Rev Genet 23:637.

Bernardi G et al (1985) The mosaic genome of the warm-blooded vertebrates. Science 228:953.

Bernardi G et al (1988) Compositional patterns in vertebrate genomes: Conservation and change in evolution. J Mol Evol 28:7.

Bickmore WA, Sumner AJ (1989) Mammalian chromosome banding—an expression of genome organization. Trends Genet 5:144.

Bodmer WF et al (1986) Gene clusters and the evolution of the major histocompatibility system. Philos Trans R Soc London B 312:303.

Bodner M et al (1988) The pituitary-specific transcription factor GHF-1 is a homeobox-containing protein. Cell 55:505.

Boncinelli E et al (1989) Organization of human class I homeobox genes. Genome 31:745.

Bornstein P, Sage H (1980) Structurally distinct collagen types. Annu Rev Biochem 49:957.

Bowman BH, Yang F (1987) DNA sequencing and chromosomal locations of human plasma protein genes. Plasma Proteins 5:1.

Brink RA (1932) Are the chromsomes aggregates of groups of physiologically inter-dependent genes? Am Nat 46:444.

Carson HL (1983) Chromosome sequences and interisland colonizations in Hawaiian Drosophila. Genetics 103:466.

Case ME, Giles NH (1971) Partial enzyme aggregates formed by pleiotropic mutants in the arom gene cluster of Neurospora crassa. Proc Natl Acad Sci USA 68:58.

Coe EH et al (1990) Linkage map of corn (maize) (Zea mays L.) (2N = 20). In O'Brien SJ (ed): "Genetic Maps," 5th ed. Cold Spring Harbor, NY: Cold Spring Laboratory Press.

Costa M et al (1988) Posterior pattern formation in C elegans involved position-specific expression of a gene containing a homeobox. Cell 55:747.

Creau-Goldberg N et al (1990) Primate genetic maps. In O'Brien SJ (ed): "Genetic Maps," 5th ed, book 4: "Nonhuman Vertebrates." Cold Spring Harbor, NY: Cold Spring Harbor Laboratory Press.

Darlington DC, Mather K (1949) "The Elements of Genetics." London: Allen and Unwin.

De Robertis EM et al (1985) The Xenopus homeo boxes. Cold Spring Harbor Symp Quant Biol 50:271.

Demerec MA, Hartman PE (1959) Complex loci in microorganisms. Annu Rev Microbiol 13:377.

Dressler GR, Gruss P (1988) Do multigene families regulate vertebrate development? Trends Genet 4:214.

Dure LS, Chan CA (1984) Cotton seed storage proteins: products of three gene families. In Vloten-Doying L van et al (eds): "Molecular Form and Function in the Plant Genome."

Eckard J (1990) The gene map of the pig (Sus scrofa domestica L.) (2N = 38). In O'Brien SJ (ed): "Genetic Maps," 5th ed, Book 4: "Nonhuman Vertebrates," Cold Spring Harbor, NY: Cold Spring Harbor Laboratory Press.

Eddington AS (1929) The Nature of the Physical World. New York: Macmillan.

Fincham JRS (1966) "Genetic Complementation." New York, WA: Benjamin.

Finney M et al (1988) The C. elegans cell lineage and differentiation gene UNC-86 encodes a protein with a homeodomain and extended similarity to transcription factors. Cell 55:757.

Foster GG et al (1981) Autosomal genetic maps of the Australian sheep blowfly, Lucilia cuprina dorsalis R.-D. (Diptera: Calliphoridae) and possible correlations with the linkage maps of Musca domestica L. and Drosophila melanogaster (MG) Genet Res 37:55.

Fox RR et al (1990) Linkage map of the rabbit (Oryctolagus cuniculus) (2N = 44). In O'Brien SJ (ed): "Genetic Maps," 5th ed, Book 4: "Nonhuman Vertebrates." Cold Spring Harbor, NY: Cold Spring Harbor Laboratory Press.

Gaertner FH, Cole KW (1977) A cluster-gene: Evidence for one gene, one polypeptide, from enzymes. Biochem Biophys Res Commun 75:259.

Gallagher DS, Womack JE (1992) Chromosome conservation in the Bovidae. J Hered 83:287.

Gaunt SJ et al (1986) Homeobox gene expression in mouse embryos varies with position by the primitive streak stage. Nature 324:662.

Gehring WJ (1985) Homeotic genes, the homeo box, and the genetic control of development. Cold Spring Harbor Symp Quant Biol 50:243.

Gehring WJ (1987) Homeo boxes in the study of development. Science 236:1245.

Gehring WJ, Hiromi Y (1986) Homeotic genes and the homeobox. Annu Rev Genet 20:147.

Giles NH (1978) Gene clusters in eukaryotes. Amer Nat 112:641.

Goldschmidt R (1955) "Theoretical Genetics." Berkeley, CA: University of California Press.

Gould SJ (1977) "Ontogeny and Phylogeny." Cambridge, MA: Belknap Harvard University Press.

Green MM (1961) Phenogenetics of the Lozenge loci in Drosophila melanogaster. II. Genetics of Lozenge-Krwshenko. Genetics 46:1169.

Green MM (1990) The foundation of genetic fine structure: A retrospective from memory. Genetics 124:793.

Green MM, Green KC (1949) Crossing over between alleles at the lozenge locus in Drosophila melanogaster. Proc Natl Acad Sci USA 35:586.

Green MM, Green KC (1956) A cytogenetic analysis of the lozenge pseudoalleles in Drosophila. Z Indukt Abstammungs Vererbungsl 87:708.

Haeckel E (1866) Generelle Morphologic der Organismen: Allgemeine Grundzuge der Organisieren Formen-Wissenschaft mechanisch gegrundet durch du von Charles

Darwin reformierte Descending-Theorio (2 vols). Berlin: George Reimer.

Hafen E et al (1983) An improved in situ hybridization method for the detection of cellular RNAs in Drosophila tissue sections and its application for localizing transcripts of the homeotic antennapedia gene complex. EMBO J 2:617.

Hahn WE et al (1977) One strand equivalent of Escherichia coli genome is transcribed: Complexity and abundant classes of mRNA. Science 197:582.

Hake S (1992) Unraveling the knots in plant development. Trends Genet 8:109.

Han K et al (1989) Synergistic activation and repression of transcription by Drosophila homeobox proteins. Cell 56:573.

Hart G (1987) Genetic and biochemical studies of enzymes. In "Wheat and Wheat Improvement." Agronomy (Monogr No 13, 2nd ed).

Hart GE (1979) Genetical and chromosomal relationships among the wheats and their relations. Stadler Genet Symp 11:9.

Hart GE, Gale MD (1990) Biochemical/molecular loci of hexaploid wheat. In O'Brien SJ (ed): "Genetic Maps," 5th ed, Book 6. Cold Spring Harbor, NY: Cold Spring Harbor Laboratory Press.

Harvey RP, Melton DA (1988) Microinjection of synthetic Xhox-1A homeobox mRNA disrupts somite formation in developing Xenopus embryos. Cell 53:687.

Hogness DS et al (1985) Regulation and products of the UBX domain of the bithorax complex. Cold Spring Harbor Symp Quant Biol 50:181.

Holmquist G (1989) Evolution of chromosome bands: Molecular ecology of noncoding DNA. J Mol Evol 28:469.

Hood L et al (1985) T cell antigen receptors and the immunoglobulin super gene family. Cell 40:225.

Hulbert SH et al (1990) Genetic mapping and characterization of sorghum and related crops. Proc Natl Acad Sci USA 87:4251.

Hunkerpiller T, Hood L (1986) The growing immunoglobulin gene superfamily. Nature 323:15.

Jarry B (1976) Isolation of a multifunctional complex containing the first three enzymes of pyrimidine biosynthesis in Drosophila melanogaster. FEBS Let 70:71.

Johnson A, Herskowitz R (1985) A repressor (MAT alpha 2 product) and its operator control: Expression of a set of cell-type specific genes in yeast. Cell 42:237.

Johnson KR et al (1987) Linkage relationships reflecting anestral tetraploidy in salmonid fish. Genetics 116:579.

Johnstone ME (1985) Three purine auxotrophic loci on the second chromosome of Drosophila melanogaster. Biochem Genet 23:539.

Kahn A et al (1971) Differences in two red-cell populations in erythroleukemia. Lancet 2:933.

Kappes D, Strominger JL, (1988) Human class II major histocompatibility complex genes and proteins. Annu Rev Biochem 57:991.

Karin M et al (1990) Growth hormone gene regulation: A paradigm for cell-type-specific gene activation. Trends Genet 6:92.

Kaufman JF et al (1984) The class II molecules of the human and murine major histocompatibility. Cell 36:1.

Kaufman TC et al (1980) Cytogenetic analysis of chromosome 3 in Drosophila melanogaster: The homeotic gene complex in polytene chromosome interval 841-B. Genetics 94:115.

Kiritani K (1962) Linkage relationships among a group of isoleucine and valine requiring mutants of neurospora crassa. Jpn J Genet 37(1):42.

Klein J (1986) "Natural History of the Major Histocompatibility Complex." New York: John Wiley.

Klein J et al (1990) The major histocompatibility complex and human evolution. Trends Genet 6:7.

Kourilsky P, Claverie J-M (1989) MHC restriction, alloreactivity, and thymic education: A common link? Cell 56:327.

Kües U, Casselton LA (1992) Homeodomains and regulation of sexual development in basideomycetes. Trends Genet 8:154.

Lalley PA et al (1989) Report of the committee on comparative mapping. Cytogenet Cell Genet 51:503.

Levan G et al (1990) Genetic map of the rat (Rattus norvegicus) (2N = 42). In O'Brien SJ (ed): "Genetic Maps," 5th ed, Book 4: "Nonhuman Vertebrates." Cold Spring Harbor, NY: Cold Spring Harbor Laboratory Press.

Levine M et al (1984) Human Dna sequences homologous to a protein coding region conserved between homeotic genes of *Drosophila*. Cell 38:667.

Lewis EB (1950) The phenomenon of position effect. Adv Genet 3:75.

Lewis EB (1954) The theory and application of a new method of detecting chromosomal rearrangements in Drosophila melanogaster. Am Nat 88:225.

Lewis EB (1955) Some aspects of position pseudoallelism. Am Nat 89:73.

Lewis EB (1967) Genes and gene complexes. In Brink RA (ed): "Heritage from Mendel." Madison, WI: University of Wisconsin Press.

Lewis EB (1978) A gene complex controlling segmentation in Drosophila. Nature 276:565.

Maizels N (1987) Diversity achieved by diverse mechanisms: Gene conversion in developing B cells of the chicken. Cell 48:359.

Maizels N (1989) Might gene conversion be the mechanism of somatic hypermutation of mammalian immunoglobulin genes? Trends Genet 5:4.

Martin GR (1987) Nomenclature for homeobox-containing genes. Nature 325:21.

May B, Johnson KR (1990) Composite linkage map of salmonid fishes (*Savelinus, Salmo, Onchorhynchus*). In O'Brien SJ (ed): "Genetic Maps." Cold Spring Harbor, NY: Cold Spring Harbor Press.

McGinnis W (1985) Homeo box sequences of the antennapedia class are conserved in higher animal genomes. Cold Spring Harbor Symp Quant Biol 50:263.

McGinnis W et al (1984a) A homologous protein coding sequence in Drosophila homoeotic genes and its conservation in other metazoans. Cell 37:403.

McGinnis W et al (1984b) Molecular cloning and chromosome mapping of a mouse DNA sequence homologous to homoeotic genes of Drosophila. Cell 38:675.

McKusick VA (1986) The gene map of homo sapiens: Status and Prospectus. Cold Spring Harbor Symp Quant Biol 51:15.

Meera Kahn P et al (1990) Domestic dog (Canis familiaris) (12-78). In O'Brien SJ (ed): "Genetic Maps," 5th ed, Book 4: "Nonhuman Vertebrates." Cold Spring Harbor, NY: Cold Spring Harbor Laboratory Press.

Milne DL, McIntosh RA (1990) Triticum aestivum (common wheat). In O'Brien SJ (ed): "Genetic Maps," 5th ed. Cold Spring Harbor, NY: Cold Spring Harbor Laboratory Press.

Mischina M et al (1986) Molecular distinction between fetal and adult forms of muscle acetylcholine receptors. Nature 321:406.

Morgan TH et al (1915) "The Mechanism of Mendelian Heredity." New York: Henry Holt.

Morizot DC et al (1990) Assignment of six enzyme loci to multipoint linkage groups in fishes of the genus Poeciliopsis (Poecillidae): Designation of linkage groups III–V. Biochem Genet 28:83.

Morizot DC et al (1991) Genetic linkage map of fishes of the genus Xiphophorus (Teleostei: Poecillidae). Genetics 127:399.

Mueller MM et al (1984) A homeo-box-containing gene expressed during oogenesis in Xenopus. Cell 39:157.

Nadeau JH (1989) Maps of linkage and synteny homologies between mouse and man. Trends Genet 5:82.

Nadeau JH, Kosowsky M (1991) Mouse map of paralogous genes. Genome 1:S433.

Nadeau JH et al (1991) Comparative map for mice and humans. Mamm Genome 1:S461.

Nash WG, O'Brien SJ (1982) Conserved regions of homologous G-banded chromosomes between orders in mammalian evolution: Carnivores and primates. Proc Natl Acad Sci USA 79:6631.

Nebert DW, Gonzales FJ (1987) P450 genes: Structure, evolution and regulation. Annu Rev Biochem 56:945.

O'Brien SJ (1986) Molecular genetics in the domestic cat and its relatives. Trends Genet 2:137.

O'Brien SJ (1990) Domestic cat (Felis catus) (2N = 38). In O'Brien SJ (ed): "Genetic Maps," 5th ed, Book 4: "Nonhuman Vertebrates." Cold Spring Harbor, NY: Cold Spring Harbor Laboratory Press.

O'Brien SJ et al (1988) Mammalian genome evolution: An evolutionary view. Annu Rev Genet 22:323.

Okajima K et al (1987) Thymidine kinase activity in individuals with galactokinase deficiency. Am J Hum Genet 41:503.

Oliver CP (1940) A reversion to wild type associated with crossing over in Drosophila melanogaster. Proc Natl Acad Sci USA 26:452.

Pinsker W, Sperlich D (1984) Cytogenetic mapping of enzyme loci on chromosomes J and U of Drosophila subobscura. Genetics 108:913.

Poschl E et al (1988) The genes for the alpha 1 (IV) and alpha 2 (IV) chains of human basement membrane collagen type IV are arranged head to head and separated by a bidirectional promoter of unique structure. EMBO J 7:2687.

Ptashne M (1986) "A Genetic Switch." Palo Alto, CA: Blackwell.

Race RR, Sanger R (1975) "Blood Groups in Man," 6th ed. Oxford: Blackwell Scientific.

Ramirez F (1989) Organization and evolution of the fibrillar collagen genes. In Olsen BR, Nimni EM (eds): "Collagen: Molecular Biology," vol 4.

Ramirez F, Di Liberto M (1990) Complex and diversified regulatory programs control the expression of vertebrate collagen genes. FASEB J 4:1616.

Rawls JM, Fristrom JW (1975) A complex locus that controls the first three steps of pyrimidine biosynthesis in Drosophila. Nature 255:738.

Roberts WM et al (1988) Tandem linkage of human CSF-1 receptor (c-fms) and PDGF receptor genes. Cell 55:655.

Rossant J, Joyner AL (1989) Towards a molecular-genetic analysis of mammalian development. Trends Genet 5:277.

Schoen RC et al (1984) Thymidine kinase activity of cultured cells from individuals with inherited galactokinase deficiency. Am J Hum Genet 36:815.

Scofield VL et al (1982) Protochordate allorecognition is controlled by an MHC-like gene system. Nature 295:499.

Scott MP et al (1983) The molecular organization of the antennapedia locus of Drosophila. Cell 35:763.

Searle AG et al (1989) Chromosome maps of man and mouse. Ann Human Genet 53:89.

Seeger MA et al (1988) Characterization of amalgam: A member of the immunoglobulin superfamily from Drosophila. Cell 55:589.

Serov OL, Pack SD (1990) American mink (Mustela vison) (2N = 30). In O'Brien SJ (ed): "Genetic Maps," 5th ed, book 4: "Nonhuman Vertebrates." Cold Spring Harbor, NY: Cold Spring Harbor Laboratory Press.

Soriano P et al (1983) The distribution of interspersed repeats is nonuniform and conserved in the mouse and human. Proc Natl Acad Sci USA 80:1816.

Stallings RL et al (1990) The Chinese hamster gene map (Cricetalus griseus) (2N = 22). In O'Brien SJ (ed): "Genetic Maps," 5th ed, Book 4: "Nonhuman Vertebrates." Cold Spring Harbor, NY: Cold Spring Harbor Laboratory Press.

Steinmetz M et al (1986) Gene organization and recombinational hotspots in the murine major histocompatibility complex. Cell 44:895.

Strominger JL (1989) Developmental biology of T cell receptors. Science 244:943.

Sturtevant AH, Novitski E (1941) The homologies of the chromosome elements of Drosophila. Genetics 26:517.

Tanksley SD, Mutschler MA (1990) Linkage map of the tomato (Lycopersicon esculentum) (2N = 24). In O'Brien SJ (ed): "Genetic Maps," 5th ed. Cold Spring Harbor, NY: Cold Spring Harbor Laboratory Press.

Threadgill DW, Womack JE (1990) Syntenic Conservation between humans and cattle. I: Human Chromosome 9. Genomics 8:22.

Threadgill DW et al (1990) Syntenic conservation between humans and cattle. II. Genomics 8:29.

Vogeli G et al (1988) The collagen gene family and its evolution: Molecular Biology of the Eye: Genes, Vision, and Ocular Disease. xx:169.

Wagner RP, Bergquist A (1963) Synthesis of valine and isoleucine in the presence of a particulate cell fraction of neurospora. Proc Natl Acad Sci USA 49:892.

Wagner RP et al (1960) Gene structure and function in *Neurospora*. Proc Natl Acad Sci USA 46:708.

Wagner RP et al (1985) A coordinate relationship between the GALK and TK1 genes of the Chinese hamster. Biochem Genet 23:677.

Weeden NF et al (1988) Application of isozyme analysis in pulse crops. In Summerfield RJ (ed): "World Crops: Cool Season Food Legumes." Dordrecht, The Netherlands: Kluwer Academic Publishers.

Weeden NF (1992) Extensive conservation of linkage relationships between pea and lentil genetic maps. J Heredity 83:123.

Weitkamp LR, Sandburg K (1990) Horse (Equus caballus) (2N = 64). In O'Brien SJ (ed): "Genetic Maps," 5th ed, Book 4: "Nonhuman

Vertebrates." Cold Spring Harbor, NY: Cold Spring Harbor Laboratory Press.

Williams AF, Barclay AN (1988) The immunoglobulin superfamily-domains for cell surface recognition. Annu Rev Immunol 6:381.

Wirz J et al (1986) Localization of the antennapedia protein in Drosophila embryos and imaginal discs. EMBO J 5:3327.

Womack JE (1987) Comparative gene mapping. Dev Genet 8:281.

Womack JE (1990) Gene map of the cow (Bos taurus) (2N = 60). In O'Brien SJ (ed): "Genetic Maps," 5th ed, Book 4: "Nonhuman Vertebrates." Cold Spring Harbor, NY: Cold Spring Harbor Laboratory Press.

Wright CE et al (1989) Interference with function of a homeobox gene in Xenopus embryos produces malformations of the anterior spinal cord. Cell 59:81.

Wurster-Hill DJ, Centerwall WR (1982) The interrelationships of chromosome banding patterns in canids, mustelids, hyena and felids. Cytogenet Cell Genet 34:178.

Zuckerkandl E (1986) Polite DNA: Functional density and functional compatibility in genomes. J Mol Evol 24:12.

CHAPTER 10

Angerer RC, Davidson EH (1984) Molecular indices of cell lineage specification in sea urchin embryos. Science 226:1151.

Arnheim N (1983) Concerted evolution in multigene families. In Nei M, Koehn RK (eds): "Evolution of Genes and Proteins." Sunderland, MA: Sinauer.

Belanger FC, Hepburn AG (1990) The evolution of CpNpG methylation in plants. J Mol Evol 30:26.

Belfaiza J et al (1986) Evolution in biosynthetic pathways: Two enzymes catalyzing consecutive steps in methionine biosynthesis orginate from a common ancestor and possess a similar regulatory region. Proc Natl Acad Sci USA 83:867.

Bird AP (1980) DNA methylation and frequency of CpG in animal DNA. Nucleic Acid Res 8:1499.

Bird AP (1986) CpG-rich islands and the function of DNA methylation. Nature 321:209.

Bird AP (1987) CpG islands as gene markers in the vertebrate nucleus. Trends Genet 3:432.

Bishop MJ, Friday AE (1988) Estimating the interrelationships of tetrapod groups on the basis of molecular data. In Benton MJ (ed): "The Phylogeny and Classification of the Tetrapods," vol 1. Oxford: Clarendon Press.

Bodmer WF, Parsons PA (1962) Linkage and recombination in evolution. Adv Genet 11:1.

Bodmer WF et al (1986) Gene clusters and the evolution of the major histocompatibility system. Philos Trans R Soc London B 312:303.

Bogart JP (1980) Evolutionary implications of polyploidy in amphibians and reptiles. In Lewis WH (ed): "Polyploidy." New York: Plenum.

Brink RA (1932) Are the chromosomes aggregates of groups of physiologically interdependent genes? Am Nat 46:444.

Brown DD et al (1971) A comparison of the ribosomal DNAs of *Xenopus laevis* and *Xenopus mulleri*: The evolution of tandem genes. J Mol Biol 63:57.

Bush GL et al (1977) Rapid speciation and chromosomal evolution in mammals. Proc Natl Acad Sci USA 74:3942.

Caccone A, Powell JR (1989) DNA divergence among hominoids. Evolution 43:925.

Caccone A, Powell JR (1990) Extreme rates and heterogeneity in insect DNA evolution. J Mol Evol 30:273.

Campbell A (1983) Transposons and their evolutionary significance. In Nei M, Koehn RK (eds): "Evolution of Genes and Proteins." Sunderland, MA: Sinauer.

Charlesworth B et al (1986) The evolution of restricted recombination and accumulation of repeated DNA sequences. Genetics 112:947.

Chen TL, Manuelidis L (1989) SINEs and LINEs cluster in distinct DNA fragments of Giemsa band size. Chromosoma 98:309.

Collins FS, Weissman SM (1984) The molecular genetics of human hemoglobin. Prog Nucleic Acid Res Mol Biol 31:317.

Contreras LC et al (1990) The largest known chromosome number for a mammal, in a South American desert rodent. Experientia 46:506.

Crosland MWJ, Crozier RH (1986) *Myrmecia pilosula*, an ant with only one pair of chromosomes. Science 231:1278.

Darlington CD, Mather K (1949) "The Elements of Genetics." London: Allen and Unwin.

Davidson EH et al (1985) Lineage-specific gene expression in the sea urchin embryo. Cold Spring Harbor Symp Quant Biol 50:321.

Deininger PL, Daniels GR (1986) The recent evolution of mammalian repetitive DNA elements. Trends Genet 2:76.

Di Nocera PP, Sakaki Y (1990) LINES: A superfamily of retrotransposable ubiquitous DNA elements. Trends Genet 6(2):29.

Dobzhansky T (1947) A directional change in the genetic constitution of a natural population of *Drosophila pseudoobscura*. Heredity 1:53.

Dobzhansky T (1963) Genetics of natural populations XXXIII. A progress report on genetic changes in populations of *Drosophila pseudoobscura* and *Drosophila persimilis* in a locality in California. Evolution 17:333.

Dobzhansky T (1970) "Genetics of the Evolutionary Process." New York: Columbia University Press.

Dobzhansky T, Epling C (1944) Contributions to the genetics, taxonomy and ecology of *Drosophila pseudoobscura* and its relatives. Carnegie Inst Wash Pub 554.

Doolittle RF (1979) Protein evolution. In Neurath H, Hill RL (ed): "The Proteins," vol 4. New York: Academic Press.

Doolittle RF et al (1989) Estimating the prokaryote–eukaryote divergence time from protein sequences. In Fernholm B, Bremer K, Jornvall H (eds): "The Hierarchy of Life." Amsterdam: Elsevier, Chap 6, p 73.

Doolittle RF et al (1990) A naturally occurring horizontal gene transfer from a eukaryote to a prokaryote. J Mol Evol 31:383.

Doolittle WF (1989) What ever happened to the progenote? In Fernholm B, Bremer K, Jornvall H (eds): "The Hierarchy of Life." Amsterdam: Elsevier.

Dorit RL et al (1990) How big is the universe of exons? Science 250:1377.

Dover CA, Tautz D (1986) Conservation and divergence in multigene families: Alternative to selection and drift. Philos Trans R Soc London B 312:275.

Dover G (1982) Molecular drive: A cohesive mode of species evolution. Nature 299:111.

Dover GA (1986) Molecular drive in multigene families: How biological novelties arise, spread and are assimilated. Trends Genet 2:159.

Dyer BD, Obar R (1985) "The Origin of Eukaryotic Cells." New York: Van Nostrand Reinhold.

Elbetieha A, Kalthoff K (1988) Anterior determinants in embryos of *Chironomus samoensis*: Characterization by rescue bioassay. Development 104:61.

Epplen JT et al (1979) Contrasting DNA sequence organization patterns in sauropsidian genomes. Chromosoma 75:199.

Erwin DH, Valentine JW (1984) "Hopeful monsters," Transposons and metazoan radiation. Proc Natl Acad Sci USA 81:5482.

Erwin DH et al (1987) A comparative study of diversification events: The early paleozoic versus the mesozoic. Evolution 41:1177.

Ferris PJ (1985) Nucleotide sequence of the central nontranscribed spacer repeats of *Physarum polycephalum*. Gene 39:203.

Field KG et al (1988) Molecular phylogeny of the animal kingdom. Science 239:748.

Fincham JRS et al (1979) "Fungal Genetics." Berkeley, CA: University of California Press.

Fisher RA (1930) "The Genetical Theory of Natural Selection." Oxford: Clarendon Press.

Flavell RB (1980) The molecular characterization and organization of plant chromosomal DNA sequences. Annu Rev Plant Physiol 31:569.

Flavell RB (1986) Repetitive DNA and chromosome evolution in plants. Philos Trans R Soc London Ser B 312:227.

Flavell RB et al (1974) Genome size and the proportion of repeated nucleotide sequences in DNA in plants. Biochem Genet 12:257.

Franklin I, Lewontin RC (1970) Is the gene a unit of selection? Genetics 65:707.

Gilbert W (1978). Why genes in pieces? Nature 271:501.

Gilbert W (1985) Genes-in-pieces revisited. Science 228:823.

Gonzalez FJ et al (1990) Evolution of the P450 gene superfamily: Animal-plant "warfare," molecular drive and human genetic differences in drug oxidation. Trends Genet 6:182.

Goodman M et al (1989) Molecular phylogeny of the family of apes and humans. Genome 31:316.

Goodman M et al (1990) Primate evolution at the DNA level and a classification of hominoids. J Mol Evol 30:260.

Graf J-D (1989) Genetic mapping in *Xenopus laevis*: Eight linkage groups established. Genetics 123:389.

Groot PC et al (1989) The human alpha-amylase multigene family consists of haplotypes with variable numbers of genes. Genomics 5:29.

Groot PC et al (1990) Evolution of the human alpha-amylase multigene family through unequal, homologous, and inter- and intrachromosomal crossovers. Genomics 8:97.

Haldane JBS (1932) "The Causes of Evolution." New York: Harper.

Haldane JBS (1957) The cost of natural selection. J Genet 55:511.

Haldane JBS (1960) More precise expressions for the cost of natural selection. J Genet 57:351.

Harbers K et al (1986) High frequency of unequal recombination in pseudoautosomal region shown by proviral insertion in transgenic mice. Nature 324:682.

Hattori M et al (1986) L1 family of repetitive DNA sequences in primates may be derived from a sequence encoding a reverse transcriptase-related protein. Nature 321:625.

Hereford LM, Rosbash M (1977) Number and distribution of polyadenylated RNA sequences in yeast. Cell 10:453.

Horii A et al (1987) Primary structure of human pancreatic alpha-amylase gene: Its comparison with human salivary amylase gene. Gene 60:57.

Horowitz NH (1965) The evolution of biochemical syntheses—Retrospect and prospect. In Bryson V, Vogel HJ (eds): "Evolving Genes and Proteins." New York: Academic Press.

Hough-Evans B et al (1980) Complexity of RNA in eggs of *Drosophila melanogaster* and *Musca domestica*. Genetics 95:81.

Hwu RH et al (1986) Insertion and or deletion of many repeated DNA sequences in human and higher ape evolution. Proc Natl Acad Sci USA 83:3875.

Imai HT, Crozier RH (1980) Quantitative analysis of directionality in mammalian karyotype evolution. Am Nat 116:537.

Imai HT et al (1986) Theoretical bases for karyotype evolution. I. The minimum-interaction hypothesis. Am Nat 128:900.

Jackson RC (1973) Chromosomal evolution in *Haplopappus gracilis*: A centric transposition race. Evolution 27:243.

Jackson JA, Fink GR (1985) Meiotic recombination between duplicated genetic elements in *Saccharomyces cerevisiae*. Genetics 109:303.

Jeffreys AJ et al (1988) Spontaneous mutation rates to new length alleles at tandem-repetitive hypervariable loci in human DNA. Nature 332:278.

Johnson KR et al (1987) Linkage relationships reflecting ancestral tetraploidy in salmonid fish. Genetics 116:579.

Joseph JL et al (1990) Interspecies distribution of abundant DNA sequences in *Lilium*. J Mol Evol 30:146.

Jurka J, Smith T (1988) A fundamental division in the Alu family of repeated sequences. Proc Natl Acad Sci USA 85:4775.

Kalthoff K (1983) Cytoplasmic determinants in dipteran eggs. In Jeffrey WR, Raff RA (eds): "Time, Space and Pattern in Embryonic Development." New York: Alan R. Liss.

Kamalay JC, Goldberg RB (1980) Regulation of structural gene expression in tobacco. Cell 19:935.

Kimura M (1983) "The Neutral Theory of Molecular Evolution." Cambridge: Cambridge University Press.

Kobel HR, Du Pasquier L (1986) Genetics of polyploid *Xenopus*. Trends Genet 2:310.

Krane DE, Hardison RC (1990) Short interspersed repeats in rabbit DNA can provide functional polyadenylation signals. Mol Biol Evol 7:1.

Lande R (1979) Effective deme size during long term evolution estimated from rates of chromosomal rearrangements. Evolution 33:234.

Lewontin RC (1974) "The Genetic Basis of Evolutionary Change." New York: Columbia University Press.

Lewontin RC (1988) On measures of gametic disequilibrium. Genetics 120:849.

Lindsley DL, Grell EH (1968) "Genetic Variations of *Drosophila melanogaster*." Carnegie Inst Washington Publ No. 627.

Lyons KM et al (1988) Length polymorphisms in human proline-rich protein genes generated by intragenic unequal crossing over. Genetics 120:267.

Maeda N, Smithies O (1986) The evolution of multigene families: Human haploglobin genes. Annu Rev Genet 20:81.

Martin W, Cerff R (1986) Prokaryotic features of a nucleus-encoded enzyme—cDNA sequences for chloroplast and cytosolic glyceraldehyde-3-phosphate dehydrogenases from mustard (*Sinapis alba*). Eur J Biochem 159:323.

Matsunaga E et al (1990) The rat P450 IID subfamily: Complete sequences of four closely linked genes and evidence that gene conversions maintained sequence homogeneity at the heme-binding region of the cytochrome P450 active site. J Mol Evol 30:155.

Maynard Smith J (1977) Why the genome does not congeal. Nature 268:693.

Merriam J et al (1991) Toward cloning and mapping the genome of *Drosophila*. Science 254:221.

Merritt CM et al (1990) Evolution of human alpha1-acid glycoprotein genes and surrounding Alu repeats. Genomics 6:659.

Morescalchi A, Olmo E (1982) Single-copy DNA and vertebrate phylogeny. Cytogenet Cell Genet 34:93.

Mortimer RK, Schild D (1980) Genetic map of Saccharomyces cerevisiae. Microbiol Rev 44:519.

Nagylaki T (1988) Gene conversion, linkage, and the evolution of multigene families. Genetics 120:291.

Nelson DR, Stobel HW (1987) Evolution of cytochrome P-450 proteins. Mol Biol Evol 4:572.

Ohta T (1989) Role of gene duplication in evolution. Genome 31:304.

Panchen AL, Smithson TR (1988) The relationship of the earliest tetrapods. In Benton MJ (ed): "The Phylogeny and Classifications of the Tetrapods," vol 1. Oxford: Clarendon Press.

Patterson JT, Stone WS (1952) "Evolution in the Genus Drosophila." New York: Macmillan.

Perelson AS, Bell GI (1977) Mathematical models for the evolution of multigene families by unequal crossing over. Nature 265:304.

Prevosti A et al (1990) Clines of chromosomal rearrangements of Drosophila subobscura in South America evolve closer to Old World patterns. Evolution 44:218.

Reynaud C-A et al (1987) Hypoconversion mechanism generates the chicken light chain preimmune repertoire. Cell 49:379.

Riede I et al (1983) DNA sequence divergence in the *Drosophila virilis* group. Chromosoma 88:109.

Samuelson LC et al (1990) Retroviral and pseudogene insertion sites reveal the lineage of human salivary and pancreatic amylase genes from a single gene during primate evolution. Mol Cell Biol 10:2513.

Schmidtke J, Epplen JT (1980) Sequence organization of animal nuclear DNA. Hum Genet 55:1.

Schultz RJ (1980) Role of polyploidy in the evolution of fishes. In Lewis WH (ed): "Polyploidy." New York: Plenum.

Shapura ST, Finnerty VG (1986) The use of genetic complementation in the study of eukaryotic macromolecular evolution: Rate of *Drosophila melanogaster* spontaneous gene duplication at two loci of *Drosophila melanogaster*. J Mol Evol 23:159.

Singer MF, Scowronski J (1985) Making sense out of LINES: long interspersed repeat sequences in mammalian genomes. Trends Biochem Sci 10:119.

Skowronski J, Singer MF (1986) The abundant LINE-1 family of repeated DNA sequences in mammals: Genes and pseudogenes. Cold Spring Harbor Symp Quant Biol 51:457.

Slatkin M (1986) Interchromosomal biased gene conversion mutation and selection in a multigene family. Genetics 112:681.

Smith GP (1973) Unequal crossover and the evolution of multigene families. Cold Spring Harbor Symp Quant Biol 38:507.

Smith-White S, Carter CR (1970) The cytology of *Brachycome lineariloba*. Chromosoma 30:129.

Smith-White S et al (1970) The cytology of *Brachycome*. Aust J Bot 18:99.

Sparrow AH, Nauman AF (1976) Evolution of genome size by DNA doubling. Science 192:524.

Sturtevant AH (1925) The effects of unequal crossing over at the bar locus in Drosophila. Genetics 10:117.

Sylvanen M (1986) Cross-species gene transfer: A major factor in evolution? Trends Genet 2:63.

Tachida H (1987) Differentiation of a multi-gene family. Evolution 41:190.

Tartof KD (1988) Unequal crossing over then and now. Genetics 120:1.

Templeton AR et al (1989) Natural selection and ribosomal RNA in Drosophila. Genome 31:296.

Thompson CB, Neiman PE (1987) Somatic diversification of the chicken immunoglobulin light chain gene is limited to the rearranged variable gene segment. Cell 49:369.

Timberlake WE et al (1977) Relationship between nuclear and polysomal RNA populations of *Achlya*. A simple eukaryotic system. Cell 10:623.

Turner JRG (1967a) On supergenes. I. The evolution of supergenes. Am Nat 101:195.

Turner JRG (1967b) Why does the genome not congeal? Evolution 21:645.

Van Ness J et al (1979) Complex population of nonpolyadenylated messenger RNA in mouse brain. Cell 18:1341.

Wade MJ, Goodnight CJ (1991) Wright's shifting balance theory: an experimental study. Science 253:1015.

Walsh JB (1986) Selection and biased gene conversion in a multigene family: Consequences of interallelic bias and threshold selection. Genetics 112:699.

Werman SD et al (1990) Rapid evolution in a fraction of the Drosophila nuclear genome. J Mol Evol 30:281.

White MJD (1978) "Modes of Speciation." San Francisco: WH Freeman.

Willard C et al (1987) Existence of at least three distinct Alu subfamilies. J Mol Evol 26:180.

Williams SM (1990) The opportunity for selection in multigene families. Genetics 124:439.

Wilson AC (1975) Evolutionary importance of gene regulation. Stadler Symp 7:117.

Wilson AC et al (1974a) Two types of molecular evolution: Evidence from studies of inter-specific hybridization. Proc Natl Acad Sci USA 71:2843.

Wilson AC et al (1974b) The importance of gene rearrangements in evolution: Evidence from studies on rates of chromosomal, protein and anatomical evolution. Proc Natl Acad Sci USA 71:3028.

Wilson AC et al (1975) Social structuring of mammalian populations and rate of chromosomal evolution. Proc Natl Acad Sci USA 72:5061.

Wilson AC et al (1977) Biochemical evolution. Annu Rev Biochem 46:573.

Wright S (1931) Evolution in Mendelian populations. Genetics 16:97.

Wright S (1941) On the probability of fixation of reciprocal translocations. Am Nat 75:513.

Wright S (1948) On the roles of directed and random changes in gene frequencies in the genetics of populations. Evolution 2:279.

Wright S (1949) Population structure and evolution. Proc Am Philos Soc 93:471.

Wright S (1977) "Evolution and Genetics of Populations," vol 4: "Variability Within and Among Natural Populations." Chicago: University of Chicago Press.

Wright S (1980) Genic and organismic selection. Evolution 34:825.

Wright S (1982) The shifting balance theory and macroevolution. Annu Rev Genet 16:1.

Wright S (1988) Surfaces of selective value revisited. Amer Naturalist 131:115.

Wright S, Dobzhansky T (1946) Genetics of natural populations: XII. Experimental production of some of the changes caused by natural selection in certain populations of Drosophila pseudoobscura. Genetics 31:125.

Xiang Y et al (1988) Gene conversions can generate sequence variants in late chorion multigene families of *Bombyx mori*. Genetics 120:221.

Zuckerkandl E et al (1989) Maintenance of function without selection: Alu sequences as "cheap genes." J Mol Evol 29:504.

Index